AROMATIC AMINE OXIDES

Aromatic Amine Oxides

by

EIJI OCHIAI

Emeritus Professor of Pharmaceutical Chemistry,
University of Tokyo (Japan)

Translated by

DOROTHY U. MIZOGUCHI

Chemotherapy Information Center,
Cancer Institute, Tokyo (Japan)

ELSEVIER PUBLISHING COMPANY

Amsterdam London New York 1967

Elsevier Publishing Company
335 Jan van Galenstraat, P.O. Box 211, Amsterdam

American Elsevier Publishing Company, Inc.
52 Vanderbilt Avenue, New York, N.Y. 10017

Elsevier Publishing Company Limited
Rippleside Commercial Estate, Barking, Essex

Library of Congress Catalog Card Number 66-25764

With 31 illustrations and 87 tables

Printed in The Netherlands

E. Ochiai, *Aromatic Amine Oxides*

Elsevier Publishing Company, Amsterdam 1967

CORRIGENDUM

Page 160, second footnote, first line,
NN_2^+ should read NO_2^+

Page 388, in formula group at bottom of
page, for (35) read 135)

Preface

The purpose of this book is to provide a review of the chemistry of aromatic amine oxides. In view of the recent development of studies on the biological activities of some of these compounds, an outline of their biological properties has also been added.

Almost a quarter of a century has passed since I began to study this field of chemistry. The motive for these studies and for the decision to write this book are given in the first chapter. As will be understood from this chapter, reports of work on aromatic amine oxides have mainly appeared in Japanese scientific journals, and quite a number of them are presented in the Japanese language. Since it was thought that not many of the readers of this book would be familiar with these Japanese scientific journals, while a still smaller number would be able to read them, special attention has been given to the inclusion of Japanese work.

In writing this book, efficient communication with the reader has been sought by simplifying the explanation of chemical reactions with the aid of formulae and charts, and much of the information contained in these schematic charts is not repeated in the text. Care has also been taken to include experimental details for those reactions which are most likely to be utilized by workers in this field.

Chapter IV, on the physicochemical properties of aromatic amine oxides, was written by Dr. Chikara Kaneko of the Tokyo Medical and Dental College, who worked with me for a long time on the amine oxides of the pyridine and quinoline series. It was checked by Dr. Tanekazu Kubota of the Shionogi Research Laboratory since my knowledge of this part of the work was not sufficient to cover it completely.

The English translation of this work was undertaken by Miss Dorothy U. Mizoguchi of the Cancer Institute, Tokyo, who also worked with me for many years, and the final version as it appears in print was completed after mutual discussion. She also undertook all negotiations with the Elsevier Publishing Company, and helped me with the proof reading and the preparation of indexes.

The nomenclature of chemical compounds in accordance with the IUPAC rules was checked by Dr. Kenzo Hirayama of the Fuji Photo Film Company, who also assisted in the preparation of the index of compounds.

I express my sincere thanks to these persons without whose willing cooperation this book would not have been possible.

Thanks are also expressed to the following organizations for permission to reproduce material published in their journals: the Pharmaceutical Society of Japan for material published in the Chemical and Pharmaceutical Bulletin and in Yakugaku Zasshi; the Spectroscopic Society of Japan for material published in Bunko Kôgaku; the Chemical Society for material published in the Journal of The Chemical Society; the American Chemical Society for the material published in the Journal of Organic Chemistry and the Journal of the American Chemical Society; Verlag Chemie GmbH for material published in Chemische Berichte; the Chemical Society of Japan for material published in the Bulletin of the Chemical Society of Japan.

I am most grateful to Dr. Ken'ichi Takeda, Director of the Shionogi Research Laboratory, for his kind consideration in having the typing of the manuscript and the drawing of chemical formulae done in his office; to Professor Masatomo Hamana of the Kyushu University, Professor Eisaku Hayashi of the Shizuoka College of Pharmacy, Dr. Manabu Fujimoto and Mr. Norio Tokutake of the Shionogi Research Laboratory for valuable discussions in going over the original Japanese manuscript; to Mr. Makoto Takahashi of the ITSUU Laboratory, who prepared final tracings of all the graphs printed in this book; to my former co-workers Drs. Ikuo Suzuki, Michitaka Natsume, Yutaka Kawazoe, Akihiro Ohta, and Masaaki Hirobe, for their kind cooperation in preparing the index; and also to Mrs. Yûko Ohta, Miss Michiko Katayama, and Mr. Shigeo Isono for typing the manuscript and drawing chemical formulae.

Finally, I am greatly indebted to Professor Morizo Ishidate, an old colleague of mine, who first tempted me to write this book.

Tokyo, February 1967 EIJI OCHIAI

Contents

Chapter 1

Introduction

My career in chemistry can be said to have started in 1922, as one of the research students under Professor H. Kondo of the Pharmaceutical Institute of Tokyo Imperial University, and in fact, with a theme on the hydrogenation of isoquinoline. At that time, Kondo was working on matrine, the main alkaloid of *Sophora flavescens* and my work was related to it. Kondo's research group had obtained a volatile amine of the composition $C_{10}H_{19}N$ by dry distillation of matrine with zinc dust. This base was later determined as β-lupinane[1] but, at that time, Kondo assumed it to be a decahydro derivative of methylquinoline or methylisoquinoline, and had been attempting to elucidate its structure by synthesis. As my work in preparing the starting materials took a long time and apparatus for high-pressure hydrogenation was not easily obtainable at that time, the work did not show any noticeable progress. On the opposite side of my laboratory bench, Dr. S. Yamaguchi was preparing a number of quinoline homologs and following up with their reduction. Yamaguchi, who was very deft with work in glass, discovered that by heating the salt of these quinolines with ice-cold saturated hydriodic acid and red phosphorus in a sealed glass tube at 230–240°, the corresponding 5,6,7,8-tetrahydroquinolines were formed in a good yield[2]. I was greatly impressed with this result because, from what could be learned from the literature then available, the pyridine portion of quinoline should be more easily reduced than the benzene portion and reduction with ordinary reducing agents would give the 1,2,3,4-tetrahydro derivative. I was unable, at that time, to understand why the hydrogenation of the pyridine portion was hindered by such a reduction method but the fact remained in my mind as an unforgettable and impressive experiment.

In the mean time, I was to take part directly in the work on matrine. The potassium salt of matrinic acid, obtained by hydrolysis of the lactam ring in matrine by treatment with ethanolic potassium hydroxide, was submitted to dry distillation with soda-lime, and two bases were obtained as the main products. For these bases, α-matrinidine (1-1) and dehydro-α-matrinidine

References p. 5

(1-2), the structural formulae below were proposed[3]. Of these, the structure of dehydro-α-matrinidine was later confirmed by its total synthesis by the research group under Tsuda[4], who was my colleague at the time. I was at a loss to explain the great resistance of the pyridine ring in this dehydro-α-matrinidine against reduction under drastic conditions, such as treatment with concentrated hydriodic acid and red phosphorus at a high temperature, or catalytic reduction over platinum.

This fact and other evidence gave me the idea[5] that hydrogen would behave as a nucleophilic reagent in this hydrogenation reaction and that the polar effect of a substituent might either promote or hinder the nucleophilic addition of hydrogen to the double bond system. In quinoline itself (1-3), the $-T$ effect* of the ring nitrogen would facilitate the nucleophilic addition of the polarised hydrogen molecule or atomic hydrogen on its pyridine portion and the 1,2,3,4-tetrahydro compound (1-4) would result. If a substituent with a large $+T$ effect was introduced into the pyridine portion (1-5), hydrogenation of the pyridine ring would be hindered and the 5,6,7,8-tetrahydro derivative (1-6) would result. Besides, it must be considered that the benzene portion would be more easily reduced by virtue of its naphthoid activity. Added to

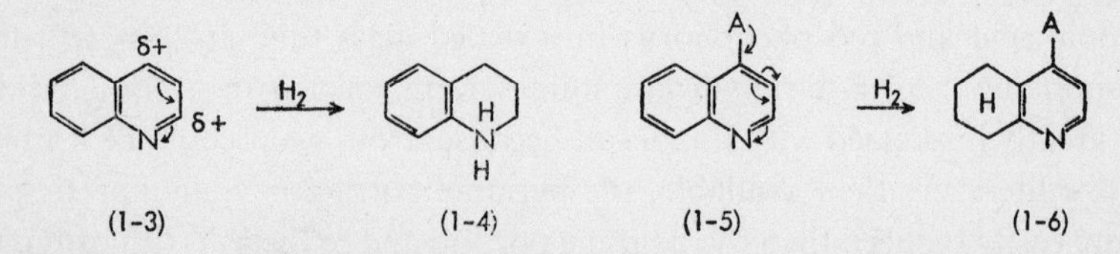

this idea, steric interference might also be taken into account. By such considerations, resistance of dehydro-α-matrinidine to hydrogenation was thought to be explicable by the mechanism shown in formula (1-2).

If this is the case, why should the quinoline salt form 5,6,7,8-tetrahydroquinoline instead of the 1,2,3,4-tetrahydro derivative on reduction with hydriodic acid and red phosphorus? Dr. Yamaguchi's experiments of those days came to my mind. This reaction had been carried out by sealing the compo-

* The + and − signs attached to *T*, *E*, *M*, and *I* effects are used throughout this book in accordance with those employed by C. K. Ingold, *Chem. Rev.*, 15 (1934) 225.

nents in a thick-walled, hard, glass tube and heating the tube. The reaction mixture in the tube had formed a crystalline mass of iodine color. The crystalline mass was transferred into a flask and steam-distilled. First, a large amount of iodine distilled over and solidified to crystals. This fact suggests that the crystalline mass in the reaction tube had consisted mainly of a polyiodide of the base. Heating of a quinoline salt with hydrogen iodide to above 200° might be the condition for formation of a polyiodide. In the polyiodide (1-7), the lone-pair electrons of iodine might act to decrease the $-T$ effect of nitrogen. Consequently, 2- and 4-positions on the pyridine ring would have less reactivity to nucleophilic addition of hydrogen and the benzene portion would be preferentially hydrogenated, as in the case of naphthalene, and the polyiodide of 5,6,7,8-tetrahydroquinoline (1-8) would result.

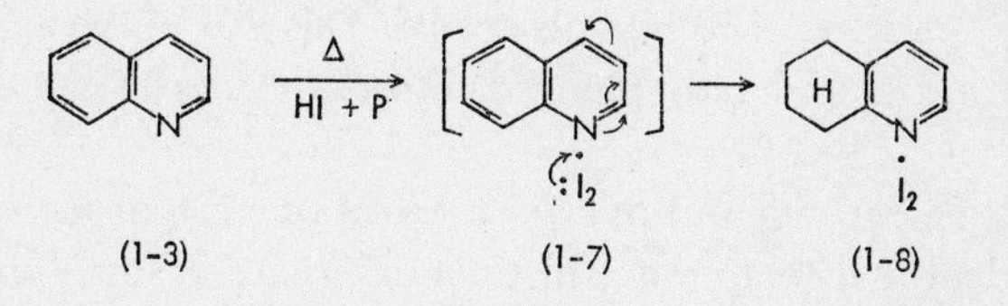

The Δ sign is used throughout this book to express "heating". The formula for polyiodide (1-7) is a tentative one.

This line of explanation has not been proved even to this day, but has offered an important clue to the question of electrophilic substitution of pyridines, which was rather difficult at that time. Pyridine was known to show a great resistance to electrophilic substitution owing to the strong polar effect of its ring nitrogen but, if the $-T$ effect of its ring nitrogen could be lessened or even cancelled by introduction of a group having lone-pair electrons onto the nitrogen atom, the position *ortho* and *para* to nitrogen, *i.e.*, 2- and 4-positions of the pyridine ring, would become active to electrophilic substitution.

However, if this fact were to be proved by experiments, polyiodide would be unsuitable as a material for electrophilic substitution, since it would easily liberate iodine and revert to pyridine. While searching for some appropriate compounds to try out this theory, I came across an important paper by Linton[6] of the U.S.A., who found (*cf.* Section *2.1.2*) that the dipole moment of pyridine 1-oxide (4.24 × D) was markedly smaller than the value calculated from the moments of pyridine and the *N*-oxide group, and pointed

References p. 5

out the contribution of the following resonance system (1-9 to 1-12) in pyridine 1-oxide.

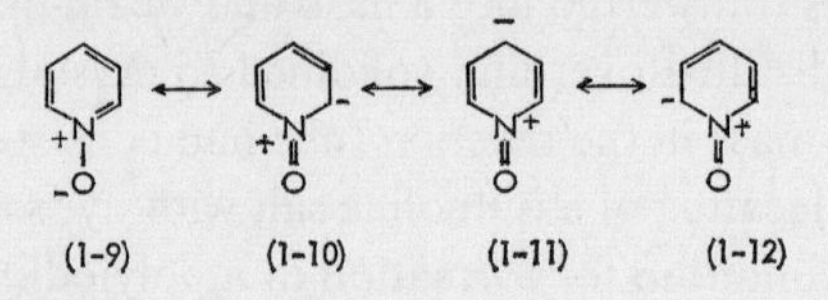

I then saw that pyridine 1-oxide itself was just the compound I had been looking for and that nitration would be the most promising electrophilic substitution for it, because under that reaction condition there was little likelihood that the oxygen in the *N*-oxide group would be liberated. Results of experiments supported my assumption, and formation of 4-nitropyridine l-oxide in a very good yield was observed[7]. This was in the autumn of 1941, when the world had already plunged into the throes of the Second World War.

We had been carrying out our work based on such observations, when it became evident that there was a distinct difference in the *N*-oxide functions between *N*-oxides of the pyridine series and those of trimethylamine and dimethylaniline, the difference being comparable to that of hydroxyl functions between phenols and alcohols, or of amino functions between aniline and alkylamines. It was therefore proposed, in June, 1943, that the *N*-oxide group in tertiary amine oxides should be classified into aliphatic and aromatic *N*-oxides[8].

This work was developed haltingly throughout the war period, without being known to other parts of the world. Since 1950, Professor den Hertog of the Netherlands, independent of our work but with similar ideas, started to publish papers on the nitration of pyridine l-oxide[9]. At about this time abstracts of our papers, published only in Japanese journals during the war years, began to appear in Chemical Abstracts. Through these abstracts, Professor Bunnett of the U.S.A. became interested in my work and kindly undertook the task of rewriting the German translation of a review of our work written in 1948[10]. An account of this work was published in Professor Bunnett's neat English in the Journal of Organic Chemistry[11] under my name.

From about this time, papers on aromatic amine oxides began to multiply gradually and many reviews on the subject were published in various countries[12]. In Japan, the work was developed mainly by my former collaborators, and especially by the co-operative research program subsidized by the Grant-in-Aid for Scientific Research from the Ministry of Education of Japan for 3 years from 1960. As a result, the chemistry of aromatic amine oxides made a great progress. For this reason, a number of papers dealing with investigations on aromatic amine oxides will be found predominantly in Japanese scientific journals.

The present book is a review of studies made on the chemistry of aromatic amine *N*-oxides up to the beginning of 1964, including the biological properties of some of these compounds, and also including bibliography published prior to 1941.

REFERENCES

1 H. Kondo, E. Ochiai, K. Tsuda, and S. Yoshida, *Ber.*, 68 (1935) 570.
2 S. Yamaguchi, *Yakugaku Zasshi*, 46 (1926) 556, 749.
3 H. Kondo, E. Ochiai, and K. Tsuda, *Ber.*, 68 (1935) 1899.
4 K. Tsuda, S. Saeki, S. Imura, S. Okuda, Y. Sato, and H. Mishima, *J. Org. Chem.*, 21 (1956) 1481.
5 E. Ochiai, *Yakugaku Zasshi*, 61 (1941) 298.
6 E. P. Linton, *J. Am. Chem. Soc.*, 62 (1940) 1945.
7 (a) E. Ochiai and M. O. Ishikawa, *Proc. Imp. Acad. (Tokyo)*, 18 (1942) 561; (b) E. Ochiai, M. O. Ishikawa, and K. Arima, *Yakugaku Zasshi*, 63 (1943) 79.
8 E. Ochiai, *Proc. Imp. Acad. (Tokyo)*, 19 (1943) 307.
9 (a) H. J. den Hertog and J. Overhoff, *Rec. Trav. Chim.*, 69 (1950) 468; (b) H. J. den Hertog and W. P. Combé, *Rec. Trav. Chim.*, 70 (1951) 581.
10 E. Ochiai, *Kagaku-no-Kenkyu*, 2 (1948) 1.
11 E. Ochiai, *J. Org. Chem.*, 18 (1953) 534.
12 (*a*) C. C. J. Culvenor, *Rev. Pure Appl. Chem.*, 3 (1953) 83; (*b*) H. J. den Hertog, *Chem. Weekblad*, 52 (1956) 387; (*c*) F. E. Cislak, *Ind. Eng. Chem.*, 47 (1955) 800; (*d*) A. R. Katritzky, *Quart. Revs.*, 10 (1956) 395; (*e*) M. Colonna, *Boll. Sci. Fac. Chim. Ind., Bologna*, 15 (1957) 1; (*f*) K. Thomas and D. Jerchel, *Angew. Chem.*, 70 (1958) 719; (*g*) E. N. Shaw in E. Klingsberg (Editor), *Pyridine and its Derivatives*, Interscience Publishers, New York, 1961, Part 2, p. 97; (*h*) D. V. Ioffe and L. S. Efros, *Usp. Khim.*, 30 (1961) 1325; (*i*) T. Kubota, *Bunkô Kenkyû*, 10 (1962) 97.

Chapter 2

General Survey of Amine Oxides

2.1 Amine Oxides

2.1.1 Amine Oxides of Aliphatic Tertiary Amines

The chemistry of amine oxides started with aliphatic tertiary amine oxides. These are usually prepared by the application of peroxide compounds, such as hydrogen peroxide or percarboxylic acids, to tertiary amines, but hydriodides of tertiary amine oxides can be formed by the application of alkyl iodide to hydroxylamine.

The structure of amine oxide is represented as in (2-1), where the lone-pair electrons of the nitrogen atom of the tertiary amine are bonded to the oxygen atom by co-ordinative covalence, and is usually expressed by the formula (2-2) or (2-3).

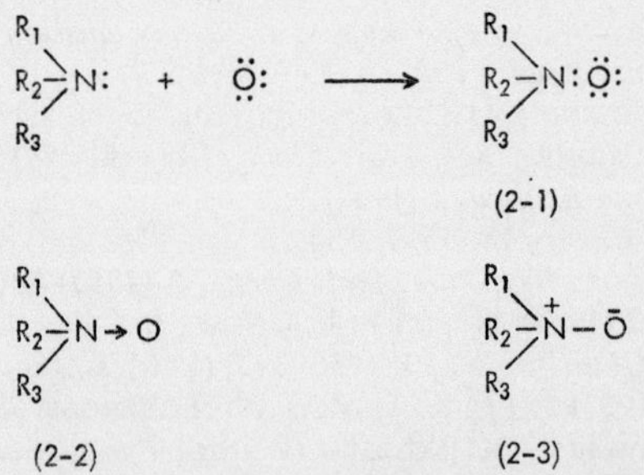

The first evidence for the single linkage between nitrogen and oxygen was furnished by the resolution of optical isomers. Meisenheimer[1] revealed that a compound can be resolved into optically active isomers when the three ligands to the nitrogen atom are different.

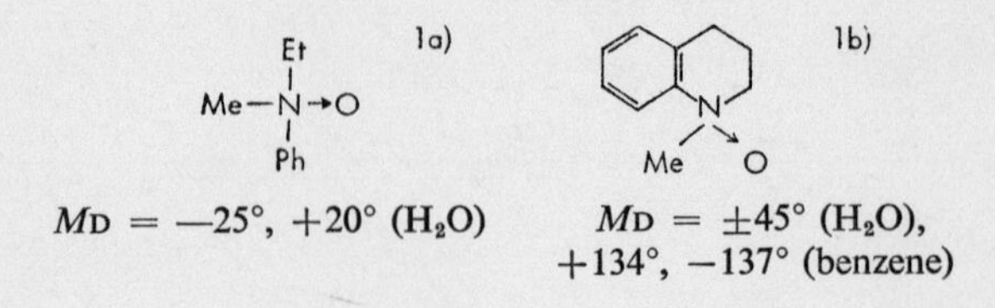

*M*D = —25°, +20° (H_2O) *M*D = ±45° (H_2O), +134°, −137° (benzene)

These experimental facts indicate that oxygen and three ligands take up a symmetric tetrahedral position in relation to the nitrogen atom, and that the linkage between nitrogen and oxygen is a single bond.

Further evidence was also furnished by Meisenheimer[2], who succeeded in the synthesis of two isomers (2-4 and 2-5) by the following route. The two isomers differ only in the product of pyrolytic decomposition.

$$R_3N{\rightarrow}O \xrightarrow{CH_3I} (R_3N\text{-}OCH_3)^+I^- \xrightarrow{Ag_2O} (R_3N\text{-}OCH_3)^+OH^- \quad (2\text{-}4)$$

$$(R_3N\text{-}OCH_3)^+OH^- \xrightarrow{\Delta} R_3N + CH_2O + H_2O$$

$$R_3N{\rightarrow}O \xrightarrow{HI} (R_3N\text{-}OH)^+I^- \xrightarrow{NaOCH_3} (R_3N\text{-}OH)^+(OCH_3)^- \quad (2\text{-}5)$$

$$(R_3N\text{-}OH)^+(OCH_3)^- \xrightarrow{\Delta} R_3N{\rightarrow}O + CH_3OH$$

Properties known to be common to these amine oxides can be explained to a certain extent by the presence of a semipolar N→O bonding but they can be summarized in the following two points.

(i) Large polarity. In general, amine oxides have large dipole moments, high melting points, and are difficult to volatilize. They are sparingly soluble in lipoids, such as ether or benzene, but at the same time are easily soluble in water, and have a tendency to form hydrates (formation of hydrogen bonding). Amine oxides generally have a lower basicity than the corresponding tertiary amines themselves.

(ii) Instability of nitrogen–oxygen bonding. The bonding between these two electronegative atoms is comparatively unstable, and constitutes one of the factors that govern the chemical properties of amine oxides. Principal phenomena that may be traced to this are the following:

(a) Facile reduction to the corresponding tertiary amines, either by catalytic reduction or by a mild reducing agent like sulfurous acid.

(b) Tendency to undergo decomposition into an aldehyde and a secondary amine on being heated.

$$(CH_3)_3N{\rightarrow}O \xrightarrow{\Delta} CH_2O + (CH_3)_2NH$$

(c) Alkyl halide adducts of amine oxides have a great tendency to be de-

References p. 17

TABLE 2-1

PROPERTIES OF ALIPHATIC TERTIARY AMINES AND THEIR *N*-OXIDES

	b.p. (°C)	m.p. (°C)	μ (D)	pKa	Solubility in water	Solubility in ether	SO_2	Heating
Trimethylamine *N*-oxide	—	212 (anhyd.)	5.02	4.65	easily sol.	diff.sol.	reduced to the amine	dec. into CH_2O + Me_2NH
Trimethylamine	70	liquid	0.65	9.74	easily sol.	easily sol.	—	—
Dimethylaniline *N*-oxide	—	152–153 (anhyd.)	4.79	4.21	easily sol.	diff.sol.	reduced to the aniline	dec. into CH_2O + PhNHMe + $PhNMe_2$ + $PhNH_2$
Dimethylaniline	194	2	1.58	5.06	limitedly sol.	easily sol.	—	—

composed into an aldehyde and a tertiary amine via the corresponding hydroxides.

$$(CH_3)_3N{\rightarrow}O \xrightarrow[RCH_2I]{} [(CH_3)_3NOCH_2R]^+I^- \xrightarrow[Ag_2O]{} [(CH_3)_3OCH_2R]^+OH^-$$

$$\longrightarrow (CH_3)_3N + RCHO + H_2O$$

These properties are compared in Table 2-1 for two of the representative amine oxides, trimethylamine *N*-oxide and dimethylaniline *N*-oxide, and the corresponding tertiary amine.

This co-ordinative covalent bond becomes more stable when phosphorus is substituted for nitrogen, by increasing the difference in electronegativity. While tertiary amines form amine oxides on reaction with peroxide compounds, tertiary phosphines tend to form phosphine oxides by treatment with weak oxidizing agents. Moreover, the phosphines may even form bonds with sulfur to give phosphine sulfides. A corresponding reaction has not been observed for amines.

$$Me_3P \xrightarrow{air} Me_3P{\rightarrow}O \quad \text{(ref. 3)}$$

$$Ph_3P \xrightarrow[S(CS_2)]{} Ph_3P{\rightarrow}S \quad \text{(ref. 4)}$$

2.1.2 *Amine Oxides of the Pyridine Series*

Properties of the amine oxides of the pyridine series, such as pyridine l-oxide and quinoline l-oxide, will be found to differ considerably from the common properties of the foregoing tertiary amine oxides. The most prominent difference is the marked decrease in polarity. Table 2-2 compares the dipole moments of pyridine l-oxide and aliphatic amine oxides with those of the corresponding amines.

TABLE 2-2

COMPARISON OF THE DIPOLE MOMENTS[5]

Amine *N*-oxide	μ (D)	Tertiary amine	μ (D)	Diff.
$Me_3N{\rightarrow}O$	5.02	Me_3N	0.65	4.37
$PhMe_2N{\rightarrow}O$	4.79	$PhNMe_2$	1.58	2.71
Pyridine 1-oxide	4.24	Pyridine	2.20	2.22

Pyridine and quinoline *N*-oxides are easily soluble in water and are both sparingly soluble in ether in the presence of water. These properties are similar to those of the foregoing amine oxides, but differ in the fact that pyridine and quinoline *N*-oxides are soluble in ether in the anhydrous state and are more volatile. As will be seen from Table 2-3, they can be purified by low-pressure distillation[6].

TABLE 2-3

BOILING POINTS OF AMINE OXIDES OF PYRIDINE SERIES[6]

N-Oxide	b.p. (°C/mm Hg)	*N*-Oxide	b.p. (°C/mm Hg)
Pyridine	138–140/15	4-Picoline	116–120/4
2-Picoline	107–109/4	Quinoline	171–172/4
3-Picoline	113–114/4	Isoquinoline	175–180/3

The second feature is the increased resistance to reduction. As will be described in a separate chapter (*cf.* Section *5.2.1*), *N*-oxides of pyridine and

References p. 17

quinoline are resistant to reduction with sulfurous acid or catalytic reduction over palladium. Their polarographic reduction potential is markedly lower than that of the *N*-oxides of aliphatic tertiary amines.

On the other hand, *N*-oxides of the pyridine series do have some properties common with the amine oxides (Section *2.1.1*). These are:

(a) When the alkyl halide adducts of such *N*-oxides are converted into the corresponding hydroxides and warmed, they undergo decomposition into an aldehyde, tertiary amine, and water (*cf.* Section *5.1.4. a*).

(b) When heated, these *N*-oxides do not form an aldehyde but are deoxygenated to tertiary amines.

These facts indicate that the nitrogen–oxygen bonding is still somewhat unstable.

Other important features of the *N*-oxides of the pyridine series are the great activity of the positions *ortho* and *para* to the nitrogen to nucleophilic substitution (compared with the corresponding tertiary amines) and the appearance of a marked tendency to undergo electrophilic substitution.

This tendency to electrophilic substitution is shown by the nitration reaction. As will be described in detail elsewhere (*cf. Section 6.2.1*), heating of pyridine 1-oxide with potassium nitrate in fuming sulfuric acid solution results in the formation of 4-nitropyridine 1-oxide in a good yield, with a minute amount of 2-nitropyridine as a by-product. Further heating of 4-nitropyridine 1-oxide with conc. sulfuric acid and potassium nitrate at 100° results in conversion of a part of the *N*-oxides into 4-nitropyridine*. This fact suggests that 2-nitropyridine obtained during nitration of pyridine 1-oxide must have been formed as 2-nitropyridine 1-oxide, which was subsequently deoxygenated; this occurs easily under these reaction conditions. These phenomena can be considered as the marked effect of *N*-oxidation, and contrasted with the fact that pyridine is highly resistant to nitration and forms only a small amount of 3-nitropyridine when treated at above 300°, giving a very poor yield of below 15–20%, in spite of many efforts to find better conditions for the reaction[7]. Similar activation of the position *para* to the *N*-oxide group can be seen in the nitration of quinoline 1-oxide.

These results of nitration indicate that the polar effect of the *N*-oxide function on the pyridine nucleus is much greater than has been assumed. The 4-position of the ring is greatly activated by the tautomeric effect of the

* Recently, F. Kröhnke proved that this deoxygenation is due to the nitrogen monoxide formed by the reaction (*cf. Section 5.2.5.c*).

N-oxide group. The reactivity of the 2-position is noticeable but much less so, owing to the simultaneous intervention of the inductive effect of the *N*-oxide function. This situation may be compared to the effect of the chlorine atom in chlorobenzene. The chlorine atom deactivates the ring but directs nitration to the *ortho* and *para* positions, with a *ortho*:*para* ratio of approximately 1:5 (ref. 8). Direction to the *ortho* and *para* positions is a consequence of its tautomeric effect, but deactivation, particularly of the *ortho* position, is caused by its inductive effect. A particularly striking observation is the fact that pyridine itself, which is very resistant to nitration, becomes very reactive when converted into the *N*-oxide; the *para* isomer greatly predominates. These facts show that both the tautomeric and inductive effects of the *N*-oxide function are greater than those of the chlorine function. The $+M$ and $-I$ effects of chlorine and the *N*-oxide group, which were estimated from their group moments, are presented in Table 2-4. They show good agreement with the experimental observations quoted above[9].

The nitro group in the 4-position of pyridine series *N*-oxides so formed is extremely active to nucleophilic substitution and undergoes facile reaction with alkoxides, phenoxides, and thiophenoxides to form the corresponding 4-alkoxy, 4-aryloxy, and 4-arylthio derivatives. Reaction with acyl chlorides gives a 4-chloro derivative. These substituents in the 4-position are also active to nucleophilic substitution and easily undergo exchange reactions with primary or secondary amines to form the corresponding 4-amino derivatives (*cf. Sections 8.3.2* to *8.6*).

Such a tendency to nucleophilic substitution also appears in the position *ortho* to the *N*-oxide group. For example, the reaction of pyridine l-oxide and

TABLE 2-4

POLAR EFFECTS OF CHLORINE AND *N*-OXIDE GROUPS[9]

	μ (D)	$-I$	$+M$
RCl	2.0	2.0	0
C_6H_5Cl	1.55	2.0	0.45
$R_3N \rightarrow O$	5.04	4.38	0
$C_5H_5N \rightarrow O$	4.24	4.38	2.34[a]

[a] Estimated from the value of the dipole moment of pyridine as 2.2 × D (Ref. 5).

phenylethynylsodium by warming in dimethyl sulfoxide gives 2-(phenylethynyl)-pyridine 1-oxide (*cf. Section 7.2*). These S_N reactions are often accompanied by deoxygenation of the *N*-oxide group, and heating of pyridine 1-oxide with phosphoryl chloride gives 2- and 4-chloropyridine (*cf. Section 7.4.1*). Such activation of *ortho* and *para* positions of the *N*-oxide group to nucleophilic substitution can be observed in many amine oxides of the pyridine series.

This experimental evidence indicates that the *N*-oxide function in amine oxides of the pyridine series imparts, in addition to the tautomeric effect comparable to that of the ring nitrogen atom of the corresponding tertiary amines, a strong inductive effect to the ring, so that its electron-withdrawing effect becomes far greater than that of the corresponding tertiary amines, and the activity of the positions *ortho* and *para* to the *N*-oxide group to S_N reaction is increased.

In other words, the *N*-oxide group in amine oxides of the pyridine series is characteristic in that it exerts both the electron-releasing and electron-withdrawing effect simultaneously.

Robinson[10] drew attention to the fact that he had already pointed out this possible overlapping of "push–pull" double polarization in nitrosobenzene[11] and suggested that the position *para* to the *N*-oxide group has both *(i)* an unshared electron (or electrons) and *(ii)* an incomplete electron configuration during the course of this reaction, giving it a free radical character.

$O{=}\overset{\oplus}{N}-C_6H_5{}^{\ominus}$ $\quad\quad$ $\overset{\ominus}{O}{=}N-C_6H_5{}^{\oplus}$

2.2 Aromatic Amine Oxides

2.2.1 Classification of Amine Oxides into Aliphatic and Aromatic Amine Oxides[9,12,13]

Consideration of *N*-oxides of the pyridine series shows that they are actually tertiary amine oxides in that the oxygen atom is co-ordinatively bonded by a single linkage to the tertiary nitrogen atom of the pyridine ring but, since the N→O group is bonded to an α-situated methine group by a double bond, they may be regarded as a kind of nitrones. Colonna[14], who at first had chosen to consider *N*-oxides of the pyridine series as aldonitrones or isoximes, pointed out the transformation of pyridine or quinoline 1-oxide into 2- and 4-chloro derivatives of pyridine or quinoline by the action of sulfuryl chloride, and assumed the formation of 2- and 4-hydroxy derivatives of pyridine or quinoline in the first stage of the reaction by a kind of Beckmann rearrangement. This agrees with the fact observed by Katada, Ochiai, and Okamoto that heating of pyridine or quinoline 1-oxide with acetic anhydride produced respectively α-pyridone[15a] or carbostyril[15b]. In this case the reaction was considered to be accompanied by 1,3-addition of the reagent (*cf. Section 7.4.6*).

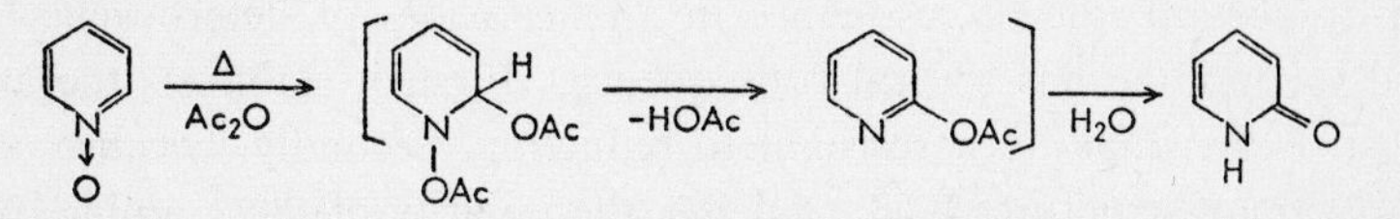

A similar 1,3-addition can be presumed in the Reissert reaction of quinoline 1-oxides, in which application of potassium cyanide in the presence of benzoyl chloride gives 2-cyanoquinolines (*cf. Section 7.4.2*).

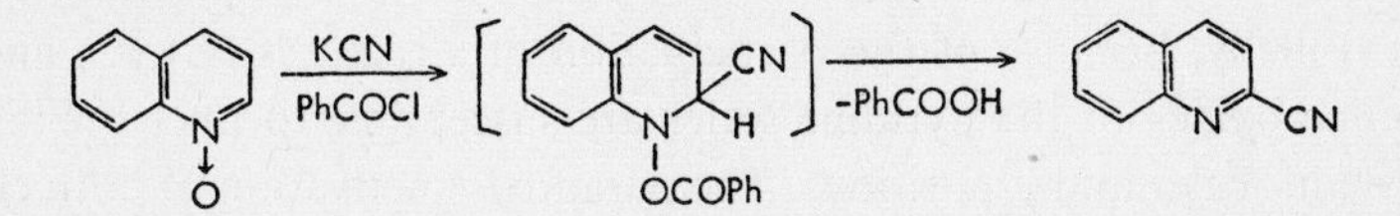

Such 1,3-addition is common in aldonitrones. A rearrangement similar to the Beckmann reaction is also observed in the case of dimethylaniline *N*-oxide. Heating of dimethylaniline *N*-oxide with concentrated sulfuric acid at 65–75° results in its rearrangement to *o*- and *p*-hydroxydimethylanilines[16]. On the other hand, nitrones in general are known to be easily reduced.

From these considerations, it is not appropriate to regard amine oxides of the pyridine series as mere nitrones. The double bond connecting the *N*-oxide group with the α-methine group of the pyridine ring is not the same as that in aldonitrones, but is a double bond of the pyridine ring with great aromaticity, which can only be represented as that of the resonance hybrid of the two principal resonance systems.

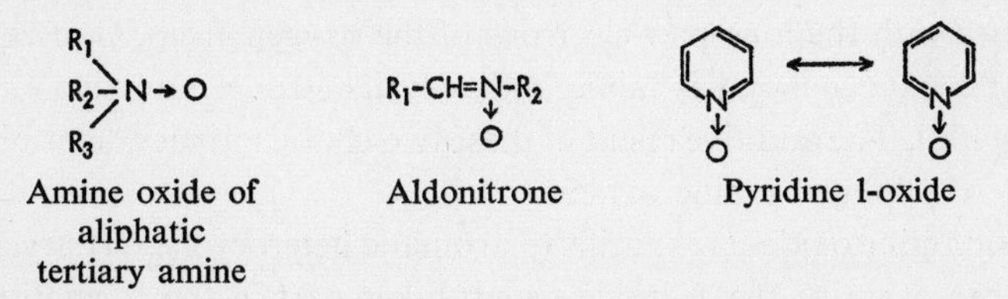

The lone-pair electrons of the oxygen atom in pyridine 1-oxide is in intimate resonance with the electron sextet of the aromatic ring, forming a resonance system between the following types of structure (A to C).

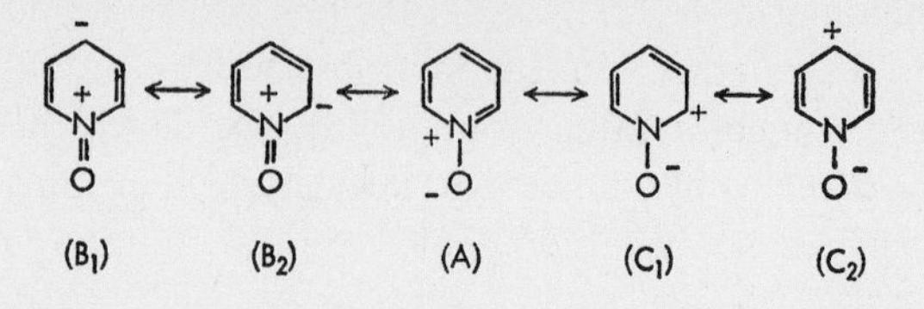

This important rôle of resonance in amine oxides of heterocyclic bases related to pyridine has several consequences. First, it stabilizes the amine oxide structure, making it resistant to reduction. Secondly, because of the contribution of structures B (B_1 and B_2), the polarity of these amine oxides is less than expected. Thirdly, by virtue of the participation of structures (B), these amine oxides are apt to be attacked by electrophilic substitution in the α- and γ-positions. As mentioned above, the predominance of γ-reactivity over α-reactivity is a consequence of the strong inductive effect of the *N*-oxide group. Finally, because of the participation of structures (C) (C_1 and C_2), the amine oxides of the pyridine series are susceptible to nucleophilic substitution in the α- and γ-positions. The paradox of activation of both electrophilic and nucleophilic substitutions in the same structure is resolved by recalling that it is not only the electron distribution in the isolated molecule that determines the rate of substitution reactions, but also the extent to which the molecule is polarized by the approaching reagent. The aromatic heterocyclic *N*-oxide function has the remarkable property of being strongly polarizable in both ways[13].

Ochiai and his collaborators recognized that the above conclusion can be applied to aromatic heterocyclic tertiary amine oxides in general, and they proposed the classification of amine oxides into aromatic and aliphatic amine *N*-oxides[9,12].

The properties of amine oxides reported in the literature had chiefly been those of aliphatic amine oxides, which lacked electron pairs that would be in resonance with the lone-pair electrons of the oxygen atom. Consequently, the nitrogen–oxygen bond is labile and only its strong inductive effect becomes manifest. The additive result of these two characteristics determines the properties of aliphatic amine oxides.

Aromatic amine oxides are specific to aromatic heterocyclic tertiary amines. The nitrogen atom of the *N*-oxide group takes part in the formation of an aromatic ring, and the nitrogen–oxygen linkage becomes markedly stable owing to the resonance of the lone-pair electrons of the oxygen atom with the

π-electron sextet of the aromatic ring, exerting both tautomeric effects of electron release and electron withdrawal simultaneously. Summation of these characteristics controls the properties of aromatic amine oxides. Attention should be paid to the fact that the characteristics of the aromatic *N*-oxide group are due to its inclusion in the aromatic ring rather than being a substituent. For example, the two lone-pairs of electrons on the oxygen of dimethylaniline *N*-oxide are not conjugated with the double bond of the ring, and the *N*-oxide group itself should be classified as an aliphatic one.

Studies on aromatic amine oxides, which developed with *N*-oxides of pyridine and benzopyridine series, such as pyridine, quinoline, isoquinoline, and acridine *N*-oxides as first representatives, have been extended step by step to the *N*-oxides of azines and benzoazines, such as pyrazine, pyridazine, pyrimidine, quinazoline, quinoxaline, phenazine, and cinnoline, and further to azoles and benzoazoles like imidazole, thiazole, benzimidazole, and benzothiazole, and to their condensed ring systems like purine. Further extension of the work may be anticipated.

Various hypotheses on the characteristic properties of the aromatic amine oxides described above have been confirmed by examination of the results of physical measurements, such as dipole moments, ultraviolet and infrared spectra, dissociation constants, nuclear magnetic and electron spin resonance spectra, and through X-ray crystallography, as will be described in Chapter 4.

2.2.2 Natural Occurrence of Aromatic Amine Oxides

The amine oxide most widely distributed in nature is trimethylamine *N*-oxide which was isolated by Suwa[17] in 1909 from the muscle of a shark (*Acanthus vulgaris*). Henze[18] revealed that this is also a constituent common to all muscles of the Cephalopods, and it was subsequently found to be a common constituent of muscles, organs, and sexual glands of Crustaceans, Cephalopods, and marine vertebrates.

Amine oxides have also been found in the alkaloid field. The first evidence appeared when in 1917 Polonovski[19] showed that geneserin, one of the alkaloids of calabar bean, was a mono-*N*-oxide (2-6) of another of its alkaloids, eserine. Later many strongly basic alkaloids with quinolizidine and pyrrolizidine rings were found to be accompanied by their *N*-oxides in nature. These included isatidine (2-7) from some *Senecio* species[20], oxymatrine (2-8)

References p. 17

from *Sophora flavescens*[21], and nupharidine (2-9) from *Nuphar japonicum*[22].

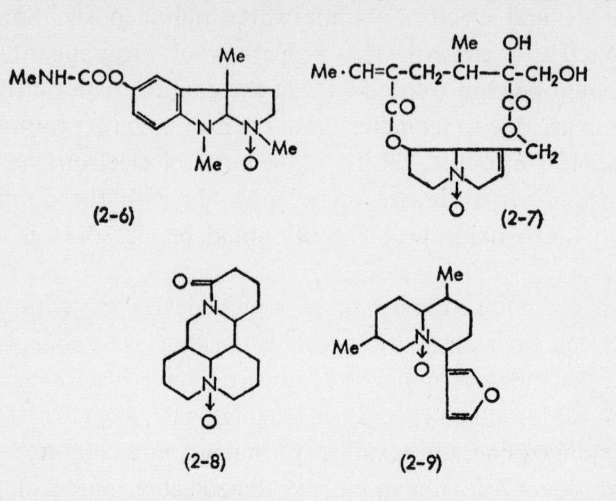

However, these are all aliphatic amine oxides, and only a few aromatic amine oxides have been found, mainly of the diazine series, as the metabolites of a comparatively small number of micro-organisms. The first of those discovered seems to be iodinin (2-10), a violet crystalline product of *Chromobacterium iodinum*, isolated by Davies[23] in 1938 during the examination of milk. Its structure was determined as 1,6-dihydroxyphenazine 5,10-dioxide by Clemo and others[24]. Iodinin was found to possess an antibacterial activity, and this discovery promoted studies on diazine *N*-oxides in the search for compounds with good antibacterial properties[25].

In 1943, White and Hill[26] isolated an antibiotic, aspergillic acid, from *Aspergillus flavus* and its structure was elucidated by the work of White and Hill[26], Dutcher[27], and of Spring and others[28] as 6-sec-butyl-2-hydroxy-3-isobutylpyrazine 1-oxide (2-11).

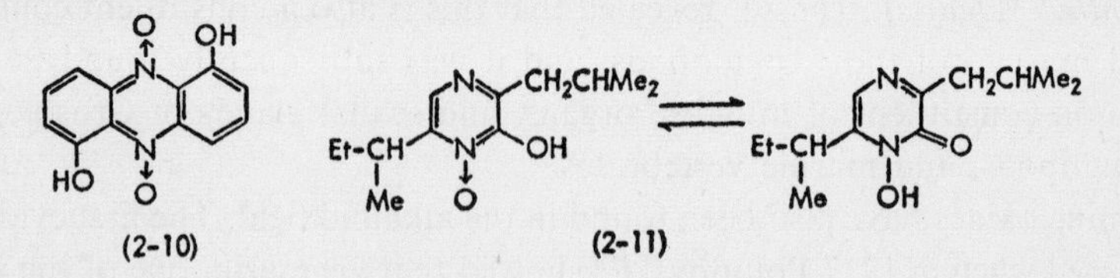

Aspergillic acid has antibiotic activity and shows a marked inhibitory effect *in vitro* on the growth of certain gram-positive and gram-negative bacteria, and on tubercle bacilli[29], but its toxity is so severe that it cannot be used for therapeutic purposes. However, the compound has opened a way for synthesis of antibacterial compounds with this acid as the model[30].

In connection with the work on aspergillic acid, antibiotics having similar cyclic hydroxamic acid skeletons have been isolated from various fungi. For example, hydroxyaspergillic acid (2-12) from *Aspergillus flavus*[31], muta-aspergillic acid (2-13) from *Asp. oryzae*[32], neohydroxyaspergillic acid (2-14) and 3-hydroxy-2,5-diisobutylpyrazine 4-oxide (2-15) from *Asp. sclerotiorum*[33] and pulcherrimine from *Candida albicans*[34].

Of these compounds, muta-aspergillic acid inhibits the growth of the Hiochibacterium and neohydroxyaspergillic acid antagonizes some bacteriophages. Pulcherrimine is an iron(III)-complex of pulcherrimic acid (2-16), and it is characterized as the *N,N'*-dioxide of 2,5-dihydroxypyrazine derivative.

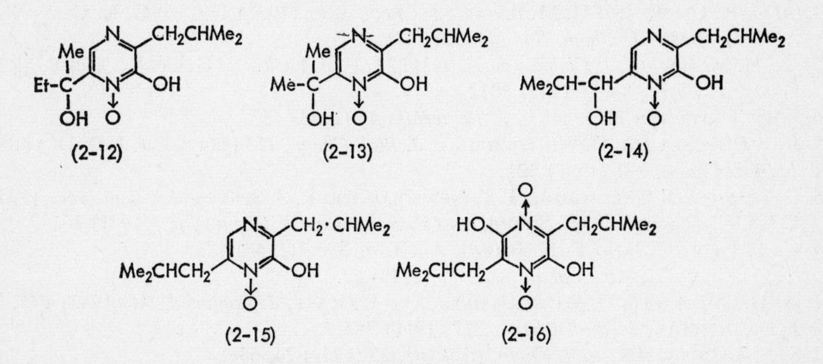

REFERENCES

1 (*a*) J. MEISENHEIMER, *Ber.*, 41 (1908) 3966; (*b*) J. MEISENHEIMER, *Ann.*, 385 (1911) 117; (*c*) J. MEISENHEIMER, *Ann.*, 428 (1922) 252; (*d*) J. MEISENHEIMER, H. GLAWE, H. GREESKE, A. SCHORNING, AND E. VIEWEG, *Ann.*, 449 (1926) 191.
2 J. MEISENHEIMER, *Ann.*, 397 (1913) 273; 399 (1913) 371.
3 A. CAHOURS AND A. W. HOFMANN, *Ann.*, 104 (1857) 23.
4 A. MICHAELIS AND L. GLEICHMANN, *Ber.*, 15 (1882) 803.
5 E. P. LINTON, *J. Am. Chem. Soc.*, 62 (1940) 1945.
6 E. OCHIAI, M. IKEHARA, T. KATO, AND N. IKEKAWA, *Yakugaku Zasshi*, 71 (1951) 1385.
7 (*a*) F. FRIEDL, *Ber.* 45 (1912) 428; F. FRIEDL, *Monatsh.*, 34 (1913) 759; (*b*) A. KIRPAL AND E. REITER, *Ber.*, 58 (1925) 699; (*c*) P. P. SCHORGGIN AND A. W. TOPTSCHIJEW, *Zh. Obshch. Khim.*, 7 (1937) 193 [*Chem. Zentr.*, II (1937) 4039].
8 M. L. BIRD AND C. K. INGOLD, *J. Chem. Soc.*, (1938) 918.
9 E. OCHIAI, *Proc. Imp. Acad. (Tokyo)*, 19 (1943) 307.
10 R. ROBINSON, *Tetrahedron*, 1 (1957) 170.
11 R. ROBINSON, *Chemistry and Industry*, 44 (1925) 456.
12 E. OCHIAI, *Yakugaku Zasshi*, 69 (1949) 1.
13 E. OCHIAI, *J. Org. Chem.*, 18 (1953) 534.
14 M. COLONNA, *Boll. Sci. Fac. Chim. Ind., Bologna,* (1940) 134 [*Chem. Abstr.*, 34 (1940) 7290].

15 (*a*) M. Katada, *Yakugaku Zasshi*, 67 (1947) 51; (*b*) E. Ochiai and T. Okamoto, *Yakugaku Zasshi*, 68 (1948) 88.
16 E. Bamberger and P. Leyden, *Ber.*, 34 (1901) 12.
17 A. Suwa, *Arch. Ges. Physiol.*, 128 (1909) 421; 129 (1909) 231 [*Chem. Abstr.*, 5 (1911) 1928].
18 M. Henze, *Z. physiol. Chem.*, 91 (1914) 230.
19 M. Polonovski, *Bull. Soc. Chim. France*, 21 (1917) 191.
20 E. C. Leisegang and F. L. Warren, *J. Chem. Soc.*, (1949) 486; (1950) 702; S. M. H. Christie, M. Kropman, E. C. Leisegang, and F. L. Warren, *J. Chem. Soc.*, (1949) 1700.
21 E. Ochiai and Y. Ito, *Ber.*, 71 (1938) 938.
22 M. Kotake, T. Kubota, and H. Hagitani, *Nippon Kagaku Zasshi*, 62 (1941) 442; T. Ukai and Y. Arata, *Yakugaku Zasshi*, 66B (1946) 138.
23 J. G. Davies, *Zentr. Bakteriol. Parasitenk.*, *Abt. II*, 100 (1939) 273 [*Chem. Abstr.*, 33 (1939) 6383].
24 (*a*) G. R. Clemo and H. McIlwain, *J. Chem. Soc.*, (1938) 479; (*b*) G. R. Clemo and A. F. Daglish, *J. Chem. Soc.*, (1950) 1481.
25 *e.g.* H. McIlwain, *J. Chem. Soc.*, (1943) 322; F. E. King, N. G. Clark, and P. M. H. Davis, *J. Chem. Soc.*, (1949) 3012.
26 E. C. White and H. J. Hill, *J. Bacteriol.*, 45 (1943) 433.
27 J. D. Dutcher and O. Wintersteiner, *J. Biol. Chem.*, 155 (1944) 359; J. D. Dutcher, *J. Biol. Chem.*, 171 (1947) 321.
28 G. Dunn, J. J. Gallagher, G. T. Newbold, and F. S. Spring, *J. Chem. Soc.*, (1949) 127; G. T. Newbold, W. Sharp, and F. S. Spring, *J. Chem. Soc.*, (1951) 2679.
29 G. T. Newbold and F. S. Spring, *J. Chem. Soc.*, (1948) 1865.
30 *e.g.* E. Shaw, *J. Am. Chem. Soc.*, 71 (1949) 67.
31 (*a*) G. T. Menzel, O. Wintersteiner, and G. Rake, *J. Bacteriol.*, 46 (1943) 109; (*b*) J. D. Dutcher, *J. Biol. Chem.*, 232 (1958) 785.
32 S. Nakamura, *Agr. Biol. Chem. (Tokyo)*, 25 (1961) 74, 658.
33 (*a*) U. Weiss, F. Strelitz, H. Flon, and I. N. Asheshov, *Arch. Biochem. Biophys.*, 74 (1958) 150 [*Chem. Abstr.*, 52 (1958) 9311]; (*b*) R. G. Micetich and J. C. MacDonald, *J. Chem. Soc.*, (1964) 1507.
34 (*a*) A. J. Kluyver, J. P. van der Walt, and A. J. van Triet, *Proc. Natl. Acad. Sci. U.S.*, 39 (1953) 583; (*b*) A. H. Cook and C. A. Slater, *J. Chem. Soc.*, (1956) 4133; (*c*) J. C. MacDonald, *Can. J. Chem.*, 41 (1963) 165; (*d*) A. Ohta, *Chem. Pharm. Bull. (Tokyo)*, 12 (1964) 125.

Chapter 3

Preparation of Aromatic Amine Oxides

3.1 By Direct N-Oxidation

In 1925, Meisenheimer and Stotz[1] carried out the reaction of quinaldine (3-1) with perbenzoic acid in benzene solution at room temperature and obtained a product identical with the substance derived from β-hydroxy-β-(*o*-nitrophenyl)ethyl methyl ketone (3-2) through reduction with zinc in acetic acid by Heller and Sourlis[2], and which was assumed by them to be 4-oxo-1,4-dihydroquinaldine (3-3). Meisenheimer and Stotz showed that this substance was not (3-3) but quinaldine 1-oxide (3-4).

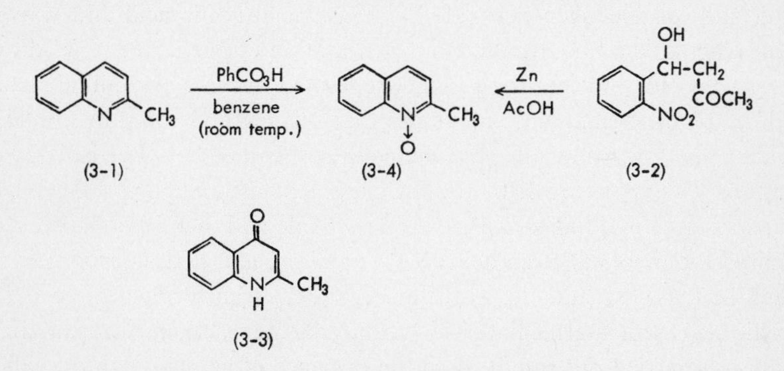

Soon after this, Meisenheimer[3] applied this reaction with perbenzoic acid further to quinoline, pyridine, isoquinoline, and 2,4,6-triphenylpyridine to transform them into their corresponding *N*-oxides. Since then, this method has developed as the standard procedure for the preparation of aromatic amine oxides.

This method with perbenzoic acid often necessitated isolation and purification of the *N*-oxide as its picrate. Later, Böhme[4] recommended the use of more stable monoperphthalic acid in place of perbenzoic acid as the oxidation agent. Bobranski *et al.*[5] utilized this acid for *N*-oxidation by reaction with the amine in ether, and showed that the *N*-oxide thereby formed would

References p. 70

separate out as its crystalline phthalate, thereby facilitating later purification.

The use of monoperphthalic acid is suitable for synthesis on a small scale, and its use has increased. For preparing a large amount of *N*-oxide, this method may be somewhat disadvantageous, because a large quantity of ether would be required for the preparation of the acid, and it is necessary to carry out the reaction at a low temperature. For this reason, it is more desirable to have a method whereby hydrogen peroxide can be used directly for preparing a large quantity of *N*-oxide, in order to avoid the use of the peracid requiring ether. Aliphatic tertiary amines are strongly basic and can be easily converted to the corresponding *N*-oxide in a comparatively good yield by allowing them to stand at room temperature with hydrogen peroxide solution[6]. Heteroaromatic tertiary amines, such as pyridines and quinolines, have a much weaker basicity and do not undergo this reaction.

In 1938, Clemo and McIlwain[7] found that whereas the application of hydrogen peroxide or benzoyl peroxide to phenazine in a neutral medium resulted in the recovery of the starting material, the use of Caro's acid or, better still, of hydrogen peroxide (5%) in glacial acetic acid and warming of the reaction mixture, resulted in the formation of phenazine *N*,*N'*-dioxide in a good yield, and that the product separated out in a crystalline state on cooling. Later, Ochiai and Sai[8] found that this method could be applied to pyridine and quinoline, and that the *N*-oxides formed could be isolated and purified by low-pressure distillation. Ochiai, Katada, and Hayashi[9] examined the reaction of pyridine with hydrogen peroxide and found that the reaction required the presence of an acid in the medium, and that a carboxylic acid would be the most suitable, as the acid and was transformed into a percarboxylic acid as an intermediate which catalyzed the reaction. Accordingly, the reaction is carried out smoothly in the presence of an acid, like phthalic or benzoic acid, which is sparingly soluble in water, by warming with hydrogen peroxide, acidification of the reaction mixture with hydrochloric acid, removal of the precipitated carboxylic acid, and then concentration of the filtrate to afford the hydrochloride of the *N*-oxide in a high yield.

Such an *N*-oxidation with a percarboxylic acid or with hydrogen peroxide in glacial acetic acid is the result of a bond formation between lone-pair electrons of the amine nitrogen and electron-deficient hydroxyl groups polarized to produce electrophilic activity. Progress of the reaction should depend mainly on the basicity of the nitrogen atom and on the ability of the oxidizing agent to form δ^{+}(OH). Consequently, the activity of the dicarboxylic acid would be larger than that of the monocarboxylic acid, and increase with the electron-withdrawing activity of its acyl group.

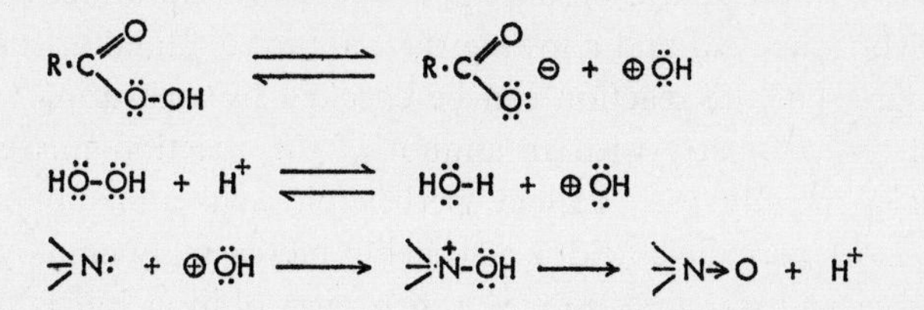

Den Hertog *et al.*[10] revealed that, while 2,6-dibromopyridine (3-5), whose basicity is low due to the overlapping of the steric hindrance and the inductive effect of bromine atoms in the two *ortho* positions, resisted *N*-oxidation with perbenzoic acid or peracetic acid, it formed the corresponding *N*-oxide (3-6) in 70–75% yield if warmed with 30% hydrogen peroxide in trifluoroacetic acid. Huisman *et al.*[11] also succeeded by the use of hydrogen peroxide in trifluoroacetic acid in converting 2-pyridyl 2,4,5-trichlorophenyl sulfide (3-7) to the corresponding sulfone *N*-oxide (3-8).

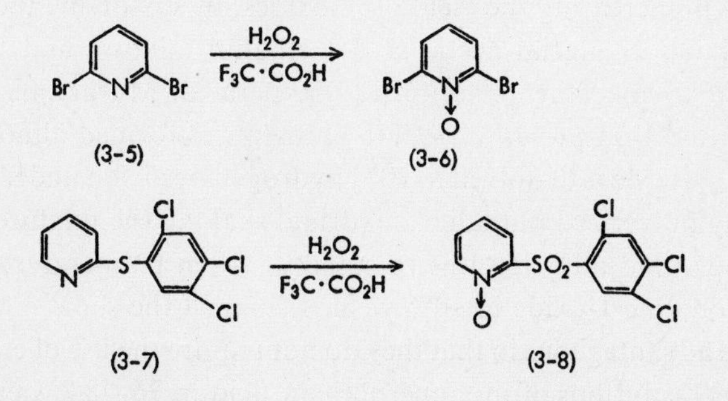

N-Oxidation of 2-chloroquinoline (3-9) is usually difficult and the starting material is recovered when using either monoperphthalic acid or hydrogen peroxide in glacial acetic acid, but Hamana and others[12] succeeded in obtaining 2-chloroquinoline l-oxide (3-10) in 60–65% yield by the application of monopermaleic acid[13].

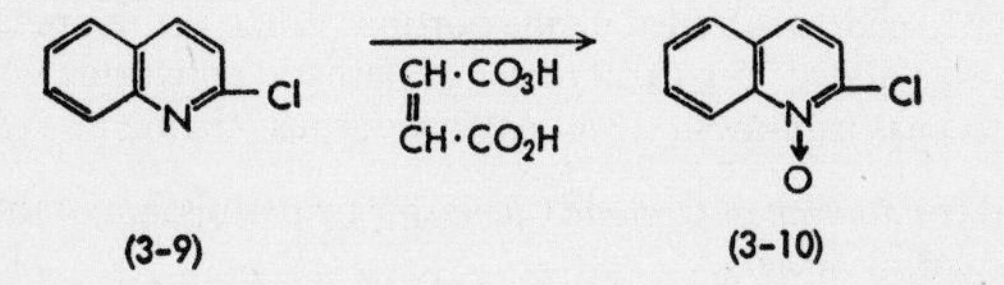

3.1.1 N-Oxidation with Percarboxylic Acid

The reaction takes place on adding a solution of an amine to a solution of a percarboxylic acid and allowing the mixture to stand at room temperature. Progress of the reaction can be checked by measuring the decrease of peracid by iodometry with an aliquot of the reaction mixture. Percarboxylic acids mostly used include perbenzoic acid and monoperphthalic acid, and monopermaleic acid is also being increasingly employed.

Perbenzoic acid was first used as a benzene solution but is now usually used as a chloroform solution, since a method has been elaborated for the preparation of such a solution[14].

When using monoperphthalic acid, the amine is added, directly *per se* or as an ether solution, to the ether solution of the peracid prepared by the method of Böhme[4,15,16]. In this case, the *N*-oxide formed will separate out as its crystalline phthalate, which facilitates subsequent treatment. If the amine is resistant to *N*-oxidation with this reagent, crystalline phthalate of the unreacted base may separate out and impede further reaction. Ochiai *et al.*[17] succeeded in preventing the precipitation of the phthalate of unreacted base, and thereby in increasing the yield of *N*-oxides, by dissolving the amine in a suitable volume of glacial acetic acid–methanol mixture and then adding this solution to the ether solution of the peracid. Murakami and Matsumura[18] reported that pyridine 1-oxide phthalate is formed almost quantitatively when pyridine is dissolved in 30% hydrogen peroxide under ice-cooling, adding finely pulverized phthalic anhydride, shaking the mixture to dissolve the anhydride, and allowing this to stand at room temperature. Colonna[19] obtained quinoline 1-oxide in 70% yield by almost the same method. These methods are advantageous in that they do not require the use of ether. Payne[16] obtained ether solutions of monoperphthalic acid in 76–78% yield in a short time by substituting the sodium hydroxide used as the basic agent in Böhme's formula by sodium carbonate.

Example 1. Solution of Monoperphthalic Acid*[16]

To a 1-l, round-bottomed flask, equipped with a mechanical stirrer and a thermometer, and cooled in an ice-salt bath, a solution of 62 g (0.5 mole) of sodium carbonate ($Na_2CO_3 \cdot H_2O$) in 250 ml of water was added. This was cooled to 5° and 70 g (0.6 mole) of 30% hydrogen peroxide was added in one portion. When the temperature of the mixture reached 0°, there was added 75 g (0.50 mole) of phthalic anhydride which had previously been pulverized to pass a 14-mesh sieve. After vigorous stirring at −5° to 0° for 30 min,

* Reprinted from the *Journal of Organic Chemistry* by permission of the copyright owners, the American Chemical Society.

all but a trace of the anhydride had dissolved. The solution was poured into a 2-l separatory funnel, covered with 350 ml of ether, and treated with an ice-cold solution of 30 ml of concentrated sulfuric acid in 150 ml of water. The liberated monoperphthalic acid was extracted with ether and separated completely from the aqueous layer by extraction with three more 150-ml portions of ether. The combined ether extract was washed with 200 ml of 40% ammonium sulfate solution and dried over anhydrous magnesium sulfate. Analysis for both hydrogen peroxide and monoperphthalic acid indicated the presence of less than 0.02 mole of the former and 0.39 mole (78% yield) of the latter (the yield based on phthalic anhydride).

Example 2. Dihydrocinchonine N,N′-Dioxide[17]

To a solution of 1 g of dihydrocinchonine dissolved in a mixture of 0.5 ml of acetic acid and 3 ml of methanol, 8 ml of ether solution of monoperphthalic acid (0.1077 mole/100 ml) was added by which the solution colored orange with evolution of heat to slight simmering. After 4 h, 5 ml more of the monoperphthalic acid solution was added, and prismatic crystals of m.p. 108–110° began to separate out after 3 h. The whole mixture was allowed to stand for 24 h at room temperature, the ether solution was decanted, and the crystalline residue was washed with ether. Yield, 1.670 g of prisms, melting at 108–110° (decomp.). The crystals were dissolved in 10% sodium hydroxide solution and extracted with chloroform. The chloroform solution was evaporated after drying over anhydrous sodium sulfate, and then the syrupy residue was recrystallized from methanol–acetone to give 745 mg of prisms, m.p. 184–185°.

The afore-mentioned decanted ether solution was shaken with 10% sodium hydroxide solution and the aqueous layer was shaken with chloroform. The chloroform solution was evaporated after drying over anhydrous sodium sulfate and the syrupy residue, which solidified on being wetted with acetone, was recrystallized from methanol–acetone mixture to give 50 mg of prisms, m.p. 184–185°. Total yield of dihydrocinchonine *N,N′*-dioxide, 795 mg (66%).

Example 3. 2-Methoxyquinoxaline 4-Oxide[20]

To a solution of 1 g of 2-methoxyquinoxaline dissolved in 15 ml of ether, 22 ml of ether solution of monoperphthalic acid (active oxygen, 10 mg/ml) was added slowly under ice-cooling and the mixture was allowed to stand in an ice box (below 10°) over night. The ether solution was decanted, the residue was basified with 40% potassium carbonate solution, and the mixture was extracted with chloroform. The chloroform solution was dried over anhydrous sodium sulfate and evaporated. The residue was dissolved in benzene and purified through a column of activated alumina. The crystalline residue thus obtained was recrystallized from petroleum ether to give white to slightly yellowish needles, m.p. 105–107°. Yield, 0.77 g (70%).

Recently, White and Emmons[13] reported that permaleic acid has stronger oxidative power than perbenzoic or perphthalic acid, and is more stable and well adapted as a reagent for *N*-oxidation. To prepare it, freshly crushed maleic anhydride is added in one portion to the methylene chloride solution of 90% hydrogen peroxide*, with stirring under ice-cooling. If the amount of

* If 90% hydrogen peroxide is not available, 30% hydrogen peroxide can be substituted for it. Compare the *example 4* (Ref. 12).

permaleic acid exceeds 10–15%, a second layer may be formed, and this layer will be found to contain a larger quantity of the peracid than the first layer by iodometric titration. This can be avoided by thorough agitation during the reaction. The oxidative activity of permaleic acid is comparable to that of trifluoroperacetic acid, but the former is cheaper than the latter and can be prepared more easily.

Example 4. 2-Chloroquinoline 1-Oxide (3-10) (Ref. 12)

To a stirred solution of maleic anhydride (14 g) in chloroform (30 ml), 30% hydrogen peroxide (2.8 g) was added under ice-cooling. After 2 hours' stirring, 2-chloroquinoline (1.6 g) was added to this mixture and the whole was kept in a refrigerator for 5 days. Deposited maleic acid was treated with a small amount of conc. potassium carbonate solution, dried over potassium carbonate, and the solvent was evaporated cautiously at water-pump pressure (below 60°). The residue was taken up with chloroform and chromatographed on alumina (elution with chloroform and acetone successively). The chloroform eluate gave an oil, which solidified on addition of a few drops of water. Recrystallization from acetone–petroleum benzine afforded 2-chloroquinoline 1-oxide, m.p. 85–92° (0.8–1.1 g). Drying of the sample over phosphorus pentoxide raised the m.p. to 105°. The acetone eluate furnished a small amount of carbostyril.

3.1.2 N-Oxidation with Hydrogen Peroxide and Glacial Acetic Acid

In this process, 30% hydrogen peroxide is added to a solution of the amine in glacial acetic acid and the mixture is warmed. Usually, 1.2–1.7 equivalents of hydrogen peroxide solution are added in two to three portions, and the reaction temperature is raised from 50–60° to 80–95°. After completion of the reaction, the solution is concentrated in vacuum, but when the experiment is performed on a large scale, it is necessary to decompose the excess of hydrogen peroxide with manganese dioxide or palladium on carbon and filter the mixture, otherwise the latter may occasionally explode on concentration. The concentrated residue is basified with sodium carbonate and extracted with chloroform. The chloroform extract is dehydrated and evaporated, and the residue is recrystallized or distilled in vacuum. Chumakov and Poluchenya[21] obtained a good result by adding the amine directly to hydrogen peroxide solution, warming this during addition of acetic anhydride, and decomposing the excess of hydrogen peroxide with formaldehyde.

Example 5. Quinoline 1-Oxide[22]

To a solution of 129 g (1 mole) of quinoline in 300 ml of acetic acid (of ca. 70% purity, recovered from a similar reaction), 90 ml of 29% aqueous hydrogen peroxide was added and the mixture was warmed at 65–70° for 3 h. Further 80 ml (total 1.5 moles) of hydrogen peroxide solution was added and the whole mixture was warmed at 65–70° for

6 h. The reaction mixture was concentrated in vacuum, the residue was basified by the cautious addition of hot, saturated solution of sodium carbonate, and the reddish brown oily substance that separated out was extracted with chloroform. The chloroform extract (*Note 1*) was evaporated on a water bath of *ca.* 80° as much as possible. The inorganic salts that separated out were removed while hot by suctional filtration and the filtrate was further evaporated under vacuum. The extraction with chloroform was carried out in four portions of 200, 150, 100, and 100 ml and each extract was treated separately. To each residue of these extracts, 40, 10, 5 and 2 ml of water was added, respectively, and when they were allowed to stand, quinoline 1-oxide crystallized as its hydrate. After allowing the mixture to stand over night, the crystals were triturated with ether, collected by suctional filtration, washed with ether, and dried. The first product was a little brownish, but was pure enough to be used as a material for nitration. Second and later products were almost colorless. Yield from each fraction, 138, 20, 7, and 2 g, respectively, and total yield was 167 g (92%); b.p. 171–172°/4 mm, m.p. 60–62°.

Note 1: Washing of the extract with water to remove the inorganic salt will result in precipitation of quinoline 1-oxide hydrate. The extract should be processed without washing.

Example 6. 2,6-Lutidine 1-Oxide Hydrochloride[23]

To an intimate mixture of 50 g of 2,6-lutidine and 150 g of glacial acetic acid, about one-third of 90 g of 30% hydrogen peroxide was added, while the initial mixture was maintained at 80° on a water bath, and the remaining portions were added gradually every 2 h, the total time of heating being 8 h. The reaction mixture was concentrated on a water bath under a reduced pressure, 20 ml of water was added when the residue became small, and the resulting solution was further concentrated. This concentration procedure was repeated three times and, finally, 40 ml of 30% hydrochloric acid was added and the acid solution was also concentrated. This concentration was repeated three times again with addition of 10 ml of water each time to remove acetic acid as much as possible. White hydrochloride precipitated out when the solution was evaporated almost to dryness; the crystalline mass was dried on a porcelain plate and recrystallized from 300 ml of ethanol, with decolorization by *ca.* 2 g of carbon. Yield, 68.5 g (92%) of colorless needles, m.p. 218–218.5°. Further recrystallization from ethanol–ethyl acetate mixture raised the melting point to 219.5°.

This method of *N*-oxidation with hydrogen peroxide and glacial acetic acid is simple and adapted for treatment of a large quantity with wide application, but is attended with various side reactions such as oxidation and hydrolysis due to the use of acid and oxidizing agent, and to heating for a long period of time. In order to prevent such side reactions, substituents may be converted into a safer form, as will be explained in the following section on these side reactions (*cf. Section 3.1.3*).

As has already been stated, a modification of this method has been devised in which a mixture of the base and hydrogen peroxide is warmed with a carboxylic acid sparingly soluble in water, such as benzoic or phthalic acid, as a catalyst[9].

A method similar in idea to the foregoing appeared in a patent[24] on the manufacture of *N*-oxides. In this process, pyridine, quinoline, or their methyl homologs are dissolved in *ca.* 1.2 moles of 35–90% hydrogen peroxide solution and the mixture is heated directly or in a solvent of glacial acetic acid or tert-butanol at 60–65°, using as catalyst the peroxide of the oxy-acid of Group VI elements in the periodic table, such as tungstic and molybdic acids. The reaction progresses most favorably in glacial acetic acid as a solvent with tungstic acid as a catalyst, but the yield is generally 63–85%, no significant difference being found from that in the absence of tungstic acid.

As will be shown later (*cf. Section 3.1.3.a*), it has been known that *N*-oxidation of diazine bases sometimes resulted in the formation of two kinds of isomeric *N*-monoxide. In such a case, the product should be examined by gas chromatography and, if the presence of two isomers is found, the product should be submitted to separatory purification.

Example 7. 3-Methylpyridazine 1-Oxide (3-11) and 2-Oxide (3-12) (Ref. 25)

A mixture of 2.4 g of 3-methylpyridazine, 15 ml of glacial acetic acid, and 7 ml of 30% hydrogen peroxide was heated at 70° for 3 h, further 7 ml of 30% hydrogen peroxide was added, and the whole kept at the same temperature for 3 h. To this solution, 15 ml of water was added, acetic acid was evaporated under a reduced pressure, and this procedure was repeated twice. After neutralization of the residue with sodium carbonate, the solution was extracted with chloroform, the chloroform layer was dried over anhydrous sodium sulfate, and evaporated. The residue was dissolved in benzene and chromatographed over alumina (27 g). The column was eluted with benzene and the benzene eluate was recrystallized from benzene to colorless prisms (2-oxide), m.p. 83–84°. Yield, 630 mg.

The crystalline residue from the fraction eluted with chloroform was recrystallized from benzene to colorless prisms (1-oxide), m.p. 68.5–69.5°. Yield, 230 mg.

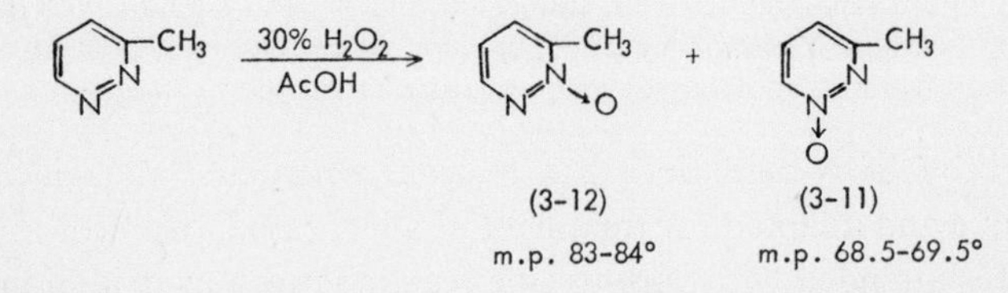

3.1.3 Side Reactions

Direct oxidation of amines is often accompanied by other oxidation reactions, such as nucleophilic substitution with a hydroxyl group. This is especially marked in the case of *N*-oxidation with hydrogen peroxide and glacial acetic acid, since it involves heating over a long period of time, and the reaction is also complicated by hydrolysis due to the acid.

It is natural that, if an easily oxidizable group is present in the molecule besides tertiary amine-nitrogen, this group may be oxidized at the same time. Peracids and hydrogen peroxide also act as a nucleophilic reagent, and the oxidation reaction may often be accompanied by substitution of the ring with hydroxyl, by cleavage of the ring, or by change in the ring system.

(a) Oxidation of Other Substituents

A primary amino group in the molecule can sometimes be oxidized to a nitro group by treatment with hydrogen peroxide in glacial acetic acid or the like, as shown below[26].

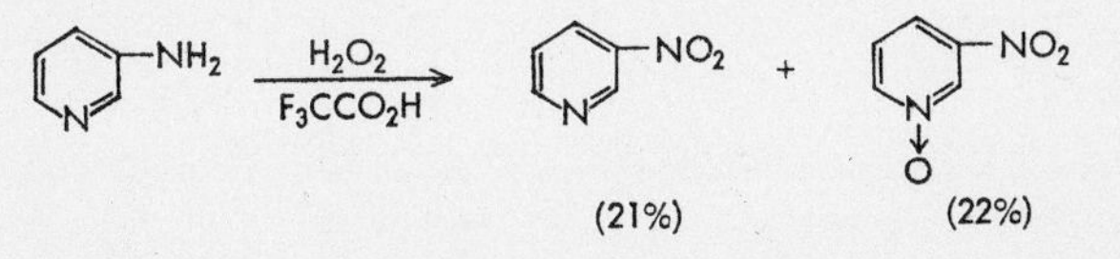

Brown[27] treated 2-aminopyridine 1-oxide and five of its methyl homologs with a mixture of 30% fuming sulfuric acid and 27% hydrogen peroxide at 10–20° and obtained the corresponding 2-nitropyridine 1-oxide in 45–60% yield, while Herz and Murty[28] carried out the same reaction on 4-amino-3-picoline and obtained 4-nitro-3-picoline 1-oxide as the main product.

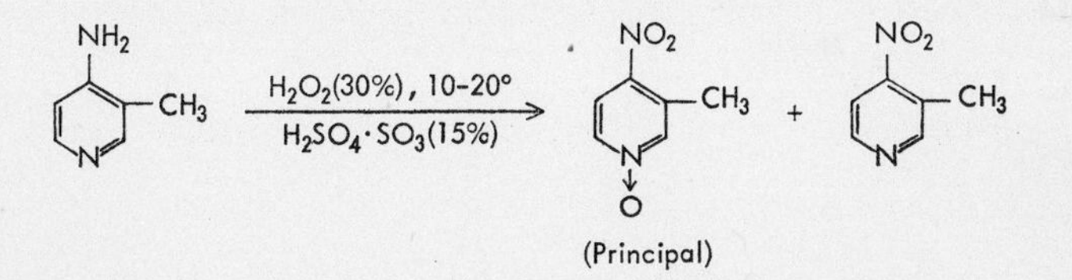

In such a case, it is usual to protect the primary amino group, if present, by acylation and hydrolyze the acylamino group after the *N*-oxidation.

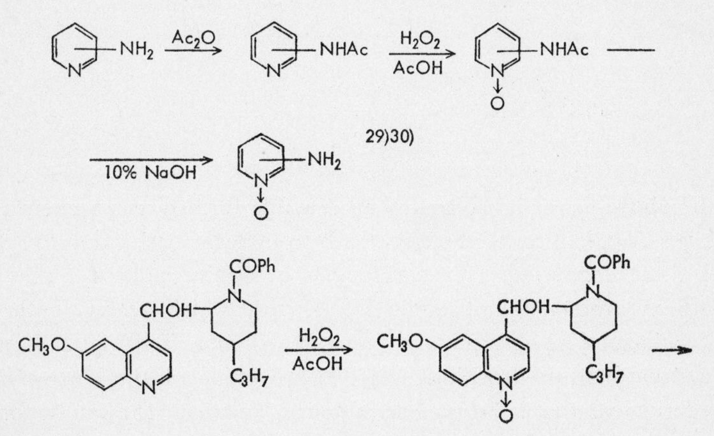

References p. 70

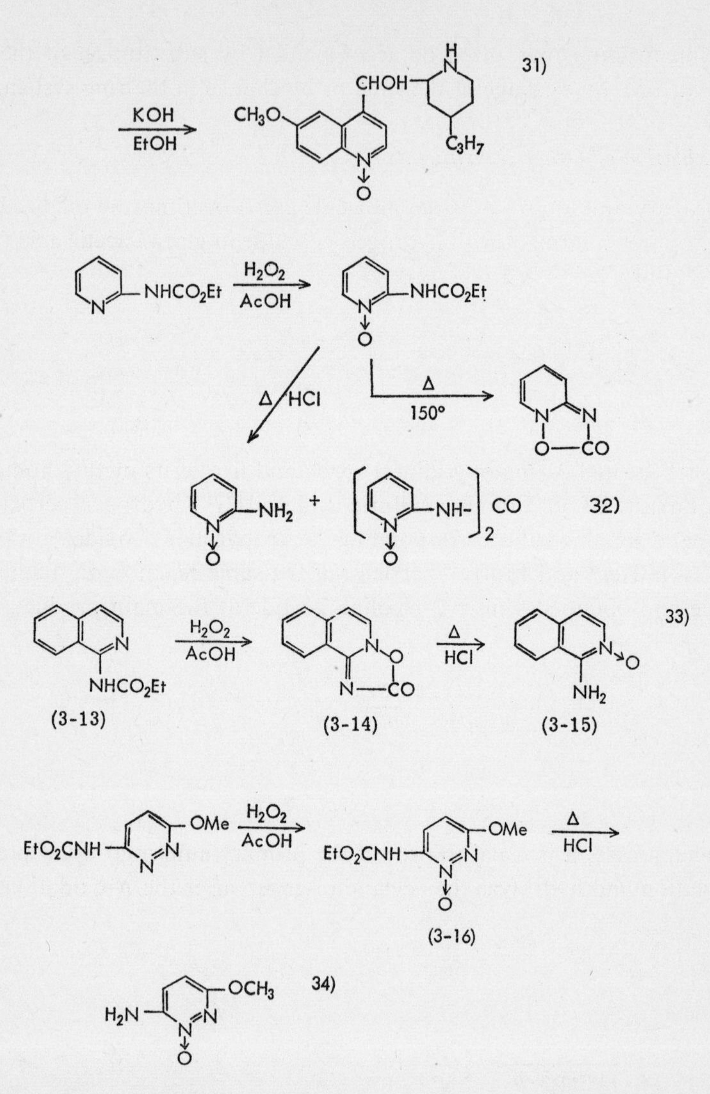

In the above examples, the urethan *N*-oxide formed as an intermediate during the *N*-oxidation of urethans tends to undergo cyclization to a 1,2,4-oxadiazolone derivative. This tendency is especially marked in the case of 1-(ethoxycarbonylamino)isoquinoline 2-oxide which readily cyclizes to [1,2,4]oxadiazolo[3,2-*a*]isoquinolin-2-one (3-14), but, if the latter is heated with concentrated hydrochloric acid, the *N*-oxide of the corresponding amino derivative, that is 1-aminoisoquinoline 2-oxide (3-15), can be formed.

In contrast, 6-(ethoxycarbonylamino)-3-methoxypyridazine 1-oxide (3-16) is resistant to the cyclization to oxadiazolone.

It has been found in *N*-oxidation of sulfonamides that a part of the imide group is oxidized to hydroxylamine[35].

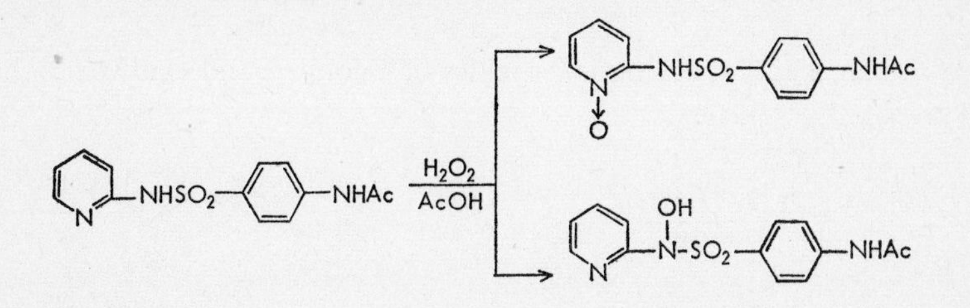

Pentimalli[36] revealed that oxidation of 2-(4-dimethylaminostyryl)pyridine (3-17) with perbenzoic acid results in oxidation of the dimethylamino group alone without oxidation of the ring nitrogen.

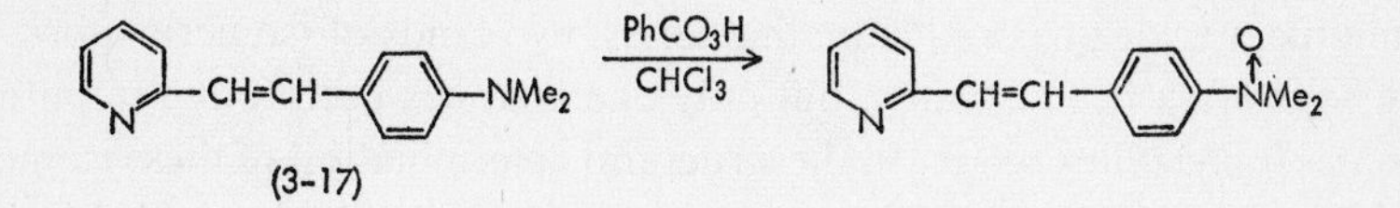

When an aliphatic tertiary amino group is also present in the molecule of an aromatic tertiary amine, the formation of an aliphatic and aromatic double *N*-oxide is generally observed. In such a case, if the di-*N*-oxide is allowed to stand in aqueous sulfurous acid solution at room temperature, only the aliphatic *N*-oxide group is reduced and the aromatic *N*-oxide group remains, in general, intact, resisting such a reduction (*cf. Section 5.2.1*). If the original diamine is allowed to stand in dilute aqueous acidic solution of hydrogen peroxide at room temperature, only the nitrogen of the aliphatic tertiary amine, with greater basicity, is oxidized to form the aliphatic *N*-oxide. For example, this may be seen in the oxidation of cinchona bases (3-18).

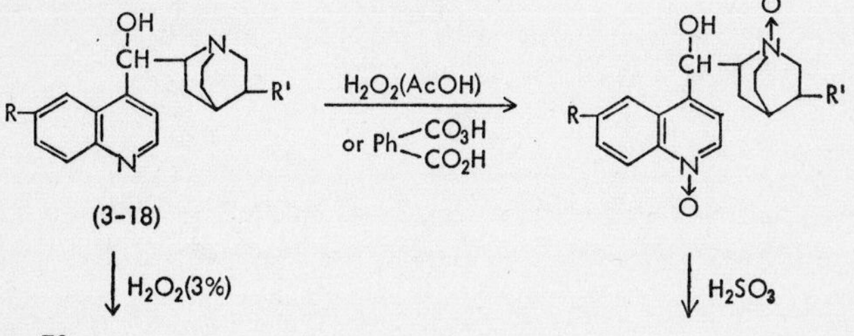

References p. 70

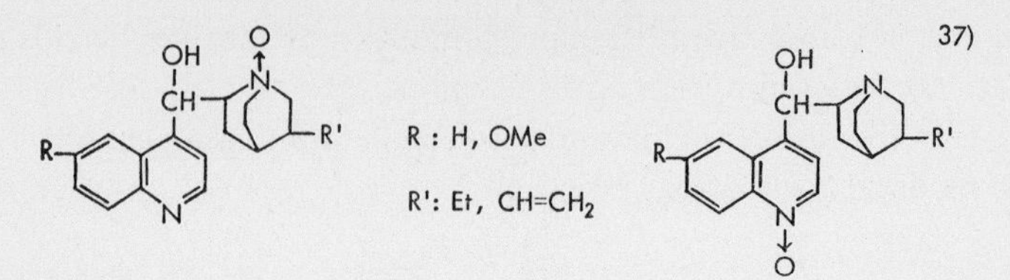

In a similar manner, aromatic *N*-oxides of nicotine[38] and anabasine[39] have been prepared.

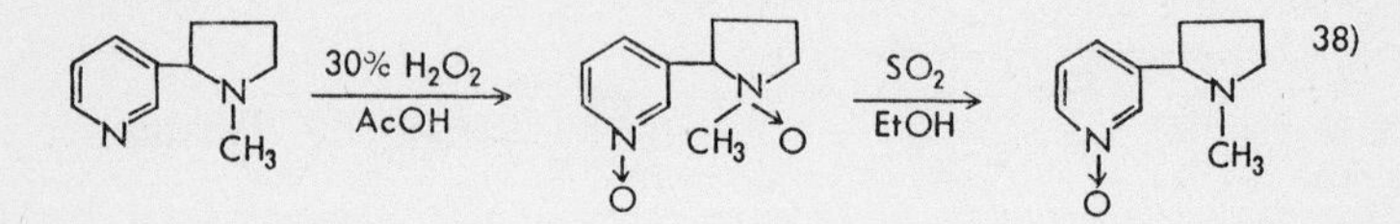

In diazine series having two tertiary amino-nitrogen atoms in the heterocycle, simultaneous formation of mono- and di-*N*-oxides has been shown to occur in pyrazine and benzopyrazine series (*i.e.*, quinoxaline and phenazine), and only monoxide has been obtained in other heterocyclic series. In pyrimidine, pyridazine, and cinnoline series, two kinds of isomeric mono-*N*-oxide have often been isolated, but only one kind has been found in phthalazine and quinazoline series. In the structural determination of these isomeric mono-*N*-oxides, besides reaction with acetic anhydride or with acid chloride, and so on, measurement of dipole moment and of NMR spectrum has been of considerable help (*cf. Section 3.3* p. 67).

In a series of compounds with pyrimidine fused with imidazole in its 5-6 position, *i.e.* in purine systems, the imidazole ring remains intact, while the nitrogen atom in 1-position of the pyrimidine nucleus is found to be oxidized.

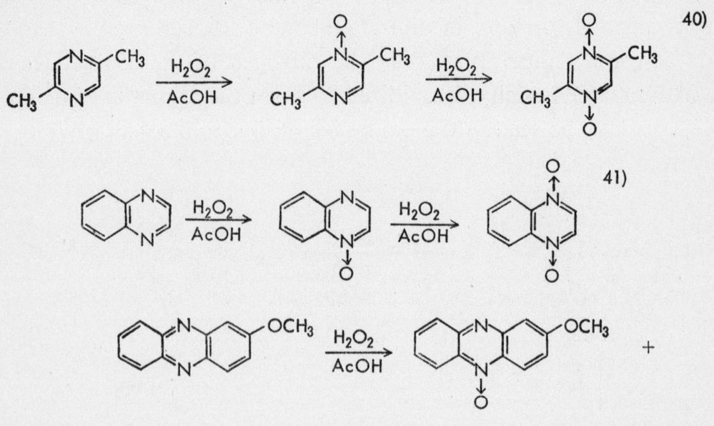

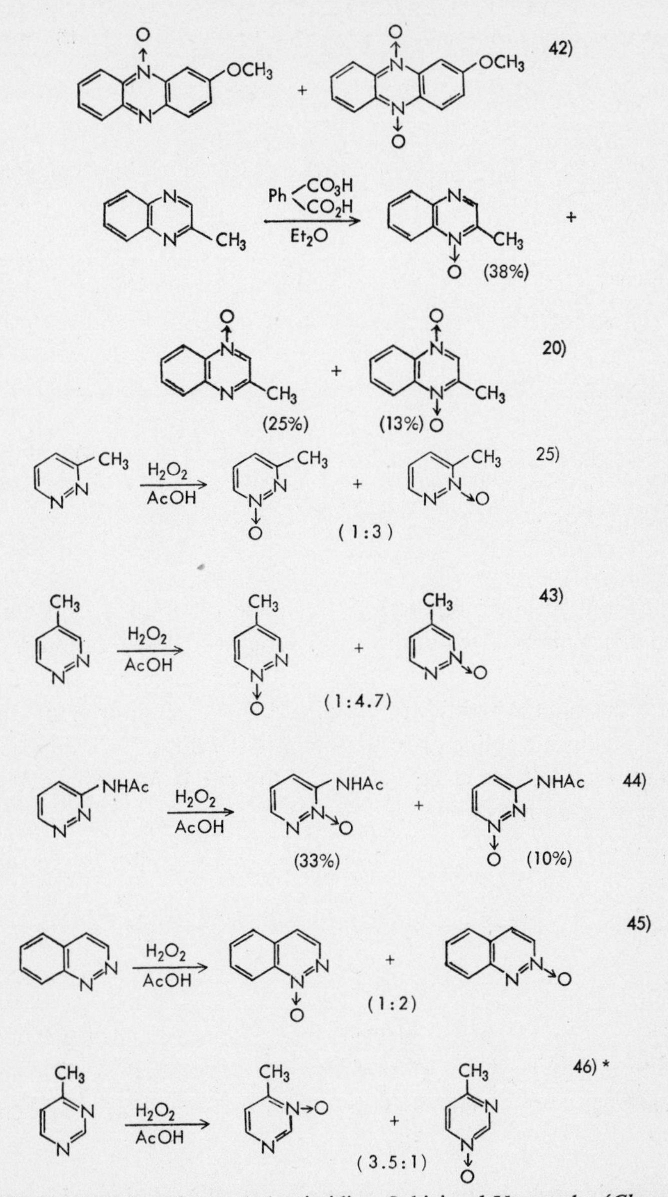

* In a similar *N*-oxidation of 4-methylpyrimidine, Ochiai and Yamanaka (*Chem. Pharm. Bull. (Tokyo)*, 3 (1955) 175) obtained mono-*N*-oxide (m.p. 74–78°) which was mistakenly assumed at that time to be 1-oxide in view of the steric hindrance of the methyl group. The presumed 1-oxide of earlier workers should consequently be corrected to the 3-oxide as a result of this work.

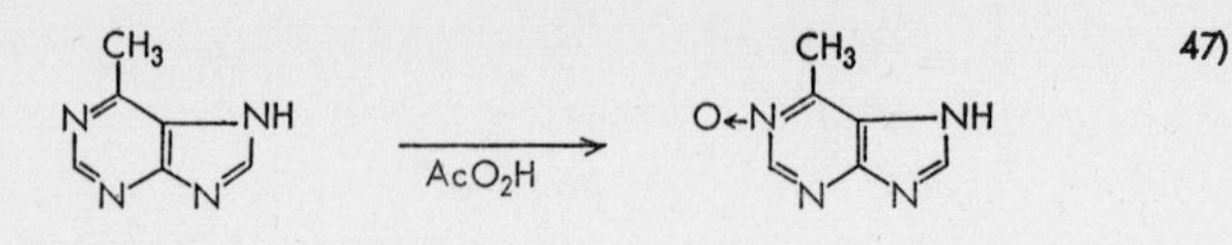

In naphthyridines formed by fusion of two pyridine rings, the two nitrogen atoms are oxidized similarly.

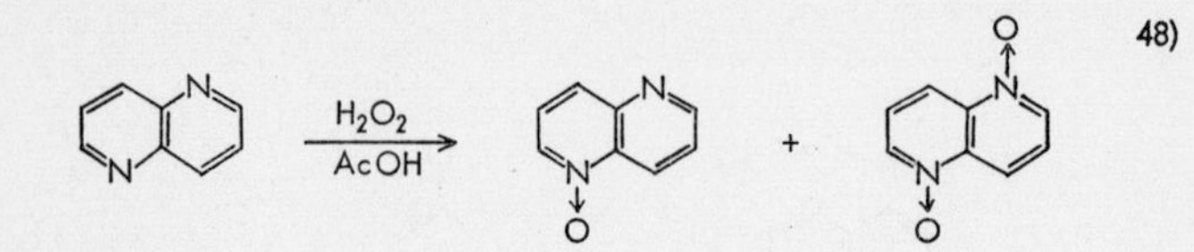

Analogously,

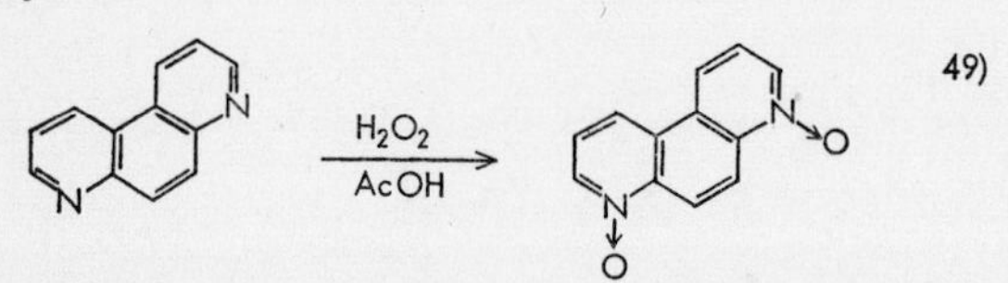

Colonna and Risaliti[50] reported that the azo group in 2- or 4-position of the pyridine ring is oxidized to the azoxy group with *N*-oxidation of the ring, and that the oxygen is first introduced onto the nitrogen atom of the azo group distant from the pyridine ring. Itai and Kamiya[51] showed that the azido group in 4-azido-2-picoline (3-20) is decomposed during *N*-oxidation of the compound, and 4,4′-azoxy-2-picoline 1,1′-dioxide is formed via the intermediate having an azo group.

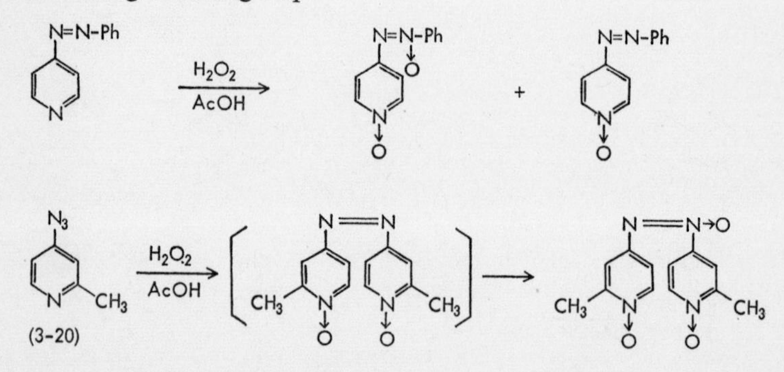

The foregoing has shown some examples of *N*-oxidation of compounds having nitrogen-containing substituents in the molecule. As examples of compounds with a sulfur-containing group, which also has active lone-pair electrons like nitrogen, Kobayashi and others[52] carried out the *N*-oxidation of compounds with dihydrothiophene ring condensed at 3-4 or 2-3 position of the quinoline ring, *i.e.*, 2,3-dihydrothieno[3,2-*c*]quinaldine (3-21) and 2,3-dihydrothieno[2,3-*b*]quinoline (3-22). They found that the sulfur atom was oxidized prior to nitrogen and that the sulfone-amine type of product (3-23) was formed via the sulfoxide by the use of monoperphthalic acid, and that sulfone *N*-oxide (3-24) was isolated only from (3-21).

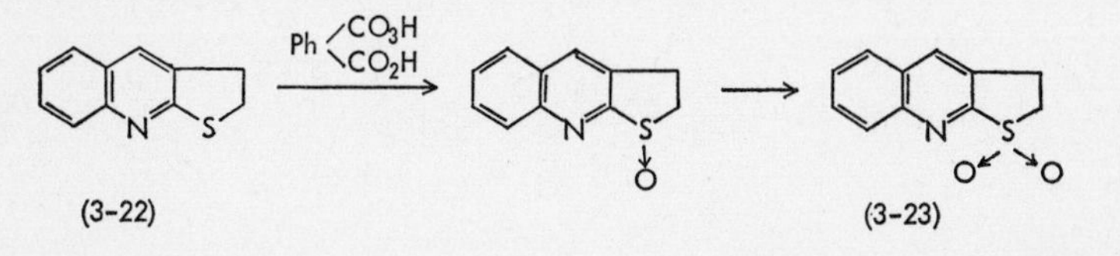

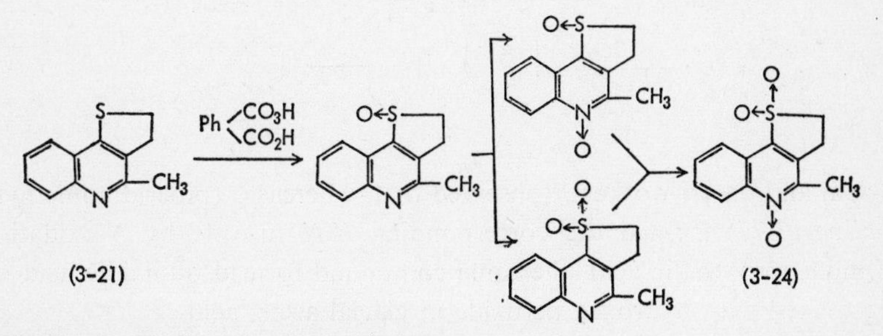

Similar examples are shown below.

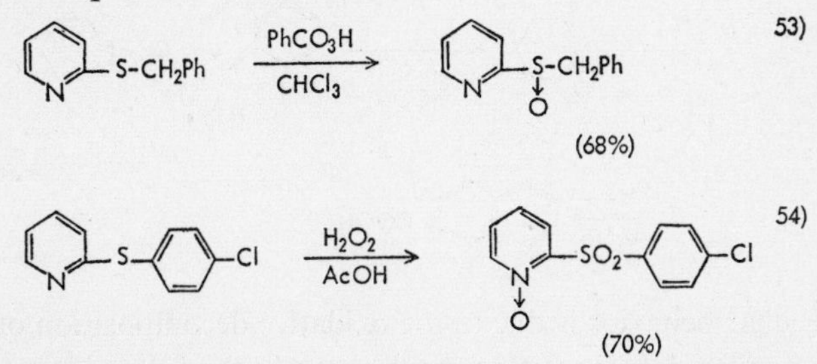

Mann and Watson[55] also found this to be true in the case of tripyridylphosphine (3-25) and tripyridylarsine (3-26).

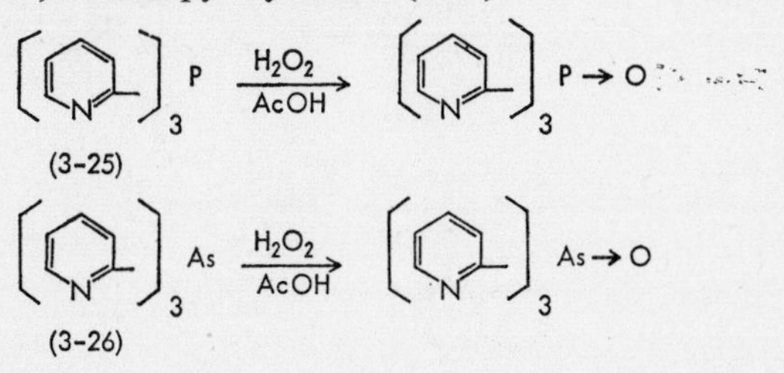

References p. 70

When a hydroxyl, aldehyde, active methylene, or active methine group is present in the molecule, these groups are also often oxidized at the same time.

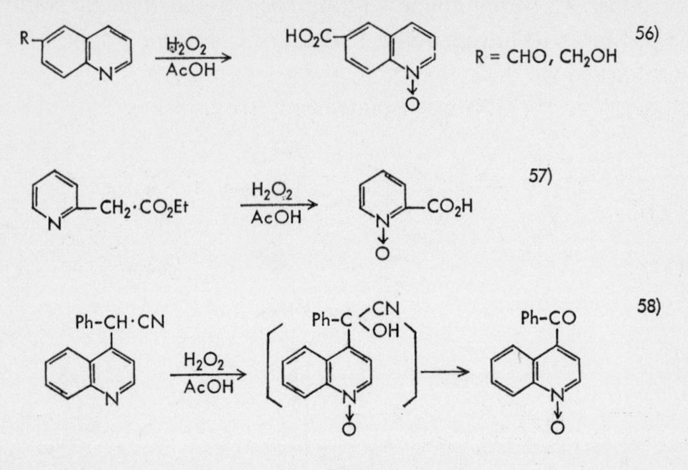

Ochiai and his co-workers[56] observed that, whereas 2-(piperidinomethyl)-quinoline (3-27) formed the corresponding *N*,*N'*-dioxide by *N*-oxidation with monoperphthalic acid, the same compound formed quinaldic acid on being treated with hydrogen peroxide in glacial acetic acid.

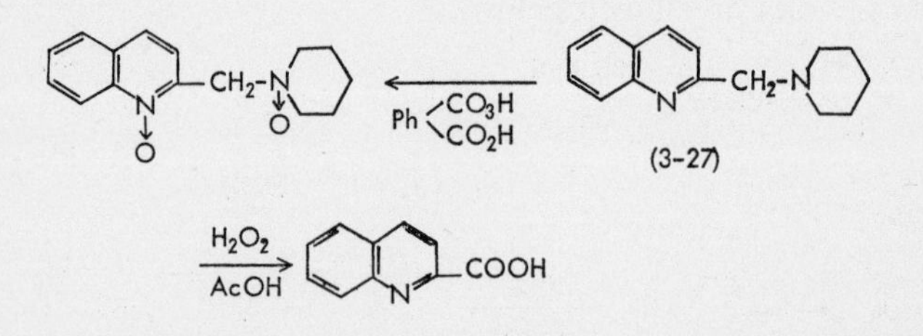

Such a peculiar behavior is due to the oxidative decomposition of the side chain and a similar decomposition can be seen in the following example.

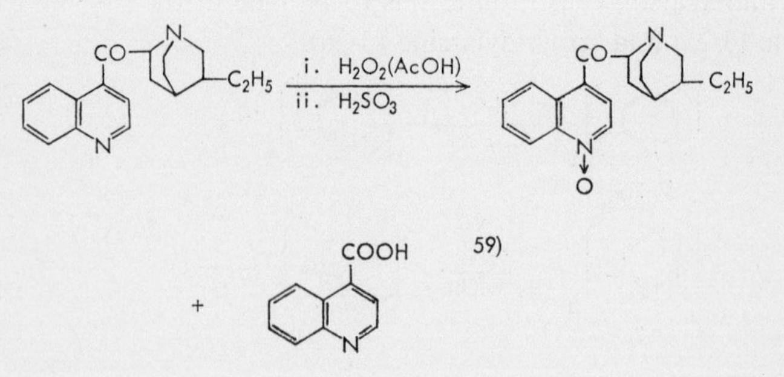

Analogously, while 11*H*-indeno[1,2-*b*]quinoline (3-27) gives the corresponding *N*-oxide in a crude yield of 85%, its isomer, 11*H*-indeno[2,1-*b*]-quinoline (3-28) is oxidized to the corresponding 11-oxo compound[60].

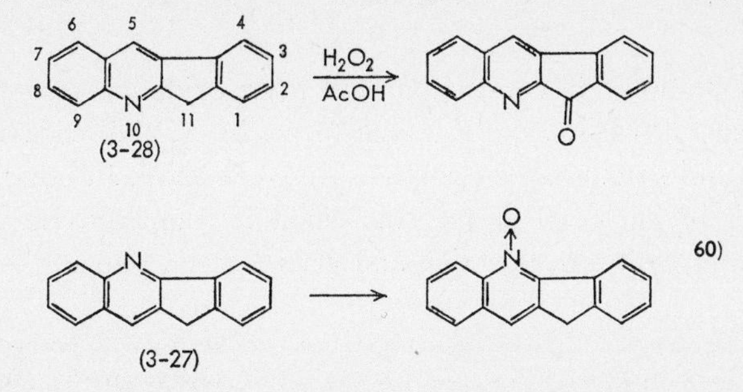

To suppress such oxidation, acylation of the hydroxyl prior to *N*-oxidation was carried out, and found to have increased the yield of the *N*-oxide slightly. For example, *N*-oxidation of pyridoxine without acetylation of the hydroxyl groups gives the *N*-oxide in only 9% yield, while previous acetylation of its three hydroxyl groups followed by *N*-oxidation increases the yield of pyridoxine *N*-oxide (3-29) to 23%[61].

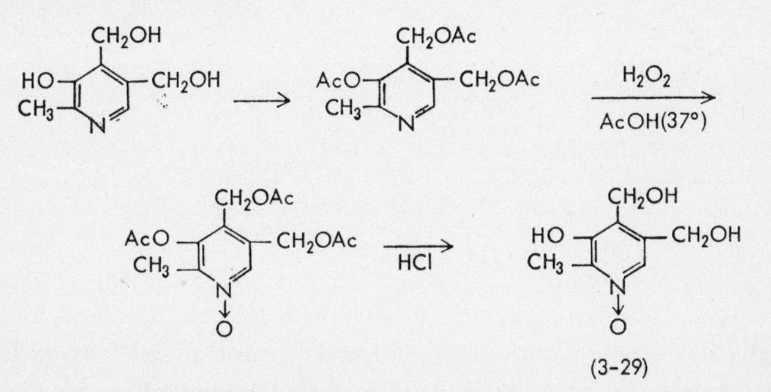

Analogously, it has been noted that 2-(hydroxymethyl)-3-methylquinoxaline 1,4-dioxide (3-30) can be derived successfully from the original quinoxaline via the intermediate acetate (3-31)[62].

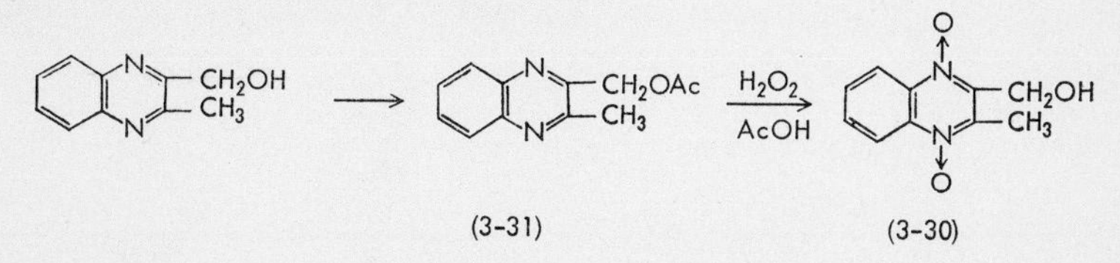

Transformation of an aldehyde to the corresponding acetal was found to effect protection to some extent in the *N*-oxidation[63].

References p. 70

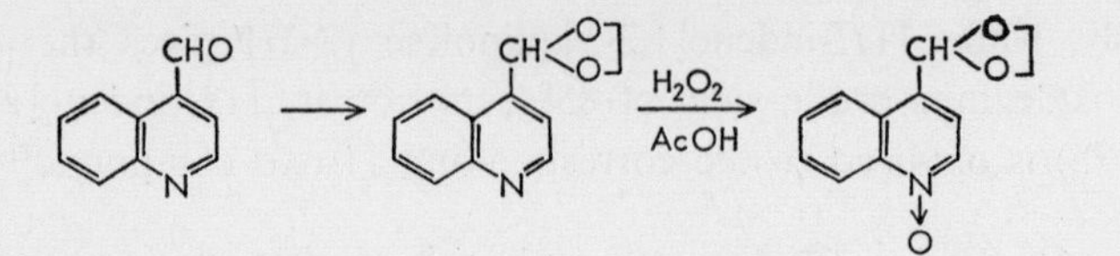

It should be noted in connection with these oxidation reactions, as has been pointed out earlier, that quinine forms its *N,N'*-dioxide without the vinyl group in its quinuclidine portion being affected[37]. In other words, the unsaturated double bond in the side chain is comparatively resistant to treatment with hydrogen peroxide and glacial acetic acid.

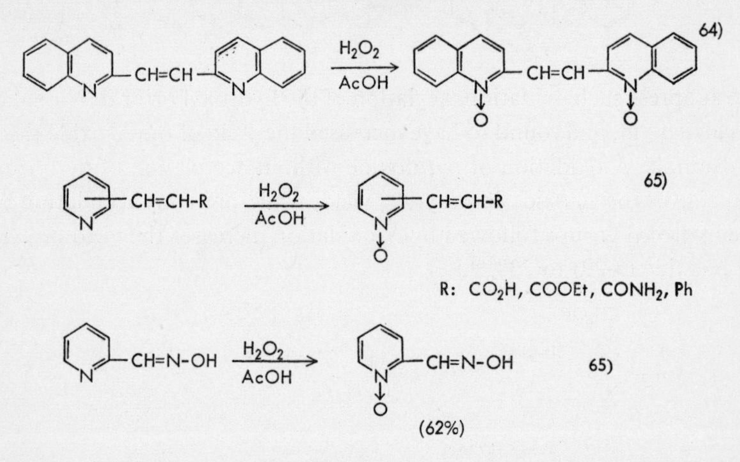

(b) Replacement with Hydroxyl Group

The peracid used as a reagent for direct oxidation acts as a nucleophilic reagent by the following manner of polarization, and as an *N*-oxidation agent. It may in some cases cause substitution with a hydroxyl group.

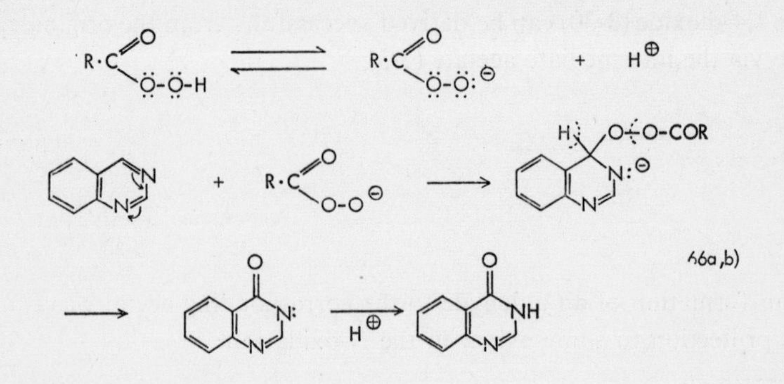

This reaction is more liable to occur in alkaline than in acid medium, but it also occurs in acid medium if the molecule has a position highly reactive to nucleophilic reagents, as shown in the above example. Similar examples are given below.

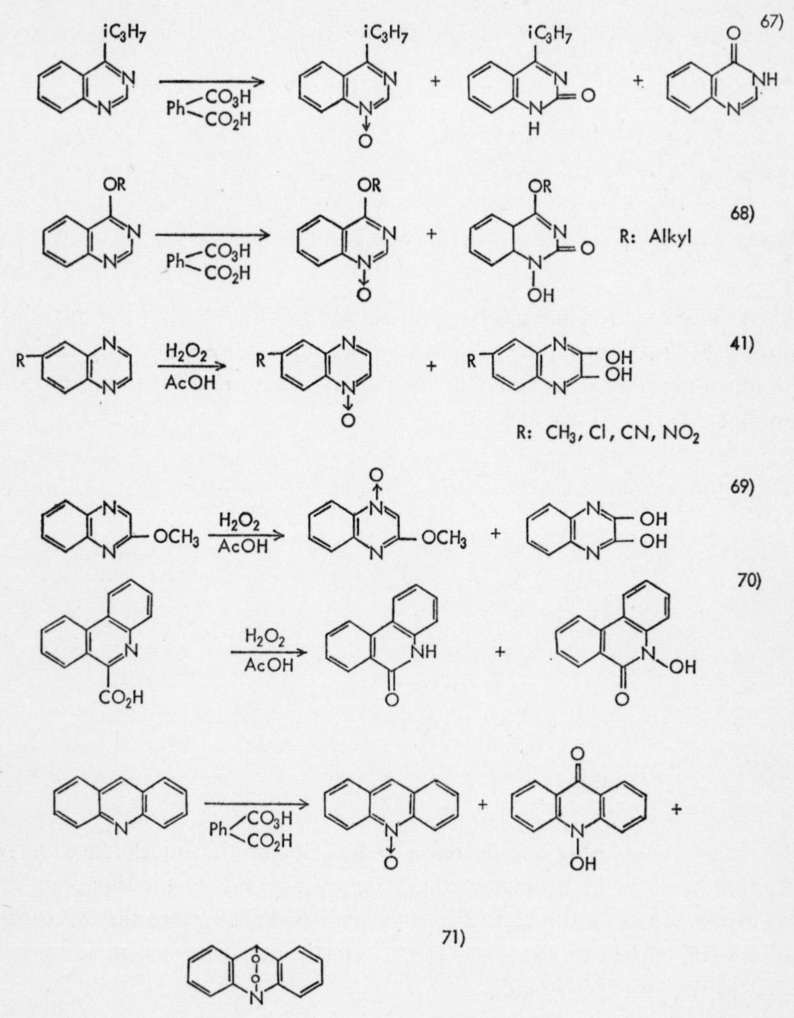

In these reactions, the position substituted by the hydroxyl group is invariably the methine group at *o*- or *p*-position activated by the ring nitrogen, and this reaction should be considered to be a kind of nucleophilic substitution by a peracid.

References p 70

On the other hand, Ochiai and his co-workers showed that the *N*-oxidation of quinaldine[72] and lepidine[73] with hydrogen peroxide and glacial acetic acid is accompanied by the formation of a small amount of 3-hydroxyl derivative.

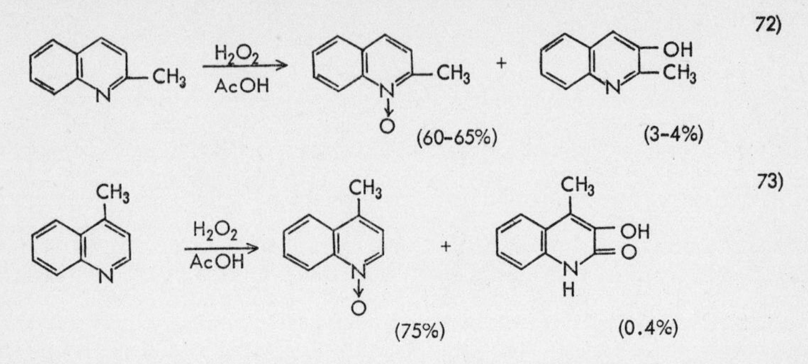

Such a side reaction had not been observed in quinoline, but the introduction of a hydroxyl group into 3-position becomes more marked in the case of 6-nitroquinoline, and only the 3-hydroxy compound is obtained from quinoline-8-carboxylic acid[74].

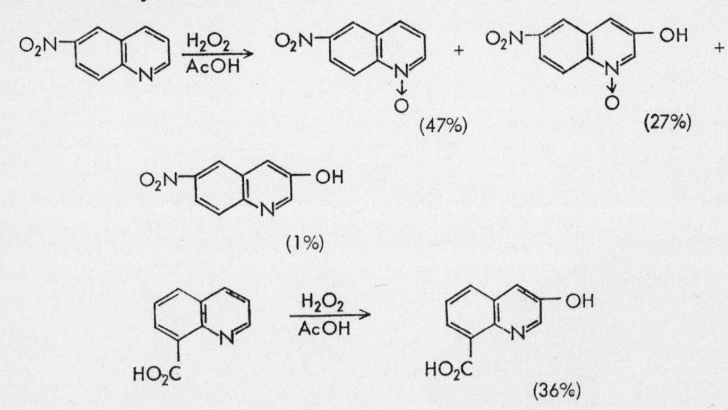

In these cases, either the decrease in basicity or the interference of *N*-oxidation by steric hindrance would probably account for the side reaction. It seems probable that introduction of a hydroxyl group into the 3-position may also be caused by the electrophilic substitution with a peracid, as observed in the case of *N*-oxidation.

(c) Oxidative Cleavage of the Ring

Ochiai and Hayashi[75] found that, while 2,4-dimethylthiazole formed 2,4-dimethylthiazole 3-oxide in *ca.* 60% yield by *N*-oxidation with hydrogen peroxide and glacial acetic acid, 4-methylthiazole, in which the 2-position of the thiazole ring is not occupied, only forms the corresponding *N*-oxide in 14% yield and the thiazole ring is decomposed by the oxidation, giving a large quantity of ammonium sulfate. The following examples show that the intermediate products can be isolated because they have a low resistance to oxidation or undergo a second cyclization.

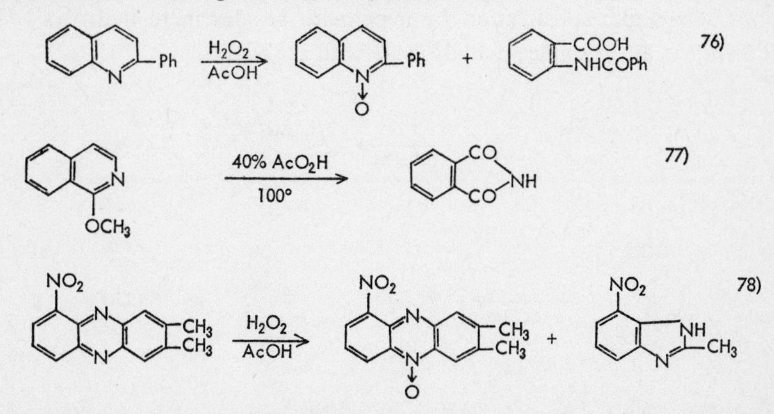

Von Euler[79] found that the imidazole ring in benzimidazole was resistant to *N*-oxidation with hydrogen peroxide and glacial acetic acid, and the reaction resulted in cleavage of the benzene portion, forming imidazole-4,5-dicarboxylic acid.

$$\text{benzimidazole} \xrightarrow[\text{AcOH}]{H_2O_2} \text{imidazole-4,5-}(HO_2C)_2$$

Kobayashi and others[52] found that *N*-oxidation of the afore-mentioned 2,3-dihydrothieno[3,2-*c*]quinaldine (3-21) and 2,3-dihydrothieno[2,3-*b*]-quinoline (3-22) by hydrogen peroxide and glacial acetic acid differs from that with monoperphthalic acid, and the oxidation itself progresses further to effect cleavage of the sulfur-containing ring, resulting in the formation of a sulfonic acid derivative.

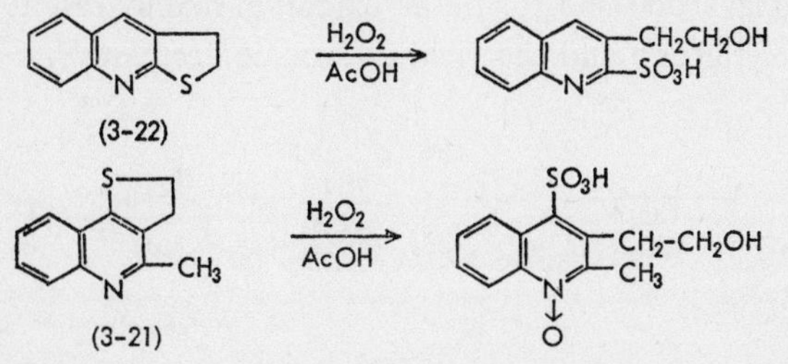

(d) Hydrolysis

Hydrolysis during *N*-oxidation frequently occurs by the use of hydrogen peroxide and glacial acetic acid, especially when the reaction does not progress smoothly and heating in glacial acetic acid solution extends over a long period of time. Consequently, it often seems possible to regard the reaction as a nucleophilic substitution by a peracid, besides mere hydrolysis. The following is an example of hydrolysis with an acid.

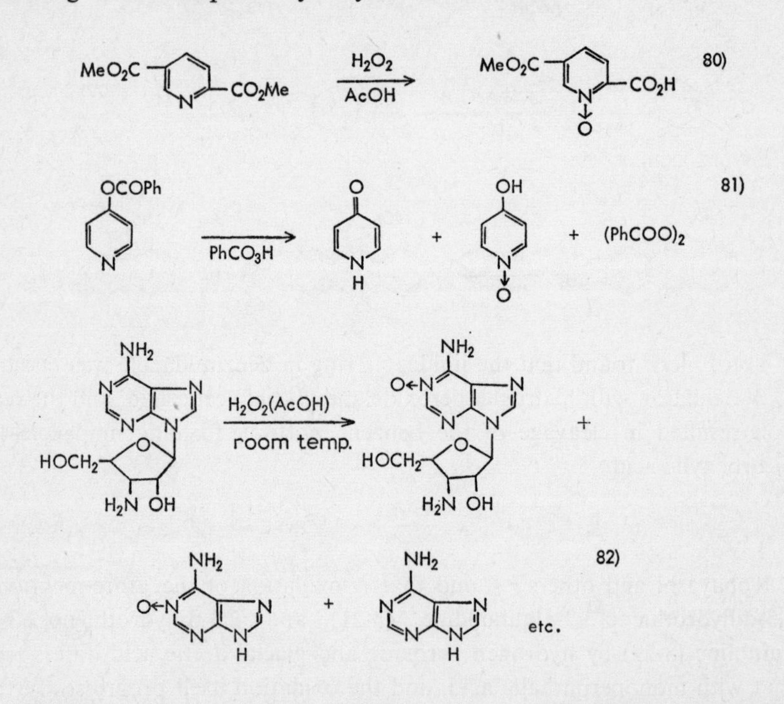

1-Phenoxyphthalazine (3-32) easily forms its 3-oxide (3-33) by the application of monoperphthalic acid. On the other hand, the use of hydrogen peroxide and glacial acetic acid for the *N*-oxidation mainly results in hydrolysis of the phenoxyl group and the yield of 3-oxide decreases[83].

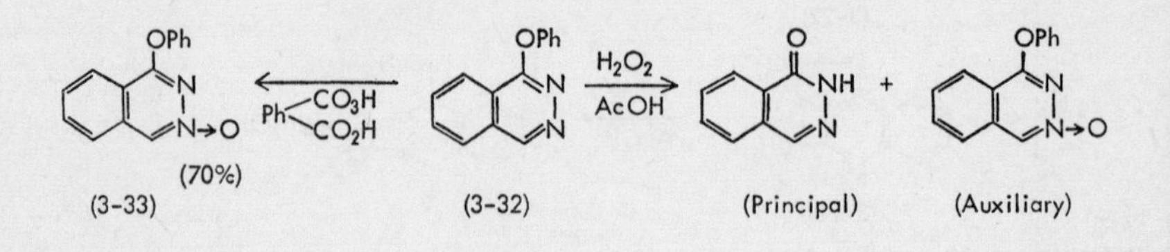

The following reactions are examples of substitution of chlorine, alkoxyl, or carbamyl groups with hydroxyl, when the *N*-oxidation reaction with hydrogen peroxide and glacial acetic acid does not progress smoothly.

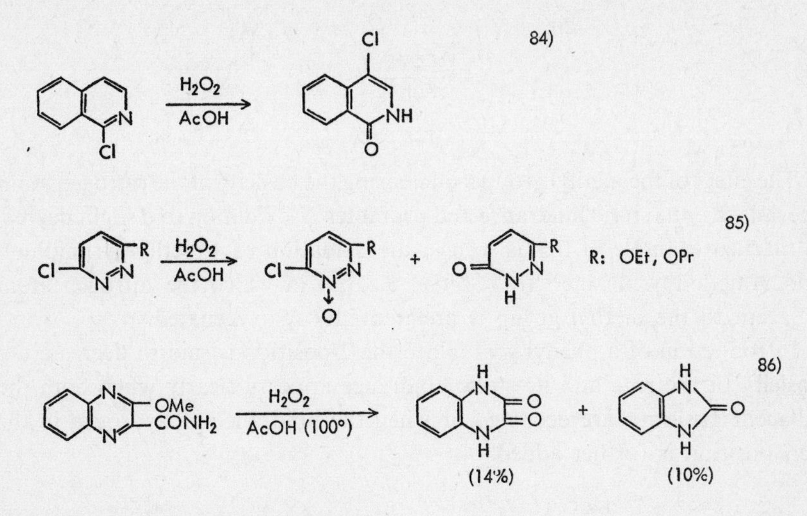

3.1.4 Steric or Electronic Hindrance

As described earlier, *N*-oxidation with peracid may be considered as due to the bonding of hydroxyl, polarized by electrophilic influence, with the lone-pair electrons of the amine-nitrogen in which case this reaction would be greatly affected not only by the nucleophilic reactivity of the nitrogen atom, *i.e.* its inherent basicity, but also by the steric or polar effect of the adjacent or conjugated substituent.

Modena and Todesco[87] examined the *N*-oxidation of pyridine and its methyl homologs with perbenzoic acid and stated that this is a second-order reaction and gave its rate constants (25°, $K \times 10^3$ sec^{-1}/mole^{-1}/l) as follows:

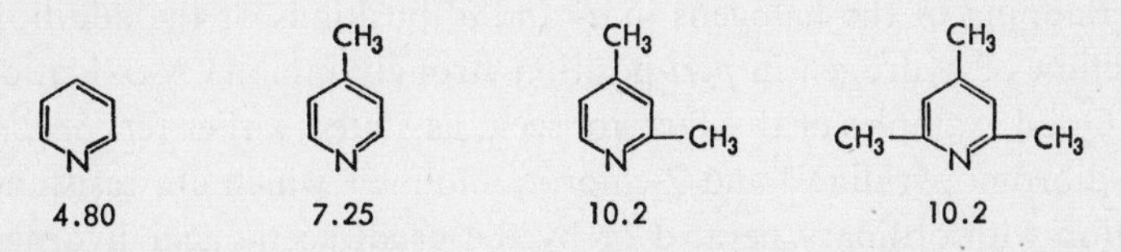

These values indicate that the polar effect of the methyl groups in 2- and 4-positions increases the basicity of nitrogen and promotes *N*-oxidation, while the spatial hindrance of the methyl groups overlapping in 2- and 6-positions exerts a slight hindrance. This steric hindrance of an alkyl group becomes more marked with an alkoxyl group. For example, the following reaction progresses slightly when R is methyl but does not progress at all when R is a benzyl group[88].

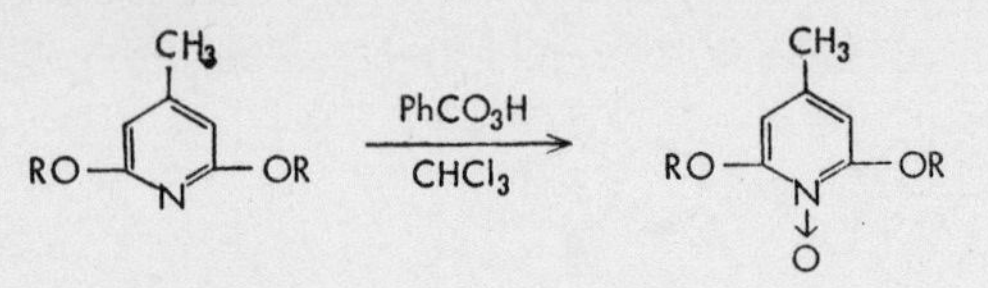

The effect of the methyl group in increasing the basicity of the nitrogen atom prevails over its steric hindrance and promotes *N*-oxidation to a slight degree. A marked example of this is seen in the oxidation of 4-methylpyrimidine[46] and 3-methylpyridazine[25] (*cf. Section 3.1.3.a*) in which the nitrogen atom adjacent to the methyl group is preferentially *N*-oxygenated.

Introduction of a phenyl group into the 2-position is said to decrease the basicity of the ring and its steric hindrance appears clearly when both the adjacent positions are occupied or when the hindrance of hydrogen in the *peri* position is further added.

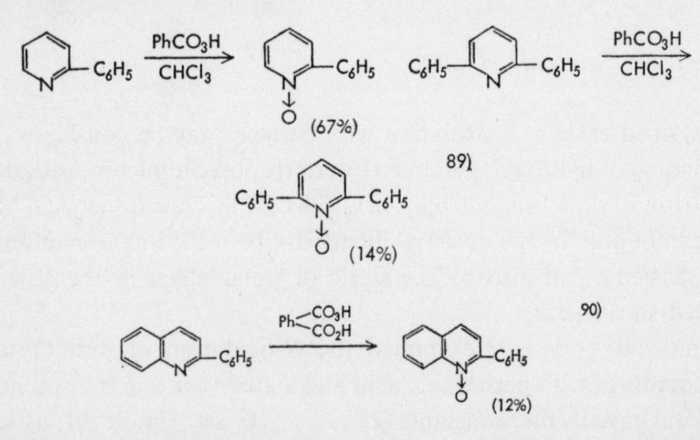

Overlapping of the halogens in α- and α'-positions or the addition of an interaction of hydrogen in *peri*-position strongly inhibits *N*-oxidation reactions. Good examples of this fact are seen, as stated earlier (*cf. Section 3.1*), in 2,6-dibromopyridine[10] and 2-chloroquinoline[12] which are resistant to *N*-oxidation with ordinary peracid or by the usual acetic acid–hydrogen peroxide process, and the *N*-oxidation is effected only by the use of trifluoroacetic acid–hydrogen peroxide or with permaleic acid. Similar hindrance of the polar effect of substituents in α- or *peri*-position has been recognized in sulfone[11], nitro, and carbonyl groups[74,60,91].

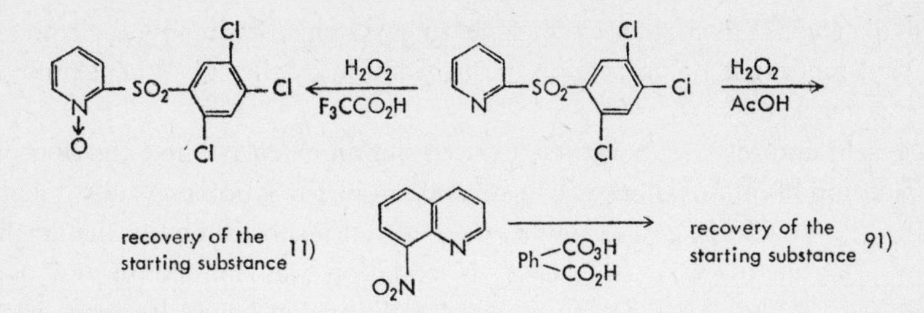

Such hindrance by the polar effect of a substituent group is sometimes seen in the 3-position.

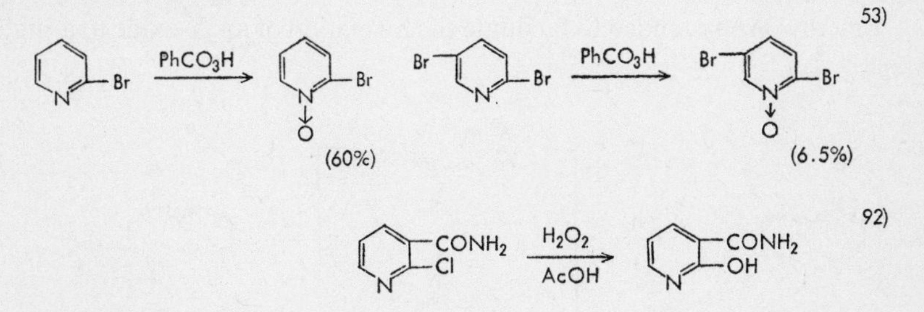

The above examples are both instances of overlapping of interference by a halogen group in 2-position. The mechanism of interference by a substituent in 3-position cannot be explained from its polar effect on the nitrogen atom alone and there have been some examples that seem to suggest the presence of an unknown factor. Such examples are 2- and 4-cyanoquinolines of low basicity, in which the nitrogen atom of the ring lies in the direct field of the polar effect of the cyano group and is further under its steric effect; these are *N*-oxygenated in a quantitative yield[93], while 3-cyano- and 3-carbamoyl-quinolines, whose 3-substituent only indirectly influences the ring nitrogen, are difficult to be converted to their *N*-oxides[94]. In addition, 6-cyanophenanthridine (3-34), whose basicity is so weak that the substance can be recrystallized from glacial acetic acid, forms an *N*-oxide in a quantitative yield[95].

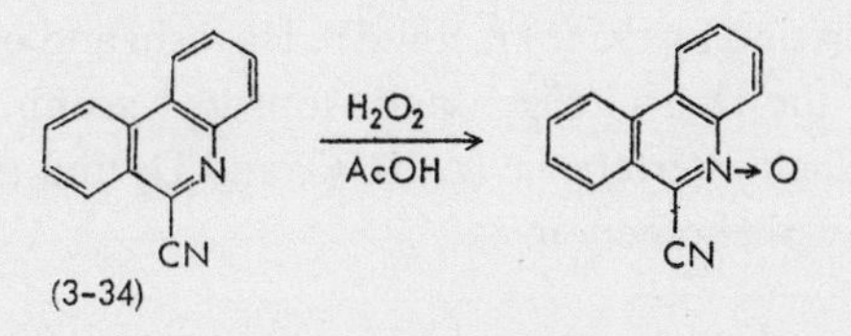

In diazines and condensed diazine ring systems, owing to their weak basicity, hindrance or promotion of *N*-oxidation by the substituent present appears markedly.

Hayashi and his collaborators[86] carried out an extensive investigation on *N*-oxidation of quinoxalines with a substituent in 2-position and substituents in 2- and 3-positions, and examined the effect of a substituent on the neighboring ring-nitrogen. In this case, *N*-oxidation was limited to the two methods of (*a*) application of monoperphthalic acid at below 10° in ether or benzene and (*b*) application of hydrogen peroxide at a suitable temperature between 60° and 100° in acetic acid as a solvent. Results of these experiments may be summarized as follows:

(i) A methyl group tended to facilitate the formation of an *N*-oxide to a slight extent.

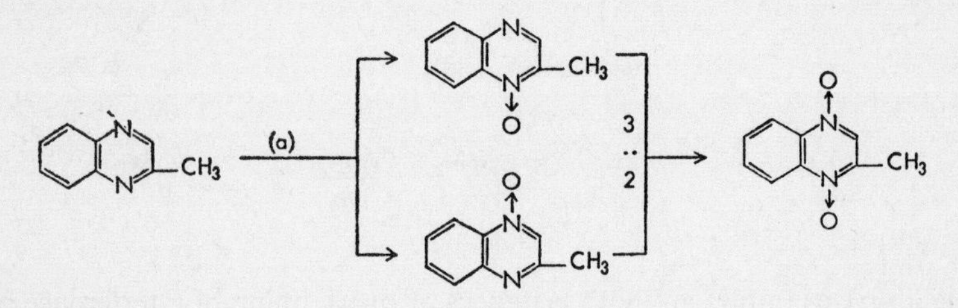

(ii) Primary alkyl groups like ethyl and propyl do not interfere with the formation of an *N*-oxide, but it is still not certain whether such groups facilitate its formation.

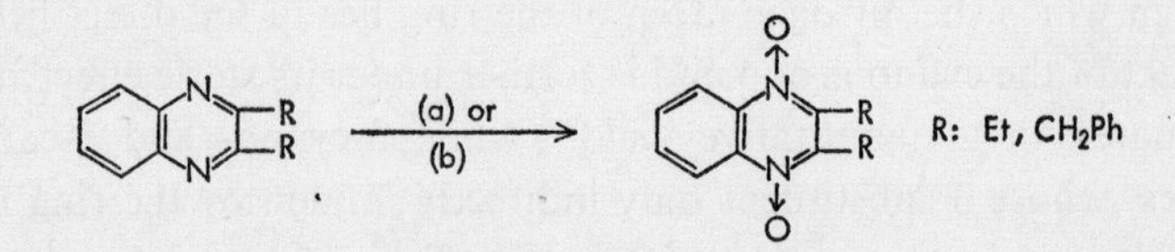

(iii) The isopropyl group (and probably secondary alkyl groups in general) exert a considerable hindrance on the formation of an *N*-oxide. For example, it has already been reported that 2,3-diisopropylquinoxaline does not form an *N*-oxide by oxidation by the (*b*) method[78]. Hayashi and others, considering the steric state of the ring nitrogen and isopropyl group, noted that there are four kinds of such interference (A, B, C, and D) and gave the following explanation to this phenomenon.

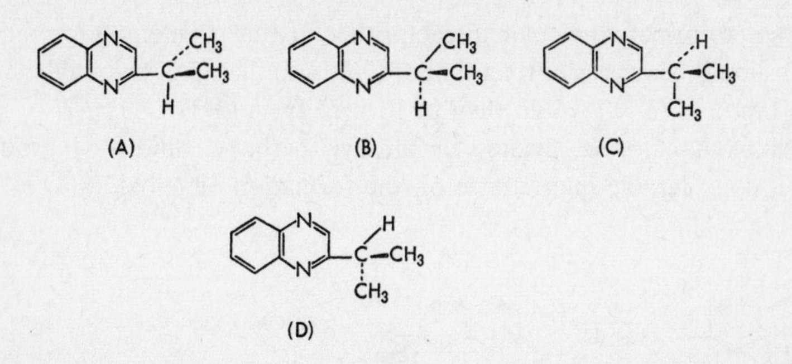

In these four states, steric hindrance is comparatively small in (A) and (B), and *N*-oxidation is possible in these states. In (C) and (D), steric hindrance is great and *N*-oxidation is thereby greatly inhibited. If two isopropyl groups are in adjacent positions, they will each take a position with least steric hindrance, *i.e.*, as in (C) and (D) as regards the neighboring nitrogen atom, since it would be difficult to take the conformation as in (A) and (B), so that *N*-oxidation of the compound becomes difficult.

(iv) The phenyl group has a slight effect on the formation of *N*-oxides, but not decisive enough to prevent it.

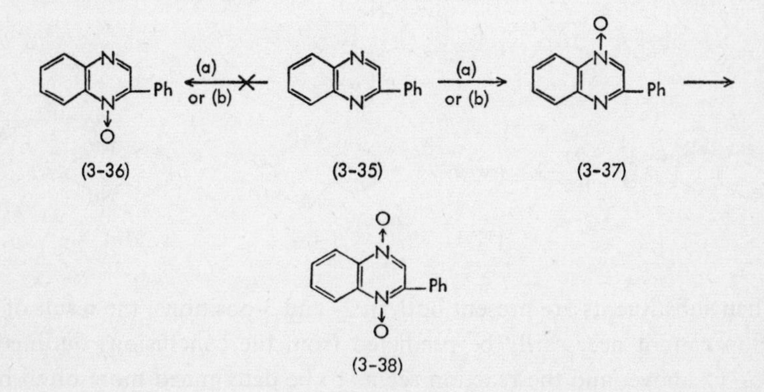

In a normal state, the quinoxaline ring and phenyl group in (3-35) are conjugated and both are assumed to be in the same plane. In such a state, *N*-oxidation of the ring-nitrogen adjacent to the phenyl group is affected by steric hindrance, but it is possible that this conjugation may be interrupted during the reaction, in which case the plane in which the quinoxaline is present and that in which the phenyl group is present undergo revolution and steric interference of the phenyl group will be decreased. It may be assumed that the state in which this interference becomes smallest is that in which the

References p. 70

two planes are perpendicular to each other. If this is true, the formation of (3-37) should be far easier than that of (3-36), so that (3-37) would be formed alone first and (3-38) would then be formed from (3-37).

(v) Alkoxyl, chlorine, cyano, carbamoyl, carboxyl, and acyl groups all exert a considerable interference on the formation of *N*-oxides.

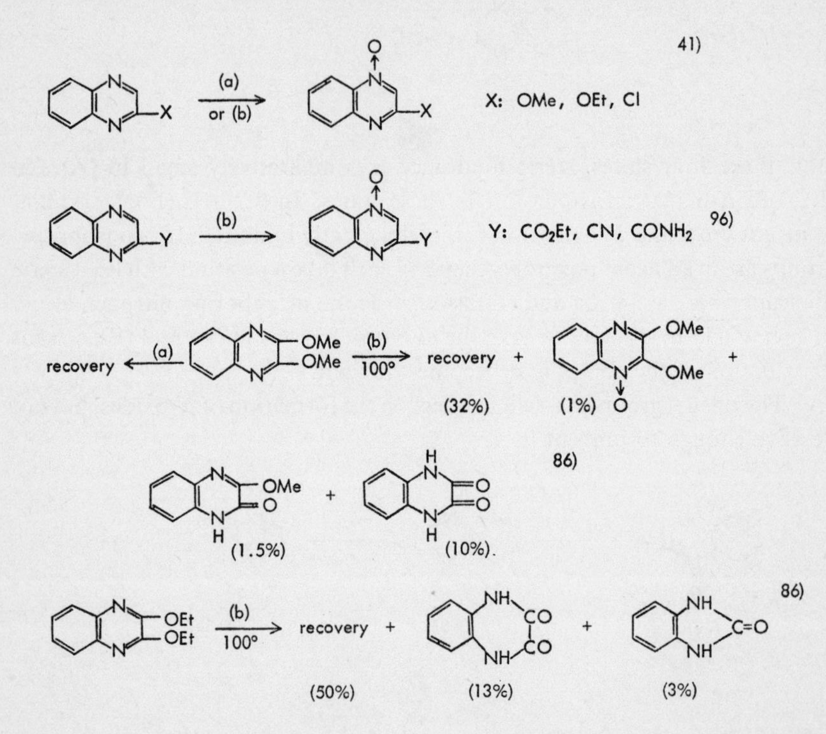

When substituents are present both in 2- and 3-positions, the result of the reaction cannot necessarily be predicted from the conclusions outlined in *(i)* to *(v)* above, and the reaction seems to be determined more often by a correlation between the two substituents. For example, the following reactions can be predicted from the foregoing conclusions.

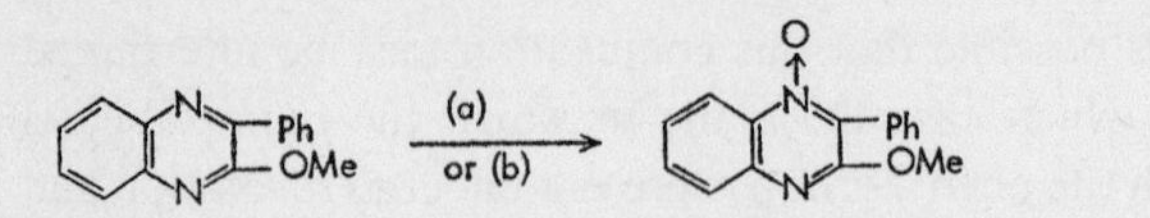

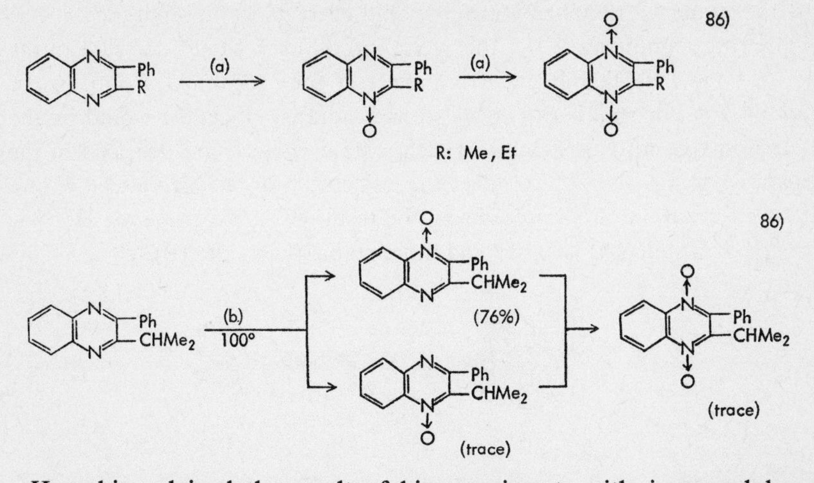

Hayashi explained the result of his experiments with isopropylphenylquinoxaline (3-39) by assuming that the bulky isopropyl group takes up so much space that it becomes easier for the molecule to assume the state (B) in which the phynyl group is not conjugated with the quinoxaline ring, than state (A) in which the phenyl group is in conjugation.

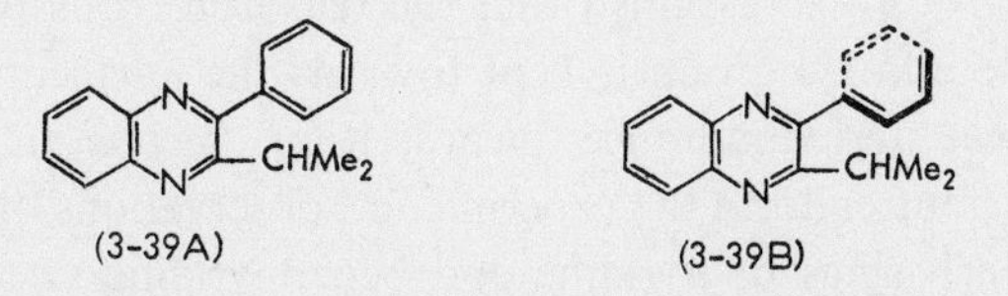

On the other hand, 3-phenyl-2-quinoxalinecarboxamide (3-40) forms only the 1-oxide (3-41) and not 4-oxide or 1,4-dioxide. This is not in conformity with the fact, outlined in (*v*) above, that 2-quinoxalinecarboxamide forms only the 4-oxide and not 1-oxide or 1,4-dioxide.

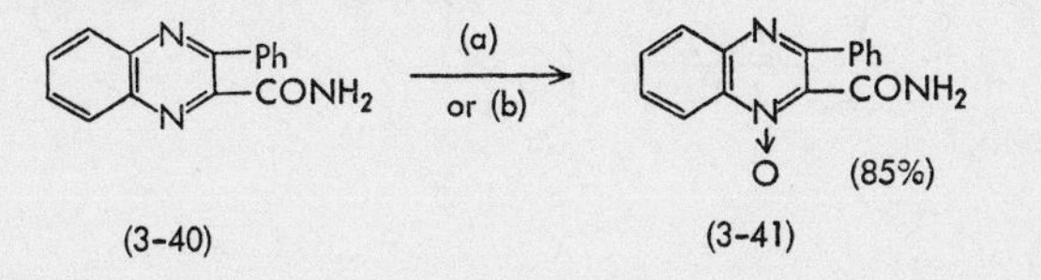

Hayashi gave the following explanation for this fact. The bulkiness of the phenyl and carbamoyl groups makes it impossible for these groups to be conjugated with the quinoxaline ring at the same time. Consequently, (A) if the phenyl group is in conjugation, this cannot apply to the carbamoyl group, whilst the converse (B) is also true.

References p. 70

In the state (A), the phenyl group would exert its steric hindrance on the neighboring ring-nitrogen, but this steric interference would decrease in the state (B). On the other hand, if the effect of the carbamoyl group on the adjacent ring-nitrogen is due to its steric hindrance, its effect would be the greatest in the state of (B) and in such a state, π-electron attraction of the carbamoyl group would also affect the adjacent ring-nitrogen and interfere with the formation of *N*-oxide. From the result of his experiments, Hayashi assumed that state (A) is easier and more stable than state (B).

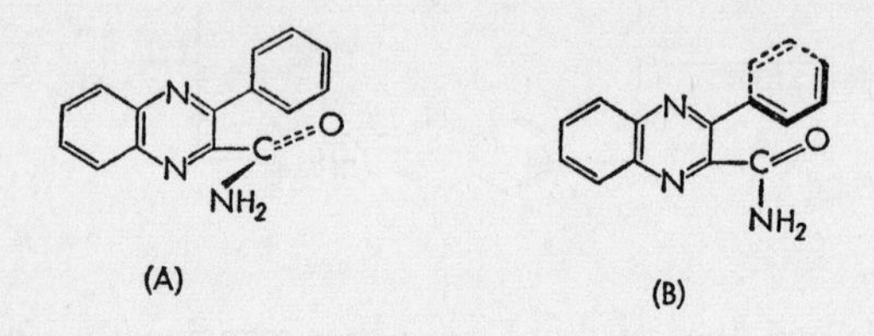

Otomasu and his collaborators determined the structure of mono-*N*-oxides obtained by the *N*-oxidation of alkoxyl derivatives of quinoxaline[97], pyrimidine[98], and pyridazine[98] from the result of their dipole moment measurements, and revealed the oxidation of nitrogen in the position not adjacent to the alkoxyl group. He assumed that the reason for this phenomenon is the fact that the alkoxyl group is bent towards the adjacent ring-nitrogen and exerts a great steric hindrance towards it (*cf. Section 4.2*).

These conclusions endorse the experimental observations on the *N*-oxidation of the hetero rings of pyridine and benzopyridine systems described earlier and offer an important clue when considering steric hindrance as encountered during *N*-oxidation. Some of the following examples can also be explained by the above considerations.

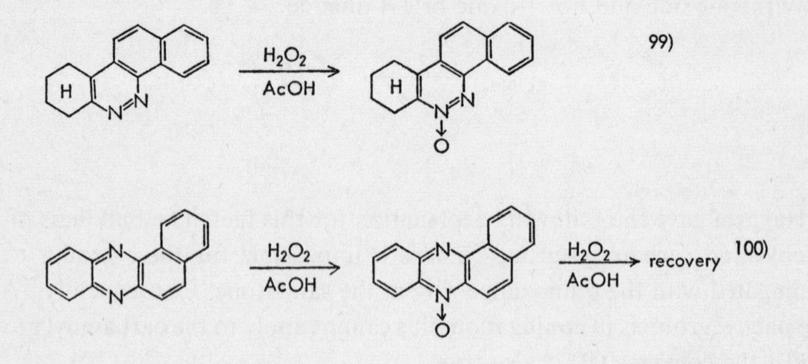

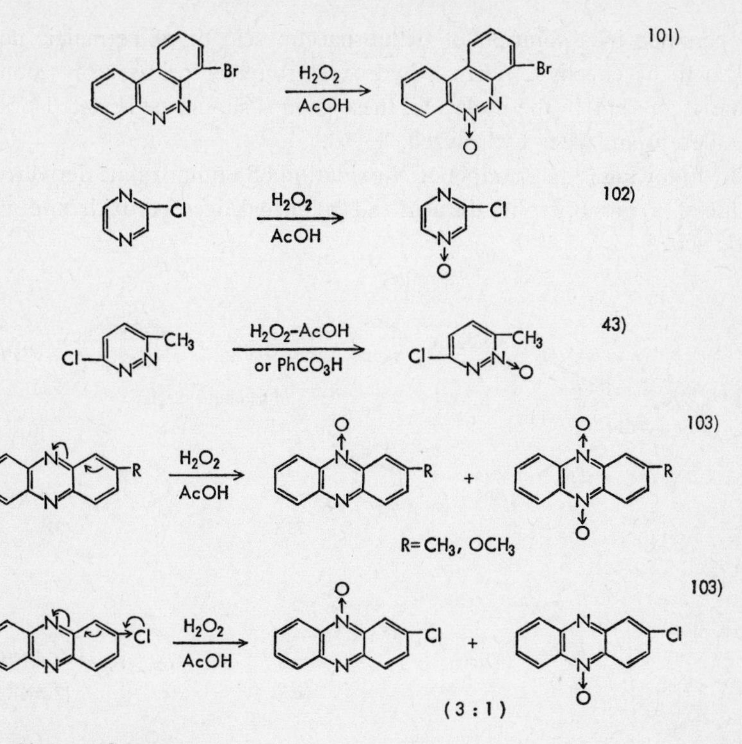

3.1.5 *Choice of Reagent*

In summarizing the result of various methods of *N*-oxidation described in the foregoing sections, it may be said that when the amine has sufficient basicity for *N*-oxidation and is comparatively stable to side reactions like oxidation and hydrolysis, then the use of hydrogen peroxide and glacial acetic acid is the most simple and convenient. Acetic acid is generally a good solvent for most of the amines because reaction conditions can be made stronger by heating and the solvent is adapted for treating a large quantity. However, the relatively aggressive reaction conditions and heating for a long period of time in acidic medium tend to increase the possibility of side reactions like hydrolysis, oxidation, and hydroxylation. The use of peroxycarboxylic acid diminishes the danger of such side reactions, because the solvent used is generally a neutral solvent like ether and chloroform, and the temperature is from about 0° to room temperature; this process, however, is somewhat difficult when a large quantity is to be treated in one batch. When the amine has a low basicity and is resistant to *N*-oxidation, the use of hydro-

References p. 70

gen peroxide in a solution of trifluoroacetic acid or of permaleic acid is known to be effective. When a hydroxyl group, or primary or secondary amine is present in the molecule, these groups should be acylated prior to *N*-oxidation and later hydrolyzed.

The following is an example of *N*-oxidation of a quinoxaline derivative by the use of (a) monoperphthalic acid and (b) of hydrogen peroxide and glacial acetic acid[86].

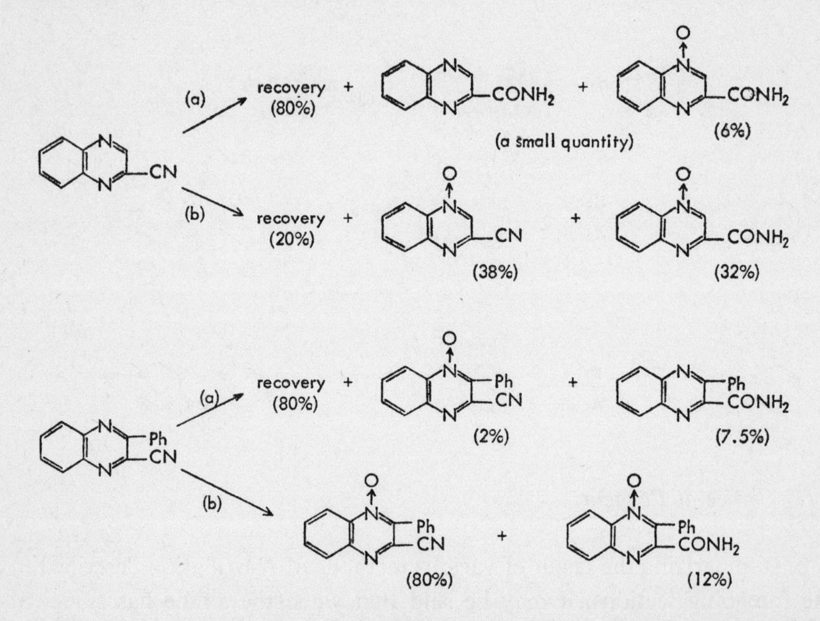

The above example shows that whereas *N*-oxidation failed to proceed by the (a) method, it progressed favorably by (b), so that it is advisable to try another method even if one method does not give a satisfactory result.

If the base has both aliphatic and aromatic tertiary amine-nitrogen, application of 3% hydrogen peroxide in a solution of dilute acid will result in *N*-oxidation of the aliphatic nitrogen, forming aliphatic *N*-oxide. If only the aromatic nitrogen is to be oxidized to form aromatic *N*-oxide, the amine should be *N*-oxidized with hydrogen peroxide in glacial acetic acid or with peracid, and the aliphatic-aromatic double *N*-oxide thereby formed should be reduced with sulfurous acid at room temperature whereby only the aliphatic *N*-oxide will be reduced, leaving the aromatic *N*-oxide intact.

N-Oxidation of diazines, having two aromatic tertiary amine-nitrogens in the molecule, is possible. While the pyrazine system compounds form mono-

N-oxide and *N,N*-dioxide, pyridazine and pyrimidine series compounds form only mono-*N*-oxide, often in two isomeric forms which can be detected by gas chromatography and separately purified.

N-Oxidation of azoles has been effected to some extent in the thiazole series, but is difficult with imidazoles, whose *N*-oxides must be synthesized in a different way. Kuhn and Blau[104] proposed the structure of the pigment Radziszewski Blue, which is formed by the oxidation of 2,2′-bisimidazole with hydrogen peroxide in sulfuric acid, as its *N,N*′-dioxide. Similarly, oxidation of 2,2′-bisbenzimidazole with hydrogen peroxide in sulfuric acid gives *N,N*′-dioxide (NO-Indigo). In general, however, imidazoles are resistant to *N*-oxidation.

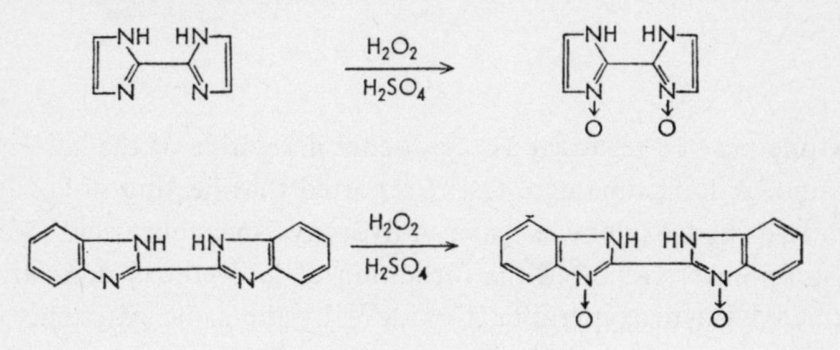

3.2 By Cyclization Reactions

3.2.1 Cyclization of Hydroxylamine Derivatives

Baumgarten and his co-workers[105] showed that when glutaconic aldehyde and hydroxylamine are heated in methanol in acid medium, the dioxime formed as an intermediate undergoes cyclization with liberation of one mole of hydroxylamine, and pyridine 1-oxide is formed.

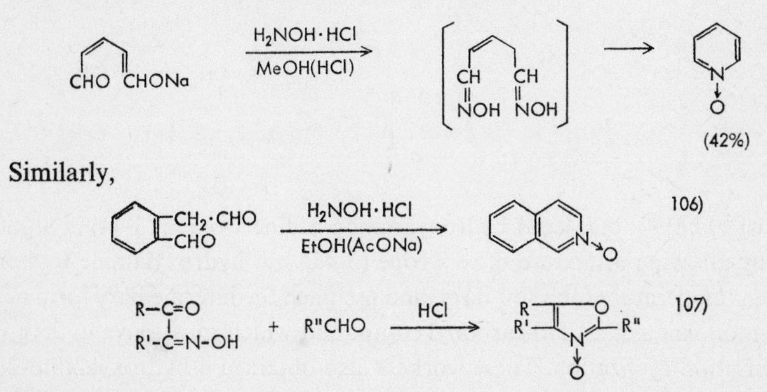

Similarly,

References p. 70

Nielsen *et al.*[108] synthesized the 2,5-dihydroxypyridine 1-oxide derivative in a good yield by ketalization of α-acylfuran, electrolytic reduction of the ketal in methanol solution, and leaving the methoxylated dihydrofuran derivatives thereby obtained in aqueous solution of hydroxylamine hydrochloride at room temperature. They assumed the intermediate formation of the 1,5-dicarbonyl compound, as in the foregoing example.

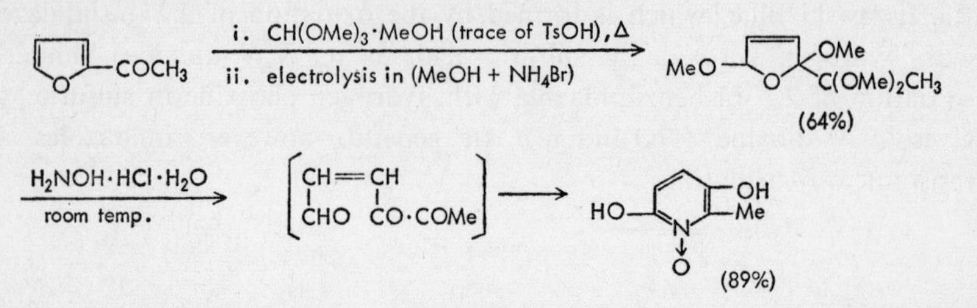

γ-Pyrone may be regarded as the di-enol anhydride of the 1,5-dicarbonyl compound. A long time ago, Ost[109] reported that heating of comanic acid (3-42) with hydroxylamine gave 4-hydroxy-2-picolinic acid 1-oxide (3-43). Later workers reported the formation of its 5-ethoxy derivative[110] and 2,6-diphenyl-4-hydroxypyridine 1-oxide[111] by the same reaction.

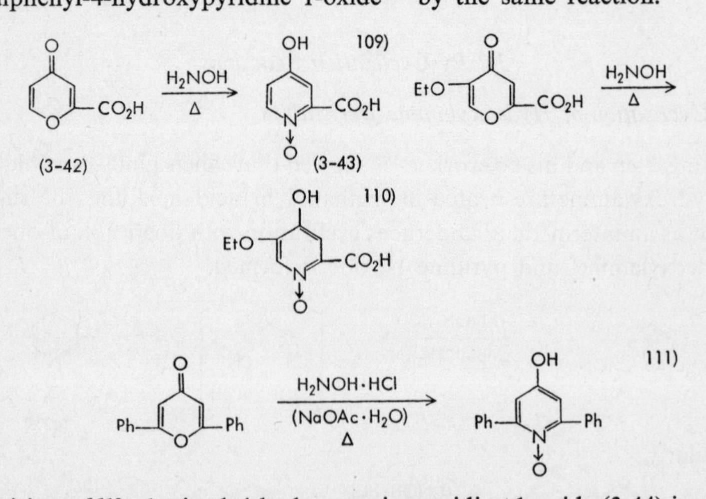

Parisi *et al.*[112] obtained 4-hydroxyaminopyridine 1-oxide (3-44) in a good yield by allowing a mixture of γ-pyrone (3-45) and hydroxylamine to stand at room temperature for a few days, and assumed the intermediary formation of the monoxime of a 1,5-dicarbonyl compound which must have undergone condensation-cyclization. These workers also obtained 4-hydroxyamino-2,6-

lutidine 1-oxide (3-46)[113] by leaving a mixture of the barium salt of the dienol compound of diacetylacetone and hydroxylamine hydrochloride in ethanol in a dark place. The above *N*-oxide has also been derived from 2,6-dimethyl-γ-pyrone by the same method[114].

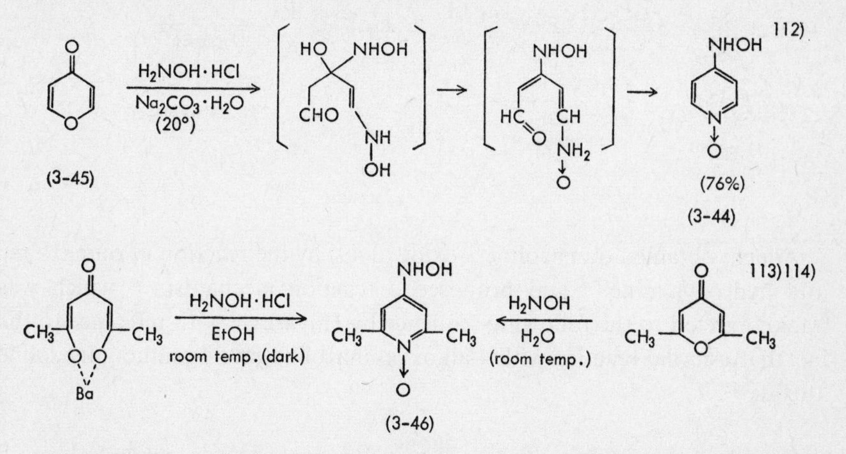

This reaction is similar to the formation of aldonitrone by the condensation of arylhydroxylamine and aldehyde. If a substituent that reacts with the hydroxyamino group or oxime is present in its γ- or δ-position, the compound is known to undergo cyclization to form an *N*-oxide compound.

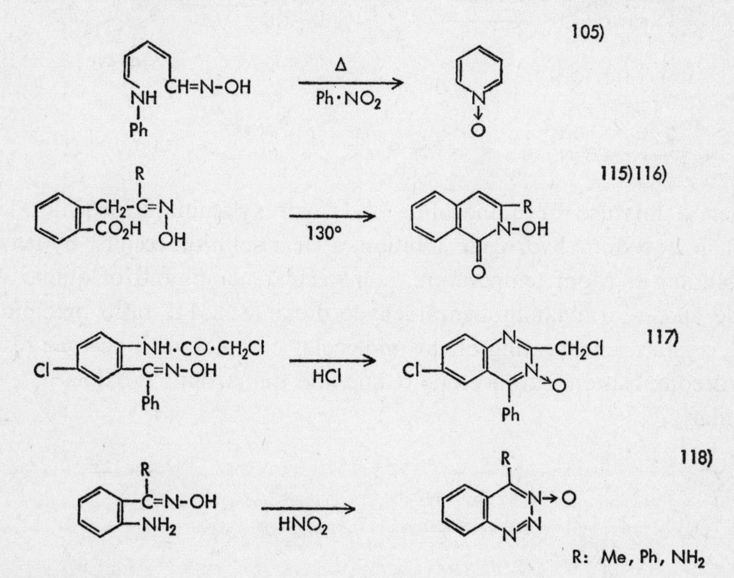

References p. 70

Shaw[119] synthesized 2,6-diamino-*s*-triazine 1-oxide derivatives (3-47) by the reaction of alkyl- and aryl-dicyanoamidine salts (3-48), derived from amidines or imido esters, with hydroxylamine.

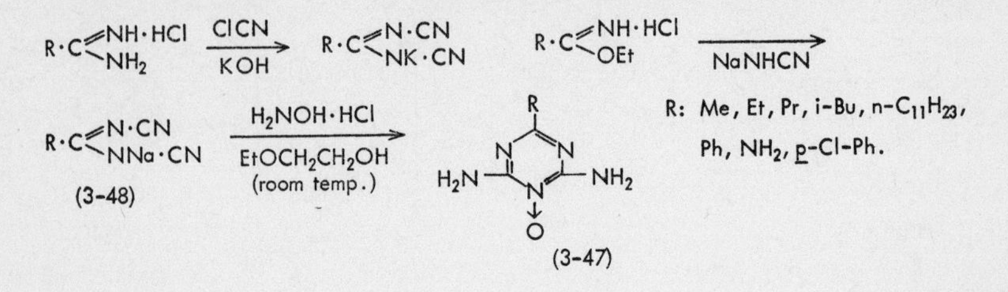

Adachi obtained quinazoline 3-oxide (3-49) by the reaction of quinazoline and hydroxylamine[66a] and proposed a reaction mechanism[120] which was later corrected to the following sequence by Hayashi[121] with reference to the fact that a similar reaction with 4-alkoxyquinazoline gave 4-aminoquinazoline 3-oxide[122].

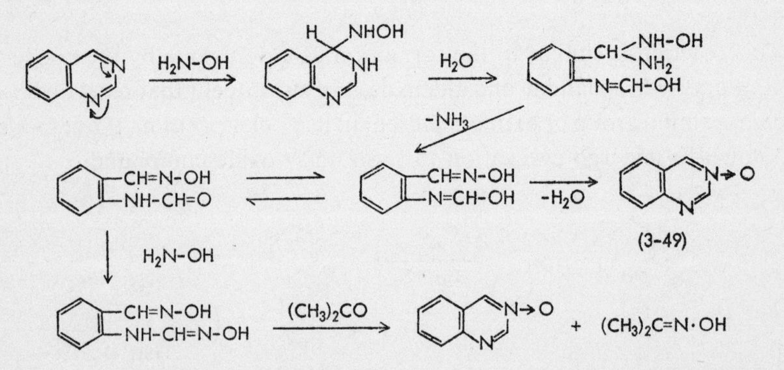

When a mixture of quinazoline and hydroxylamine hydrochloride is shaken in 2N sodium hydroxide solution, a clear solution results. By leaving this solution at room temperature, a molecular compound of quinazoline 3-oxide and *o*-formylaminobenzaldehyde dioxime in 1:1 ratio precipitates out as crystals, and heating of this molecular complex with acetone results in its decomposition into acetone oxime and quinazoline 3-oxide.

Similarly,

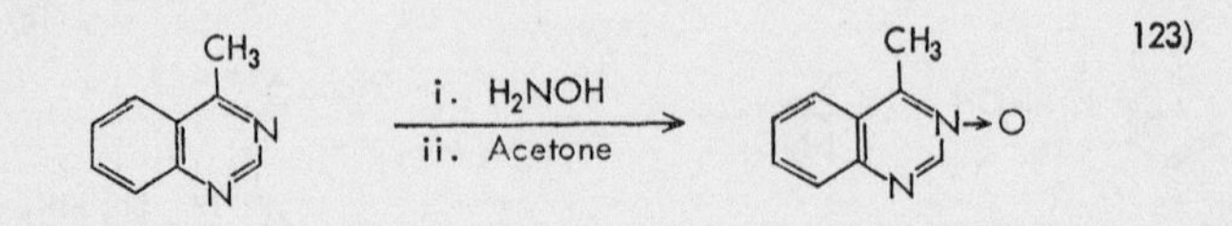

If the temperature is raised during this reaction, amination by hydroxylamine is likely to occur, while the presence of a methoxyl or phenoxyl group in the 4-position is known to facilitate the reaction[122].

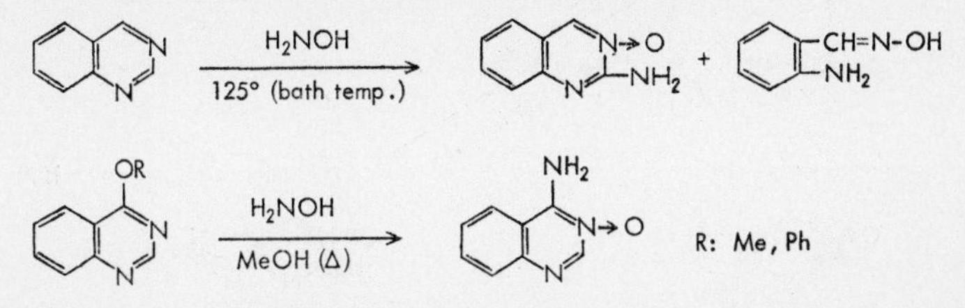

This method of hydroxylamine cyclization is useful for the preparation of quinazoline 3-oxide, since *N*-oxidation of quinazoline with hydrogen peroxide and glacial acetic acid gives 4-quinazolinol[66a], and *N*-oxidation of quinazoline with a substituent in the 4-position, such as 4-alkoxy-[68] and 4-isopropylquinazoline[67], with monoperphthalic acid results in the formation of 1-oxide (3-50).

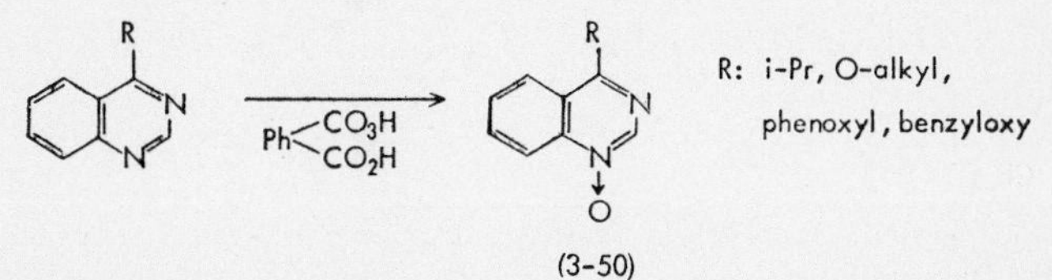

(3-50)

In order to verify the structure of the foregoing quinazoline 3-oxide and its 4-methyl derivative, Adachi synthesized these *N*-oxides by the condensation of orthoformic acid ester with *o*-aminobenzaldoxime[120] or with *o*-aminoacetophenone oxime[123], respectively.

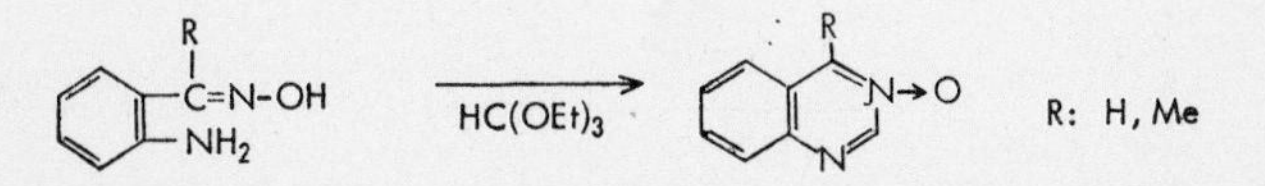

Syntheses of purine *N*-oxides using the same reaction have been devised as shown below.

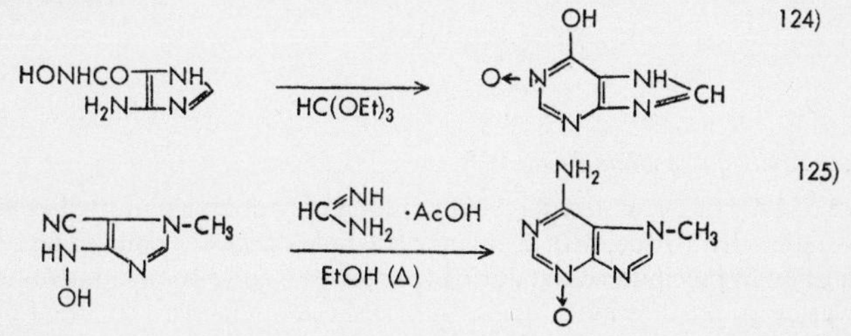

References p. 70

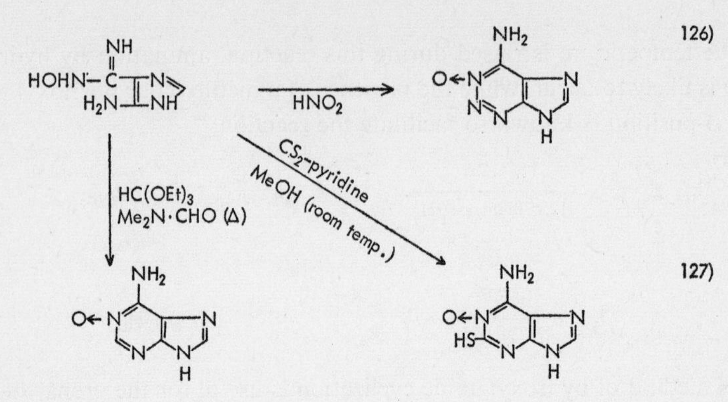

The following is an example of the same reaction, carried out as intermolecular condensation, to obtain mono-*N*-oxides of pyrazine or imidazole derivatives.

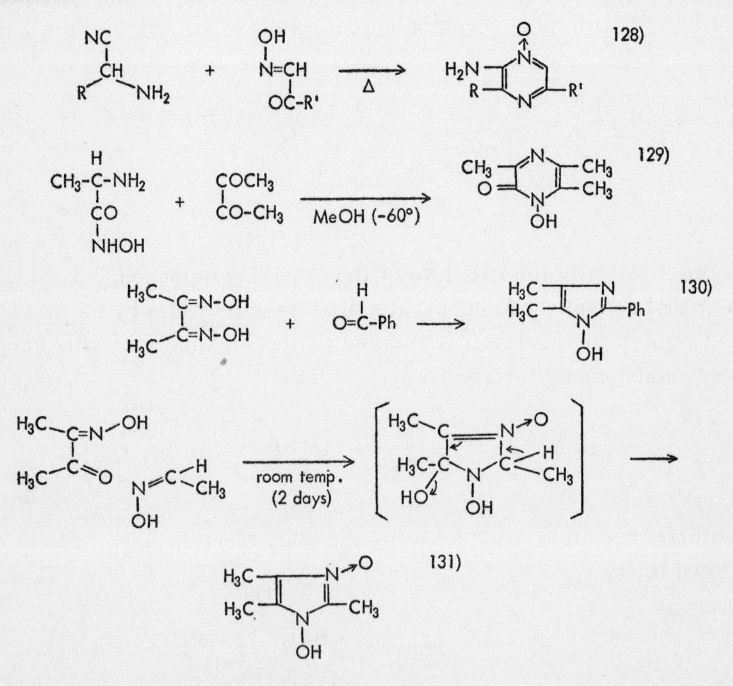

Example 1. 4-Aminoquinazoline 3-Oxide[122]

A mixture of 1.6 g of 4-methoxyquinazoline, 1.4 g of hydroxylamine hydrochloride, and 10 ml of 2N sodium hydroxide, with 20 ml of methanol added, was refluxed on a water bath for 2 hours. Crystals precipitated out during the reaction. After completion of the reaction, methanol was evaporated as much as possible and the residue was collected by filtration when cooled. Crude crystals of m.p. 220° (decomp.) amounted to 2.1 g, which were recrystallized from methanol, m.p. 227° (decomp.).

Example 2. Quinazoline 3-Oxide[66a]

A mixture of 1.5 g of *o*-aminobenzaldoxime and 5 ml of ethyl orthoformate was refluxed at 140° (bath temp.) for 1 h, excess orthoformate was evaporated under a reduced pressure, and 1.6 g of crystals of m.p. 135–146° was collected from the residue. Recrystallization from acetone gave 1.1 g of quinazoline 3-oxide, m.p. 150–152°.

3.2.2 Cyclization through Reduction of Nitro Compounds

Reduction of an aromatic nitro compound with an active substituent in the *ortho* position results in the condensation of the hydroxyamino compound formed as an intermediate with the active group in the *ortho* position, producing an aromatic *N*-oxide compound. In a classic work, Heller and Sourlis[132] carried out the reduction of methyl 2-hydroxy-2-(*o*-nitrophenyl)ethyl ketone (3-2) with zinc and acetic acid, and assumed the product to be a ketone compound (3-3), but Meisenheimer *et al.*[1] corrected this and proved it to be quinaldine 1-oxide (3-4), as was shown earlier (*cf. Section 3.1*).

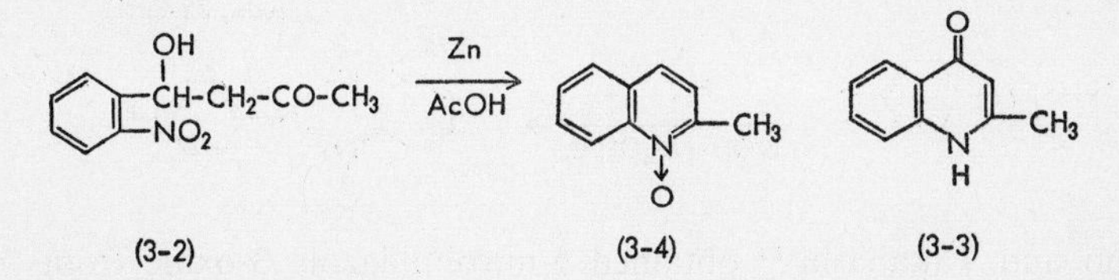

Similarly, Heller and others[133] assumed the product from the reduction of *o*-nitrobenzalcyanoacetamide to be a dihydroindole derivative (3-51) but this was later corrected by Taylor and others[134] to 2-amino-3-carbamoylquinoline 1-oxide (3-52).

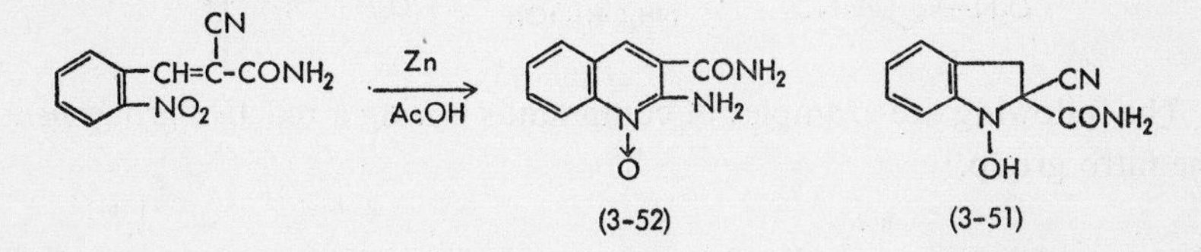

Since then, many of the derivatives of quinoline 1-oxide have been synthesized by this method, but various modifications have also been reported other than reduction with zinc and acetic acid; *e.g.* heating an ethanolic ammonia solution while bubbling hydrogen sulfide[135,136], catalytic reduction of an ethanolic solution over palladium–kieselguhr[137], heating an ethanolic

References p. 70

solution with addition of palladium–carbon, using cyclohexene as the hydrogen donor[138]. However, *N*-oxides of the quinoline series can be prepared by direct oxidation, so that these methods are little used for them and have been developed more for the syntheses of benzoazole series *N*-oxides, for which such methods are indispensable.

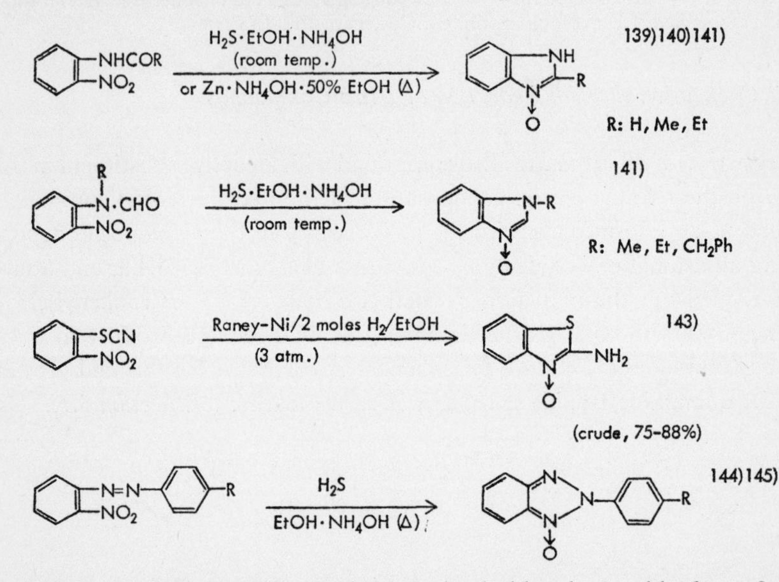

Kanô and Takahashi[142] obtained 5-nitroimidazole 3-oxide from 2,4-dinitroformanilide by bubbling hydrogen sulfide through its saturated solution in ethanolic ammonia at 0° and allowing the solution to stand at room temperature, by which the nitro group in 4-position was not reduced.

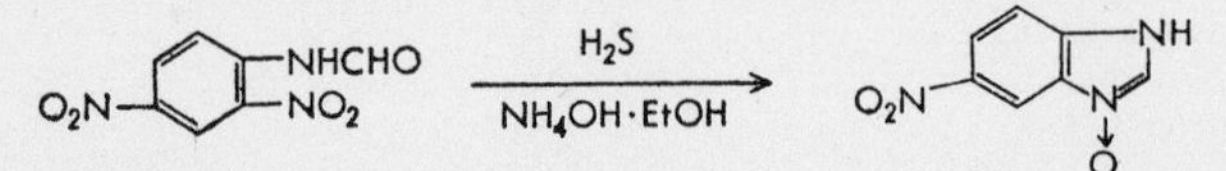

The following are examples of compounds having a reactive group *peri* to the nitro group.

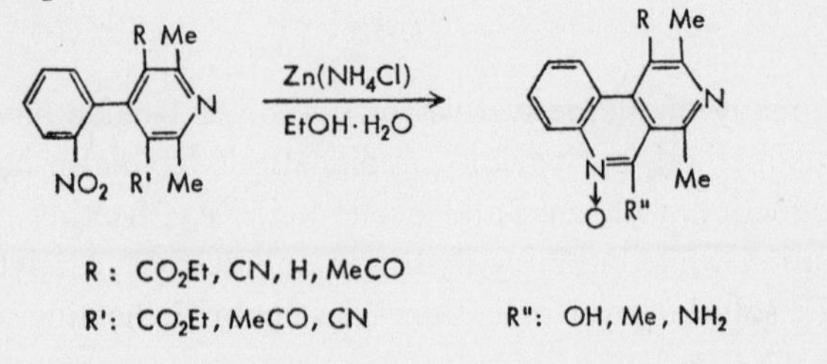

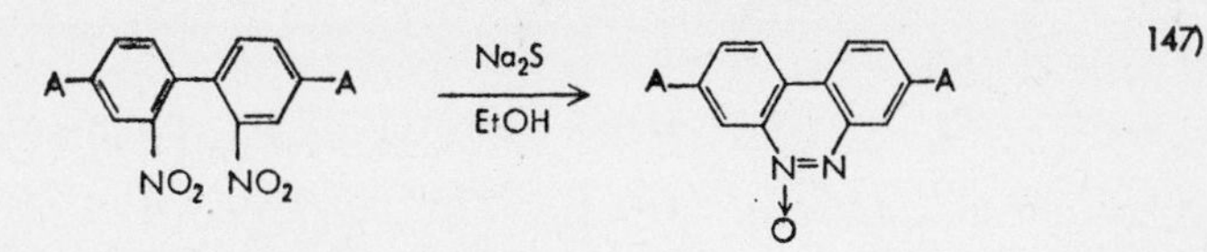

Example 3. Benzimidazole N-Oxide[141]

To a solution of 2-nitroformanilide (2.0 g) in 50% ethanol (60 ml), NH_4Cl (1.0 g) was added, followed by zinc dust (5.0 g) in small portions with stirring. The mixture was heated under reflux for 3 h. The resulting grey mixture was filtered while hot and the residue was washed with hot ethanol. The filter cake was broken up into a powder and suspended in a dilute ammonia solution. Hydrogen sulfide was passed through this suspension to decompose the zinc complex of benzimidazole *N*-oxide and the precipitate was removed by filtration. The filtrate was evaporated and the crystalline residue was recrystallized from ethanol to 0.7 g of benzimidazole *N*-oxide, m.p. 215° (decomp.).

The filtrate from the grey reaction mixture was evaporated and the residue was recrystallized from acetone to give 2-benzimidazolone (0.4 g), m.p. above 300°.

Example 4. 5-Nitrobenzimidazole 3-Oxide[142]

Hydrogen sulfide was passed through a mixture of 2,4-dinitroformanilide (7.0 g) in ethanol (500 ml) containing ammonia (saturated at 0°, 20 ml) for 3 h. After leaving the solution overnight at room temperature, the resulting brown solution was concentrated to *ca.* 50 ml, the precipitated sulfur was filtered off, and washed with ethanol. Evaporation of ethanol from the combined filtrate and washings gave brown crystals which were recrystallized from ethanol (*ca.* 1 l) to pale yellow plates (1.5 g), m.p. 274° (decomp.).

3.2.3 *Intramolecular Condensation-Cyclization of Nitro Compounds*

The nitro group in some cases undergoes dehydrative cyclization with a substituent in the molecule having active hydrogen, in the presence of alkali or acid, to form an aromatic *N*-oxide compound. Usually, alkali hydroxide is used as the condensation agent. The following are examples of compounds having CH_2, CH, or NH_2 as a substituent with active hydrogen.

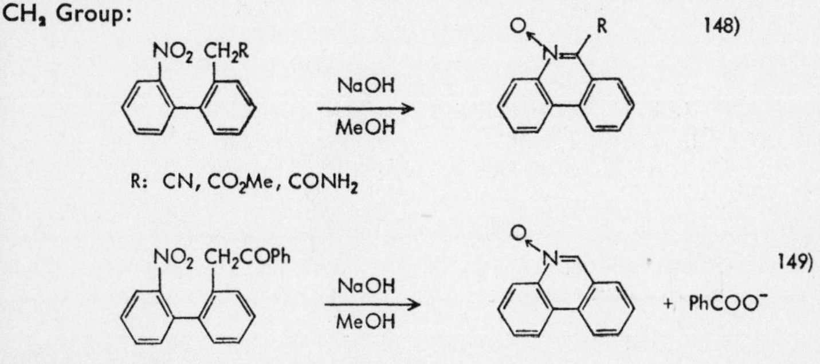

References p. 70

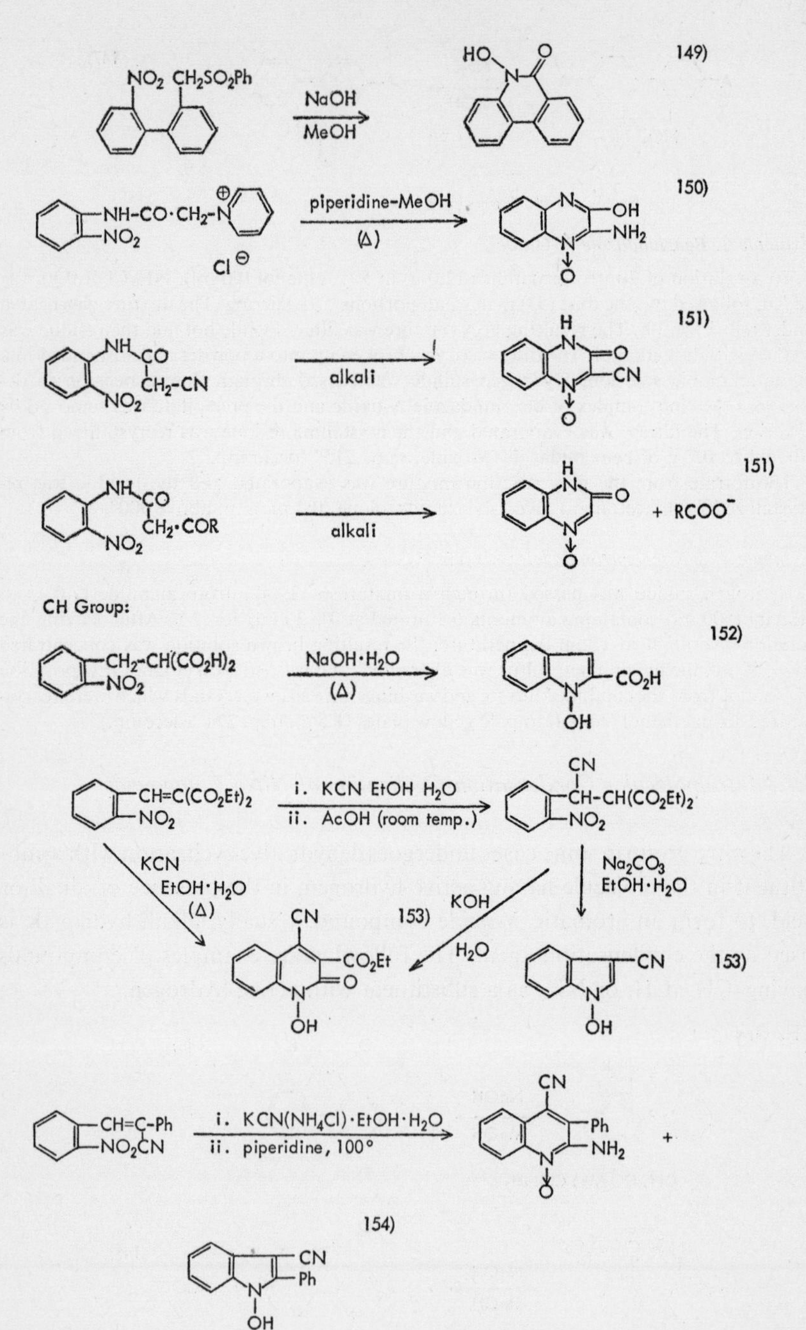
149)
NaOH
MeOH
150)
piperidine-MeOH
(Δ)
151)
alkali
151)
alkali
+ RCOO⁻
CH Group:
152)
NaOH·H₂O
(Δ)
i. KCN EtOH H₂O
ii. AcOH (room temp.)
KCN
EtOH·H₂O
(Δ)
153)
KOH
H₂O
Na₂CO₃
EtOH·H₂O
153)
i. KCN(NH₄Cl)·EtOH·H₂O
ii. piperidine, 100°
154)

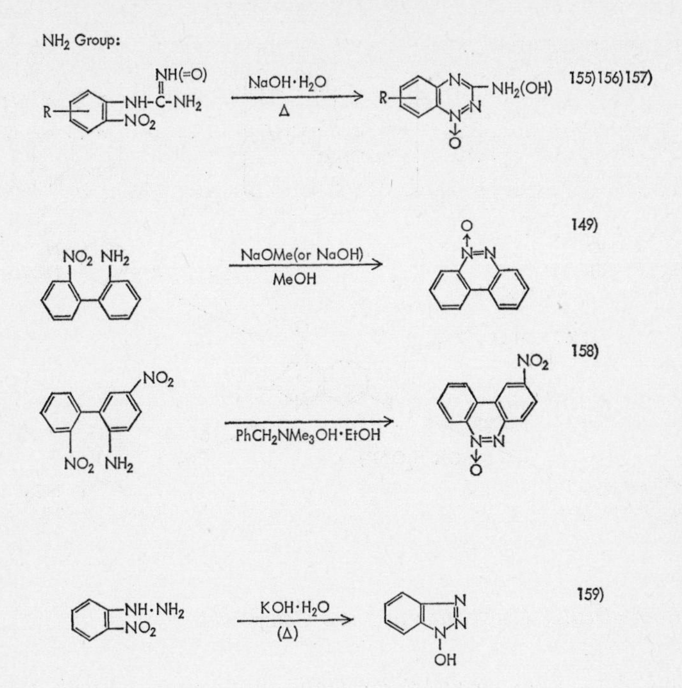

The foregoing reactions used alkali as the condensation agent and the following is an example using an acid catalyst.

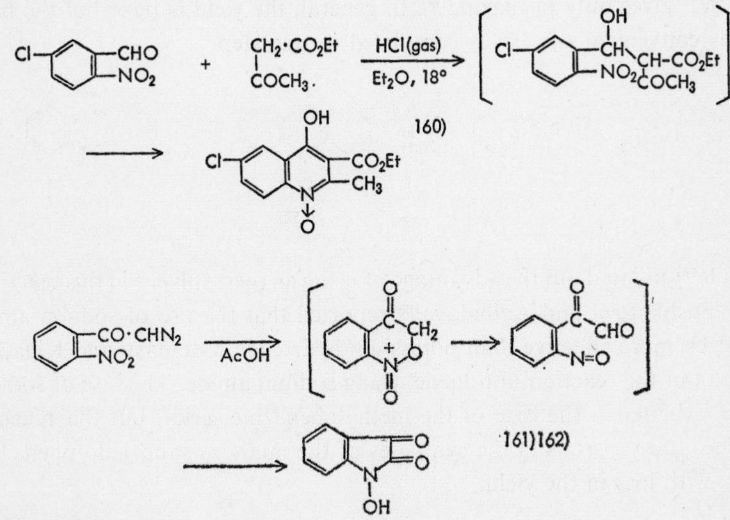

References p. 70

Other reactions are:

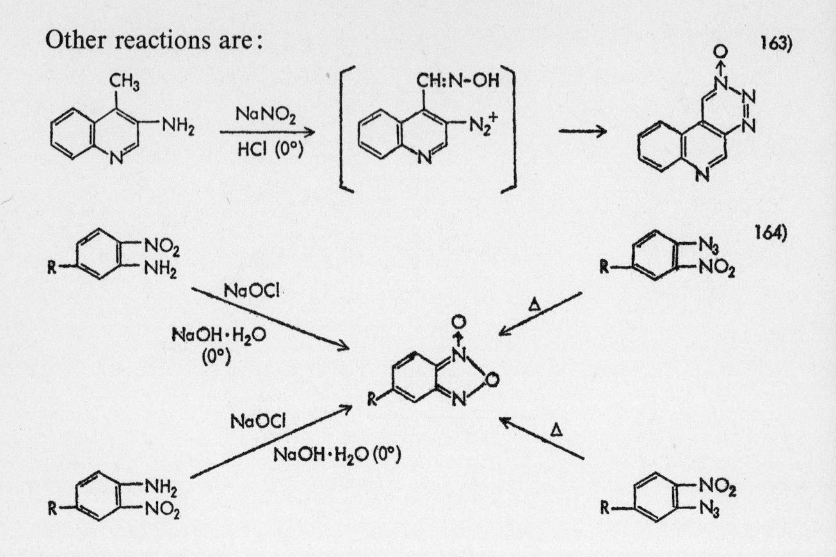

3.2.4 Intermolecular Cyclization of Nitro Compounds

Heating of an arylamine and an aromatic nitro compound with powdered potassium hydroxide at 110–160° results in the formation of phenazine and phenazine 5-oxide (3-53); this is known as the Wohl–Aue reaction[165]. While heating at 120–125° results in formation of an *N*-oxide compound, that at 140–160° gives only phenazine[166]. In general, the yield is poor but the reaction is convenient since it is completed in one step.

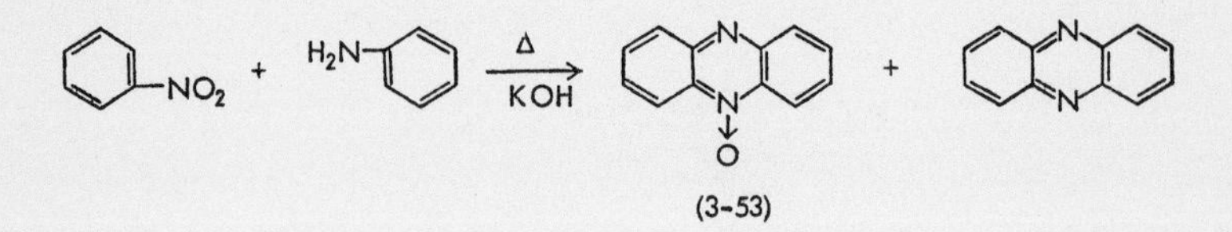

Soule[167] pointed out the advantage of using an inert solvent in this reaction, while Pushkareva and Agibalova[168] reported that the use of sodium amide would be more effective than potassium hydroxide. Otomasu and Kidani[169] carried out the reaction in toluene, using sodium amide. This use of sodium amide is better in the case of the methylphenazine series, but the reaction becomes too vigorous in the syntheses of the methoxy- and halo-phenazine series, with loss in the yield.

The mechanism of this reaction is thought to be the following. The reaction begins with nucleophilic attack of the anionic imine from arylamine on the aromatic nitro compound, and the *o*-nitrodiarylamine thereby formed is reductively cyclized by excess of aromatic amine.

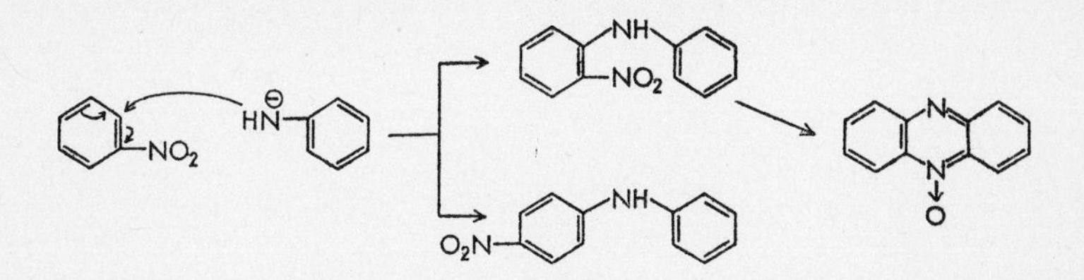

This hypothesis is based on the fact that Wohl[170] had obtained *p*-nitrodiphenylamine as the by-product of this reaction, while Yosioka and Otomasu[171], and Vivian and others[172] synthesized phenazine derivatives by the same treatment of *o*-nitrodiphenylamine and its derivatives with potassium hydroxide. Other examples in support of the above assumption are the formation of 1-chlorophenazine by the Wohl–Aue reaction of 1,2-dichloro-3-nitrobenzene and aniline by Yosioka and Otomasu[171], and by Chernetzkii and Kiprianov[173], and of 1-methoxyphenazine from 3-nitroveratrol and aniline by Yosioka and Otomasu[171], which may be taken as the substitution of chlorine or methoxyl in the position *ortho* to the nitro group by the phenylimino ion.

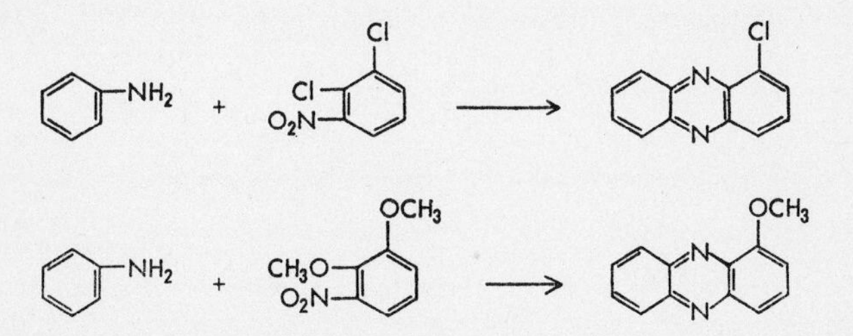

Yosioka and Otomasu[174], however, found that the condensation of nitrobenzene and 4-aminoresorcinol dimethyl ether produced 2-methoxyphenazine and its 10-oxide instead of dimethoxyphenazine, and that 2-methoxyphenazine 5-oxide was formed in the main, with a small amount of 1,8-dimethoxyphenazine, on heating *p*-nitroanisole and *o*-anisidine in toluene with potassium hydroxide[175]. These facts indicate that the liberation of methoxyl from the position *ortho* to the imino group also occurs under other influences.

References p. 70

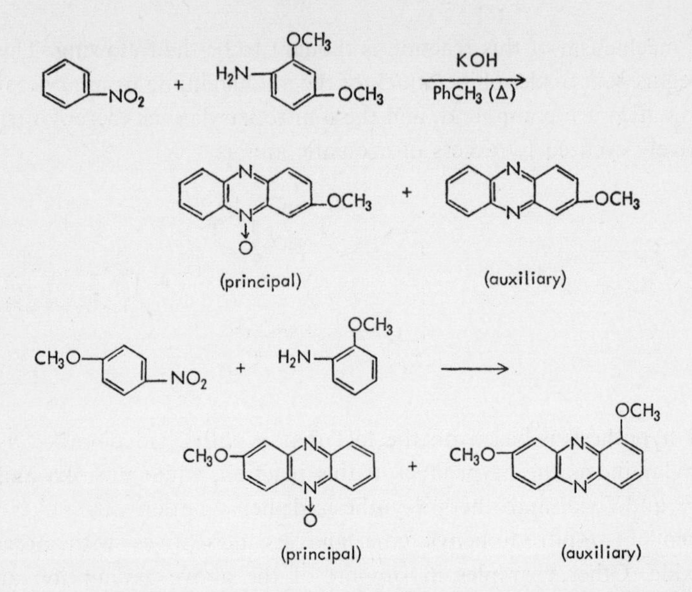

In short, the Wohl–Aue reaction is not very well adapted for the preparation of phenazine *N*-oxides because of the concurrent formation of phenazine and the poor yield. Direct oxidation of phenazine would be better, but when there is steric hindrance in the phenazine skeleton, this Wohl–Aue reaction is advantageous. For example, direct *N*-oxidation of benzo[*a*]phenazine (3-54) gives only its 7-oxide (3-55), while by the Wohl–Aue reaction 12-oxide (3-56) can be synthesized, although its yield is not so good[100].

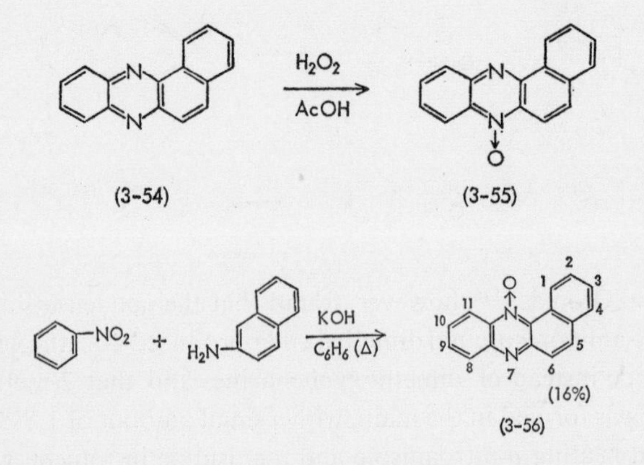

The above are examples of condensation of aromatic nitro compounds with alkali. Kliegel *et al.*[176] synthesized 9-hydroxyacridine 10-oxide (3-57), though in a small amount, by the condensation of *o*-nitrobenzaldehyde and benzene, with the use of concentrated sulfuric acid.

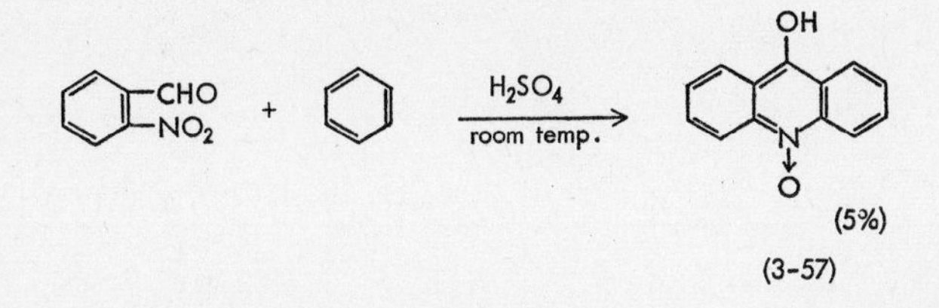

Example 5. 2,7-Dimethoxyphenazine 5-Oxide[176]

A mixture of *p*-nitroanisole (10 g), *p*-anisidine (10 g), and well-powdered potassium hydroxide (30 g) was heated in toluene (150 ml) under a reflux in an oil bath for 6 h. After the reaction, toluene was distilled off under a reduced pressure and 300 ml of water was added. Unreacted substance was removed by steam distillation and the residue was extracted with benzene. The benzene solution was extracted once with 10% hydrochloric acid and the acid layer was neutralized with ammonia water, then filtered, leaving a yellowish brown substance, which weighed 1.35 g (7.1% yield). The product was dissolved in benzene and purified by chromatography on alumina. The column was developed with benzene and a yellow band containing 2,7-dimethoxyphenazine and its mono-*N*-oxide was gradually eluted. The first eluate gave 0.5 g of 2,7-dimethoxyphenazine as yellow needles, m.p. 240°, which was recrystallized from ligroin. The last eluate gave 0.55 g of 2,7-dimethoxyphenazine mono-*N*-oxide, m.p. 236° (decomp.), as orange yellow needles, which was recrystallized from chloroform.

3.2.5 *Cyclization of Nitrone Derivatives*

Aromatic *N*-oxides may be considered as a kind of cyclic nitrone[177], so that aromatic *N*-oxides should be formed if a nitrone molecule can be obtained as an intermediate and if the latter molecule contains a reactive group which will undergo condensation-cyclization. Based on such ideas, synthesis of aromatic amine oxides via nitrones by the condensation of nitrosoaryl and active methylene has been devised, as shown below.

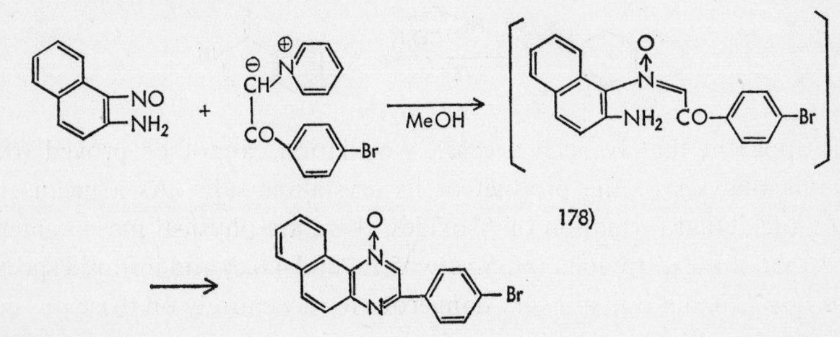

References p. 70

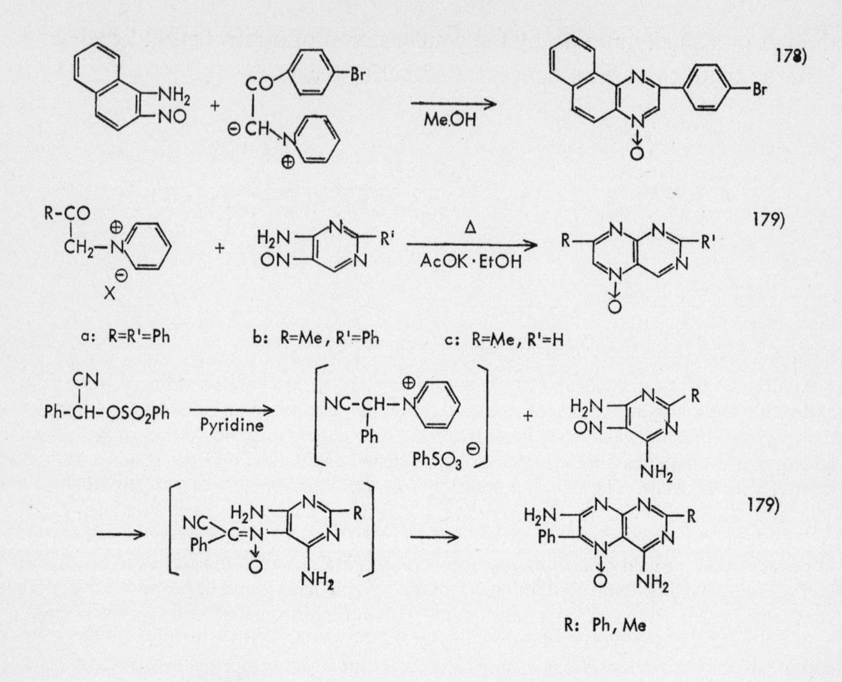

3.3 Characterization of the N-Oxide Group

The aromatic *N*-oxides synthesized by the methods outlined in the preceding sections must be examined to prove the presence of the *N*-oxide group. This is especially true in the case of direct oxidation because a side reaction often occurs to bring about oxidation of the α-position and a possibility for the formation of a lactam, an isomer of the expected *N*-oxide. For example, *N*-oxidation of quinazoline with monoperphthalic acid in ether results in quantitative formation of 4-quinazolone, an *N*-oxide compound not being formed at all[66a,b].

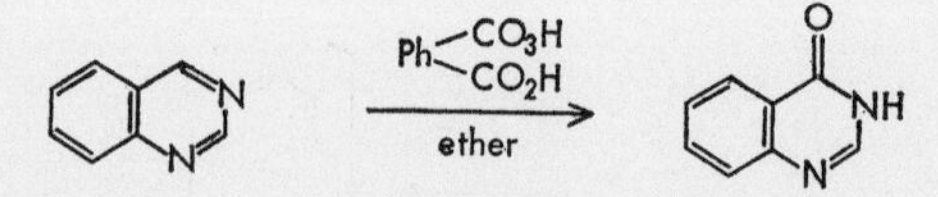

It is apparent that in such a case, *N*-oxidation cannot be proved from elemental analyses of the product or its crystalline salts. As a means for proving the actual formation of *N*-oxides, there are physical measurements such as that of the ultraviolet (*cf. Section 5.1.3* and 4.8.2) and infrared spectra (*cf. Section 4.7*), but it is still very dangerous to rely entirely on these physical means. It is more desirable to convert such products to the corresponding tertiary amines by deoxygenation, such as by catalytic reduction over Raney nickel (*cf. Section 5.2.2*) or by the action of phosphorus trichloride (*cf. Section 5.2.3*).

Coats and Katritzky[181] devised a method of testing *N*-oxides in which aromatic *N*-oxides when heated with dimethylaniline in hydrochloric acid become blue. This was assumed to be due to the oxidative action of the *N*-oxide group and formation of a pigment of the Crystal Violet series by condensation with formaldehyde formed as a result of this reaction. This coloration reaction is known to take place with methyl homologs of pyridine 1-oxide, also those substituted with cyano, hydroxyl, amino, and nitro groups, and in quinoline 1-oxides, but also takes place in the case of nitrobenzene and *m*-dinitrobenzene, and is not specific for aromatic *N*-oxides.

Example 1. Color Reaction of Aromatic N-Oxides[181]

In a test tube, 0.2 ml of dimethylaniline, 0.05 ml of conc. hydrochloric acid, and 0.1 g of the test compound are placed and the mixture is boiled for 1 min. When cooled, 1 ml of ethanol is added. If an *N*-oxide group is present, the solution will become intensely blue.

In the case of *N*-oxides of diazine series, in which there is a possibility of *N*-oxidation of two nitrogen atoms, a simple mono-*N*-oxide or two kinds of isomeric mono-*N*-oxides may have been formed. Consequently, it is necessary to determine which of the two tertiary nitrogen atoms has been oxidized. For this purpose, some physical methods have been devised, such as the measurement of dipole moment (*cf. Section 4.2*) and NMR spectrum (*cf. Section 4.6.2*), and readers are referred to these sections.

Various chemical methods of determination have been suggested, but there is no set formula and the method must depend upon the specific case. Two of the representative reactions will be given here.

Example 2. N-Oxidation of 4-Methoxyquinazoline

Yamanaka[68] obtained two kinds of mono-*N*-oxides (A and B) by the application of an ether solution of monoperphthalic acid to 4-methoxyquinazoline (3-58).

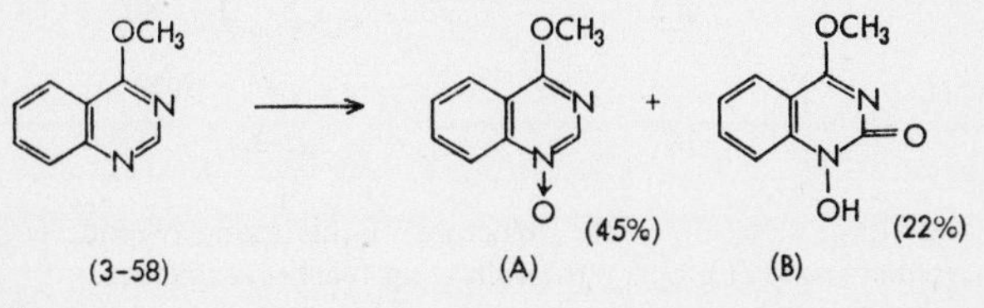

References p. 70

The product (A) was converted to 3-methyl-4-quinazolone (3-59) by the reaction sequence shown below and was determined as 1-oxide by identification with the product synthesized from anthranilic acid methylamide.

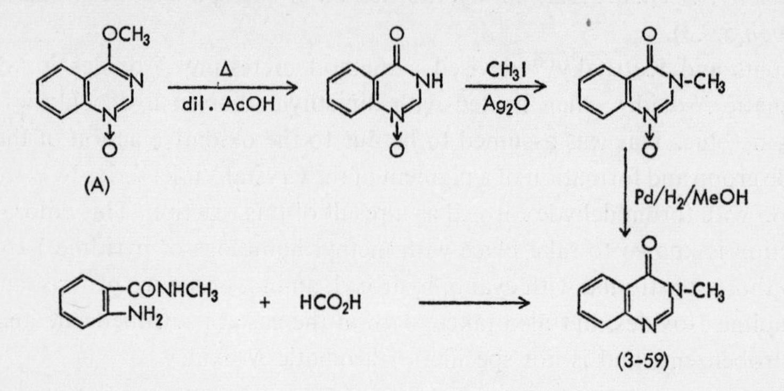

In this case, the nitrogen atom which was not oxidized was first protected and then the oxygen atom removed from the *N*-oxide group. Compound (B) was determined as 1-hydroxy-2-oxo-4-methoxy-1,2-dihydroquinazoline by converting the *N*-oxide group of (A) to the lactam with tosyl chloride and alkali, and identifying it with the reduction product of the methylated compound of (B).

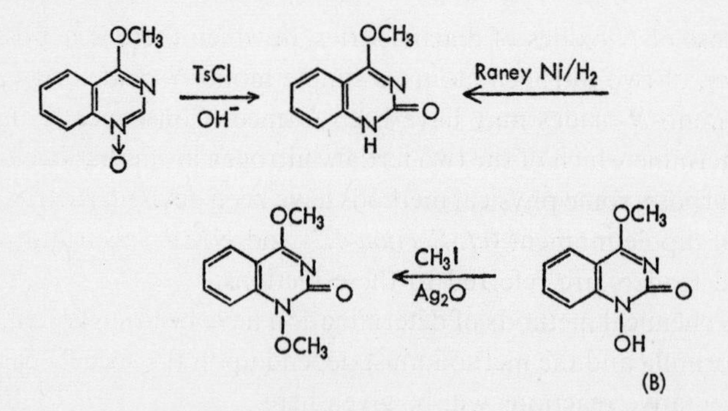

Example 3. N-Oxidation of 2-Phenylquinoxaline

N-Oxidation of 2-phenylquinoxaline (3-36) with hydrogen peroxide and glacial acetic acid or with monoperphthalic acid gives only one kind of mono-*N*-oxide[182].

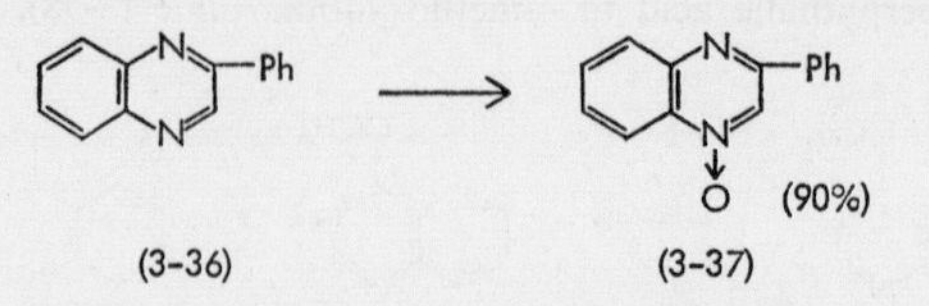

Hayashi and Iijima determined the structure of this mono-*N*-oxide compound as 2-phenylquinoxaline 4-oxide (3-37) by the following reaction sequence.

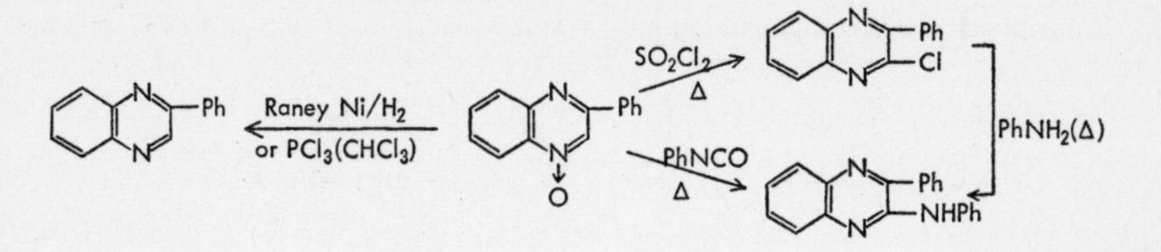

Deoxygenation of this product would revert it to 2-phenylquinoxaline, which shows that the bonded oxygen forms an *N*-oxide group and not the lactam-oxygen, and formation of phenyl-chloroquinoxaline with sulfuryl chloride indicates that there is no phenyl group in the position α to the *N*-oxide group. This point can further be proved by the formation of 2-anilino-3-phenylquinoxaline by heating the mono-*N*-oxide with phenyl isocyanate.

Example 4. N-Oxidation of 2-Methylquinoxaline

N-Oxidation of 2-methylquinoxaline with monoperphthalic acid results in the formation of 1-oxide (A) and 4-oxide (B) in about 3:2 ratio[183], and Hayashi and his co-workers determined their structure in the following manner.

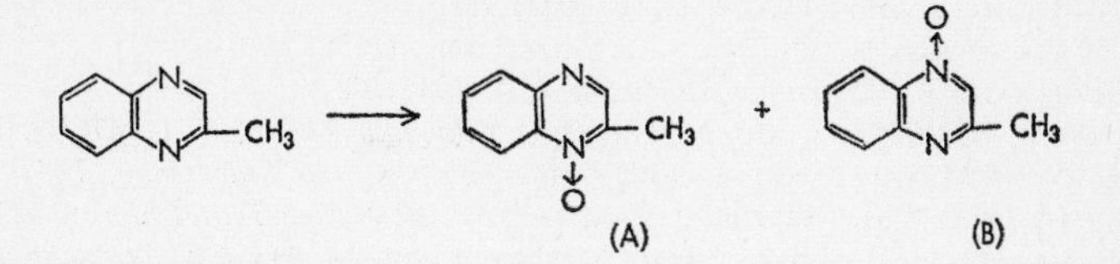

Oxidation of the methyl group in (A) with selenium dioxide to an aldehyde and heating of its oxime with acetic anhydride gave a nitrile. This nitrile did not agree with 2-cyanoquinoxaline 4-oxide, which had been obtained by the *N*-oxidation of 2-cyanoquinoxaline and whose structure had been proved by another method. This fact indicates that the compound (A) is the 1-oxide.

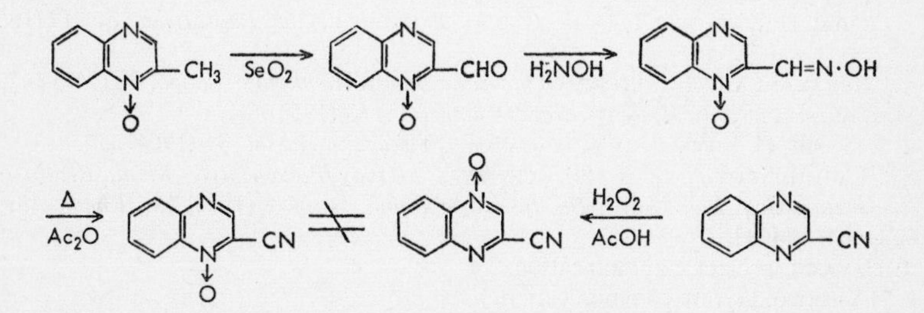

Similar oxidation of the methyl group in (B) to the aldehyde and heating of its oxime with acetic anhydride failed to form the expected nitrile. This fact shows that while the α-position of the *N*-oxide group is occupied in the oxime of (A) and only dehydration of the oxime to nitrile occurs, this position is vacant in the compound (B) and reaction of its oxime and acetic anhydride causes rearrangement to the lactam type (*cf. Section 7.4.4*) at the same time.

References p. 70

Consequently, this compound (B) must be the 4-oxide.

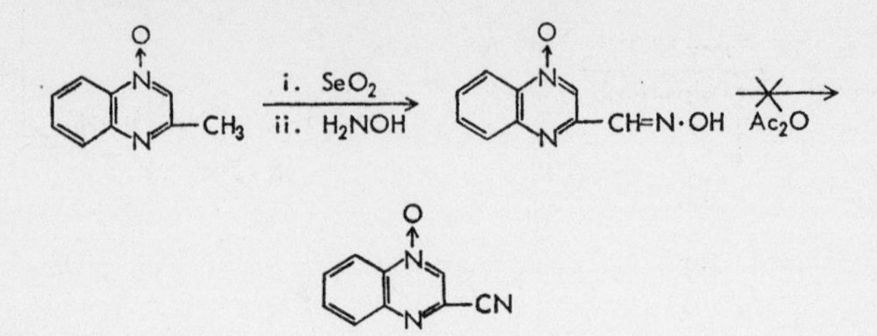

REFERENCES

1 J. Meisenheimer and E. Stotz, *Ber.*, 58 (1925) 2334.
2 G. Heller and A. Sourlis, *Ber.*, 41 (1908) 2692.
3 J. Meisenheimer, *Ber.*, 59 (1926) 1848.
4 H. Böhme, *Ber.*, 70 (1937) 379.
5 B. Bobranski, L. Kochańska, and A. Kowalewska, *Ber.*, 71 (1938) 2385.
6 *e.g.* E. Ochiai and Y. Itô, *Ber.*, 71 (1938) 938.
7 G. R. Clemo and H. McIlwain, *J. Chem. Soc.*, (1938) 479.
8 E. Ochiai and Z. Sai, *Yakugaku Zasshi*, 65A (1945) 73.
9 E. Ochiai, M. Katada, and E. Hayashi, *Yakugaku Zasshi*, 67 (1947) 33.
10 R. F. Evans, M. van Ammers, and H. J. den Hertog, *Rec. Trav. Chim.*, 78 (1959) 408.
11 P.A. van Zwieten, M. Gerstenfeld, and H.O. Huisman, *Rec. Trav. Chim.*, 81 (1962) 604.
12 M. Yamazaki, N. Honjo, K. Noda, Y. Chono, and M. Hamana, *Yakugaku Zasshi*, 86 (1966) 749.
13 R. W. White and W. D. Emmons, *Tetrahedron*, 17 (1962) 31.
14 *Organic Syntheses, Collective Volume* (2nd Ed.), John Wiley & Sons, New York–London 1956, p. 431.
15 *Organic Syntheses, Collective Volume* III, John Wiley & Sons, New York–London, 1955, p. 619.
16 G. B. Payne, *J. Org. Chem.*, 24 (1959) 1354.
17 E. Ochiai, H. Kataoka, T. Dodo, and M. Takahashi, *Ann. Rept. Itsuu Lab.*, 12 (1962) 19.
18 M. Murakami and E. Matsumura, *Osaka Sangyô Kenkyûjo Nempô*, 5 (1947) 147.
19 M. Colonna and A. Risaliti, *Gazz. Chim. Ital.*, 83 (1953) 61.
20 E. Hayashi, C. Iijima, and Y. Nagasawa, *Yakugaku Zasshi*, 84 (1964) 177.
21 Yu. I. Chumakov and M. Polucheniya, *Metody Polucheniya, Khim. Reaktivovji Preparatov, Gos. Kom. Sov. Min. SSSR po Khim.*, No. 4-5 (1962) 59 [*Chem. Abstr.*, 60 (1964) 14468].
22. E Hayashi, private communication.
23 M. Fujimoto, private communication.
24 R. C. Witman (Shell Oil Co.), U. S. Pat. 3,047,579 (July 31, 1962).
25 M. Ogata and H. Kanô, *Chem. Pharm. Bull. (Tokyo)*, 11 (1963) 32.
26 E. C. Taylor and J. S. Driscoll, *J. Org. Chem.*, 25 (1960) 1716.
27 E. V. Brown, *J. Am. Chem. Soc.*, 79 (1957) 3565.
28 W. Herz and D. R. K. Murty, *J. Org. Chem.*, 26 (1961) 122.
29 R. Adams and S. Miyano, *J. Am. Chem. Soc.*, 76 (1954) 2785.
30 F. Leonard and A. Wajngurt, *J. Org. Chem.*, 21 (1956) 1077.

31 Y. Kobayashi, *Chem. Pharm. Bull. (Tokyo)*, 6 (1958) 273.
32 A. R. Katritzky, *J. Chem. Soc.*, (1957) 191, 4385.
33 H. Tanida, *Yakugaku Zasshi*, 79 (1959) 1063.
34 T. Horie and T. Ueda, *Chem. Pharm. Bull. (Tokyo)*, 11 (1963) 114.
35 S. J. Childress and J. V. Scudi, *J. Org. Chem.*, 23 (1958) 67.
36 L. Pentimalli, *Tetrahedron*, 14 (1961) 151.
37 E. Ochiai, G. Kobayashi, and J. Hasegawa, *Yakugaku Zasshi*, 67 (1947) 101; E. Ochiai and M. Y. Ishikawa, *Chem. Pharm. Bull. (Tokyo)*, 6 (1958) 208; E. Ochiai, M. Y. Ishikawa, and Y. Oka, *Chem. Pharm. Bull. (Tokyo)*, 7 (1958) 744; *Ann. Rept. Itsuu Lab.*, 12 (1962) 29; E. Ochiai, H. Kataoka, T. Dodo, and M. Takahashi, *Ann. Rept. Itsuu Lab.*, 12 (1962) 11.
38 A. W. Johnson, T. J. King, and J. R. Turner, *J. Chem. Soc.*, (1958) 3230; E. C. Taylor and N. E. Boyer, *J. Org. Chem.*, 24 (1959) 275; Ya. L. Goldfarb and V. K. Zvorykina, *Izv. Akad. Nauk. SSSR, Otd. Khim. Nauk* (1958) 748 [*Chem. Abstr.*, 52 (1958) 20219].
39 Ya. L. Goldfarb, F. D. Alashev, and V. K. Zvorykina, *Izv. Akad. Nauk. SSSR, Otd. Khim. Nauk*, (1958) 788 [*Chem. Abstr.*, 52 (1958) 20151].
40 G. T. Newbold and F. S. Spring, *J. Chem. Soc.*, (1947) 1183.
41 J. K. Landquist, *J. Chem. Soc.*, (1953) 2816.
42 Y. S. Rozum and A. I. Kiprianov, *Zh. Obshch. Khim.*, 29 (1959) 1306 [*Chem. Abstr.*, 54 (1960) 9938].
43 M. Ogata and H. Kanô, *Chem. Pharm. Bull. (Tokyo)*, 11 (1963) 35.
44 T. Itai and T. Nakashima, *Chem. Pharm. Bull. (Tokyo)*, 10 (1962) 936.
45 H. Kanô, M. Ogata, and K. Tôri, *Chem. Pharm. Bull. (Tokyo)*, 10 (1962) 1123; 11 (1963) 1527.
46 M. Ogata, H. Watanabe, K. Tori, and H. Kanô, *Tetrahedron Letters*, (1964) 19.
47 M. A. Stevens, A. Giner-Sorolla, H. W. Smith, and G. B. Brown, *J. Org. Chem.*, 27 (1962) 567.
48 E. P. Hart, *J. Chem. Soc.*, (1954), 1879.
49 F. Linsker and R. L. Evans, *J. Am. Chem. Soc.*, 68 (1946) 403.
50 M. Colonna and A. Risaliti, *Gazz. Chim. Ital.*, 85 (1955) 1148; 86 (1956) 1067.
51 T. Itai and S. Kamiya, *Chem. Pharm. Bull. (Tokyo)*, 9 (1961) 87.
52 G. Kobayashi, Y. Kuwayama, and S. Okamura, *Yakugaku Zasshi*, 83 (1963) 234.
53 E. Shaw, J. Bernstein, K. A. Losee, and W. A. Lott, *J. Am. Chem. Soc.*, 72 (1950) 4362.
54 P. A. van Zwieten, M. Gerstenfeld, and H. O. Huisman, *Rec. Trav. Chim.*, 81 (1962) 611.
55 F. G. Mann and J. Watson, *J. Org. Chem.*, 13 (1948) 502.
56 E. Ochiai, S. Suzuki, Y. Utsunomiya, T. Ohmoto, K. Nagatomo, and M. Itoh, *Yakugaku Zasshi*, 80 (1960) 339.
57 R. Adams and S. Miyano, *J. Am. Chem. Soc.*, 76 (1954) 3160.
58 E. Ochiai and Y. Tamai, *Ann. Rept. Itsuu Lab.*, 13 (1963) 5.
59 E. Ochiai, H. Kataoka, T. Dodo, and M. Takahashi, *Ann. Rept. Itsuu Lab.*, 12 (1962) 14.
60 N. H. Cromwell and R. A. Mitsch, *J. Org. Chem.*, 26 (1961) 3812, 3817.
61 T. Sakuragi and A. Kummarow, *J. Org. Chem.*, 24 (1959) 1032.
62 J. Francis, J. K. Landquist, A. A. Leri, J. A. Silk, and J. M. Thorp, *Biochem. J.*, 63 (1956) 455.
63 H. Kataoka, private communication.
64 M. Colonna, *Gazz. Chim. Ital.*, 90 (1960) 1179.
65 A. R. Katritzky and A. M. Monro, *J. Chem. Soc.*, (1958) 150.

66 (*a*) K. ADACHI, *Yakugaku Zasshi*, 77 (1957) 507. (*b*) J. K. LANDQUIST, *J. Chem. Soc.*, (1956) 1885.
67 E. HAYASHI AND T. HIGASHINO, *Chem. Pharm. Bull. (Tokyo)*, 12 (1964) 43.
68 H. YAMANAKA, *Chem. Pharm. Bull. (Tokyo)*, 7 (1959) 152.
69 G. H. W. CHEESEMAN, *J. Chem. Soc.*, (1961) 1246.
70 E. HAYASHI AND H. OHKI, *Yakugaku Zasshi*, 81 (1961) 1033.
71 K. LEHMSTEDT AND H. KLEE, *Ber.*, 69 (1936) 1514.
72 E. OCHIAI, H. TANIDA, AND S. UYEDA, *Pharm. Bull. (Japan)*, 5 (1957) 188.
73 E. OCHIAI AND H. TANIDA, *Pharm. Bull. (Japan)*, 5 (1957) 621.
74 E. OCHIAI, CH. KANEKO, I. SHIMADA, Y. MURATA, T. KOSUGE, S. MIYASHITA, AND CH. KAWASAKI, *Chem. Pharm. Bull. (Tokyo)*, 8 (1960) 126.
75 E. OCHIAI AND E. HAYASHI, *Yakugaku Zasshi*, 67 (1947) 34.
76 A. RISALITI, *Ricerca Sci.*, 24 (1954) 2351 [*Chem. Abstr.*, 49 (1955) 15902].
77 M. M. ROBISON AND B. L. ROBISON, *J. Am. Chem. Soc.*, 80 (1958) 3446.
78 J. K. LANDQUIST AND G. J. STACEY, *J. Chem. Soc.*, (1953) 2822.
79 H. VON EULER, H. HASSELQUIST, AND O. HEIDENBERG, *Arkiv Kemie*, 14 (1957) 237.
80 M. L. PETERSON, *J. Org. Chem.*, 25 (1960) 565.
81 J. I. G. CADOGAN, *J. Chem. Soc.*, (1959) 2844.
82 N. N. GERBER, *J. Med. Chem.*, 7 (1964) 204.
83 E. HAYASHI, T. HIGASHINO, CH. IIJIMA, Y. KÔNO, AND T. DOIHARA, *Yakugaku Zasshi*, 82 (1962) 584.
84 M. M. ROBISON AND B. L. ROBISON, *J. Org. Chem.*, 23 (1958) 1017.
85 T. ITAI AND S. SAKO, *Chem. Pharm. Bull. (Tokyo)*, 10 (1962) 989.
86 E. HAYASHI, CH. IIJIMA, AND Y. NAGASAWA, *Yakugaku Zasshi*, 84 (1964) 163.
87 G. MODENA AND P. E. TODESCO, *Gazz. Chim. Ital.*, 90 (1960) 702.
88 D. E. AMES AND T. F. GREY, *J. Chem. Soc.*, (1955) 631.
89 H. GILMAN AND J. T. EDWARD, *Can. J. Chem.*, 31 (1953) 457.
90 N. K. USHENKO AND T. E. GORIZDRA, *Ukr. Khim. Zh.*, 15 (1949) 296 [*Chem. Abstr.*, 48 (1954) 3366].
91 R. W. GOULEY, G. W. MOERSCH, AND H. S. MOSHER, *J. Am. Chem. Soc.*, 69 (1947) 303.
92 E. C. TAYLOR AND A. J. CROVETTI, *J. Am. Chem. Soc.*, 78 (1956) 214.
93 J. DRUEY AND H. V. DAENIKER, *Helv. Chim. Acta*, 41 (1958) 2148.
94 E. HAYASHI, private communication.
95 E. HAYASHI AND H. OHKI, *Yakugaku Zasshi*, 81 (1961) 1035.
96 A. S. ELINA AND O. YU. MAGIDSON, *Zh. Obshch. Khim.*, 25 (1955) 161 [*Chem. Abstr.*, 50 (1956) 1839].
97 H. OTOMASU, R. YAMAGUCHI, K. ISHIGÔ-OKA, AND H. TAKAHASHI, *Yakugaku Zasshi*, 82 (1962) 1434.
98 H. OTOMASU, T. TAKAHASHI, AND M. OGATA, *Chem. Pharm. Bull. (Tokyo)*, 12 (1964) 714.
99 R. S. W. BRAITHWAITE AND G. K. ROBINSON, *J. Chem. Soc.*, (1962) 3671.
100 I. J. PACHTER AND M. C. KLOETZEL, *J. Am. Chem. Soc.*, 73 (1951) 4958.
101 J. F. CORBETT AND P. F. HOLT, *J. Chem. Soc.*, (1961) 5029.
102 B. KLEIN AND J. BERKOWITZ, *J. Am. Chem. Soc.*, 81 (1959) 5160.
103 H. OTOMASU, H. TAKAHASHI, AND K. YOSHIDA, *Yakugaku Zasshi*, 84 (1964) 1080.
104 R. KUHN AND W. BLAU, *Ann.*, 615 (1958) 105.
105 P. BAUMGARTEN, R. MERLÄNDER, AND J. OLSHAUSEN, *Ber.*, 66 (1933) 1803.
106 C. SCHÖPF AND K. KOCH, *Ber.*, 69 (1936) 2766.
107 J. W. CORNFORTH AND R. H. CORNFORTH, *J. Chem. Soc.*, (1947) 96.
108 J. T. NIELSEN, N. ELMING, AND N. CLAUSON-KAAS, *Acta Chem. Scand.*, 9 (1955) 14,30.
109 H. OST, *J. prakt. Chem.*, [2], 29 (1884) 378.

110 A. Peratoner and A. Tamburello, *Gazz. Chim. Ital.*, 41 (1911) 666.
111 G. Soliman and I. El-Sayed El-Kholy, *J. Chem. Soc.*, (1954) 1755.
112 F. Parisi, P. Bovina and A. Quilico, *Gazz. Chim. Ital.*, 90 (1960) 903.
113 F. Parsii, P. Bovina, and A. Quilico, *Gazz. Chim. Ital.*, 92 (1962) 1138.
114 P. Yates, M. J. Jorgenson, and S. K. Roy, *Can. J. Chem.*, 40 (1962) 2146.
115 M. M. Robison and B. L. Robison, *J. Am. Chem. Soc.*, 80 (1958) 3443.
116 J. Gottlieb, *Ber.*, 32 (1899) 958.
117 L. H. Sternbach, S. Kaiser, and E. Reeder, *J. Am. Chem. Soc.*, 82 (1960) 475.
118 J. Meisenheimer, O. Senn, and P. Zimmermann, *Ber.*, 60 (1927) 1736.
119 J. T. Shaw, *J. Org. Chem.*, 27 (1962) 3890.
120 K. Adachi, *Chem. Pharm. Bull. (Tokyo)*, 7 (1959) 479.
121 E. Hayashi, *Collection of Papers, 10th Anniversary of the Shizuoka College of Pharmacy*, (1963) 31.
122 K. Adachi, *Yakugaku Zasshi*, 77 (1957) 510.
123 K. Adachi, *Yakugaku Zasshi*, 77 (1957) 514.
124 E. C. Taylor, C. C. Cheng, and O. Vogl, *J. Org. Chem.*, 24 (1959) 2019.
125 E. C. Taylor and P. K. Löffler, *J. Org. Chem.*, 24 (1959) 2035.
126 M. A. Stevens, H. W. Smith, and G. B. Brown, *J. Am. Chem. Soc.*, 82 (1960) 3189.
127 R. M. Cresswell and G. B. Brown, *J. Org. Chem.*, 28 (1963) 2560.
128 W. Sharp and F. S. Spring, *J. Chem. Soc.*, (1951) 932.
129 G. Dunn, J. A. Elvidge, G. T. Newbold, D. W. C. Ramsay, F. S. Spring, and W. Sweeny, *J. Chem. Soc.*, (1949) 2707.
130 G. La Parola, *Gazz. Chim. Ital.*, 75 (1945) 216.
131 (*a*) J. B. Wright, *J. Org. Chem.*, 29 (1964) 1620. (*b*) O. Diels and R. van Leeden, *Ber.*, 38 (1905) 3363.
132 G. Heller and A. Sourlis, *Ber.*, 41 (1908) 2692.
133 G. Heller and P. Wunderlich, *Ber.*, 47 (1914) 1617.
134 E. C. Taylor and N. W. Kalenda, *J. Org. Chem.*, 18 (1953) 1755.
135 P. Friedländer and H. Ostermeier, *Ber.*, 14 (1881) 1916; 47 (1914) 3369.
136 K. G. Cunningham, G. T. Newbold, F. S. Spring, and J. Stark, *J. Chem. Soc.*, (1949) 2091.
137 K. H. Bauer, *Ber.*, 71 (1938) 2226.
138 R. T. Coutts and D. G. Wibberley, *J. Chem. Soc.*, (1962) 2518.
139 St. von Niementowski, *Ber.*, 43 (1910) 3012.
140 R. Kuhn and W. Blau, *Ann.*, 615 (1958) 99.
141 S. Takahashi and H. Kanô, *Chem. Pharm. Bull. (Tokyo)*, 11 (1963) 1375.
142 S. Takahashi and H. Kanô, *Chem. Pharm. Bull. (Tokyo)*, 12 (1964) 286.
143 T. A. Liss, *Chem. Ind. (London)*, (1964) (9) 368 [*Chem. Abstr.*, 60 (1964) 10666].
144 K. Elbs, O. Hirschel, F. Wagner, K. Himmler, W. Türk, A. Henrich, and E. Lehmann, *J. prakt. Chem.*, 108 (1924) 209.
145 W. C. J. Ross and G. P. Warwick, *J. Chem. Soc.*, (1956) 1728.
146 S. B. Hansen and V. Petrow, *J. Chem. Soc.*, (1953) 350.
147 E. Ullmann and P. Dieterle, *Ber.*, 37 (1904) 24.
148 C. W. Muth, .J C. Ellers, and O. F. Folmer, *J. Am. Chem. Soc.*, 79 (1957) 6500.
149 C. W. Muth, N. Abraham, M. L. Linfield, R. B. Wotring, and E. A. Pacofsky, *J. Org. Chem.*, 25 (1960) 736.
150 G. Tennant, *J. Chem. Soc.*, (1963) 2429.
151 R. Fusco and S. Rossi, *Chim. Ind. (Milan)*, 45 (1963) 834 [*Chem. Abstr.*, 60 (1964) 10683].
152 A. Reissert, *Ber.*, 29 (1896) 639.

153 J. D. LOUDON AND I. WELLINGS, *J. Chem. Soc.*, (1960) 3462.
154 J. D. LOUDON AND G. TENNANT, *J. Chem. Soc.*, (1960) 3467.
155 F. ARNDT, *Ber.*, 46 (1913) 3522.
156 J. A. CARBON, *J. Org. Chem.*, 27 (1962) 185.
157 F. J. WOLF, R. M. WILSON, JR., K. PFISTER, AND M. TISHLEV, *J. Am. Chem. Soc.*, 76 (1954) 4611.
158 J. W. BARTON AND J. F. THOMAS, *J. Chem. Soc.*, (1964) 1265.
159 T. ZINCKE AND P. SCHWARZ, *Ann.*, 311 (1909) 329.
160 J. D. LOUDON AND I. WELLINGS, *J. Chem. Soc.*, (1960) 3470.
161 F. ARNDT AND W. PARTALE, *Ber.*, 60 (1927) 446.
162 J. A. MOORE AND D. H. AHLSTROM, *J. Org. Chem.*, 26 (1962) 5255.
163 D. W. OCKENDEN AND K. SCHOFIELD, *J. Chem. Soc.*, (1953) 1915.
164 D. BRITTON AND W. E. NOLAND, *J. Org. Chem.*, 27 (1962) 3218.
165 A. WOHL AND A. AUE, *Ber.*, 34 (1901) 2442.
166 E. I. ARAMOVA AND I. YA. POSTOVSKII, *Zh. Obshch. Khim.*, 22 (1952) 502.
167 E. E. SOULE (to Mathieson Alkali Works, Inc.), U. S. Pat. 2,332,179 (October, 1944) [*Chem. Abstr.*, 38 (1944) 1534].
168 Z. V. PUSHKAREVA AND G. I. AGIBOLOVA, *Zh. Obshch. Khim.*, 7 (1938) 151.
169 Y. KIDANI AND H. OTOMASU, *Pharm. Bull. (Japan)*, 4 (1956) 391.
170 A. WOHL, *Ber.*, 36 (1903) 4135.
171 I. YOSIOKA AND H. OTOMASU, *Yakugaku Zasshi*, 76 (1956) 1051.
172 H. C. WATERMANN AND D. L. VIVIAN, *J. Org. Chem.*, 14 (1949) 289; D. L. VIVIAN, *J. Am. Chem. Soc.*, 73 (1951) 457; D. L. VIVIAN, J. L. HARTWELL, AND H. C. WATERMANN, *J. Org. Chem.*, 19 (1954) 1136.
173 P. V. CHERNETZKII AND A. I. KIPRIANOV, *Zh. Obshch. Khim.*, 23 (1953) 1743 [*Chem. Abstr.*, 48 (1954) 13695].
174 I. YOSIOKA AND H. OTOMASU, *Pharm. Bull. (Japan)*, 2 (1954) 56.
175 I. YOSIOKA AND H. OTOMASU, *Pharm. Bull. (Japan)*, 1 (1953) 67.
176 A. KLIEGEL AND A. BRÖSAMLE, *Ber.*, 69 (1936) 197; A. KLIEGEL AND A. FEHRLE, *Ber.*, 47 (1914) 1629.
177 M. COLONNA, *Boll. Sci. Fac. Chim. Ind. Bologna*, (1940) 134.
178 K. GERLACH AND F. KRÖHNKE, *Ber.*, 95 (1962) 1124.
179 I. J. PACHTER, P. E. NEMETH, AND A. J. VILLANI, *J. Org. Chem.*, 28 (1963) 1197.
180 N. J. LEONARD AND W. V. RUYLE, *J. Org. Chem.*, 13 (1948) 903.
181 N. A. COATS AND A. R. KATRITZKY, *J. Org. Chem.*, 24 (1959) 1836.
182 E. HAYASHI AND CH. IIJIMA, *Yakugaku Zasshi*, 82 (1962) 1093.
183 E. HAYASHI, CH. IIJIMA, AND Y. NAGASAWA, *Yakugaku Zasshi*, 84 (1964) 165.

Chapter 4

Physico-chemical Properties of Aromatic Amine N-Oxides

4.1 General

As has already been stated in Chapter 2 (*cf. Section 2.2.1*), there are two kinds of tertiary amine *N*-oxide compounds, aliphatic and aromatic. The four atomic orbitals of the nitrogen atom forming the *N*-oxide group in aliphatic compounds are close to sp^3, and have an approximately regular tetrahedral structure. This had already been pointed out by Meisenheimer[1] from the fact that when the three groups bonded to the nitrogen are different, the compound can be resolved into optically active isomers (*cf. Section 2.1.1*), and the fact has been confirmed from X-ray diffraction data (*cf. Section 4.3*). In the case of aromatic *N*-oxide compounds, however, the lone-pair electrons of the nitrogen before formation of a nitrogen–oxygen bond are on the sp^2 orbital, so that the nitrogen–oxygen bond is in the same plane with the hetero ring, and the $2p\pi$-atomic orbital of its oxygen atom is in parallel with those of the atoms constituting the hetero ring. As its result, a large interaction between the nitrogen–oxygen bond and the remaining π-electron system can be postulated, and this constitutes the main reason for the characteristic behavior of aromatic *N*-oxide compounds, which differs from that of aliphatic *N*-oxides in chemical reactions. For example, the following variety of resonance structures (4-1 to 4-3) can be considered for pyridine 1-oxide.

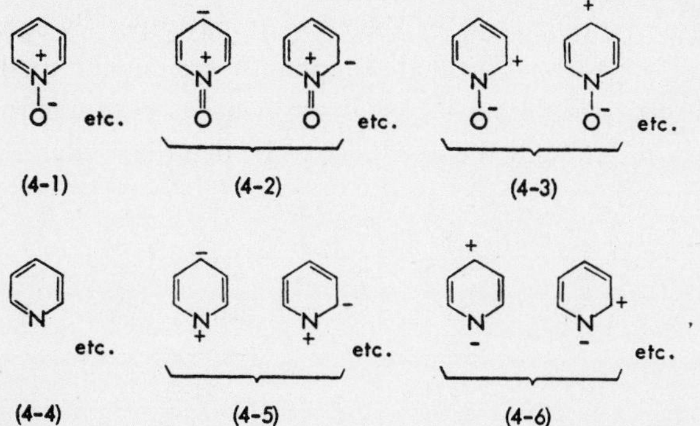

References p. 167

Consequently, these aromatic *N*-oxide compounds in general are expected to show a double-bond character in their nitrogen–oxygen bond, which means that they not only differ from the afore-mentioned aliphatic *N*-oxide compounds in which the nitrogen–oxygen bond is single, but also show various chemical properties different from those of their parent heterocyclic nitrogen compounds.

For the sake of comparison with resonance structures (4-1 to 4-3) of pyridine 1-oxide, those (4-4 to 4-6) of pyridine are also shown above. As will be expected from the lability of (4-5) corresponding to (4-2), the ring system in aromatic *N*-oxide compounds in general has a greater π-electron density than that of the corresponding parent compounds. In this point, pyridine 1-oxide is isoelectronic with a phenoxide anion and this fact alone suggests the increased activity of *N*-oxide compounds to electrophilic reagents. This does not mean, however, that *N*-oxide compounds are necessarily less reactive to nucleophilic reagents than their parent *N*-heterocycles because the negative charge in (4-3) is in the oxygen, which is more electronegative than nitrogen, and the negative charge in (4-6) is in the nitrogen.

The foregoing has been stated on the assumption that pyridine 1-oxide reacts as a neutral molecule but, actually, as will be discussed later, *N*-oxide compounds are bases, though weak, as are their parent hetero-aromatic compounds, and in most cases undergo reactions as $\geqslant N^+$–O-substituted pyridinium ion (including the hydroxypyridinium ion). In such a case, electronegativity of the oxygen atom increases, even if the configuration of the atomic orbital in the oxygen does not change, and the drift of electrons from the oxygen to the ring is inhibited, resulting in the nitrogen–oxygen bond becoming more like a single bond than that in a non-polar solvent. In other words, the contribution of the structure (4-8) in the resonance system of the hydroxypyridinium ion shown below becomes much smaller than that of (4-2) in the free pyridine 1-oxide. Therefore, it may generally be said that the properties of $\geqslant N^+$–O-substituted compounds are similar to those of the $\geqslant N^+$-substituted pyridinium ion, and the fact has been confirmed by the various physical data and the reactivity to be described later.

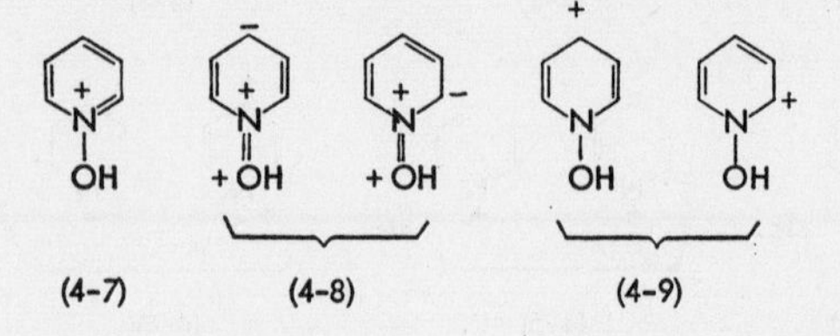

When the six-membered ring monoazines like pyridine are changed to diazines and triazines, the π-electron deficiency of their ring system increases, and the contribution of the structure (4-2) in the resonance system of their *N*-oxides would be expected to become larger than that of (4-3). The same phenomenon may be expected when these *N*-oxide compounds are substituted with electron-withdrawing groups (designated as W in some cases). On the other hand, *N*-oxide compounds having electron-donating groups (designated as D) would give larger contribution of (4-3) structure than the non-substituted *N*-oxides. In the *N*-oxides of five-membered aromatic heterocyclic nitrogen compounds, the contribution of (4-3) type would be larger than that of (4-2) from the excess of π-electron density of their ring, compared to the six-membered π-deficient nitrogen-heterocycles, such as pyridine 1-oxide.

On the other hand, change of ring properties in *N*-oxides of azanaphthalene (or polyazanaphthalene) or their higher benzenoid homologs, formed by the condensation of an extra benzene ring to the azabenzene (or polyazabenzene) *N*-oxides, results in the localization of double-bond characters of their parent ring, as would be expected. Consequently, while the contribution of (4-1a) and (4-1b), corresponding to (4-1) of pyridine 1-oxide, is equivalent, that of (4-10a) is larger than that of (4-10b) in quinoline 1-oxide, and, as a result, the benzene portion of azanaphthalene *N*-oxide retains the naphthoid activity (activity of the α-position) as in naphthalene.

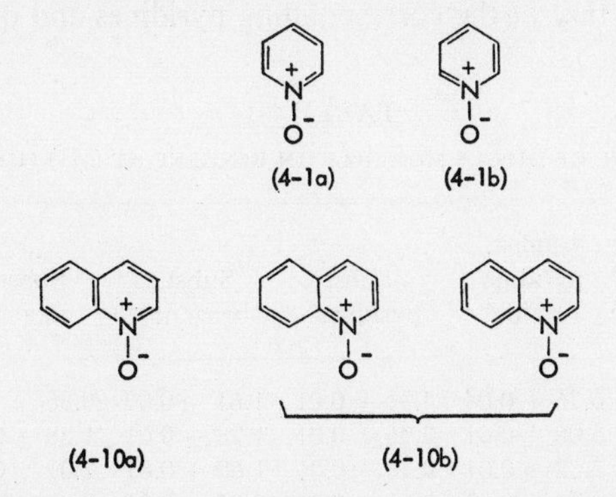

The present chapter has been based on the facts revealed by physicochemical examination of the characteristics of aromatic amine *N*-oxides, and is divided into sections on dipole moment, X-ray diffraction, polarography, dissociation constant, nuclear magnetic resonance spectra, infrared spectra, and ultraviolet spectra. In addition, in order to summarize these physical data and to correlate them to the reactivity of these compounds, some considerations on the molecular orbital methods carried out on these aromatic amine *N*-oxide compounds will be discussed.

4.2 Dipole Moment

Linton[2] pointed out that the difference (2.02D) in dipole moments of pyridine 1-oxide (4.24D) and of pyridine (2.22D) is far smaller than that (4.37D) between trimethylamine *N*-oxide (5.02D) and trimethylamine (0.65D), and concluded that the contribution of a resonance structure like (4-2) is important in pyridine 1-oxide (*cf. Section 1*). Ochiai was the first to point out that this conclusion by Linton means that the susceptibility of the ring in pyridine 1-oxide to substitution by electrophilic reagents should be increased[3], and actually showed that nitration of pyridine 1-oxide chiefly occurs at 4-position, comparatively easily in contrast to the inactivity of pyridine to this reaction. From further experiments on the chemical changes of 4-nitropyridine 1-oxide[4], Ochiai and others suggested that there must be further contribution of an additional resonance structure like (4-3) in the resonance system of pyridine 1-oxide (*cf. Section 2.2.1*). This assumption was endorsed by kinetic experiments[5], which showed that substitution of 2- or 4-halogenated pyridine and quinoline 1-oxides by nucleophilic reagents is far easier than that of the corresponding pyridines and quinolines. These

TABLE 4-1

MAGNITUDE OF DIPOLE MOMENTS (IN BENZENE AT 25°) (UNIT IN D)

Substituent	4-Subst. pyridine 1-oxide[6]	4-Subst. pyridine[6]	Subst. benzene[6]	Subst. alkane[6]	4-Subst. pyridine-BCl_3[7]
$N(CH_3)_2$	6.76 ± 0.04	4.31 ± 0.01	1.61 ± 0.02	0.86	
OCH_3	5.08 ± 0.01	2.96 ± 0.01	1.28 ± 0.01	1.28 ± 0.01	8.86 ± 0.02
Cl	2.82 ± 0.01	0.78 ± 0.01	1.60 ± 0.01	2.01 ± 0.01	6.71 ± 0.03
CH_3	4.74 ± 0.01	2.61 ± 0.01	0.35 ± 0.05	0	8.37 ± 0.03
H	4.24 ± 0.02	2.22 ± 0.02	0	0	7.70 ± 0.02
$COOC_2H_5$	3.80 ± 0.01	2.53 ± 0.04	1.85	1.8	
$COCH_3$	3.19 ± 0.02	2.41 ± 0.01	2.96 ± 0.02	2.75 ± 0.05	7.74 ± 0.07
NO_2	*ca.* 0.7	1.58 ± 0.01	4.01 ± 0.02	3.25 ± 0.05	
CN		1.65 ± 0.03	4.05 ± 0.01	3.60 ± 0.05	4.20 ± 0.05

results were also supported by the measurement of dipole moments of 4-substituted pyridines and their *N*-oxides by Katritzky and others[6,7], whose values are given in Table 4-1.

These workers analyzed their values by applying the treatment of mesomeric moment, defined by Sutton and others[8] as giving the degree of interaction between the π-electron system and its substituent (Z) in aromatic compounds (Ar–Z), to heteroaromatic compounds and calculated the mesomeric moments ($\vec{\mu}_m$) of 4-substituted pyridines, their boron trichloride adducts, and their *N*-oxides. The mesomeric moment of each of these compounds is defined by the following equations (*1* to *4*).

$$\vec{\mu}_m \text{ (subst. benzene)} = \vec{\mu}(\text{Ph–Z}) - [\vec{\mu}(\text{benzene}) + \vec{\mu}(\text{alk–Z})] \quad (1)$$

$$\vec{\mu}_m \text{ (4-subst. pyridine)} = \vec{\mu}\,(\text{Z–C}_5\text{H}_4\text{N}) - [\vec{\mu}(\text{pyridine}) + \vec{\mu}(\text{alk–Z})] \quad (2)$$

$$\vec{\mu}_m \text{ (4-subst. pyridine N–O)} = \vec{\mu}(\text{Z–C}_5\text{H}_4\text{N–O}) - [\vec{\mu}(\text{Py–N–O}) + \vec{\mu}(\text{alk–Z})] \quad (3)$$

$$\vec{\mu}_m \text{ (4-subst. pyridine–BCl}_3) = \vec{\mu}(\text{Z–C}_5\text{H}_4\text{N}\cdot\text{BCl}_3) - [\vec{\mu}(\text{Py–BCl}_3) + \vec{\mu}(\text{alk-Z})] \quad (4)$$

The mesomeric moments of pyridines, pyridine 1-oxides, and pyridine-boron trichloride adducts so obtained are compared with those of benzene derivatives in Table 4-2.

The data given in Table 4-2 are summarized in Fig. 4-1, which shows that,

TABLE 4-2

MAGNITUDE OF MESOMERIC MOMENT (IN BENZENE AT 25°) (UNIT IN D)

Z	Subst. benzene[8]	4-Subst. pyridine[6]	4-Subst. pyridine 1-oxide[6]	4-Subst. pyridine–BCl_3[7]
$N(CH_3)_2$	1.66	2.27	2.74	
OCH_3	0.96	1.16	1.39	1.35
Cl	0.41	0.57	0.59	1.02
CH_3	0.35	0.39	0.50	0.67
H	0.00	0.00	0.00	0.00
$COOC_2H_5$	0.50	0.47	0.93	0.17
$COCH_3$	0.56	0.21	0.62	
NO_2	0.76	0.55	0.99	
CN	0.45	0.27		0.10

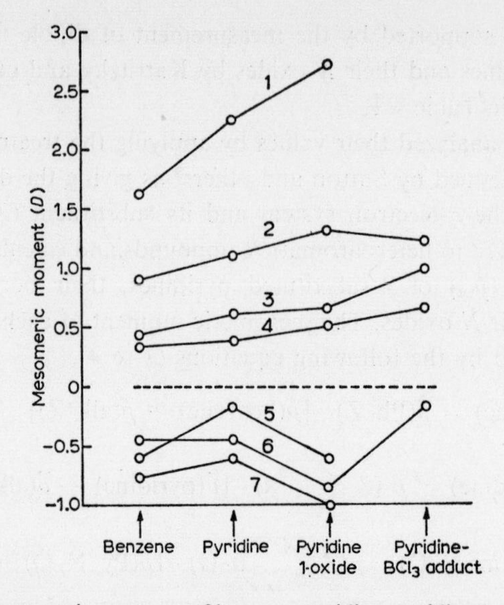

Fig. 4-1. Plot of mesomeric moments of benzene, pyridine, pyridine 1-oxide, and pyridine–boron trichloride adduct systems with various substituents[6,7]. Substituents: 1, NMe_2; 2, OMe; 3, Cl; 4, Me; 5, COMe; 6, COOEt; 7, NO_2. (Reprinted from the *Journal of the Chemical Society*, by permission of the copywright owners, The Chemical Society.)

in contrast to substituted benzenes and pyridines, the mesomeric moment shifts more to the positive value when Z is an electron-donating substituent (D) and more to the negative value when Z is an electron-withdrawing substituent (W) in 4-substituted pyridine 1-oxides (4-Z-pyridine 1-oxide).

From these results Katritzky[9] concluded that both structures (4-2) and (4-3) take part in the resonance system of the pyridine 1-oxide series, that the degree of their contribution is dependent on the nature of the substituent (Z) in the 4-position, that the resonance contribution of (4-11) corresponding to (4-2) is larger when Z is an electron-withdrawing group and that of (4-12) corresponding to (4-3) is larger when Z is an electron-donating group.

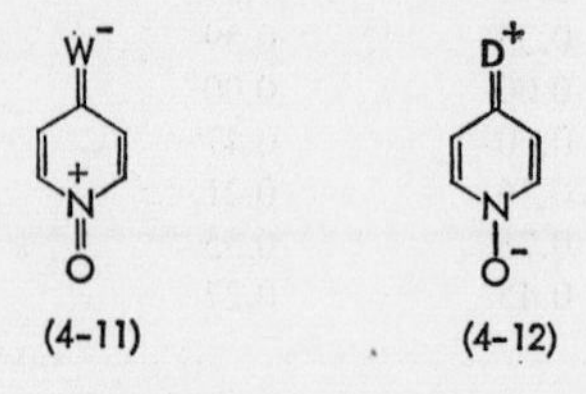

As for the dipole moment of pyridine 1-oxide itself, Bax and Katritzky[9] made a qualitative examination of the relative contribution of (4-2) and (4-3). They compared the dipole moment of pyridine 1-oxide with that of a pyridine–BX_3 type complex, which is thought to be in the same hybrid state relative to nitrogen and in which the contribution of the canonical form comparable to (4-2) can be neglected. For the same reason, they compared the moments of trimethylamine *N*-oxide and the trimethylamine–boron trihalide complex, and calculated the mesomeric moment, $\vec{\mu}_m$, involved in the drift of electrons from the oxygen to the ring (due to the contribution of a structure such as (4-2)), from the following formula.

$$\begin{aligned}\vec{\mu}_m &= \{\vec{\mu}(\text{Py–N–O}) - \vec{\mu}[(CH_3)_3\text{N–O}]\} - \{\vec{\mu}(\text{Py–}BX_3 \text{ complex}) - \\ &\qquad \vec{\mu}[(CH_3)_3\text{N–}BX_3 \text{ complex}]\} \\ &= [(4.24) - (5.02)] - (1.31)^* \\ &= (-0.78) - (1.31) \\ &= -2.09 \text{ (direction of moment, O}\rightarrow\text{ring)}\end{aligned}$$

The calculated moment is $\vec{\mu}_m = 2.09$D. It is thus concluded that in non-substituted pyridine 1-oxide, the inward drift of electrons originating from the oxygen atom to the ring is slightly greater. This value of 2.09D is of approximately the same order as the value of 2.34D** obtained as the difference between the moment (6.58D) predicted by Linton[2] for pyridine 1-oxide and the observed moment (4.24D).

Sharpe and Walker[10] experimentally obtained the $\mu_{C-Z} - \mu_{C-H}$ moments for the four ring systems indicated in Table 4-3 (mono-substituents in benzene, and 3- and 4-substituents in pyridine, nitrobenzene, and pyridine 1-oxide), and showed that these values indicate the degree of relative interaction between the substituent (Z) and the ring system.

In calculating the values shown in Table 4-3, these workers measured the dipole moments of various *N*-oxide compounds and obtained values different from those of Katritzky and his associates. Sharpe explained this by the probable lack of sufficient care to protect the compounds from moisture during the measurement by Katritzky and others. The values found by

* This value is almost unchanged with the kind of halogen atom (X) in BX_3 and is in the range of 1.24 to 1.39D (*cf.* Ref. 9).

** In this case, the predicted dipole moment of pyridine 1-oxide was evaluated from the difference between the dipole moment of pyridine and of those of trimethylamine and trimethylamine *N*-oxide. Therefore, Linton's data did not take into consideration the difference (if any) in the hybridization of atomic orbitals in reference to nitrogen between pyridine and trialkylamine series.

References p. 167

TABLE 4-3

$\mu_{C-Z}-\mu_{C-H}$ MOMENTS (IN D)[10]

	Benzene	Pyridine[a]	Nitrobenzene	Pyridine 1-oxide[a]
Substituent in 3-position				
CH_3	+0.37	+0.35	+0.37	+0.27
Cl	−1.58	−1.73	−(1.71, 1.53)	−1.43
Br	−1.56	−1.64	−(1.71, 1.53)	−1.43
I	−1.40	−1.35	−(1.16, 1.02)	−1.00
Substituent in 4-position				
CH_3	+0.37	+0.39	+0.41	+0.31
tert-C_4H_7	+0.45	+0.52	+0.60	+0.44
Cl	−1.58	−1.43	−1.39	−1.36
Br	−1.56	−1.44	−1.35	−1.29
I	−1.40	—	−1.11	—
Phenyl	0	+0.34	+0.35	+0.42

[a] Calculated by assuming μ (pyridine 1-oxide) = 4.19D; μ (pyridine) = 2.21D

TABLE 4-4

DIPOLE MOMENTS OF 3- AND 4-SUBSTITUTED PYRIDINE 1-OXIDES[10]

Substituent	4-Subst. compound	3-Subst. compound	Substituent	4-Subst. compound	3-Subst. compound
CH_3	4.50	4.33	Cl	2.83	3.68
C_2H_5	4.54		Br	2.90	3.69
C_3H_7	4.56		I	—	3.79
iso-C_3H_7	4.59		Phenyl	4.61	—
tert-C_4H_9	4.63				

Sharpe and Walker are given in Table 4-4, since these values seem to be more correct.

The fact that the value of $\mu_{C-Z}-\mu_{C-H}$ reflects the mutual mesomeric and inductive interactions between Z and the ring was confirmed by plotting these values for 3- and 4-substituted pyridine 1-oxide against the σ-values (substituent constants taken from the work of Jaffé[11]); this gave a good linear relationship. (Similar linear relationships are obtained for substituted pyridines and nitrobenzenes.)

It is interesting to compare the values of $\mu_{C-Z}-\mu_{C-H}$ for 4-halogen-substituted compounds shown in Table 4-3 between different ring systems. In this case, the smaller the absolute value of the $\mu_{C-Z}-\mu_{C-H}$ moment, the greater should be the interaction between the substituent halogen and the ring system*. It may, therefore, be concluded that, from the values of $\mu_{CZ}-\mu_{CH}$ in 4-halogenated derivatives, the electron-accepting power of the ring decreases in the order of pyridine 1-oxide, nitrobenzene, pyridine, and benzene**. This conclusion of Sharpe and Walker is in good agreement with that of Katritzky and his associates.

These discussions were based on the vector analysis of dipole moments obtained experimentally. By using the molecular orbital method (*cf. Section 4.9*), the contribution of the resonance systems (4-2) and (4-3) can be discussed theoretically and also compared with the known results. Kubota and Watanabe[12] made a detailed examination along these lines for pyridine 1-oxide and *N*-monoxides of pyridazine, pyrimidine, and pyrazine, and the values they calculated are in good agreement with those observed. They concluded that the π-moment of (4-2) type is higher than that of (4-3) type. The same calculations were also made by Kubota[13] for other basic heterocyclic *N*-oxide compounds.

As shown above, the dipole moment is a valuable means not only for judging the nature of pyridine 1-oxides but also for structural elucidation of compounds like diazine-series *N*-oxides, which can form two kinds of *N*-monoxide.

Ogata and others[14] carried out *N*-oxidation of 4-methylpyrimidine (4-13) and isolated two kinds of *N*-monoxide in 3.5:1 ratio. They measured the

* When the substituent Z is a halogen (X), comparison of the ring systems between A and B gives

$$(\vec{\mu}_{CX}-\vec{\mu}_{CH})_A-(\vec{\mu}_{CX}-\vec{\mu}_{CH})_B=[\vec{\mu}_{CX}]_A-[\vec{\mu}_{CX}]_B$$

and $\vec{\mu}_{CX}$ can be divided into σ and π portions. The σ portion has a moment (the bond moment of C–X) in the direction of C→X and this moment is fairly large. On the other hand, the π portion would be $^{+}X=\langle\text{ring}\rangle N{-}O^{-}$, and the moment would be in the direction of C←X, the reverse of that in the σ portion. When the substituent is a halogen, the magnitude of the moment would be greater in the σ portion than in the π portion. Consequently, the system with greater π portion, in other words, with greater electron-accepting power of the ring, $[\vec{\mu}_{C-X}]$ would be smaller.

** The above order is not obtained with methylated derivatives. This is considered to be due to the fact that the hyperconjugation effect of the methyl is so small that it is to about the same degree as the small difference in C–H bond moment of various unsubstituted compounds[10].

dipole moments of these *N*-monoxides and their analogs (4-14 and 4-15), compared their values with the moments calculated on the basis of a reasonable assumption, and concluded that the main product (4-16) is a 3-oxide and the minor product (4-17), a 1-oxide. These results are summarized in Table 4-5.

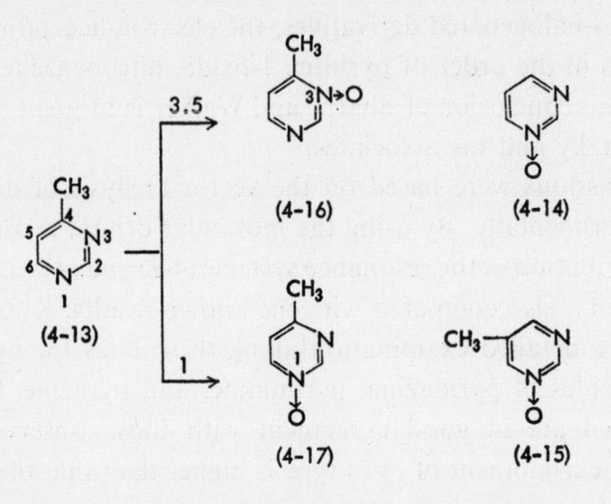

In a similar manner, Watanabe and others[15] carried out structural determination of the *N*-monoxides of methylpyridazines (4-18) by the comparison of observed and calculated values of their dipole moments (*cf.* Table 4-6).

Otomasu and others[16,17] compared the observed and calculated values of the dipole moment of *N*-oxides of diazines possessing an alkoxyl group in

TABLE 4-5

DIPOLE MOMENT OF PYRIMIDINE *N*-OXIDES (IN BENZENE AT 25°)[14]

Compound No.	Observed	Calculated[a]
4-16	3.72	3.70
4-17	4.05	3.99
4-14	3.65	3.67
4-15	4.03	4.00

[a] Each of the calculated moments was obtained as a vector sum of group moments. On the assumption of a regular hexagon form for the ring, the moments of (4-14) can be calculated as 3.67D from the observed moment of 2.22D for pyridine and that of 4.24D for its *N*-oxide in benzene solution. Taking the moment of C–CH_3 as that of 0.37D for toluene, the moments of (4-15), (4-16), and (4-17) can be calculated.

TABLE 4-6

DIPOLE MOMENT OF METHYLPYRIDAZINES AND THEIR *N*-OXIDES[15] ($\mu \times$ D)

Compd.	Subst. pyridazine			Subst. pyridazine *N*-oxide		
No.	Substituent	Obsvd.	Calcd.[a]	Substituent	Obsvd.	Calcd.[a]
	3-CH_3, 6-Cl	4.51		H	5.21	
	3-CH_3	4.13		2-Oxide		
4-18	4-CH_3	4.48	(4.48)	3-CH_3	5.23	
4-19	3-Cl, 4-CH_3	4.72	(4.72)	3-CH_3, 6-Cl	5.29	
4-20	4-CH_3, 6-Cl	4.79	(4.79)	4-CH_3, 6-Cl	5.67	5.55 (1-oxide[b] 6.04)
				4-CH_3	5.53	
				3-CH_3, 5-NO_2	2.71	
				3-CH_3, 5-NO_2, 6-Cl	3.28	
				1-Oxide		
				3-CH_3	5.23	
				3-Cl, 4-CH_3	5.67	5.58 (2-oxide[b] 6.08)
				4-CH_3	5.70	
				4-NO_2	2.28	

(4-18)

[a] First, by a simple vector analysis of observed moments of (4-18), (4-19), and (4-20), μ_{C-Cl} and μ_{C-CH_3} of pyridazine ring and μ of its parent pyridazine were calculated respectively as 1.59, 0.33, and 4.15D, and the values in this table were calculated from these figures.

[b] Figures in parentheses are values calculated by assuming that the *N*-oxide was present in the reverse position.

the position *ortho* to the ring nitrogen, and revealed that oxidation of these compounds invariably produces *N*-oxides at the nitrogen far from the alkoxyl group and that diazine *N*-oxide and its parent base possessed a moment, which indicates that the alkoxyl group is bent towards the tertiary nitrogen adjacent to the alkoxyl group. In other words, these facts support the view that such diazines are present in the *cis* form (4-23) in the non-polar solvent formed by the two conformers. The example of 3-alkoxypyridazine (4-21) is shown below.

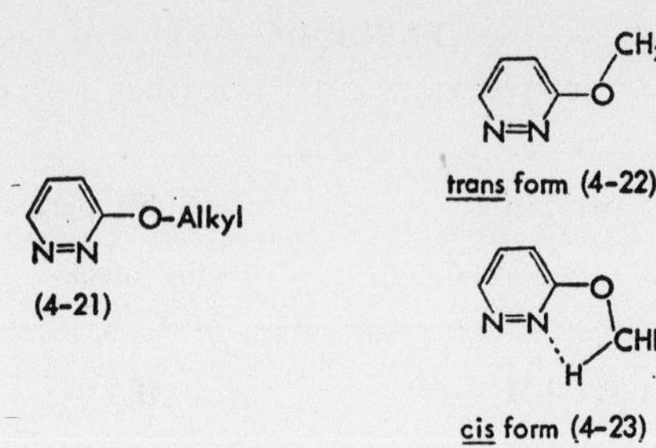

(4-21)

trans form (4-22)

cis form (4-23)

These results of Otomasu and others are very important contributions in that they provide theoretical evidence for the chemical observation that the nitrogen far from an alkoxyl group is always oxygenated in these kinds of compound. The observed and calculated values obtained by Otomasu and others for quinoxaline[16], pyrimidine, and pyridazines[17] are listed in Tables 4-7 and 4-8. Readers are referred to the literature[17] for the methods of calculation. These values also indicate that these compounds are all present in the *cis* form (4-24, taking 3-methoxyquinoxaline 1-oxide as an example) as in the case of the foregoing pyridazine.

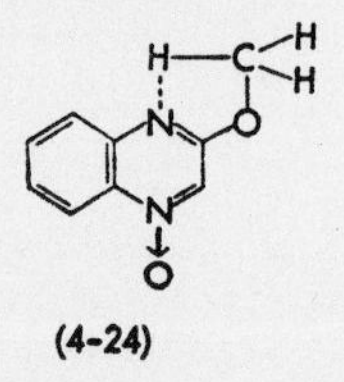

(4-24)

Similar studies have been made by others. Ogata[18] measured the dipole moments of two kinds of *N*-monoxide obtained by *N*-oxidation of 3-phenyl-6-chloropyridazine (4-25) and concluded that the compound with greater moment (5.54D) is the 1-oxide and the one with smaller moment (5.19D) the 2-oxide.

Sasaki and others[19] assumed the structure of the *N*-oxide compound, obtained by the peracid oxidation of 3-amino-5,6-dimethyl-1,2,4-triazine, to be its 2-oxide (4-26) from its dipole moment (4.13D), but their data are insufficient and it seems dangerous to make a conclusion from this fact alone.

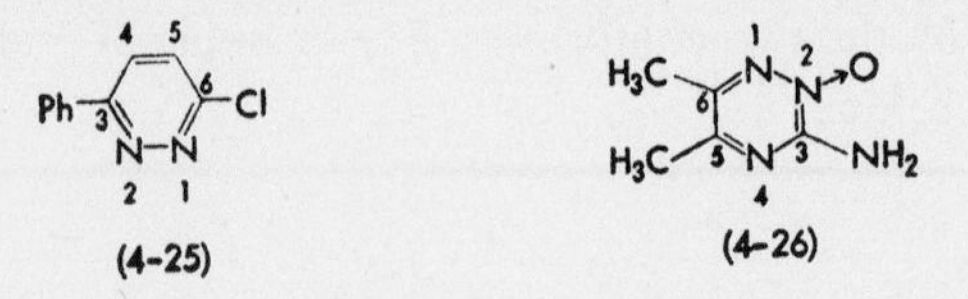

(4-25)

(4-26)

TABLE 4-7

DIPOLE MOMENT OF QUINOXALINES[16]

Compound	Conformation of alkyl	Position of *N*-oxide	Dipole moment (μ in D) Calcd.[a]	Found
quinoxaline (N-1, 2, 3, N-4)			(0.82)	0.82
N→O			1.66	1.87
$-CH_3$			(0.52)	0.52
N→O, $-CH_3$		1-oxide	**1.54**	1.55
		4-oxide	1.93	
$-OCH_3$	*cis*		**1.66**	1.47
	trans		0.87	
$-OCH_3$, N→O	*cis*		1.16	3.14
	trans	1-oxide	2.62	
	cis		**3.12**	
	trans	4-oxide	0.82	
$-OEt$	*cis*		**1.66**	1.54
	trans		0.87	
$-OEt$, N→O	*cis*		1.16	3.31
	trans	1-oxide	2.62	
	cis		**3.12**	
	trans	4-oxide	0.82	
$-OEt$, $-CH_3$	*cis*		**1.31**	1.70
	trans		1.10	
$-OEt$, $-CH_3$, N→O	*cis*		0.97	3.24
	trans	1-oxide	2.70	
	cis		**2.85**	
	trans	4-oxide	0.77	

[a] The values in bold-face type are close to observed values.

TABLE 4-8

DIPOLE MOMENT OF PYRIMIDINES AND PYRIDAZINES[17]

Compound	Conformation of alkoxyl	Position of *N*-oxide	Dipole moment (μ in D) Calcd.[a]	Found	Difference
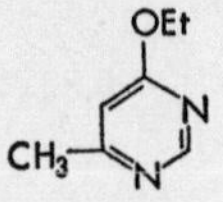	*cis*		**1.67**	2.20	0.53
	trans		3.60		−1.40
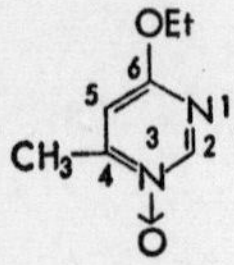	*cis*	1-oxide	2.79	3.95	1.16
		3-oxide	**3.49**		0.46
	trans	1-oxide	5.12		−1.17
		3-oxide	4.71		−0.76
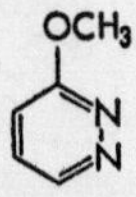	*cis*		**2.66**	2.87	0.21
	trans		5.08		−2.21
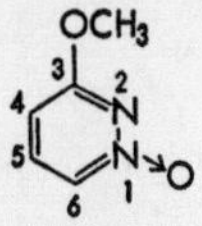	*cis*	1-oxide	**4.60**	4.80	0.20
		2-oxide	4.42		0.38
	trans	1-oxide	6.96		−2.16
		2-oxide	6.84		−2.06
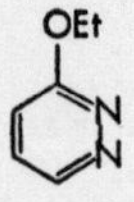	*cis*		**2.66**	3.17	0.51
	trans		5.08		−1.91
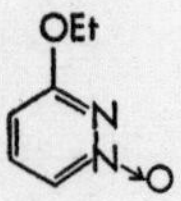	*cis*	1-oxide	**4.60**	4.76	0.16
		2-oxide	4.42		0.33
	trans	1-oxide	6.96		−2.20
		2-oxide	6.84		−2.08
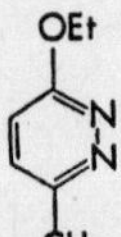	*cis*		**2.63**	2.86	0.23
	trans		5.06		−2.20
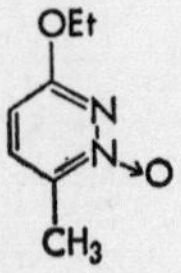	*cis*	1-oxide	**4.50**	5.11	0.61
		2-oxide	4.39		0.71
	trans	1-oxide	6.89		−1.78
		2-oxide	6.72		−1.61

[a] The values in bold-face type are close to observed values.

Dipole moments have been reported by British chemists for 4-nitrobenzo-[*c*]cinnoline and its *N*-oxide[20], acridine, 9-phenylacridine, and their *N*-oxides[21]. Dipole moments have also been reported for *N*-oxides of numerous acridines, quinoxalines, phenazines, and their substituted derivatives by Soviet chemists[22,23].

4.3 X-Ray Diffraction

Numerous reports[24–27] have already appeared on X-ray crystallographic analyses or electron diffraction studies of trimethylamine *N*-oxide. Rérat[25] carried out a crystallographic analysis of its hydrochloride and reported that the distance between nitrogen and oxygen is 1.424 Å*. He confirmed that the conformation of the four atoms attached to the nitrogen is close to a regular tetrahedron. Caron and others[26] carried out a detailed X-ray analysis of the free base of trimethylamine *N*-oxide and also confirmed the near-tetrahedral structure. However, they reported the $N^+–O^-$ bond length as 1.388 Å, a little shorter than that of its hydrochloride. According to these authors, this is due to the difference in the formal charge between $N^+–OH$ and $N^+–O^-$.

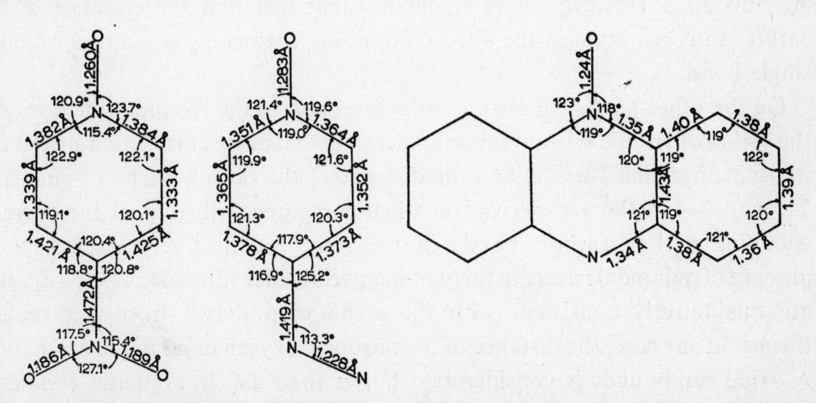

Fig. 4-2. Intramolecular bond distances and bond angles of 4-nitropyridine 1-oxide, 4,4′-*trans*-azopyridine 1,1′-dioxide, and phenazine *N*-oxide.

Several reports on similar studies with aromatic *N*-oxide compounds have been published recently. Detailed crystallographic analyses have been made on 4-nitropyridine 1-oxide[28a], 4,4′-*trans*-azopyridine 1,1′-dioxide[28b], and phenazine *N*-oxide[29]. These results are illustrated in Fig. 4-2.

* The nitrogen–oxygen bond length for the free base has been reported as 1.36 (Ref. 24) and 1.44 Å (Ref. 27).

The characteristic point in the bond data of 4-nitropyridine 1-oxide is that the distance between nitrogen and oxygen is different in the *N*-oxide and nitro groups, and this agrees, for example, with the order of wave numbers of the absorption due to the N^+–O^- stretching vibrations of the nitro and *N*-oxide groups in the infrared spectrum of this compound (*cf. Section 4.7.1*). The same may be said of the two sets of carbon–carbon bonds. In 4-nitropyridine 1-oxide, the length of the carbon–carbon bonds parallel to the molecular axis is much shorter than that of other carbon–carbon bonds and this is a proof of the quinoidal nature (as shown by 4-27) of this molecule.

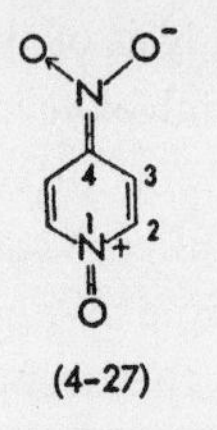

(4-27)

Eichhorn[28] assumed that the double-bond character of the carbon–carbon bonds at 2–3 and 3–4, and the nitrogen–carbon bond at 1–2, is respectively 90, 30, and 20%. However, it is an inconsistent fact that the distance of a carbon–nitrogen bond at the nitro group is approximately equal to a purely single bond.

On the other hand, the nitrogen–oxygen bond in the *N*-oxide function of the 4,4′-azopyridine 1,1′-dioxide is longer than that of the corresponding bond in 4-nitropyridine 1-oxide, and the distance of the carbon–carbon bonds at 2–3 and 3–4 in the azo derivative, when compared with that of the corresponding carbon–carbon bonds in the 4-nitro derivative, indicates the absence of quinoidal nature in this 4,4′-azopyridine 1,1′-dioxide. These results are qualitatively consistent with the assumption drawn from electronic theory. In any case, the distance of the nitrogen–oxygen bond in the aromatic *N*-oxide compounds is considerably shorter than that in aliphatic *N*-oxide compounds and this is thought to indicate the contribution of the (4-2) type resonance structure in the aromatic *N*-oxide compounds.

As stated in *Section 4.1*, the conversion of aromatic *N*-oxides to the 1-hydroxypyridinium ion results in a marked decrease of the double-bond character of the nitrogen–oxygen bond and its bond length is expected to become greater. Crystallographic analysis of the hydrochloride of pyridine 1-oxide has endorsed this assumption and Nambu, Oda, and Watanabe[30] have given 1.3 Å as the distance of the nitrogen–oxygen bond of the *N*-oxide function, while Tsoucaris[31], from further detailed X-ray analysis, gave the value of 1.37 Å. Since a nitrogen–oxygen bond of 1.37 Å in the hydrochloride of pyridine 1-oxide is considerably shorter than that of 1.424 Å for the nitrogen–oxygen bond in trimethylamine *N*-oxide hydrochloride, there is still some contribution of (4-8) type resonance structure*. On the other hand, as will be understood from the foregoing data for trimethylamine *N*-oxide, there is no great change of this magnitude between the free base and its hydrochloride in aliphatic *N*-oxide compounds.

4.4 Polarography

Ochiai and others[4,32] had indicated already in 1943 that the nitrogen-oxygen bond in pyridine 1-oxides is reduced at a far more negative potential than that in simple aliphatic *N*-oxides (*e.g.* matrine *N*-oxide, dimethylaniline *N*-oxide, etc.) in their polarographic reduction. The difference in the nitrogen–oxygen bond in these two groups of compounds indicates the larger stability of such a bond in aromatic *N*-oxide compounds and these workers concluded that, taking pyridine 1-oxide as an example, this more negative reduction potential may be attributed to the contribution of a canonical form like (4-2). Similar studies have since been carried out in various countries of the world on aromatic *N*-oxide compounds. Table 4-9 lists the half-wave reduction potential (= $E_{\frac{1}{2}}$ in V) relative to the nitrogen–oxygen bond in typical aromatic *N*-oxides, in comparison with the initial reduction potential of their parent bases.

The following facts are deduced from the half-wave reduction potentials of the nitrogen–oxygen bond in representative unsubstituted nitrogen-containing heterocycles listed in Table 4-9.

(i) Aromatic *N*-oxide compounds are generally reduced at a far more negative potential than the aliphatic *N*-oxides. In other words, aromatic *N*-oxides are more resistant to reduction.

(ii) The reduction potential, $E_{\frac{1}{2}}$, of the nitrogen–oxygen bond in aromatic *N*-oxide compounds generally varies with pH of the medium and the reduction is easier on the acid side of the pH. For example, $E_{\frac{1}{2}}$ of pyridine 1-oxide is approximately −1.4, −1.28, and −1.01, respectively, at pH 7.0,

* In this case, the difference of atomic orbitals relative to nitrogen in these two systems cannot be neglected.

References p. 167

TABLE 4-9

HALF-WAVE REDUCTION POTENTIALS OF *N*-OXIDE COMPOUNDS AND THEIR PARENT BASES ($-E_{1/2}$ V)[a]

Compound	*N*-Oxide compound						Parent bases	
	Reduction potential relative to nitrogen–oxygen bond at various pH						1st reduction potential	
pH	1.8	3.0	3.5	7.0			7.0	
Matrine			−0.4562[4,32]					
Dimethylaniline			−0.7047[4,32]	−1.018[33]				
Pyridine	−1.010[35]		−1.2786[4,32]	−1.364[33]	−1.41[36]	−1.40[37]	−1.75[36]	
Quinoline	−0.772[35] (−0.805)	−1.0692[4,32]		−1.164[33]				
Acridine				−0.854[33]			−0.618[34]	
Quinoxaline							−0.66[36]	−0.676[38]
1-oxide				−0.650[33]	−0.52[36]			
1,4-dioxide				−0.618[33]	−0.52[36]			
Phenazine							−0.364[33]	−0.36[36]
5-oxide				−0.218[33]	−0.19[36]			
5,10-dioxide				−0.218[33]	−0.23[36]			
Benzo[c]cinnoline							−0.58[36]	
5-oxide				−0.69[36]				
5,6-dioxide				−0.60[36]				

[a] Potentials relative to saturated calomel electrode.

3.5, and 1.8, and that of quinoline 1-oxide is −1.16, −1.07, and −0.8 at pH 7.0, 3.0, and 1.8, respectively.

(iii) Polarographic reduction becomes facile when a benzene ring is fused to monocyclic *N*-oxide or a second nitrogen is introduced into its ring. For example, the $E_{\frac{1}{2}}$ values of pyridine, quinoline, acridine, and phenazine *N*-oxides at pH 7.0 approach a positive value in this order, the values being approximately −1.4, −1.16, −0.854, and −0.218, respectively. This means that these compounds become more prone to reduction in this order.

Considering that the reduction is generally a nucleophilic reaction of the electrons, this increased facility of reduction may be understood as the increased electrophilic nature of the *N*-oxide group itself by the stabilization of the energy levels of the lowest empty molecular orbital in the *N*-oxide compounds, due to the increased conjugation in the case of a fusion of a benzene ring and to the introduction of a negative hetero atom in the other case. Consequently, the same tendency can be seen in the first half-wave reduction potentials of their parent heterocycles. For example, $E_{\frac{1}{2}}$ (initial) of pyridine, acridine, and phenazine at pH 7 are −1.75, −0.618, and −0.36, respectively.

However, as is seen in the case of phenazines, quinoxalines, etc., *N*-oxidation of the second nitrogen hardly has any effect on the $E_{\frac{1}{2}}$ value. This is assumed to be due to the fact that the extra electron-withdrawal by the formation of a positive charge in the second nitrogen atom by its *N*-oxidation is compensated for by the drift of the electrons from oxygen to the ring by the formation of an *N*-oxide at that nitrogen atom[36].

Dependence of $E_{\frac{1}{2}}$ on pH in the reduction of an *N*-oxide group, as stated in *(ii)* above, was already pointed out by Foffani[37] and by Kubota and others[39], together with the fact that its reduction wave rapidly decreases with increase of pH value and disappears on the alkaline side (above pH 9). Kubota and Miyazaki[39] assumed, from their detailed studies on pyridine 1-oxide and its alkyl derivatives, that their reduction progressed through the following steps.

$$\geqslant N \rightarrow O + H_3^+O \xrightarrow{\vec{k}} \geqslant N^+{-}OH \xrightarrow{2H^+ + 2e} [\geqslant N^+{-}H \leftrightharpoons \geqslant N + H_3O^+]$$

They indicated that the polarographic reduction on the alkaline side is due to the kinetic current, i, dependent on the size of $\vec{k}\cdot[H^+]$, and calculated the rate of recombination of the proton, $\vec{k}$, by using the equation $i = i_d \{1/[1 + 1.128\,(\vec{k}\cdot t\cdot[H^+]^2/Ka)^{-1/2}]\}$, obtaining the value of $\vec{k} \approx 1 \times 10^{11-12}$ $(sec)^{-1}$ $(mole/l)^{-1}$, where i is the limiting current, i_d is the diffusion current, and Ka is the acid dissociation constant. They also proved experimentally that the relationship between $E_{\frac{1}{2}}$ and pH (1.5–6.7) holds for pyridine 1-oxide at $E_{\frac{1}{2}} = -0.0865(\text{pH})-0.849$, and for 4-methylpyridine 1-oxide at $E_{\frac{1}{2}} = -0.0948(\text{pH})-0.883$.

As may be assumed from the result described in *(iii)* above, the nitrogen–oxygen bond in aromatic *N*-oxide compounds would be expected to be more easily reduced, the lower the energy potential of the lowest empty molecular orbital of the *N*-oxide compounds. In fact, $E_{\frac{1}{2}}$ (due to the nitrogen–oxygen bond) of many of the substituted *N*-oxides agrees with this assumption. As an example, $E_{\frac{1}{2}}$ values of substituted pyridine 1-oxides are listed in Table 4-10.

As will be clear from the data in Table 4-10, the reduction potential is generally more negative than that of the parent *N*-oxide when the substituent is an electron-donating group (D) and is more positive when the substituent is an electron-withdrawing group (W). This fact also supports the theory of Ochiai and others[4,32] that one of the reasons for the difficult reduction of the nitrogen–oxygen bond in aromatic *N*-oxide compounds is the co-ordination of the electrons of the oxygen to nitrogen, the reactive site for reduction (indicated by the contribution of the (4-2) formula among the resonance structures of pyridine 1-oxide).

The same conclusion has been drawn in the case of substituted phenazine and acridine *N*-oxides[33,34]. It should be noted in the reduction of substituted *N*-oxide compounds, as was pointed out by Emerson and others[36], that the

TABLE 4-10

HALF-WAVE REDUCTION POTENTIAL OF SUBSTITUTED PYRIDINE 1-OXIDES ($-E_{1/2}$ V)

	Substituent in pyridine 1-oxide	pH 1.8[34]	pH 3.5[4,32]	pH 7.0[36]
D	4-OCH_3			−1.56[36]
D	4-OC_6H_5		−1.2820[4,32]	
D	2,4,6-$(CH_3)_3$		−1.3924[4,32]	
	Pyridine	−1.010[34]	−1.2786[4,32]	−1.41[36]
W	2-COOH	−0.702[34]		
W	3-COOH	−0.835[34]		
W	4-COOH	−0.670[34]		−1.19[36]
W	2-Phenyl	−0.745[34]		
W	4-Phenyl	−0.745[34,35]		

substituent itself in some cases is reduced before the nitrogen–oxygen bond. For example, it has been revealed by Kubota[40] that, in 4-nitropyridine 1-oxide, the first reduction wave is that of the reduction of NO_2 to NHOH, and the second wave is that due to the reduction of the *N*-oxide group in the 4-hydroxyaminopyridine 1-oxide so formed (*cf.* Table 4-11).

The same fact has been pointed out for azo- and azoxypyridine 1,1′-dioxides, in which the reduction of azo and azoxy groups precedes that of the *N*-oxide group[35].

Summarizing the foregoing, it may be considered that the reduction of the *N*-oxide group progresses through two-electron reduction (nucleophilic reaction) of the nitrogen–oxygen bond, the site reactive to polarographic reduction in the rate-determining step of this reduction. Further, this is not inconsistent with the fact that *N*-oxide compounds are more easily reduced in the 1-hydroxypyridinium ion type than in their basic type because, in the pyridinium ion, protonation of the lone-pair electrons of the oxygen increases its electronegativity and the drift of the electrons from oxygen to the ring (and to nitrogen), which is the characteristic of the *N*-oxides, is markedly inhibited. In this sense, aromatic *N*-oxides become closer to aliphatic *N*-oxides in their conjugate acid form. Very recently, Kubota and Miyazaki[40] measured the reduction potentials ($E_{\frac{1}{2}}$) of 23 kinds of 3- and 4-substituted pyridine 1-oxides at various pHs and found that the reduction potential of the *N*-oxide group in these compounds can be expressed by the equation $-E_{\frac{1}{2}} = a \cdot \mathrm{pH} + b$, as had already been found for pyridine and 4-methylpyridine 1-oxides[39]. In Table 4-11 are shown their data for the reduction potential of 3- and 4-substituted pyridine 1-oxides at pH 5, together with values for a and b, and substituent constants for these substituents.

When the $E_{\frac{1}{2}}$ values at each pH are plotted against the σ-values shown in this table, a good linear relationship is obtained. The slope of this line, *i.e.* the reaction constants, ϱ', of Hammett's equation, was +0.457, +0.495, and +0.584 at pH 3, 4, and 5, respectively.

Kubota and Miyazaki concluded from these results that the initial process of the electrode reaction starts with the shift of electrons from the electrode to the reactive point, *i.e.* as a nucleophilic reagent (because ϱ is positive). They further found semi-quantitatively that the lower the energy potential of the lowest empty molecular orbital of each *N*-oxide compound, calculated by the molecular orbital theory using the second-order perturbation method, the closer the reduction potential came to the positive value. These conclusions are in good agreement with the numerous data presented above.

References p. 167

TABLE 4-11

REDUCTION WAVE OF THE *N*-OXIDE GROUP IN 3- AND 4-SUBSTITUTED PYRIDINE 1-OXIDES IN pH REGION OF 1.5 TO 7–8

Values of *a* and *b* when represented as $-E_{1/2} = a(\text{pH}) + b$, and Hammett's σ values[a]

	σ^a	*a*	*b*	$E_{1/2}$ V at pH 5
Substituent at 4-position				
NH_2	−0.660	0.1023	0.953(pH 2.6–7)	−1.465
	−0.13[b]	0.074	1.026(pH 1–2.6)	
NHOH	−0.339	0.069	1.089	−1.434
OCH_3	−0.268	0.094	0.946	−1.416
OC_2H_5	−0.250	0.0963	0.914	−1.396
CH_3	−0.170	0.0948	0.883	−1.357
C_2H_5	−0.151	0.0933	0.864	−1.331
Nil		0.0865	0.849	−1.281
$NHCOCH_3$	−0.015	0.0788	0.795	−1.189
Cl	+0.227	0.087	0.739	−1.174
Br	+0.232	0.0975	0.606	−1.094
CN	+0.660[c]	0.057	0.572(pH 4–9)	−0.857
COOEt	+0.678	0.058	0.572	−0.862
NO_2	+0.778	0.04	0.004(for 1st wave)	−0.204(NO_2 group)
	+1.270[d]	0.096	0.956(for 2nd wave)	−1.436(N→O group)
Substituent at 3-position				
NH_2	−0.161	1.095	0.75(pH 1–4.5)	−1.280
		0.074	0.91(pH 4.5–8)	
OCH_3	+0.115	0.0785	0.775	−1.168
CH_3	−0.069	0.0816	0.864	−1.272
C_2H_5	−0.043	0.081	0.841	−1.246
Nil		0.0865	0.849	−1.281
$NHCOCH_3$	+0.154[e]	0.072	0.769	−1.129
F		0.0795	0.7465	−1.144
Cl	+0.373	0.0793	0.674	−1.071
Br	+0.391	0.0778	0.651	−1.040
CN	+0.56[c]	0.0713	0.586	−0.943
$COOC_2H_5$	+0.398	0.0665	0.659	−0.992

[a] These σ values are cited from Jaffé's review[11] unless otherwise stated.
[b] P. ZUMAN, *Collection Czech. Chem. Commun.*, 25 (1960) 3225.
[c] D. H. MCDANIEL AND H. C. BROWN, *J. Org. Chem.*, 23 (1958) 420.
[d] Jaffé's σ^* ($= \sigma^-$) value[54].
[e] H. H. JAFFÉ AND G. O. DOAK[47].

4.5 Dissociation Constant

The dissociation constants, pKa, of some aromatic *N*-oxide compounds and their parent bases are given in Table 4-12, together with those of aliphatic *N*-oxide compounds and their parent bases.

As is clear from Table 4-12, the pKa values of the *N*-oxides of aliphatic series are approximately constant (pKa 4–5) in spite of the considerable difference in the basicity of their parent bases (*e.g.*, pKa of 9.74 for trimethylamine against that of 5.06 for dimethylaniline). On the other hand, pKa values of aromatic *N*-oxide compounds vary, although they are lower than that of their parent bases, and the values are far smaller than those of aliphatic *N*-oxide compounds. These facts indicate that the interaction between the *N*-oxide group and the aromatic ring in aromatic *N*-oxide compounds, indicated by the canonical form like (4-2), still exists considerably in an aqueous solution. Consequently, in *N*-oxide compounds of aromat-

TABLE 4-12

OBSERVED pKa (IN WATER AT 20–25°)

Tertiary amines	pKa	Amine *N*-oxide	pKa		
Aliphatic series					
Trimethylamine	9.74[35]		4.65[35]	4.48[48]	4.60[49]
Triethylamine	10.76[35]		5.13[35]	4.96[48]	
Dimethylaniline	5.06[35]		4.21[35]	4.04[48]	
Diethylaniline	6.56[35]		4.53[35]		
Tribenzylamine			4.70[51]		
Aromatic series					
Pyridine	5.29[35]		0.79	0.56[41]*	
	5.14[46]*		0.78[41]		
Quinoline	4.85[46]*		0.92[41]	0.70[41]*	
Isoquinoline	5.40[44]		1.05[41]	0.81[41]*	
	5.14[46]*		1.01[47]		
Acridine	5.60[46]*		1.57[41]	1.37[41]*	
Phenazine	(1.23[44])	*N*-oxide	1.32[41]	1.13[41]*	
		N,N-dioxide	1.62[41]		
Phenanthridine	4.52[45]		0.82[41]	0.56[41]*	
1,10-Phenanthroline	4.80[42]	*N*-oxide	6.6[43]		
	4.86[51]				

The pKa values with an asterisk (*) are corrected for ionic strength.

References p. 167

ic series, resonance like that formulated below, which is not present in

aliphatic *N*-oxides, results in the decreased negativity of the oxygen, with a consequent decrease in the affinity to the hydrogen ion compared to that in aliphatic *N*-oxides, and this may be the reason for the small pKa values observed. Since the pKa values increase, though to a slight degree, in the order of pyridine, quinoline, isoquinoline, and acridine *N*-oxides, it may be assumed that the contribution of a resonance structure of (4-3) type increases as the parent heterocycle becomes larger†.

It should be noted in Table 4-12 that, in the case of diazine *N*-oxide, it is still uncertain whether the first proton attacks the nitrogen in the ring or the oxygen atom in the *N*-oxide group, because the pKa value of diazine *N*-monoxide is close to those of its parent base and those of ordinary *N*-oxides, and it is difficult to determine the position at which the proton attacks (*cf.* pKa of phenazine *N*-monoxide with those of phenazine and acridine *N*-oxide in Table 4-12).

The strong basicity of *o*-phenanthroline *N*-monoxide is attributed to the stable hydrogen-bond structure (4-28) of its conjugate acid[43]:

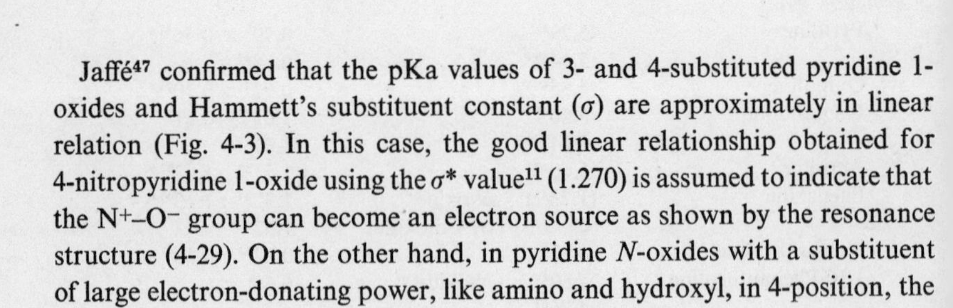

(4-28) (4-29)

Jaffé[47] confirmed that the pKa values of 3- and 4-substituted pyridine 1-oxides and Hammett's substituent constant (σ) are approximately in linear relation (Fig. 4-3). In this case, the good linear relationship obtained for 4-nitropyridine 1-oxide using the σ^* value[11] (1.270) is assumed to indicate that the N^+–O^- group can become an electron source as shown by the resonance structure (4-29). On the other hand, in pyridine *N*-oxides with a substituent of large electron-donating power, like amino and hydroxyl, in 4-position, the

† Even in the excited state, the shift of the electrons from the oxygen to the ring decreases in this order (*cf.* Section 4.8).

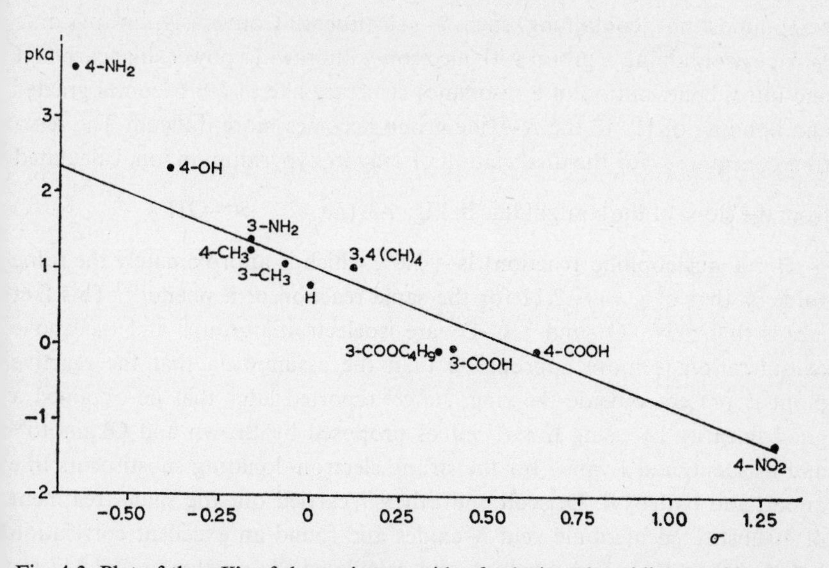

Fig. 4-3. Plot of the p*K*'s of the conjugate acids of substituted pyridine 1-oxides against substituent constants[47]. (Reprinted from the *Journal of the American Chemical Society*, by permission of the copywright owners, the American Chemical Society.)

pKa values and σ are not linear. This may be understood from the fact that the group $\geqslant N^+$–OH is the strongest electron-withdrawing group*, so that in 4-amino- and 4-hydroxy-pyridine 1-oxides, Hammett's constant of these substituents requires a more negative value**.

In any case, as is evident from this Fig. 4-3, the pKa values of pyridine 1-oxide with substituents donating electrons, like methyl, hydroxyl, amino, etc., are larger than that of pyridine 1-oxide and the fact indicates that the contribution of a resonance structure like (4-30) is greater than that in the

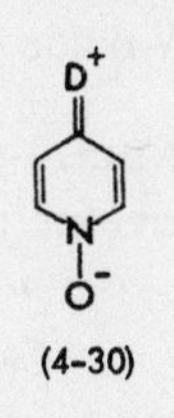

(4-30)

* The σ values of -1.28 for 4-NH_2 and -0.67 for 4-OH have been evaluated from the data in Fig. 4-3 (Ref. 47).

** H. H. Jaffé calculated the substituent constant (σ^*_{para}) of N^+–OH as 3.9 (*J. Am. Chem. Soc.*, 77 (1955) 4445).

References p. 167

compounds not containing such a substituent. Conversely, in pyridine 1-oxides containing a group with electron-withdrawing power, like carboxyl and nitro, contribution of a resonance structure like (4-29) becomes greater and bonding of H^+ to the *N*-oxide group becomes more difficult. The reaction constant, ϱ, for the dissociation of 1-hydroxypyridinium ion, calculated from the slope of the straight line in Fig. 4-3, (*i.e.* N^+–OH→ N→O + H^+, a nucleophilic reaction) is +2.09, which is approximately the same value as that of $\varrho = +2.11$ for the same reaction of a phenol[11]. This fact means that ⩾N⁺–O⁻ and ⩾C–O⁻ are isoelectronic groups and the above consideration is more appropriate than the assumption that the reactive point is present outside the ring. Jaffé[54] reported later that he obtained a good linearity by using the σ^+ values proposed by Brown and Okamoto[52] and by Deno and Evans[53] for the strong electron-donating substituents like amino and hydroxyl. Driscoll and others[55] carried out the same treatment of 4-substituted nicotinic acid *N*-oxides and found an excellent correlation between the pKa's and σ values, and calculated the ϱ value as +2.347 for the following formulae.

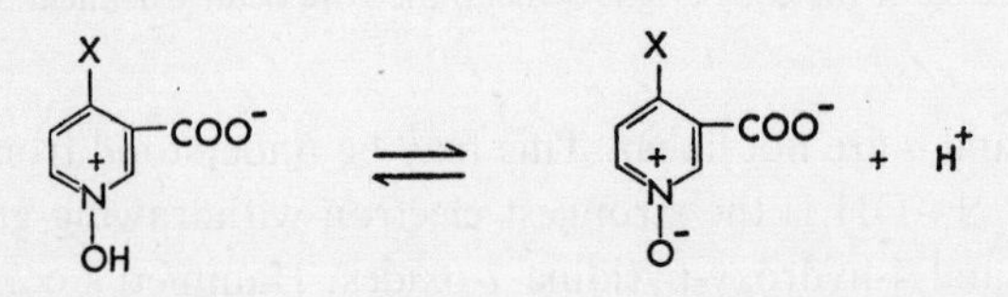

Jaffé[56] calculated σ_{meta} and σ_{para} when ⩾N⁺–O⁻ is taken as a substituent, from the result of the measurement of pKa's of substituted pyridine 1-oxides, and the values he obtained are listed in Table 4-13.

The last line in Table 4-13 gives the values obtained by Shindo[57a,b], using a non-polar solvent, and is given for comparison. In all of these, the σ values are positive, indicating that the *N*-oxide group acts rather as an electron-

TABLE 4-13

σ VALUES OF SUBSTITUTED PYRIDINE 1-OXIDES

Solvent	σ_{para}	σ_{meta}	Method	Reference
H_2O	+1.34	+1.47	pKa	56
50% EtOH	+1.37	+1.50	pKa	56
Non-polar solvent	+0.25	+1.18	I.R.	57

withdrawing group. However, the fact that the σ values, especially σ_{para}, become smaller (*i.e.* closer to minus values) as the solvent changes from water to ethanol to a non-polar solvent, seems to indicate that there is a fairly large shift of the electrons from the oxygen to the ring in a non-polar solvent.

Before concluding this section, pKa data for pyridine 1-oxides are summarized in Table 4-14. The basicity due to the *N*-oxide is indicated as the pKa value of the conjugate acid and is expressed as the "proton gained" in that sense. When there are two basic functional groups, as in aminopyridine 1-oxide, these are divided into pK(1st) and pK(2nd). The substituents (Z) are listed in the order of electron-donating to electron-withdrawing groups, but their alkylated and acylated compounds are shown together with their parent bases.

4.6 Nuclear Magnetic Resonance Spectra

Aromatic *N*-oxide compounds are interesting in that their *N*-oxide group can act both as an electron donor and an electron acceptor (*cf. Section 2.2.1*). If the increase in π-electron density on carbons in the aromatic ring results in increase of the shielding of protons attached to these carbon atoms, some information concerning localization of π-electrons on the ring should be obtained from the chemical shift of ring protons in such aromatic *N*-oxide compounds. Taking pyridine 1-oxide as an example, γ-H and α-H* should naturally be in a higher magnetic field than β-H if the contribution of (4-2) structure is larger than that of (4-3) and a reverse phenomenon should be apparent if the contribution of (4-3) is larger than that of (4-2). On the basis of this view, the electronic structure of aromatic *N*-oxides is being studied through NMR spectra by groups of scientists headed by Kawazoe and Tôri in Japan, and by Katritzky in England. It is now known that the factor determining the chemical shift of ring protons is not only the π-electron density but that there are many other factors. Consequently, in the present section, various factors determining the chemical shift of ring protons

* With regard to α-H, besides the large decrease of σ-electron density at C-2, where an inductive effect of the positive nitrogen may be expected, there are various neighboring effects of the *N*-oxide group, as will be mentioned later (*cf. Section 4.6.1.e*), and it is almost impossible at present to make any primary correlation between π-electron density and its chemical shift.

TABLE 4-14

pKa VALUES OF SUBSTITUTED PYRIDINE 1-OXIDES IN AQUEOUS SOLUTION (AT 20–25°)

(a) Mono-substituted compounds

Substituent (Z)	Proton gained				Proton lost	
	pK (1st)			pK (2nd)		
4-Amino	3.65[47]	3.54[58]	3.69[60]	−6.27[a,58]		
4-acetamido	1.59[61]					
4-Methylamino	3.85[60]					
4-(N-methylacetamido)	1.36[61]					
4-(N-methylbenzamido)	1.70[61]					
4-Dimethylamino	3.88[60]					
2-Amino	2.67[60]					
2-acetamido	−0.42[61]					
2-benzamido	−0.44[61]					
2-Methylamino	2.61[60]					
2-(N-methylacetamide)	−1.02[61]					
2-(N-methylbenzamido)	−1.39[61]					
2-Dimethylamino	2.27[60]					
3-Amino	1.47[47]			−2.1[52]		
3-acetamido	0.99[61]					
4-Hydroxy	2.36[47]	2.45[60]			5.76[56]	5.9[59]
4-methoxy	2.05[60]					
4-benzyloxy	1.99[60]					
1-methoxy-4-oxo	2.57[60]					
1-benzyloxy-4-oxo	2.58[60]					
2-Hydroxy	−0.8[60]				5.99[60]	5.9[59]
2-methoxy	1.23[60]					
2-ethoxy	1.18[60]					
1-methoxy-2-oxo	−1.3[60]					
1-benzyloxy-2-oxo	−1.7[60]					
3-Hydroxy					6.4[59]	
4-Mercapto	1.53[61]				3.82[61]	
4-benzylthio	2.09[61]					
2-Mercapto	−1.95[61]				4.67[61]	
2-benzylthio	−0.23[61]					
4-Methyl	1.29[47]					
2-Methyl	1.022[62]					
3-Methyl	1.08[47]					
Nil	0.79[47]					
4-Phenyl	0.83[63]					
2-Phenyl	0.77[63]					
3-Phenyl	0.74[63]					
4-(*p*-Aminophenyl)	3.64[b,63]					
4-(*p*-Nitrophenyl)	0.58[63]					
4-(*m*-Nitrophenyl)	0.58[63]					

For footnotes [a–d], see page 104

TABLE 4-14 *(continued)*

Substituent (Z)	Proton gained		Proton lost
	pK (1st)	pK (2nd)	
2-(*p*-Aminophenyl)	3.82[b,63]		
	0.25[b,63]		
2-(*m*-Aminophenyl)	3.92[b,63]		
	0.20[b,63]		
2-(*p*-Nitrophenyl)	0.28[63]		
2-(*m*-Nitrophenyl)	0.26[63]		
3-(*m*-Nitrophenyl)	0.47[63]		
4-COOH	−0.48[47]		2.86[56] (3.71[56])[c]
3-COOH	0.09[47]		2.73[56] (3.54[56])[c]
3-$COOC_4H_9$	0.03[47]		
4-NO_2	−1.7[47]		

(b) Disubstituted compounds

Substituent (Z_γ)	Proton gained	Proton lost
Nicotinic acid 1-oxide series (Z_β = COOH)		
4-NH_2		3.98[d,55]
4-OH		3.08[d,55]
4-OCH_3		2.88[d,55]
Nil		2.74[d,55]
4-Cl		1.80[d,55]
4-NO_2		0.50[d,55]
2-Picoline 1-oxide series (Z_α = CH_3)		
4-NH_2	4.10[62]	
4-$N(CH_3)_2$	4.37[62]	
4-OCH_3	9.414[62]	
4-OC_2H_5	2.106[62]	
Nil	1.022[62]	
4-CN	−0.674[62]	
4-NO_2	−0.968[62]	
2,6-Lutidine 1-oxide series ($Z_{\alpha,\alpha'}$ = CH_3's)		
Nil	1.442[62]	
4-COOH	−0.015[62]	
4-$COOC_2H_5$	−0.126[62]	
4-$CONH_2$	−0.317[62]	
4-CN	−0.614[62]	
4-NO_2	−0.861[62]	

For footnotes [a–d], see page 104

(Continued overleaf)

TABLE 4-14 *(continued)*

(c) Substituted quinoline 1-oxides

Substituent	Proton gained		Proton lost
	pK (1st)	pK (2nd)	
4-Amino		−6.55[a,58]	
2-Hydroxy	>1[64]		6.88[64]

[a] This pK value corresponds to $H_3\overset{+}{N}$–$C_5H_4\overset{+}{N}$–OH ⟶ H_2N–$C_5H_4\overset{+}{N}$–OH + H^+.

[b] The high pK value is due to the primary amino group and the low value to the *N*-oxide group.

[c] The value in 50% ethanol solution.

[d] Driscoll *et al.*[55] concluded that the determined pK value is not a measure of the equilibrium (a) but rather a measure of the ionization of the zwitter ionic form of nicotinic acid 1-oxide, as shown by (b).

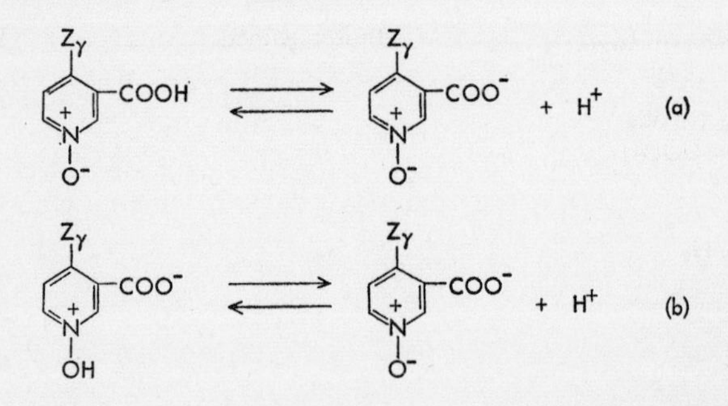

in aromatic compounds in general will be described first and studies on the electronic state of *N*-oxide compounds by the above three groups of workers will be described.

4.6.1 *Various Factors Determining the Chemical Shift of Hydrogen in an Aromatic Ring*

(a) π-Electron Density

The relationship between chemical shift of hydrogen in an aromatic ring and π-electron density on the carbon atoms in the ring to which these hydrogen atoms are attached has been discussed for some time and many studies have been made in relation to the Hammett parameter and reactivity[65]. Quantitization of these correlations has been attempted by Fraenkel and others[66]. The latter workers indicated that a proportionality $\Delta\delta H = a \cdot \Delta\varrho$ existed between $\Delta\delta$ (proton chemical shift referred to the benzene proton resonance, in p.p.m.) and $\Delta\varrho$ (local excess π-electron charge located on the aromatic carbon). The proportionality constant a (p.p.m./electron) in this equation represents the deviation of the chemical shift of hydrogen bonded to that carbon if another π-electron is assumed to exist on an optional carbon atom in the benzene ring.

Fraenkel and others proposed $a = 10$ experimentally, using aromatic anions and cations of various ring sizes. Later, Spiesecke and Schneider[67] suggested the value of 10.6, and Dailey and others[68] proposed 9.3 by carrying out corrections for the ring current effect and solvent effect. In any case, as was pointed out by Schaefer and Schneider[69], it has been found that values agreeing well with experimental results can be obtained by giving this constant, a, the value of 10 ± 1.

This relationship has been obtained with aromatic ions of known density of uniform π-electrons, or with an excess or a deficiency of π-electrons. For determining the chemical shift of aromatic hydrogen atoms in general, not only the π-electron density of the carbon atom to which this hydrogen is attached, but the following factors must also be taken into consideration.

(b) Ring Current Effect

The effect of diamagnetic ring current due to delocalized π-electrons on the chemical shift is great and its magnitude has been estimated theoretically with a fair precision[69–71]. This effect is thought to give an experimental measure for the degree of aromaticity, and Jackman and Elvidge[72] assumed that the aromaticity of 2-pyridone is only 35% of that of benzene.

(c) Magnetic and/or Electric Effect of Substituents

The effect of substituents like NO_2, COR, and COOR on adjacent hydrogen is very large and since the effect is proportional to $1/r^3$ (r being the distance between the substituent and the proton in question), its effect on a distant proton is very small. For example, the effect of the substituent in substituted benzenes is rather great on an *ortho* proton but generally negligibly small on *meta* and *para* protons. Yamaguchi[73] carried out a quantitative examination of the magnetic anisotropy effect of the nitro group on an *ortho* proton.

(d) Intermolecular Interaction with Solvent Molecules or Other Molecules in the Solution

Aromatic compounds in general are π-electron donors or n-electron donors,

References p. 167

as represented by alkylbenzenes and pyridines. They therefore tend to have an intermolecular interaction in a protic solvent by protonation of active hydrogen or by hydrogen-bond formation, or with some aprotic molecules, by electrostatic force, charge transfer interactions, etc. This affects the electronic structure of aromatic compounds; further, the magnetic anisotropy of the solvent molecules showing this property is rather large, which may also affect the chemical shift of aromatic hydrogen. Consequently, if the chemical structure of the molecules is to be deduced from the chemical shifts, experimental conditions† must be so chosen that such an effect can be neglected.

On the other hand, this effect is utilized for studies on the charge transfer interaction between alkylbenzenes and tetracyanobenzene[74], and in quantitative studies on the solvent effect of aromatic compounds by comparing the chemical shift of ring protons in substituted benzene compounds in acidic medium (such as trifluoroacetic acid) and in neutral medium (such as carbon tetrachloride)[75].

(e) Effect of Hetero Atom in Aromatic Heterocyclic Compounds (Especially those containing Nitrogen)

(i) Electronic Effect due to Nitrogen. When a hetero atom such as nitrogen, oxygen, etc. is present in the conjugated system its electronic effect brings about a disturbance of the uniform distribution of π-electrons. Consequently, it seems inappropriate to interpret the chemical shift of aromatic protons only from π-electron density.

(ii) Magnetic Anisotropic Effect due to n-π^* Transition of Nitrogen. In heterocycles of the azabenzene series, like pyridine, the magnetic effect due to an n-π^* transition of lone-pair electrons of the nitrogen (*cf. Section 4.6.2*) is anisotropic and is known to be quite large. Thus, this transition causes a large paramagnetic effect on α-CH, a phenomenon studied intensively and independently by Matsuoka and his group[76] and by Baldeschwieler and

† As a solvent for spectral measurement, neutral solvents with small dielectric constant are generally desirable. In addition, the solvent molecule should be magnetically isotropic or its anisotropic character should be as small as possible. The chemical shift of the proton in question in a dilute solution in the solvents satisfying such conditions, like carbon tetrachloride, hexane, and cyclohexane, is considered to be unaffected by intermolecular interaction with solvent molecules. On the other hand, acidic solvents like trifluoroacetic acid and sulfuric acid, and benzene and carbon disulfide are considered to have a large solvent effect.

his group[77]. Recently, Gil[78] made a refined treatment of this effect and found that there is a good agreement in the electron density deduced from the chemical shift and that from self-consistent molecular orbital calculations in pyridine.

(f) Other Factors

As is evident from the studies of Kotowycz and others[79] on the pyridinium cation, the effect of a counter ion cannot be neglected in aromatic quaternary bases. When $\geqslant$C–H changes to $\geqslant$N:, besides the afore-mentioned electronic and magnetic anisotropic effects, an electric field effect due to changes in the dipole moment may cause a change in the chemical shift of CH in the α position next to this nitrogen[78]. In such heteroaromatics, its basic center nitrogen easily receives the acid-base type interaction of the solvent, and the foregoing electronic and magnetic effects due to nitrogen change greatly[75,80]. Consequently, it still seems premature to correlate the difference in the chemical shift of α-H in the free base of pyridine and its conjugate acid primarily to their α-electron density (or to net electron density including σ-electrons).

4.6.2 Nuclear Magnetic Resonance Spectra of Aromatic N-Oxides, with Special Reference to Correlation with π-Electron Density

As will be clear from considerations given in the foregoing sections, the chemical shift of protons in pyridines, especially the α-protons adjacent to the ring nitrogen atom, is largely due to various factors related to the effect of nitrogen, but it may be concluded that the chemical shift of β- and γ-protons, to a first approximation, reflects the increase or decrease of π-electron density. Consequently, this interrelation should naturally apply in the case of pyridine *N*-oxides. In the case of quinoline *N*-oxides, however, factors other than π-electron density must be considered for the chemical shift of a proton which is present in the 8(peri)-position, besides that in the 2-position as a proton adjacent to the *N*-oxide group.

In order to carry out the analysis of the electronic structure of pyridine 1-oxide in this way, Okamoto, Kawazoe, and others[80] measured the NMR spectra of 2,5-dimethylpyridine 1-oxide in various solvents whose spectra can be easily analysed. The result they obtained is shown in Fig. 4-4, in comparison with that of its parent compound, 2,5-dimethylpyridine. In this case, as will be clear from the above discussion, the difference in the chemical shifts of at least γ- and β-protons reflects the difference in π-electron densities of the pyridine ring.

References p. 167

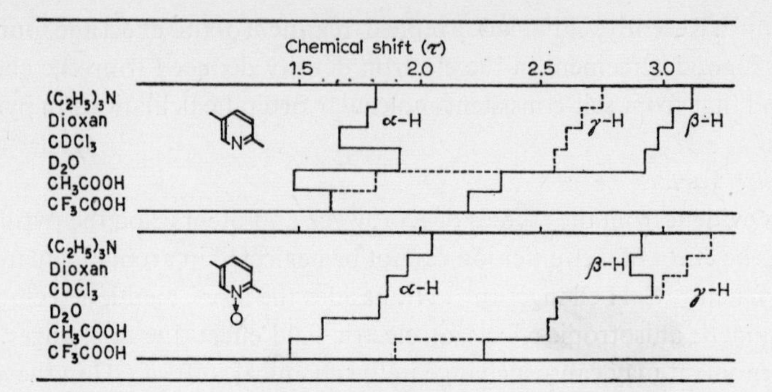

Fig. 4-4. Proton resonances of 2,5-dimethylpyridine and its oxide in various kinds of solvents[80].

The most interesting fact shown by Fig. 4-4 is that in an aprotic solvent, while γ-H is in a lower field than β-H in the parent base (hereinafter indicated as $H_\gamma < H_\beta$), the situation is the reverse ($H_\gamma > H_\beta$) in the corresponding *N*-oxide compounds. As the solvent becomes acidic, both H_γ and H_β shift to a lower magnetic field in the parent base, although their order ($H_\gamma < H_\beta$) is retained, while this relation in the corresponding *N*-oxide becomes $H_\gamma \approx H_\beta$ in a neutral protic solvent, and the order is reversed and becomes the same as that in the parent base, *i.e.* $H_\gamma < H_\beta$, in the kind of solvent in which the *N*-oxide would be present as a perfect conjugate acid (such as trifluoroacetic acid). Since the behavior of the chemical shift of H_γ and H_β in the *N*-oxide compounds may be considered as approximately reflecting the π-electron density, Okamoto, Kawazoe, and others concluded that the contribution of the (4-2) form is large in aprotic solvent, that of (4-3) becomes larger as the solvent becomes protic, and the contribution of (4-9) becomes more important than that of (4-8) in a sufficiently acid solvent like trifluoroacetic acid, in which the *N*-oxide can take the conjugate acid form (4-7). It should be noted here that Okamoto, Kawazoe, and others[80] pointed out the absence of reports giving an example of the reversion of the relative shift of protons in a molecule by the change of the solvent, as was experienced in *N*-oxides, either in aliphatic amines or aromatic amines. For example, in anilines, the relative shift of H_{meta} and H_{para} is 0.5 p.p.m. in carbon tetrachloride, with H_{para} in a higher field than H_{meta}, and even in a sufficiently acidic solution this only becomes $H_{para} \approx H_{meta}$[75]. In any case, the facts that the electronic structure of *N*-oxide compounds easily undergoes change with a slight difference in the nature of the solvent, and significant difference in electron densities occurs at various positions, such as the inversion of π-electron densities in β- and γ-positions, indicate that the mesomeric effect of the *N*-oxide group acts both in electron-withdrawing and electron-donating ways and that the mesomeric effect of the *N*-oxide group is easily changed even by the slight difference in the nature of the reagent. This is believed to correspond with the fact that the *N*-oxide compounds are active to both electrophilic and nucleophilic reagent.

Okamoto, Kawazoe, and others[80] experimentally studied the effect of *N*-oxidation on the chemical shift of protons in the parent base. They found that in aprotic solvents like carbon tetrachloride and dioxan, H_α and H_γ undergo a higher shift and H_β a lower shift* without exception (*cf.* Fig. 4-4).

The reason for this is that *N*-oxidation increases the negativity of the nitrogen, and electron density of the aromatic ring as a whole is decreased, resulting in the decrease of electron density on the β-carbon which does not receive the electron-donating resonance contribution from the oxygen atom. Similarly, Katritzky and Lagowski[81] found that the decrease in the chemical shift of the proton in *ortho*-position of monosubstituted benzenes and the β-proton in 4-substituted pyridines, their *N*-oxides (in aqueous solution), and their conjugate acids (in 20N sulfuric acid) decreases in the order of phenyl, pyridine, nitropyridine, pyridine $H^+ \sim$ pyridine N^+–OH, and concluded that this order is in parallel with the electron density of the carbon atoms on each ring. Katritzky and others drew conclusions about a similar order of electron density from the observed order of chemical shift of the α-proton in pyridine and its *N*-oxides, phenyl $\gg$ nitropyridine $>$ pyridine $\gg$ pyridine $H^+ \sim$ pyridine N^+–OH. As was pointed out above, it seems impossible at present to discuss the chemical shift of the α-proton merely from the electron density in that position. Consequently, the foregoing order of electron density relative to the α-position is still not unquestioned; the same applies to the order of the decrease of electron density.

The nuclear magnetic resonance spectra of pyridine 1-oxides may be summarized as follows. The chemical shifts of the protons in various positions in aprotic (non-polar) solvents are in the order of $H_\gamma > H_\beta > H_\alpha$ from the

* Of these, the chemical shift of H_γ and H_β can be considered as primarily due to the increase or decrease of the electron density. The higher shift of H_α of the *N*-oxide over that in the parent base is due, besides to the increase in net electron density by the contribution of the (4-2) structure, to the decreased contribution of $n \rightarrow \pi^*$ transition (paramagnetic effect, *i.e.*, effect of shifting to a lower magnetic field) by *N*-oxidation. However, quantitative analysis of the magnitude of this shift requires consideration of the paramagnetic effect of the *N*-oxide group itself on adjacent CH, as was pointed out by Tôri and others.

References p. 167

TABLE 4-15

NMR SPECTRA OF PYRIDAZINE N-OXIDE DERIVATIVES AT 60 Mc.p.s. IN $CDCl_3$[83]

Substituent	Chemical shift					
	τ_3	τ_4	τ_5	τ_6	τ_{CH_3}	τ_{OCH_3}
None	1.46	2.78	2.17	1.74		
3-Cl		2.77	2.25	1.82		
3-CH_3		3.02	2.42	1.90	7.48	
					7.64	
4-CH_3	1.65		2.43	1.83	7.64	
5-CH_3	1.62	3.01		1.92	7.63	
6-CH_3	1.63	2.88	2.27		7.49	
3,6-Cl_2		2.78	2.10			
3-Cl, 4-CH_3			2.41	1.90	7.61	
3-Cl, 5-CH_3		2.95		2.01	7.64	
3-Cl, 6-CH_3		2.83	2.27		7.51	
4-Cl, 6-CH_3	1.61		2.32		7.49	
3,4-Cl_2, 6-CH_3			2.23		7.50	
3-OCH_3		3.31	2.47	2.05		5.98
3-OCH_3, 6-CH_3		3.33	2.48		7.55	5.98
4-OCH_3, 6-CH_3	1.92		2.80		7.48	6.07
3-OCH_3, 6-Cl		3.28	2.29			5.96
						5.92
3,6-$(CH_3O)_2$		3.27	2.64			6.02
3-OCH_3, 4-Cl, 6-CH_3			2.46		7.56	5.92
3,4-$(CH_3O)_2$, 6-CH_3			3.05		7.51	5.90
						6.02
4-NO_2	0.70		1.55	1.84		
4-NO_2, 6-CH_3	0.80		1.55		7.41	
3-Cl, 4-NO_2, 6-CH_3			1.66		7.46	
3-OCH_3, 4-NO_2, 6-CH_3			1.65		7.49	5.82

higher magnetic field. The same regularity is also observed in pyridazine and its alkyl and halogen derivatives[82,83]. Therefore, this regularity can be utilized for determining the position of the *N*-oxide group in the mono-*N*-oxides of non-symmetrically substituted diazines. NMR spectra of pyridazine *N*-oxide in $CDCl_3$ measured by Tôri are given in Fig. 4-5, and the chemical shifts of ring protons in pyridine 1-oxides, measured by Tôri and Kawazoe in $CDCl_3$ and in dioxan, respectively, are given in Tables 4-15 and 4-16.

As will be seen from these Tables and Figure, the protons in each position of pyridazine 1-oxides, including those of halogen- and alkyl-substituted derivatives, without exception are in the order of $H_3 < H_6 < H_5 < H_4$ from

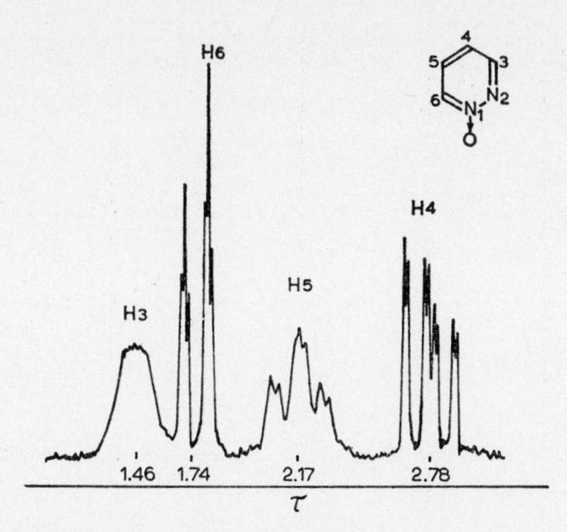

Fig. 4-5. Nuclear magnetic resonance spectrum of pyridazine 1-oxide at 60 Mc.p.s. in deuterochloroform.

the lower to the higher magnetic field. In fact, this interrelation was used to determine the structure of the two *N*-oxides of benzotetrahydrocinnoline and its substituted derivatives[84]. The data in Table 4-16 seem to indicate that H_3 of the *N*-oxide compounds is in a higher magnetic field than that of their parent base and this seems to differ from the behavior of proton chemical shifts in pyridine series compounds. Kawazoe and others explained this phenomenon by assuming that the *N*-oxide compounds have a far lower basicity than pyridazine itself and that, therefore, the $n \rightarrow \pi^*$ transition effect of the free nitrogen atom (N_2) in the *N*-oxide molecule is smaller than that in its parent base. In any case, from the chemical shift of the protons in *N*-oxides, the electron density of C_4 is larger than that of C_5 in chloroform† and in dioxan. Tôri and others[85,86] found some interesting facts by examining NMR spectra of monoaza- and diazanaphthalene *N*-oxides. They showed that the *N*-oxide group gives a magnetic anisotropy effect on a proton in 8(peri)-position of quinoline and cinnoline 1-oxides, resulting in a 1.2–1.4 p.p.m. lower shift. As an example, NMR spectra of quinoline 1-oxide in dioxan and in trifluoroacetic acid are shown in Fig. 4-6.

As will be seen from these spectra, H_2 and H_8 in quinoline 1-oxide in dioxan are in a low field, apart from other H's. In acidic medium, most of the resonance lines shift to a lower field but, relatively, the lower shift of H_4 is

† Kawazoe pointed out that chloroform has a fairly high ability for hydrogen bonding with *N*-oxides. This fact has also been confirmed by infrared data (*cf. Section 4.7.1*).

TABLE 4-16

NMR SPECTRA OF PYRIDAZINE AND ITS *N*-OXIDE DERIVATIVES[82] (at 60 Mc.p.s. in dioxan)

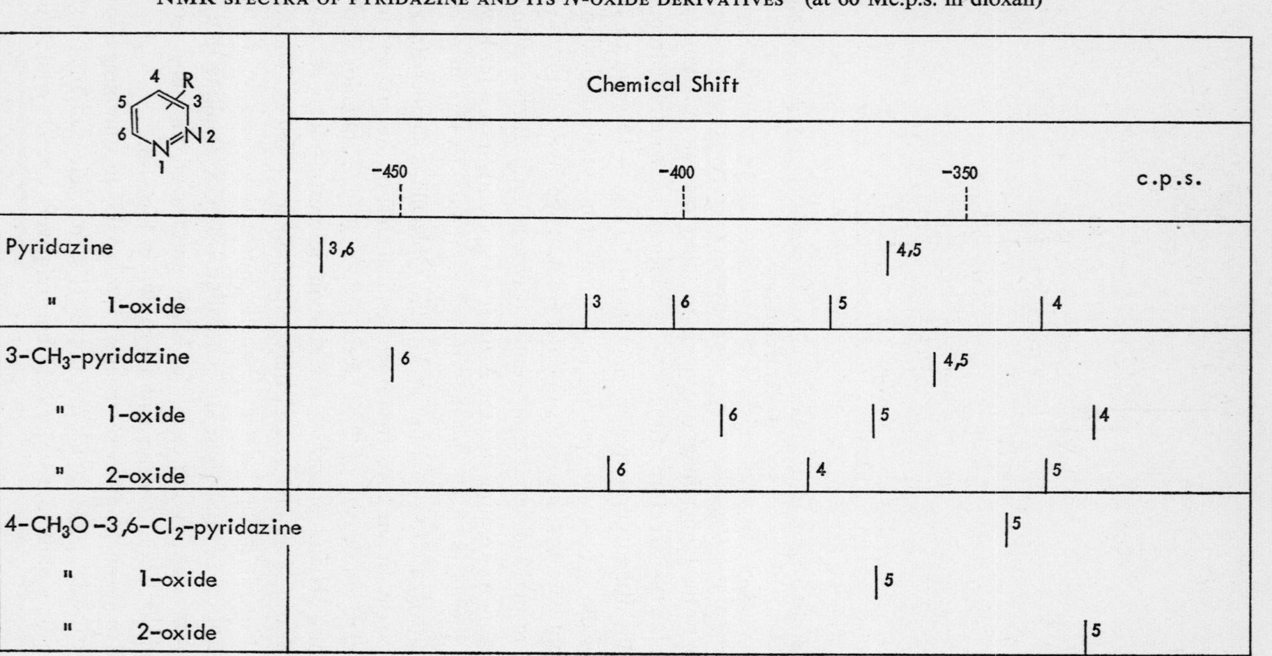

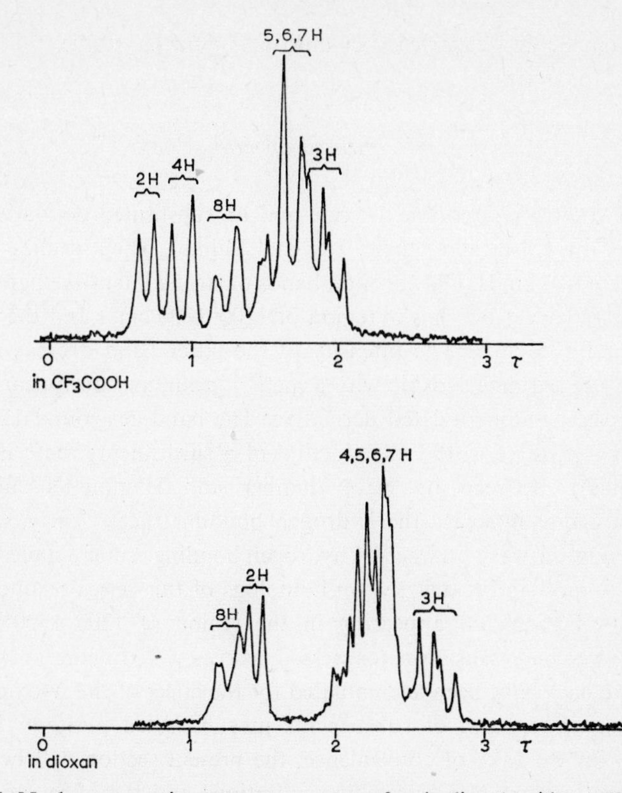

Fig. 4-6. Nuclear magnetic resonance spectra of quinoline 1-oxide at 60 Mc.p.s.

greater, as would be expected, while H_8 makes a higher shift. This is probably due to the disappearance of a large anisotropic effect of the *N*-oxide group by protonation. Such a large anisotropic effect is also expected on the *ortho* position* and therefore the electron density of the α-position in pyridine 1-oxide and pyridazine 1-oxide is not as positive as the chemical shift might indicate. This position (α), however, receives the greatest inductive effect of the *N*-oxide group and it seems impossible at present to evaluate correctly the magnitude of both the anisotropic and inductive effects of the *N*-oxide group.

As can be understood from the foregoing, quantitative studies on the electronic effect of the *N*-oxide group, especially on the neighboring protons,

* The magnetic anisotropy effect of the *N*-oxide group on the *ortho*-proton has been evaluated by Tôri[84] who gave the value of *ca.* −3 p.p.m.

and its magnetic and electric effect must await further studies, both experimental and theoretical.

4.7 Infrared Spectra

Infrared spectra of pyridine 1-oxide and its substituted derivatives were first studied by Costa and others, Ito, and Shindo. Their studies revealed that there are two kinds of absorption bands of strong intensity characteristic of pyridine 1-oxides; one in the region of 1200–1300 cm^{-1} and the other in the region of 835 cm^{-1}. The intensity of the latter band decreases by the substitution of pyridine 1-oxide with a methyl group and disappears almost entirely in the pentamethylated derivative. The band at around 1200–1300 cm^{-1} is very sensitive to the introduction of a substituent; there is a good proportionality between its wave number and Hammett's substituent constant σ, and it indicates the hydrogen bonding effect clearly, the band shifting to a lower wave number by hydrogen bonding. Later detailed studies, chiefly by Shindo and Katritzky, on homologs of this series resulted in the assignment of the band appearing in the region of 1200–1300 cm^{-1} to the N–O stretching vibration (expressed as $\nu_{N^+-O^-}$). Furthermore, some quantitative data have been accumulated for the effect of the *N*-oxide group on the absorption band of substituents in substituted pyridine 1-oxides. Therefore, for the sake of convenience, the present section will be divided into observations on N–O stretching vibrations and the effect of *N*-oxide group on the absorption of substituents.

4.7.1 N–O Stretching Vibration

The most characteristic point in the infrared spectra of the *N*-oxides of azabenzene heterocycles is that the absorption due to the N–O linkage and its stretching vibration appears as an absorption of strong intensity in the region of 1200–1300 cm^{-1}. Ito and Hata[87] assigned the band at 835 cm^{-1} (in chloroform) in pyridine 1-oxide to the N–O stretching vibration, but examination of various characteristics of these two absorption bands to date indicates that the band in the region of 1200–1300 cm^{-1} is the one due to the N–O stretching vibration[57a,88,89]

(a) Absorption Intensity

All aromatic *N*-oxide compounds exhibit absorptions of a strong intensity

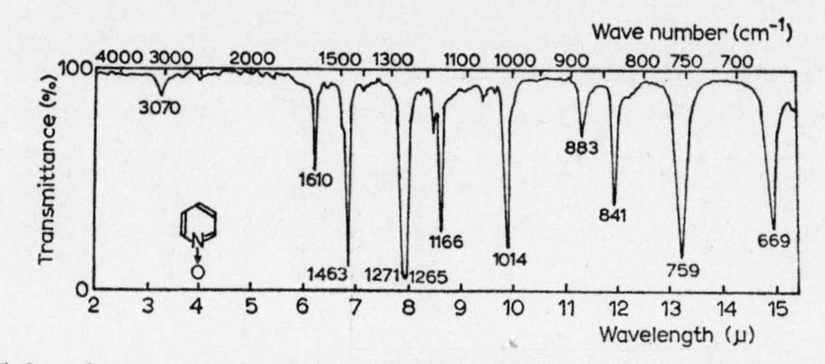

Fig. 4-7. Infrared spectrum of pyridine 1-oxide (650–1400 cm^{-1} in CS_2; 1400–5000 cm^{-1} in CCl_4).

in the region of 1200–1300 cm^{-1}. The infrared spectrum of pyridine 1-oxide measured by Shindo[57a] is given in Fig. 4-7.

This absorption band at 1265 cm^{-1} (ν_{N-O} in CS_2) is the strongest in the majority of spectra. This is naturally expected from the large polarity of the nitrogen–oxygen bond, since this vibration greatly changes the dipole moment. Shindo[57a] pointed out that there is a strong absorption at 1282 cm^{-1} in pentamethylpyridine 1-oxide, which indicates that this absorption band is independent of ring hydrogen. Consequently, Ito and Hata's assignment of this absorption to the C–H in-plane bending vibration is known to have been wrong.

(b) Solvent Effect of Methanol and Other Media

Shindo[90] examined the variation of infrared spectra of pyridine 1-oxide in dilute carbon disulfide solution during addition of a small amount of methanol and found that a marked change occurs only in the absorption at 1265 cm^{-1}. As shown in Fig. 4-8, the absorption band at 1265 cm^{-1} gradually disappears with increasing amount of methanol and a new band appears on the lower wave-number side, at around 1240 cm^{-1}. Shindo explained this phenomenon as the decreased contribution of (4-2) by the formation of a hydrogen bonding between the oxygen in the *N*-oxide group and methanol, resulting in the decreased double bond character of the nitrogen–oxygen bond and a shift of its stretching vibration to the lower wave-number side. Therefore, the absorption at 1240 cm^{-1} was assigned to the stretching vibration of the bonded N–O. At the same time, it was proved that the absorption band at 840 cm^{-1} does not show such a clear hydrogen-bonding effect.

On the other hand, the formation of a hydrogen bond by addition of

References p. 167

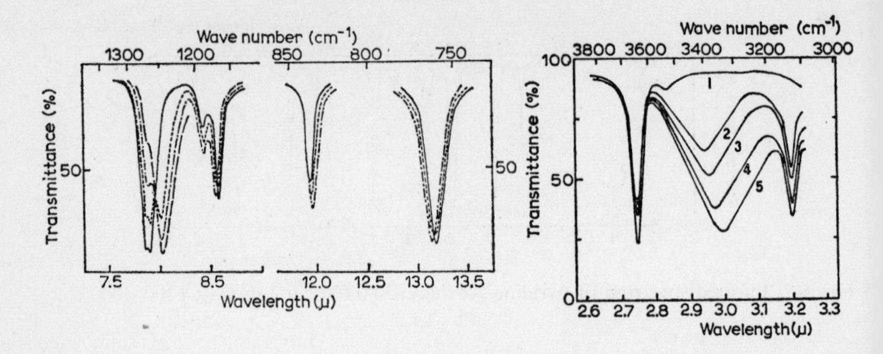

Fig. 4-8. Change in the infrared spectra of pyridine 1-oxide on addition of methanol. ———— 0.03 mole in CS_2; —·—·—·— + 0.3 mole MeOH; - - - - - - + 0.15 mole MeOH – – – – – + 0.6 mole MeOH.

Fig. 4-9. Absorption of bonded OH in the system (0.01 mole MeOH in CCl_4 + 0.02 mole substituted pyridine 1-oxide). 1, Reference; 2, 3-chloropyridine 1-oxide; 3, 4-chloropyridine 1-oxide; 4, pyridine 1-oxide; 5, 4-methylpyridine 1-oxide (LiF prism; cell thickness: 20.0 mm).

methanol to the *N*-oxide was clearly indicated by the absorption of ν_{OH} of methanol. As shown in Fig. 4-9, the stretching vibration of free hydroxyl in methanol, taken in dilute carbon tetrachloride solution to avoid intermolecular association, appears at 3650 cm^{-1}, but when pyridine 1-oxide is present in this solution, the stretching vibration of bonded hydroxyl of methanol appears as a wide band with a maximum at 3360 cm^{-1}.

The spectra shown in Fig. 4-9 seem to indicate, from the relationship between the shift (ν) and intensity (ε) of free and bonded OH in substituted pyridine 1-oxides under identical conditions, that the increase, in the electron-donating power of the substituent (Z) increases the hydrogen-bonding ability of the *N*-oxide group and, inversely, the increase in electron-withdrawing power of the substituent decreases the hydrogen-bonding ability. Shindo actually found that the wave number and intensity of the stretching vibrations of bonded OH are in linear relation with Hammett's substituent constant, σ (Ref. 11)*. This relationship also applies to 3-methylpyridine 1-oxide. As has already been stated in the case of the dissociation constant, this may be

* M. Tamres, S. Searles, E. M. Lighly, and D. W. Mohrman (*J. Am. Chem. Soc.*, 76 (1954) 3983) indicated that there is a linear relationship between basicity and absorption maximum of bonded OD by the bonding of substituted pyridine and CH_3OD.

TABLE 4-17

INFRARED ABSORPTION OF OXIMES, AMINE *N*-OXIDES, AND AROMATIC NITROSO AND NITRO COMPOUNDS

Compound	Absorption region (cm^{-1})	Reference
Simple oximes	930–960	93
Aliphatic tertiary amine *N*-oxides	950–970	94
Isatin oxime, quinone oxime	975–1015	95,96
Quinone monoxime K or Na salt	1200	96
Aromatic *N*-oxides	1200–1300	
Aromatic nitroso compounds	1260–1380	97
Aromatic nitro compounds	1300–1380 sym. 1500–1560 asym.	98

interpreted as corresponding to the difference in basicity of the *N*-oxide group by substitution.

Wiley and his group[89], and Katritzky and his group[91] indicated that the infrared spectrum of *N*-oxides in chloroform (ordinary ν_{C-H} of chloroform is at 3067 cm^{-1}) exhibits absorption of fair intensity (ε *ca.* 100) at around 2980 cm^{-1} and pointed out that this absorption is ν_{C-H} due to bonded $CHCl_3$ which had been shifted by hydrogen bonding (4-31) with the oxygen of the *N*-oxide group and which was no longer compensated by the solvent.

N
↓
O······$HCCl_3$

(4-31)

(c) Appropriateness of Absorption Region

Shindo[92] classified various kinds of N–O stretching vibrations according to their intensity and pointed out the appropriateness of the presence of stretching vibrations of N–O of aromatic *N*-oxide compounds in the region of 1200–1300 cm^{-1}.

As will be clear from the data in Table 4-17, the $\nu_{N \rightarrow O}$ frequency of aromatic *N*-oxides is higher than that expected from the pure single stretching mode of N–O. This can also be concluded from the calculation of bond order (1.5) for the $\geqslant N^{+}-O^{-}$ bond by the molecular orbital method carried out by Kubota (*cf. Section 4.9*). Such a fact may be explained as the contribution of canonical forms like (4-2) in pyridine 1-oxide. In fact, aliphatic amine *N*-

TABLE 4-18

N–O STRETCHING VIBRATION OF PYRIDINE 1-OXIDE[88,57a]

Condition	Wave number (cm^{-1})
Gas	1301
Nonpolar solvent	*ca.* 1260–1270
Chloroform	1248
Solid	1242
Methanol–carbon tetrachloride	1235

oxides, which do not have the contribution of such canonical forms, are known to show ν_{N-O} absorption at 950–970 cm^{-1} (Ref. 94). As will be shown in the section on ultraviolet absorption (*cf. Section 4.8.1*), the *N*-oxide group is very liable to change with changes in the solvent or the state of the *N*-oxide compound itself and, naturally, the position of its ν_{N-O} absorption shifts according to the conditions of measurement. Wave numbers of the ν_{N-O} absorption of pyridine 1-oxide under various conditions are given in Table 4-18.

(d) Correlation with Position and Nature of Substituents[57b]

The N–O stretching absorption of pyridine 1-oxides is closely related to the position and nature of ring substituents. As indicated in Table 4-19, the absorption shifts to a higher wave-number when the substituent (Z) is an electron-withdrawing group (W) and inversely to the lower wave-number when the substituent is an electron-donating group (D). This fact shows that the contribution of (4-2) and (4-3) type structures in each compound is affected by the electronic action of the substituent; the contribution of (4-11) precedes when Z is W, and that of (4-12) precedes the other when Z is D, and the double-bond character of the nitrogen–oxygen bond varies with these changes. Shindo[57b] plotted the Hammett's substituent constant (given in Table 4-19) against the N–O stretching vibration of substituted pyridine 1-oxides measured in dilute carbon disulfide solution and found that there is a good linear correlation between these two values, as shown in Fig. 4-10. A similar relation was found to hold for the values measured in Nujol[90].

Shindo pointed out that the majority of 3-substituted pyridine 1-oxides deviate from this linearity and assumed that the absorptions assigned to the N–O stretching frequency in these compounds contain a fair amount of frequencies other than that of ν_{N-O}. The N–O stretching frequency of pyridine 1-oxides, which shows good linearity (see Fig. 4-10), shifts to the lower wave-number side by 20–40 cm^{-1} on addition of methanol, as shown in the case of pyridine 1-oxide, while those that deviate from this linearity, such as the 3-methyl derivative, give a shift of ν_{N-O} to the lower wave number region of less than 15 cm^{-1}. This fact confirms the participation of modes other than the N–O stretching in the latter compounds. The same phenomenon has been pointed out by Katritzky and others[91,99,100].

It should be emphasized here that the N–O stretching frequency in 2- and

TABLE 4-19

N–O STRETCHING FREQUENCIES OF SUBSTITUTED PYRIDINE 1-OXIDES[57b]

Substituent	σ-Value	Diluted CS_2 solution (cm^{-1})	
		$\nu_{N\rightarrow O}$[a]	Inflexion or satellite band[b]
4-NO_2	0.778	1303	1285 w,sh
3-NO_2[c]	0.710	1298	1265 s
3-CH_3, 4-NO_2	0.709	1311	1258 m
3-CN[c]	0.678	1307	1289 s
4-CN[c]	0.628	1301	
4-$COOC_2H_5$	0.522	1267?	1297 m
4-$COCH_3$	0.516	1258	1292, 1299 m
3-$COOC_2H_5$	0.398	1302?	1284 m,sh
3-Br	0.391	1294	1263 vs, 1309 w,sh
3-Cl	0.373	1293	1265 vs, 1250, 1309 w,sh
3-$COCH_3$[c]	0.306	1302	1287 sh
4-Br	0.232	1271	
4-Cl	0.227	1269	
Nil	0.000	1265	1271 s sh
3-C_2H_5	−0.043	1276	
3-CH_3	−0.069	1285	1272 sh, 1309 m
3,5-$(CH_3)_2$	−0.138	1319	
4-C_2H_5	−0.151	1260	
4-CH_3	−0.170	1260	
4-OCH_3	−0.268	1240	1236 m
4-$OCH_2C_6H_5$	−0.416	1238	1225 w,sh
2-CH_3		1260	

[a] Intensity is very strong

[b] s: strong, m: medium, w: weak, sh: shoulder.

[c] Saturated solution in CS_2; cell thickness 5.0 mm.

Others as saturated or 0.2% solution in CS_2; cell thickness 1.0 mm.

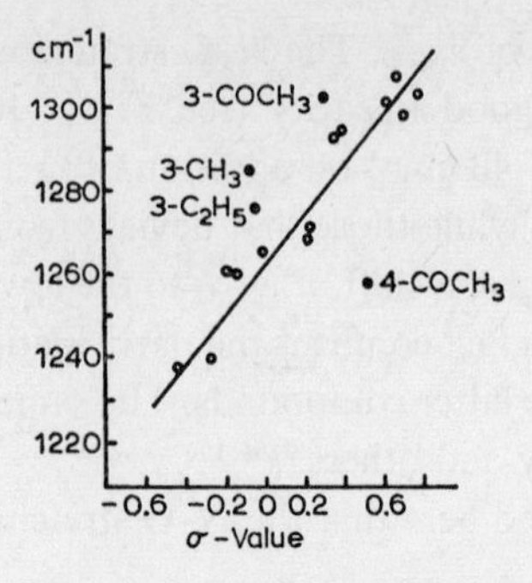

Fig. 4-10. N–O stretching frequency *vs.* Hammett's substituent constant in substituted pyridine 1-oxides.

4-methylpyridine 1-oxides is 5 cm^{-1} further towards lower wave-numbers than in pyridine 1-oxide, reflecting the electron-donating nature of the methyl group, but the absorptions in 3-methyl and 3,5-dimethyl derivatives are shifted to the higher wave-number side by 20 and 53 cm^{-1}, respectively. Such a behavior of 3-methylated derivatives seems inconsistent with the electron-donating nature of the methyl group, but the increase in basicity of the *N*-oxide group is confirmed if the hydrogen bonding with methanol (*cf.* Fig. 4-9) is evaluated by the inspection of $\nu_{OH(MeOH)}$, which shows good agreement with the Hammett's law. Consequently, even this point supports the view that the N–O stretching vibration in 3-substituted pyridine *N*-oxides is strongly coupled with other vibrational modes. Wiley and others[89], and Shindo[101] also pointed out that the N–O stretching vibration appears in a markedly higher wave-number region in 5-methylpyrimidine 1-oxide and 3-methylpyrazine 1-oxide (in both cases the methyl group is in position β to the *N*-oxide group).

Shindo[57a] showed that this relation can be utilized for the determination of the position of the substituent in alkyl-substituted pyridine 1-oxides.

(e) Absorption and Position of N–O Stretching Frequencies in N-Oxides of Nitrogen Heteroaromatics other than Pyridine

The N–O stretching frequencies and frequencies of the bands characteristic of the *N*-oxide group in *N*-oxides of other *N*-heteroaromatics are given in Table 4-20. In all these compounds, absorption bands are present in approximately the same region as that of ν_{N-O} in pyridine 1-oxides.

In general, the N–O stretching vibration in the *N*-oxides of azanaphthalene

TABLE 4-20

N–O STRETCHING VIBRATION OF *N*-OXIDE GROUP IN SOME *N*-HETEROAROMATICS

Compound	ν_{N-O} (cm^{-1})	Other bands characteristic of N–O mode (cm^{-1})	Ref.
Pyrazine 1-oxide	1350–1300	880–845, 870	100,101
Pyrazine 1,4-dioxide	1280–1250		102,103
Pyrimidine 1-oxide	1300–1200	870–845	89,101
Quinoline 1-oxide	1340–1280 [a]	*ca.* 1150	104,105
Isoquinoline 2-oxide	1260–1210 [a]	*ca.* 1350	

[a] The ν_{N-O} is split into two bands. There is evidently a strong coupling with other vibrational modes as no simple correlation can be found between the structure and frequency.

and its higher benzenoid homologs undergoes coupling with other vibrational modes, so that a clear-cut assignment, as in azabenzene *N*-oxides, here becomes difficult or almost impossible. Consequently, studies in this direction are not abundant. Clemo and others[105] reported on the infrared absorptions of the *N*-oxides of quinoline, phenazine, and their substituted derivatives, while Lüttke[106] described the infrared spectra of benzo[*c*]cinnoline dioxides.

The nature of the absorption bands in the region of 1200–1300 cm^{-1} discussed above shows the correctness of their attribution to the N–O stretching vibration. Recently, Kubota and others succeeded in assigning the vibrational spectra of pyridine 1-oxide and proved the correctness of foregoing assumptions[107]. These workers measured the infrared and Raman spectra of pyridine 1-oxide and its pentadeuterated compound and analyzed the result with fluorobenzene, whose fundamental vibrations have been well analyzed, as a model molecule. They concluded that, of the absorptions of pyridine 1-oxide, the bands at *ca.* 1270 and *ca.* 840 cm^{-1} correspond to the vibrations shown below.

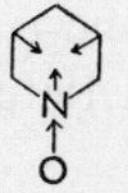

(4-32) Corresponds to the absorption at 1270 cm^{-1}

(4-33) Corresponds to the absorption at 840 cm^{-1}

As will be clear from the vibration indicated by (4-32), the absorption of the N–O stretching vibration is accompanied by a vibration of the carbon atom in the β position, which suggests that ν_{N-O} of the 3-substituent (Z) undergoes coupling with ν_{C-Z} at the 3-position; this coupling is considered to become stronger as the vibrational frequency of ν_{C-Z} approaches that of ν_{N-O}. Consequently, this conclusion of Kubota and others seems well to explain the fact that the majority of compounds which do not fall into the linearity between ν_{N-O} and σ values, as pointed out by Shindo in substituted pyridine 1-oxides, are 3-substituted compounds. This also explains the fact that, in spite of the electron-donating nature of alkyls, 3-alkylated compounds show a shift of the ν_{N-O} to a higher wave-length region.

4.7.2 Effect of the N-Oxide Group on the Absorption of Substituents[57b,108]

The position of the absorption of substituents in substituted pyridines[108] and their *N*-oxides, measured by Shindo, is listed in Table 4-21.

In any of the four instances, substituted pyridines and their *N*-oxides, especially the 2- and 4-substituted compounds, absorb in a much higher wave-number region than substituted benzenes, indicating the strong electron-withdrawing effect of the ring nitrogen. Shindo plotted these vibrational frequencies against the σ values of 0.93 (4-position, σ_{para}) and 0.62 (3-position, σ_{meta}), calculated by Jaffé[109] for the pyridine series from the rate of esterification of pyridine-carboxylic acids, and found a good linear relation between these values, as shown in Fig. 4-11.

Shindo calculated the σ-value of the 2-position, a value not determined before, from the data given in Table 4-12 and gave the value of σ_{ortho} as 1.02. On the other hand, 3-substituted compounds of the *N*-oxide series absorb in a markedly higher wave-number region and the fact confirms the strong electron-withdrawing effect of the *N*-oxide group in this position. 4-Substituted compounds adsorb in the lowest wave-number region among this series of hetero compounds, indicating that the electron-withdrawing effect of the *N*-oxide group is extremely small in this position. It was actually found that the $\nu_{C=O}$ and $\nu_{C\equiv N}$ of the esters show good linearity with σ_{para} of 0.25 and σ_{meta} of 1.18 for the *N*-oxide group, calculated from the data in Table 4-21.

In general, $\sigma_{meta}-\sigma_{para}$ is considered to show the size of the mesomeric effect[110] and the large difference of 0.93 in the above case was pointed out by Shindo[57b] as indicating the strong $+M$ effect of the *N*-oxide group, besides the strong $-I$ effect. When this value (0.93) is compared with the small value of 0.13 of $\sigma_{meta}-\sigma_{para} = 1.47–1.34$, calculated by Jaffé[56] in aque-

TABLE 4-21

ABSORPTION FREQUENCIES OF SOME DERIVATIVES OF PYRIDINE AND PYRIDINE 1-OXIDE

Compound	Stretching frequency (cm^{-1})		Compound	Stretching frequency (cm^{-1})
Acetyl derivatives	$\nu_{C=O}$ (CS_2)		Ethoxycarbonyl derivatives	$\nu_{C=O}$ (CS_2)
4-acetylpyridine 1-oxide[a]	1693		ethyl isonicotinate 1-oxide[a]	1727
3-acetylpyridine 1-oxide[b]	1706		ethyl nicotinate 1-oxide[a]	1737
4-acetylpyridine	1703		ethyl isonicotinate	1734
3-acetylpyridine	1697		ethyl nicotinate	1730
2-acetylpyridine	1704		ethyl picolinate	1722 1750
acetophenone	1692		ethyl benzoate	1724
	ν_{NO_2}	($CHCl_3$)		
Nitro derivatives	asym.	sym.	Cyano derivatives	$\nu_{C\equiv N}$ (CCl_4)
4-nitropyridine 1-oxide[c]	1531	1342	4-cyanopyridine 1-oxide[b]	2240
3-nitropyridine 1-oxide[c]	1539	1360	3-cyanopyridine 1-oxide[b]	2247
4-nitropyridine	1539	1355	4-cyanopyridine	2243
3-nitropyridine	1533	1355	3-cyanopyridine	2240
2-nitropyridine	1548	1355	2-cyanopyridine	2245
p-nitrotoluene	1517	1347	cyanobenzene	2237

[a] Approx. 0.3% solution; cell thickness, 1.0 mm.
[b] Saturated sol. in CS_2 or CCl_4; cell thickness, 5.0 mm.
[c] Approx. 3% solution; cell thickness, 0.1 mm.

References p. 167

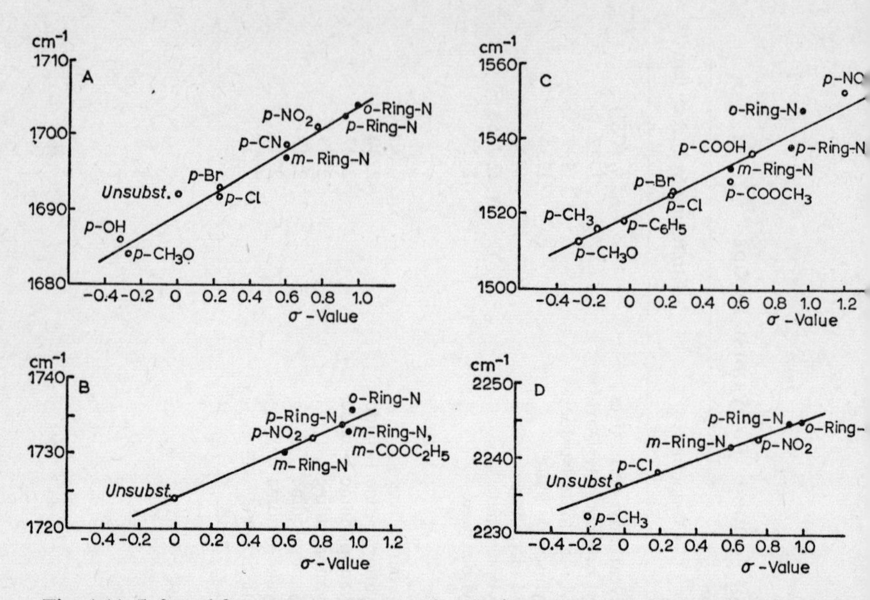

Fig. 4-11. Infrared frequency *vs.* Hammett's substituent constant for substituted benzene and pyridine derivatives. A, acetyl derivatives; B, ethoxycarbonyl derivatives; C, nitro derivatives; D, cyano derivatives; ○, substituted benzene derivatives; ●, substituted pyridine derivatives.

ous solution (*cf.* Table 4-13), it can be seen that the contribution of the (4-2) type to *N*-oxide compounds (*i.e.* pyridine *N*-oxide) in an aprotic solvent (carbon disulfide) is far larger than that in aqueous solution. Shindo measured the infrared absorptions of the foregoing compounds in methanol and in water, and pointed out from the position of absorption of substituents that the σ values (especially σ_{para} values) tend to approach Jaffé's value in these solvents[90]. In any case, the wave number of the absorption of substituents in substituted pyridines and their *N*-oxides in an aprotic solvent increases in the following order (where σ values of the N and *N*-oxide group are given in parentheses under each compound):

4-Substd. *N*-oxide < 3-substd. < 4-substd. < 2-substd. < 3-subst. *N*-oxide
(σ_{para} 0.25) (σ_{meta} 0.62)(σ_{para} 0.93)(σ_{ortho} 1.02) (σ_{meta} 1.18)

In a protic solvent, the order of the first two is reversed and the order becomes 3-pyridine < 4-substd. *N*-oxide ∼ 4-substd. < 3-substd. *N*-oxide
(σ_{meta} 0.62) (σ_{para} 0.93)

This order may be considered to show the degree of the decrease of electron density in the substituted position of aromatic compounds. It has already been revealed by Jaffé that the σ value is directly related to the electron density of the carbon atoms in the ring[110]. Similar observations have been made by Tani and Fukushima[112], and by Katritzky[111] on cyanopyridine 1-oxides from the measurement of their molecular extinction coefficients, which are listed in Table 4-22.

The fact that the intensity of the C≡N stretching vibration of 4-cyanopyridine 1-oxide is extremely high indicates the large effect of the electron-donating nature of the nitrogen–oxygen bond (shown by the contribution of the (4-2) structure) on the 4-position. On the other hand, the very low intensity of this vibration in 2-cyanopyridine 1-oxide is due presumably to the fact that the $+M$ effect expected from the *N*-oxide group by the contribution of the (4-2) structure on the 2-position is counteracted by the $-I$ effect of the adjacent N^+ and the net electron density in the 2-position becomes lower than that in the 4-position. Katritzky and Jones[113] compared the N–H stretching vibrations of 2-aminopyridine and its *N*-oxide, and of pyridinium with the data on the same substituted pyridines measured by Goulden[114] and pointed out that the ν_{N-H} frequencies increase in the order of aniline, 3-substd., 4-

TABLE 4-22

C≡N STRETCHING VIBRATIONS AND THEIR INTENSITY

Substituted pyridine	A. R. Katritzky[111] $v_{C\equiv N}$	A. R. Katritzky[111] ε_A	H. Tani, K. Fukushima[112] $v_{C\equiv N}$	H. Tani, K. Fukushima[112] E_{max} (l/mol. cm)
Pyridine 1-oxide				
4-cyano	2230	72	2235	123
3-cyano	2243	14	{2241 2249	{21.0 18.7
2-cyano	2241	14	2243	25.9
Pyridine				
4-cyano	2238	18	2241	33.5
3-cyano	2235	38	2236	65.1
2-cyano	2240	11	{2237 2245	{19.5 18.7
Cyanobenzene	2233	85	2231	122
Solvent	CHCl$_3$		CHCl$_3$	
Monochromator of infrared spectrometer	NaCl prism		gratings	

References p. 167

and 2-substd. pyridines, 2-substd. pyridine 1-oxide, and 2-pyridinium, and he interpreted this as indicating the increasing order of the electron-accepting power of each ring. However, among the compounds mentioned above, 2-aminopyridine 1-oxide is considered to have an intramolecular hydrogen bonding as indicated by (4-34) and these data do not seem to help in evaluating the net effect of the *N*-oxide group on the 2-position.

The readers are further referred to the reviews of Katritzky and Ambler[115], and of Shindo[92] for more detailed data.

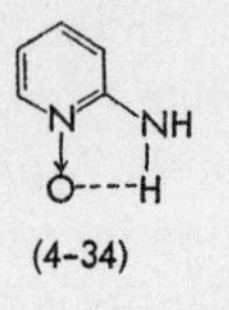

(4-34)

4.8 Ultraviolet Absorption Spectra

4.8.1 Characteristics of the Ultraviolet Spectra of Aromatic N-Oxides

In the ultraviolet spectra in an aprotic solvent, *N*-oxidation of ring nitrogen in basic aromatic heterocycles such as pyridine, quinoline, and isoquinoline produces a considerable red shift and hyperchromic effect on the π–π^* band[116–122,35,41]. On the other hand, in a hydroxylic solvent, the π–π^* bands of *N*-oxide compounds undergo a blue shift mainly due to the hydrogen-bonding effect, and in strong acid solution, in which *N*-oxides exist as their conjugate acids, the spectra agree approximately with those of the parent heterocycles[116–123,35,41].

Taking pyridine 1-oxide as an example, the π–π^* band of this compound in the longest wave-length region has been identified as that appearing at 280.5 mμ (log ε 4.14) in ether. Compared to the corresponding band of pyridine (254 mμ, log ε 3.39, in H_2O), the intensity of this band is much stronger and its wave length is much longer in pyridine 1-oxide. Although this band of pyridine, as would be expected for the usual *N*-heterocycles, shows a slight increase in its intensity (255 mμ, log ε 3.65, in 1N HCl) by its N-protonation, the corresponding band in pyridine 1-oxide shows a blue shift, together with a hypochromic shift, in the order of ether, water, and 1N hydrochloric acid. As would be understood from the pKa value of pyridine 1-oxide (*cf. Section 4.5*), this compound exists as its conjugate acid (1-hydroxypyridinium ion) in 1N hydrochloric acid solution and its absorption bands are similar to those of the protonated pyridine itself.

TABLE 4-23

ULTRAVIOLET ABSORPTION SPECTRA OF PYRIDINE 1-OXIDE

Solvent	Dielectric constant	$\pi \rightarrow \pi^*$ band of pyridine 1-oxide[116] $\lambda_{max}(m\mu)$	log ε	π–π^* band of pyridine[124] $\lambda_{max}(m\mu)$	log ε or (ε)
Ether	4.335	280.5	4.14		
Heptane	2.023			252	(2090)
Ethanol	25.8	265.5	4.10	253	(3600)
H_2O	81	255	4.10	254.5[116], 253[124]	3.39[116] (3600)
0.1N NaOH		255	4.09		
1N HCl		255	3.55	255	3.65[116]
98% H_2SO_4				252.5	(5200)

Ultraviolet spectra of pyridine 1-oxide in various solvents, reported by Kubota[116], are summarized in Fig. 4-12 and Table 4-23, together with those of pyridine[116,124].

As stated already in the beginning of this Chapter (see Section *4.1*), the ground state of pyridine 1-oxide is presumably a resonance hybrid consisting of three structures (4-1, 4-2, and 4-3). The ground state of this compound, involving such highly polar structures as (4-1), (4-2), and (4-3), is undoubtedly stabilized considerably by an electrostatic interaction due to the polarity of the solvents and by hydrogen bonding in hydroxylic solvents. Kubota and Yamakawa[125] have shown that, in non-hydrogen bonding solvents, the blue shift of the π–π^* band occurs with increasing polarity of the solvent or, to be more exact, the larger the dipole–dipole interaction between the *N*-oxide and the solvent, the larger will be the blue shift of the band in question. Kubota[117,120,126,127] also studied the effect of hydrogen bonding on the π–π^* band of aromatic *N*-oxides in a ternary system consisting of the *N*-oxide compound

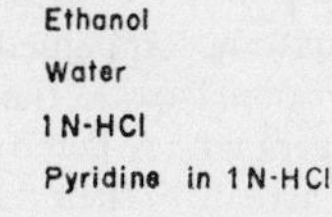

Fig. 4-12. Ultraviolet absorption spectra of pyridine 1-oxide[116].

References p. 167

and a suitable proton donor in an aprotic solvent, and confirmed that a marked blue shift† of the π–π^* bands is caused by hydrogen bonding. This shift of the π–π^* band indicates that the π^* level is considerably less stabilized by polar solvents and the ability to undergo hydrogen bonding is much more reduced at the excited π^* level than in the ground state. Thus, in the π–π^* state, the structure (4-2) (and related structures) should be more important and, in this state, the oxygen atom loses much or most of its charge. In this sense, the spectrum of pyridine 1-oxide may be classed as an intramolecular charge-transfer spectrum[41].

Kubota and Miyazaki[41,128] examined the spectra of many aromatic *N*-oxides in heptane, ethanol, water, and so on, and showed that the π–π^* band shifts to the blue with increasing polarity of the solvent used, and pointed out that this is one of the characteristics of aromatic *N*-oxides††. A similar trend to a blue shift due to such solvent changes has also been observed in the fluorescence spectra of aromatic *N*-oxides[125,127,128].†††

In addition to the foregoing π–π^* band, n–π^* transition, presumably due to the excitation of $2p$ electrons of oxygen to the π^* level of the molecule, has been identified in a longer wave-length region at 330 mμ in the vapor spectrum of pyridine 1-oxide by Ito and Hata[129]. The fact that this band is comparatively weak and is accompanied by fine vibrational structures suggests the correctness of this assignment. Similar n–π^* bands have been identified for 2- and 3-picoline 1-oxides in their vapor spectra[130,131]. It was also found that the symmetry of these n–π^* transitions is incidentally the same as that of the n–π^* transition of pyridine[132]. These n–π^* bands§ are also observed in a solution of non-polar solvents and disappear completely in hydroxylic solvents[129]. Such a fact is also in agreement with this assignment. As an example, ultraviolet spectra, together with their band analysis, for pyridine 1-oxide, measured by Ito and Hata[129,133], and for 4-methylpyrimi-

† Generally speaking, the π–π^* bands of ordinary aromatic or hetero-aromatic compounds shift to the red as the dielectric constant of the solvent increases because, in such aromatic compounds, the corresponding excited states are much more polar than their ground states.
†† Blue shifts are usually accompanied by hypochromic shifts but in some exceptional cases, such as in 4-bromopyridine 1-oxide, the reverse phenomenon has been found (*cf.* Table 4-28).
††† *N*-Oxidation of aromatic *N*-heterocycles, such as quinoline, isoquinoline, acridine, etc., usually results in the increase of their fluorescence ability but not in pyridines. This increased fluorescence is partly explained as being due to smaller interactions between π–π^* and n–π^* transitions in the *N*-oxides than in their parent heterocycles.
§ The *N*-oxides of azanaphthalene and its higher benzenoid homologs show the n–π^* bands in a longer wave-length region than those of azabenzene *N*-oxides, but in most cases these are hidden in the π–π^* bands and cannot be observed.

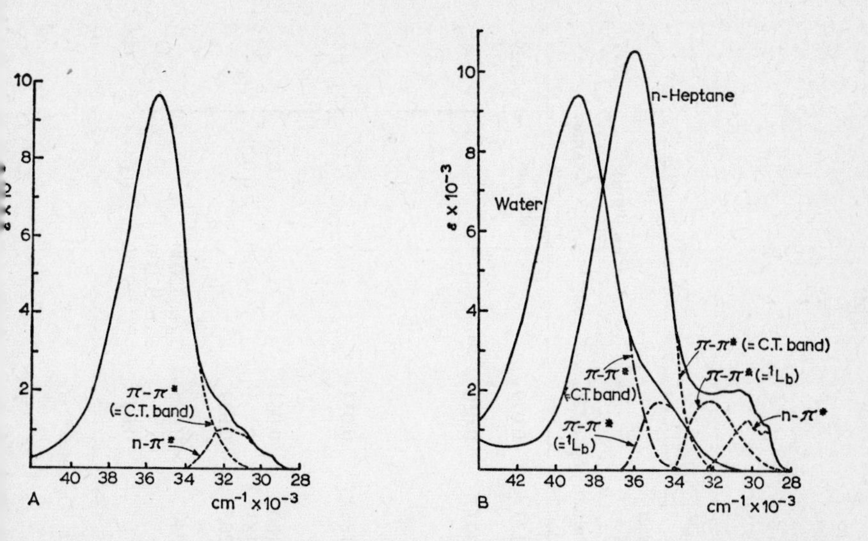

Fig. 4-13. Ultraviolet absorption spectra and band analysis. A, pyridine 1-oxide in hexane solution[133]; B, 4-methylpyrimidine 1-oxide in heptane and in water[13]. (Reprinted from the *Bulletin of the Chemical Society of Japan*, by permission of the Chemical Society of Japan).

dine 1-oxide, measured by Kubota[13], are shown in Fig. 4-13. In these spectra, the solid curves are the observed absorption lines and the dotted lines refer to the analyzed π–π^*, n–π^*, and ^1L_b bands (*v.i.*). Hata and Ito[129,133] found, after analysis of the data thus obtained, that the ratio of oscillator strength (f) between n–π^* and π–π^* bands of pyridine 1-oxide derivatives, *i.e.* the value of $f(n\text{–}\pi^*)/f(\pi\text{–}\pi^*)$, is around 0.05 and does not exceed 0.1. This also gives supporting evidence for the assignment with respect to the n–π^* and π–π^* transitions† mentioned above.

Similar data obtained so far for azabenzene *N*-oxides are summarized in Table 4-24. It will be seen from this table that, in the spectra of 4-bromopyridine 1-oxide and 4-methylpyrimidine 1-oxide, there are other bands, besides those assigned to the n–π^* and π–π^* transitions, of comparatively weak intensity and which do not disappear in a protic solvent. Kubota[13] assigned these bands as ^1L_b bands (one of the π–π^* bands, which will be explained later). Corresponding bands have not been observed in the spectra

† While the f-values for the n–π^* and π–π^* transitions are of considerably different order for different compounds, the ratio of the two f-values is in the same order of 10^{-2} for pyridine[134a], benzaldehyde, acetophenone, and p-benzoquinone[134b].

TABLE 4-24

π–π^* BANDS OF AZABENZENE *N*-OXIDES

Compound	Solvent	$\pi \to \pi^*$ band							Ref.
				Charge transfer band (=C.T.)			(^1L_b band)		
		cm⁻¹	*f*-value	cm⁻¹	*f*-value	$f_{n\to\pi^*}/f_{\pi\to\pi^*}$	cm⁻¹	*f*-value	
Pyridine 1-oxide	hexane	~32000	0.012	35520	0.173	0.079			129
Pyrazine *N*,*N*-dioxide	dioxane	27250	0.001	30900	0.303	0.050			13
		25800	0.014						
			0.015						
	water			32670					
Pyrazine mono-*N*-oxide	heptane	34050							13
		33350	0.006	36500	0.186	0.033			
		32520							
	water			37880					
4-Methylpyrimidine 1-oxide	heptane	~30000	0.011	35970	0.180	0.064	32180	0.020	13
	water			38820	0.183		34750	0.020	
4-Chloropyridine 1-oxide	CCl_4	*ca.* 30000	0.0059	34602	0.205	0.029			133
	water			39216					
4-Bromopyridine 1-oxide	CCl_4	*ca.* 30000	0.0063	34364	0.249	0.025	*ca.* 32000[a]		133
	water			37593			*ca.* 32000[a]		
Pyridine	isoöctane	*ca.* 37000	0.0030			0.073	39750	0.041	134[a]

[a] No assignment was given in the original paper, but it could be assigned as an ^1L_b band[13].

of other *N*-oxides listed in Table 4-24, presumably because of overlapping of these with the charge transfer bands of much higher intensity. As will be seen from Table 4-24 (the spectral data for pyridine given in this table are those measured in vapor phase, so that the band position for the ^1L_b state is in a shorter wavelength than that in organic solvents), the ^1L_b band of pyridine appears at 256 mμ with $\varepsilon_{max} = 2.66 \times 10^3$ in an aqueous solvent, while the conjugate acid of pyridine 1-oxide absorbs at 257 mμ ($\varepsilon = 2.9 \times 10^3$) and 217 m$\mu$ ($\varepsilon = 4.9 \times 10^3$). In addition, these two spectra are similar. The spectra of the conjugate acid of pyridine 1-oxide would be interpreted as follows. The protonation at the oxygen atom of the *N*-oxide decreases the intramolecular charge transfer from the oxygen atom to the ring system so that the charge transfer spectrum discussed above now disappears, or, in other words, the band shifts to a much shorter wavelength. In the same wavelength region as that of the ^1L_b band of pyridine, a band corresponding to ^1L_b for pyridine 1-oxide in its conjugate acid form appears which had originally been hidden under the charge transfer band of pyridine 1-oxide.

Accordingly, in the excited state of the conjugate acids of aromatic *N*-oxides, the contribution of (4-9) and its related structures should be much larger than in the ground state. In this sense, similarity between the spectrum of the conjugate acid of *N*-oxides and that of protonated parent heterocycles seems to be theoretically reasonable. These discussions on the spectra of azabenzene *N*-oxides could be extended to those of azanaphthalene and its higher benzenoid derivatives. However, it could be expected that the larger the conjugated system in parent heterocycles, the smaller will be the charge transfer character pertinent to the *N*-oxide group. Thus, the ultraviolet spectra of azanaphthalene *N*-oxides and their higher benzenoid homologs, even those of the *N*-oxides of free bases, could be interpreted by considering the *N*-oxide group as a simple substituent (*v.i.*). Fig. 4-14 shows the result of calculations made by Kubota[125,135] on the π-electron distributions at the highest occupied and lowest vacant molecular orbitals for pyridine 1-oxide and acridine 10-oxide.

Considering the fact that the π–π^* band corresponds to the one-electron transition from the highest occupied to the lowest vacant molecular orbital, Kubota's calculation given in Fig. 4-14 clearly shows that the larger the parent heterocycles, the smaller will be the electron migration from oxygen to the ring in the corresponding transition. In this sense, these data seem to support the foregoing discussions.

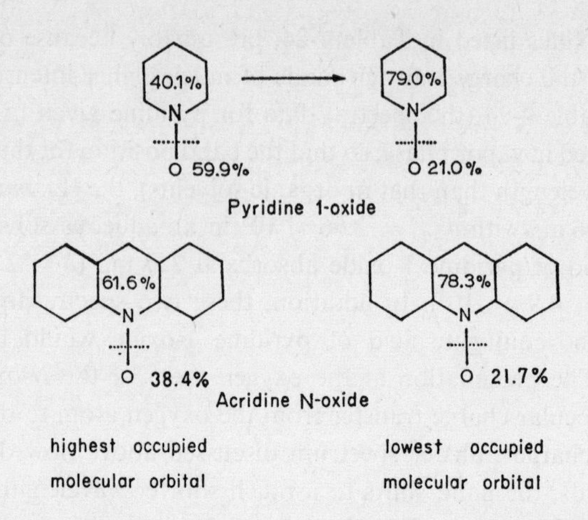

Fig. 4-14. Distribution of one electron. (Reprinted from the *Bulletin of the Chemical Society of Japan*, by permission of the Chemical Society of Japan.)

4.8.2 Assignment and Interpretation of π–π^* Bands in Aromatic N-Oxide Compounds

The free-electron model has been applied to the classification of the π–π^* absorption spectra of aromatic hydrocarbons by Platt[136a]. Various nomenclature systems proposed by Platt and others for the classification of aromatic absorption spectra are summarized in Table 4-25, using benzene and naphthalene as examples. Each band shown in this Table is due to the transition polarized along mutually perpendicular axes of the molecule, corresponding to the classical fundamental electrical oscillators suggested originally by Lewis, Calvin, and Bigeleisen[136b].

The π–π^* bands of benzenoid hydrocarbons, their aza homologs, and their conjugate acids are presented in Table 4-26. Table 4-27 shows similar data for representative aromatic *N*-oxide compounds measured by Kubota[135,141], who also made band assignments included in the Table. As an illustration, the whole pattern of absorption spectra of some basic aromatic N-oxides[41] is presented in Fig. 4-15. Comparison of the data in Tables 4-26 and 4-27 will show the marked difference in the spectra of the two systems, the *N*-oxides and their parent heterocycles. This difference is possibly due to the mutual interaction of the π-electrons between the oxygen atom and the ring system in the *N*-oxide compounds.

TABLE 4-25

NOMENCLATURE OF BENZENE AND NAPHTHALENE BAND SPECTRA

Transition			Polarization	
Platt[136a]	Clar[137]	Doub and Vandenbelt[138]		Wave length
^1L_b	*a*	Secondary	Longitudinal(X)	Longer
^1L_a	*para*	Primary	Transversal(Y)	↑
^1B_b	*β*	Second primary	Longitudinal(X)	Shorter

(Transverse) Y

Longitudinal

X

This strong interaction between the *N*-oxide group and the ring *π*-system is typical of aromatic *N*-oxides and is not (or almost not) found in aliphatic *N*-oxides. Accordingly, the absorption spectrum of tribenzylamine *N*-oxide is similar to that of tribenzylamine[50], and that of dimethylaniline *N*-oxide[35,122] is similar to that of the protonated parent base (conjugate acid of dimethylaniline). Thus, dimethylaniline *N*-oxide absorbs the light in a comparatively shorter wavelength region than dimethylaniline. However, as has already been discussed, the spectra of the conjugate acid of aromatic *N*-oxides are quite similar to those of the conjugate acids of the corresponding parent bases.

To change the aromatic *N*-oxides into their conjugate acids, Jaffé[47] recommended the use of a mixture of equal volumes of water and sulfuric acid (*ca.* 62% sulfuric acid solution) as a spectroscopic solvent, after detailed examination of the pKa values of various pyridine 1-oxide derivatives. Although this condition is sufficient to protonate the *N*-oxide function without exception, it has been confirmed by spectrometry that the amino function in 2- and 4-aminopyridine 1-oxides is not protonated in this medium (*cf. Section 4.5*).

As stated above, in the spectra of the azanaphthalene *N*-oxide series the effect of *N*-oxidation seems to be the same as that produced by the introduction of a substituent having lone-pair electrons, such as NH_2, OR, etc. into the position of nitrogen (*cf.* Tables 4-26 and 4-27). In the case of quinoline 1-oxide, in which *N*-oxidation stabilizes transition along the y-axis of the

TABLE 4-26

π–π^* ABSORPTION BANDS

Compounds	Band assignment						Ref.
	^1L_b		^1L_a		^1B_b		
	mμ	log ε	mμ	log ε	mμ	log ε	
	263	2.34	280	3.84			136
N	251	3.30	192	3.80	175	4.90	139
⊕ N H	255.7	3.72					140
	312	2.45	289	3.97	220	5.12	136
N	313	3.37	270	3.59	225	4.48	139
⊕ N H	313	3.80		*a*	233	4.54	139
N	317	3.49	266	3.61	217	4.57	139
⊕ NH	331	3.62	273.5	3.29	227.5	4.57	139
		b	379	3.95	256	5.26	136
N		*b*	356	4.00	248	5.30	139
⊕ N H	354.6 337.8	4.32 3.99	424.1 403 385.1	3.24 3.49 3.49	255	5.09	135,141

[a] ^1L_a band of quinoline shifts to the red with decreasing pH and is hidden behind the longer wave-length transition (^1L_b band).

[b] ^1L_b band is not observed in these compounds and is probably submerged under the ^1L_a band.

TABLE 4-27

ULTRAVIOLET SPECTRA OF REPRESENTATIVE AROMATIC *N*-OXIDE COMPOUNDS

Compound	Solvent	Absorption mμ	Absorption log ε	Band assignment
Pyridine 1-oxide				
Free base	Heptane	312.5~	3.44	n–π^*
		281.5	4.16	C.T. + (^1L_b)
		220.8	4.16	*a*
		216.5	4.25	
		214.0	4.20	
	dehyd. EtOH	265.5	4.10	C.T. + (^1L_b)
		212.0	4.24	*a*
	H_2O	253.8	4.09	C.T. + (^1L_b)
		206.0	4.29	*a*
Conjugate acid	2N HCl	258.1	3.51	^1L_b
		217.0	3.69	^1L_b
Quinoline 1-oxide				
Free base	Heptane	360.1	3.87	^1L_a + ^1L_b
		350.0	3.98	
		335.0	3.92	
		257.1	4.06	^1B_b
		247.0	4.21	
	dehyd. EtOH	340.0~	3.83	^1L_a + ^1L_b
		326.8	3.90	
		323.6	3.89	
		317.5	3.88	
		245.1	4.03	^1B_b
	H_2O	335.6	3.78	^1L_a + ^1L_b
		320.0	3.86	
		312.0	3.88	
		304.0	3.81	
		229.9	4.76	^1B_b
Conjugate acid	2N HCl	335.6~	3.42	
		320.5	3.78	^1L_a + ^1L_b
		308.6		
		233.1	4.61	^1B_b

[a] Actually, calculation by the ASMO–CI method under the condition of a composite system $\left(\text{N}^{-\delta+} + \text{O}^{\delta-}\right)$ shows that the band appearing in this region seems to have a character of $^1B_1$, whose symmetry is the same as those of ^1L_b and ^1B_b bands, and that another band, $^1A_1$, having the same symmetry as that of an ^1L_a band is expected in a much shorter wave-length region than that of $^1B_1$. It is also noted that the symmetry of the C.T. band and the ^1L_a band is the same, so that the ^1L_a band is also mixed in the C.T. band although the contribution from the former is much smaller than that from the latter[135].

References p. 167 *(Continued overleaf)*

TABLE 4-27 (*continued*)

Compound	Solvent	Absorption mμ	Absorption log ε	Band assignment
Isoquinoline 2-oxide				
Free base	Heptane	388.5	3.24	^1L_b
		381.8	3.08	
		375	3.13	
		367.5	3.25	
		356	3.11	
		349	3.07	
		337.5	2.99	
		314.5	4.23	^1L_a
		301.5	4.14	
		291	3.89	
		267	4.08	^1B_b
		260	4.16	
		256.5	4.12	
	dehyd. EtOH	342.6	2.94	^1L_b
		342.6	3.03	
		309.0	3.96	^1L_a
		306	3.96	
		296.5	4.06	
		288	3.98	
		257	4.47	^1B_b
	H_2O	336	2.89	^1L_b
		325	3.08	
		301.6	3.79	^1L_a
		294	3.94	
		286	3.91	
		257	4.47	^1B_b
Conjugate acid	40% H_2SO_4	331	3.58	^1L_b
		319.5	3.49	
		280	3.48	^1L_a
		235	4.71	^1B_b

parent heterocycle considerably (*cf.* Table 4-25), the red shift of the ^1L_a band is larger than that of the ^1L_b band. Accordingly, as will be seen in Tables 4-26 and 4-27, the spectrum of this *N*-oxide has a broad absorption band in the longer wavelength region. However, since the *N*-oxidation of isoquinoline produces a comparatively large red shift of the transition along the longitudinal direction (corresponding to the x-axis in Table 4-25), isoquino-

TABLE 4-27 (*continued*)

Compound	Solvent	Absorption mμ	Absorption log ε	Band assignment
Acridine *N*-oxide				
Free base	Heptane	452.5	4.05	^1L_a
		444.4	3.89	
		425.5	3.96	
		420	3.82	
		408	3.64	
		403	3.66	
		384.6	3.28	^1L_b
		367.6	3.05	
		275.5	4.75	^1B_b
	dehyd. EtOH	440.5	3.82	^1L_a
		416.7	3.92	
		397	3.75	
		372	3.52	^1L_b
		354.6	3.26	
		263.9	4.93	^1B_b
	H_2O	434.8	3.69	^1L_a
		413.2	3.86	
		393.7	3.74	
		367.6	3.70	^1L_b
		350.9	3.45	
		259.1	4.99	^1B_b
Conjugate acid	2N HCl	424.1	3.24	^1L_a
		403.2	3.49	
		385.1	3.49	
		354.6	4.32	^1L_b
		337.8	3.99	
		255.0	5.09	^1B_b

line 2-oxide exhibits two separate bands in its spectrum (*cf.* Tables 4-26 and 4-27).

The spectra of dihydrocinchonine and its 1-oxide, measured by Ochiai and Ishikawa[142], are given in Fig. 4-16. These spectra prove the correctness of assigning the 1-oxide structure and definitely exclude the alternate structure in which the nitrogen in the quinuclidine system has been oxygenated.

References p. 167

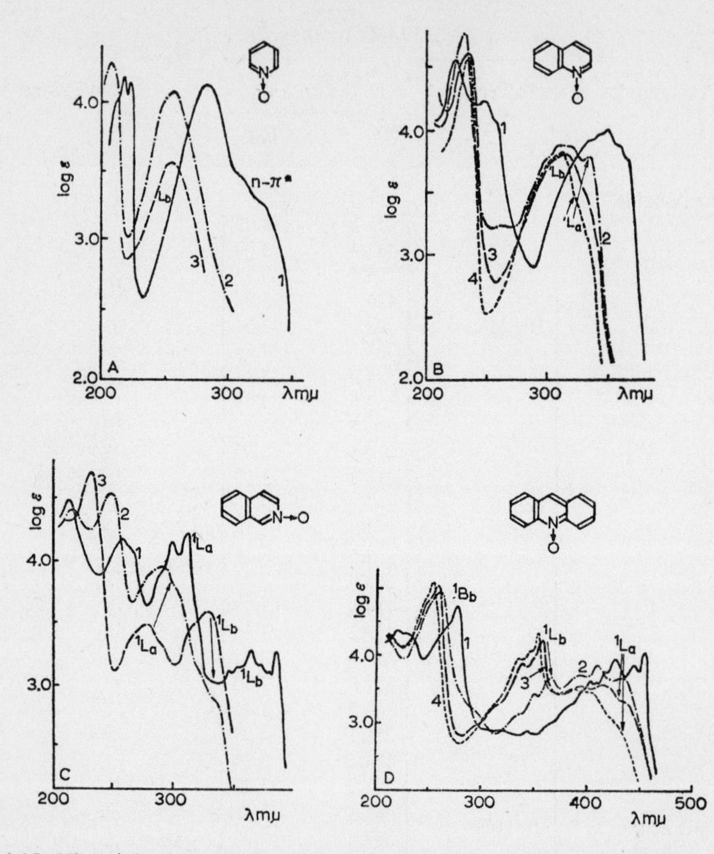

Fig. 4-15. Ultraviolet spectra of representative aromatic *N*-oxides. A, pyridine 1-oxide (1, heptane; 2, H_2O; 3, $\geqslant N^+–OH$); B, quinoline 1-oxide (1, heptane; 2, H_2O; 3, 2 N HCl; 4, quinoline in 0.1 N H_2SO_4); C, isoquinoline 2-oxide (1, heptane; 2, H_2O; 3, 40% H_2SO_4); D, acridine 10-oxide (1, heptane; 2, H_2O; 3, 2 N HCl; 4, acridine in 0.1 N H_2SO_4).

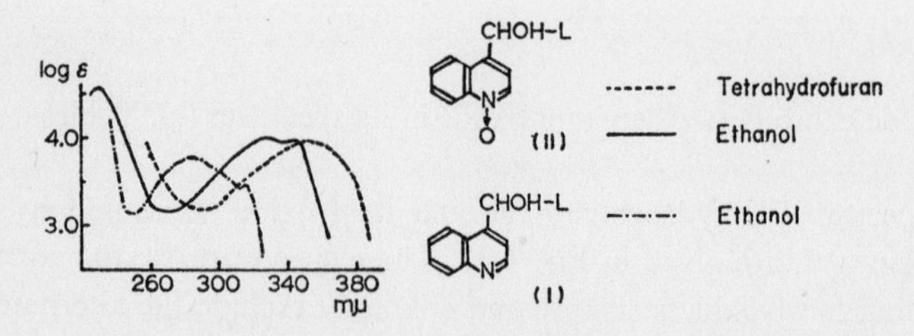

Fig. 4-16. Ultraviolet spectra of dihydrocinchonine and its *N*-oxide.

4.8.3 Ultraviolet Spectra of Substituted Pyridine 1-Oxides

Tables 4-28 to 4-30 summarize the ultraviolet absorption spectra of pyridine 1-oxide derivatives reported to date. Except where otherwise stated, these spectra were measured in an aqueous medium. In general, pyridine compounds, under appropriate conditions, exist in any one of the four distinct charged species; the neutral form, cation, anion, and the zwitter ion; the first three denoted in these Tables as N, C, and A, although in some cases N also covers the zwitter ion. In pyridine 1-oxides having a basic substituent, such as NH_2 and NR_2, the first protonation is known to occur at the *N*-oxide function (*cf. Section 4.5*).

The second protonation (to the amino group) is much retarded in those compounds in which the amino group is present in the 2- or 4-position of the pyridine ring, due to the strong electron-withdrawing power of the N^+–OH group ($\sigma_{para} = +3.9$). Thus, 2- and 4-aminopyridine 1-oxides do not form the di-cation even in 62% sulfuric acid solution (*cf.* Table 4-28). Formation of a di-cation by these amino derivatives is effected by the use of 95% sulfuric acid as a solvent, in which the spectra of 2- and 4-aminopyridine 1-oxides almost coincide with that of the conjugate acid of pyridine. Compounds with an amino group in 3-position, such as 3-aminopyridine 1-oxide, are readily converted to their di-cations in 62% sulfuric acid solution and such properties are actually observed in the amino derivatives of azanaphthalene *N*-oxides and their higher benzenoid homologs.

When the spectra of free *N*-oxides measured by different workers are to be compared, solvation effects due to the solvents used (caused mainly by hydrogen bonding) should always be kept in mind. Generally, the longest wavelength band of the π–π^* bands of free *N*-oxide, especially in azabenzene *N*-oxide derivatives, is quite susceptible to the solvent change and shifts to the blue as the polarity of the solvent increases, mainly due to hydrogen bonding, as would be expected from the charge transfer character of the band concerned. This blue shift is accompanied, in the majority of cases, by appreciable hypochromic shift. In rare cases, such as in 4-bromopyridine 1-oxide, the blue shift accompanies hyperchromic shift (*cf.* Table 4-28).

As will be understood from the remarks made so far, the spectra of substituted pyridine 1-oxides should not be compared with those of the corresponding substituted pyridines. One of the reasons for this is the consideration that in a comparatively small conjugate system, such as pyridine, conversion of the ring nitrogen to a highly polar N^+–O^- group causes too

TABLE 4-28

π–π^* BANDS OF PYRIDINE 1-OXIDE DERIVATIVES (IN H_2O)

Substituent[a]		λ(mμ)			$\varepsilon \times 10^{-4}$			Remarks	Ref.
Nil	(N)			254			1.19		121
	(C)		217	257		0.49	0.29		121
3-CH_3	(N)		209	254		1.98	1.17		121
	(C)		220	263		0.42	0.35		121
4-CH_3	(N)		206	256		1.87	1.43		121
	(C)		226	254		0.83	0.27		121
3-COOH	(N)		220	260		2.24	1.02		121
	(C)		(218)	265		(0.69)	0.21		121
	(A)		216	257		2.18	1.09		121
4-COOH	(N)		216	280		1.12	1.71		121
	(C)		232	272		0.96	0.35		121
	(A)		212	273		1.51	1.60		121
4-OH	(N)			262			1.35		121
	(C)			240			1.05		121
	(A)			272			1.77		121
1-Methoxy-	(N)			262			1.68		60
4-pyridone	(C)			243			1.12		60
4-OCH_3	(N)			261			1.63		60
	(C)			244			1.20		60
2-OH	(N)		225	296		0.64	0.52		60
	(C)		209	276		0.43	0.56	exists as 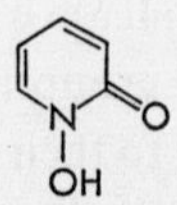	60
	(A)	210	(237)	313	2.67	(0.62)	0.62		60
2-OH	(N)		228	305		0.72	0.46	(EtOH)	59
	(N)		228	305		0.65	0.40	(EtOH)	143
	(N)		230	303		0.65	0.475	0.01N $H_2SO_4 \cdot$EtOH	60
	(A)	225	(244)	323	2.32	(0.57)	0.68	0.01N NaOEt·EtOH	60
2-OMe	(N)	213	249	293	2.50	0.76	0.45		60
	(C)	(210)		280		(0.52)	0.63		60
1-Methoxy-	(N)		224	296		0.53	0.41		60
2-pyridone	(C)		(207)	276		0.38	0.52		60
1-Benzyloxy-	(N)		214	297		1.14	0.62		60
2-pyridone	(C)		207	279		(1.16)	0.64		60
2-OEt	(N)	214	249	293	2.72	0.74	0.45		60
	(C)		(209)	280		(0.62)	0.70		60
3-OH	(N)		225	263		1.45	1.04	EtOH	59
	(N)	224.5	263.5	304				EtOH	144
3-OMe	(N)	223	262	299				EtOH	144
2-SH	(N)	235	261.5	322	1.01	0.70	0.36		145
	(C)	209	241	299.5	1.72	0.93	0.92	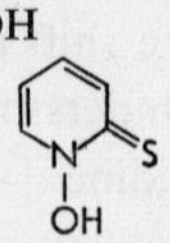	145
	(A)	249	290	329	2.51	1.14	0.46		145
2-S-CH_2-Ph	(N)	237.5	261.5	307.5	2.86	1.02	0.47		145
	(C)		251	314		1.23	1.11		145
4-SH	(N)	214(233)	285	326	0.79(0.46)	0.86	0.99		145
	(C)		222	286		0.86	1.59		145
	(A)		225	323		0.65	1.45		145

[a] (N), (C), and (A) respectively indicate neutral form, cation, and anion.

TABLE 4-28 *(continued)*

Substituent[a]		$\lambda(m\mu)$			$\varepsilon \times 10^{-4}$			Remarks	Ref.
4-S-CH_2-Ph	(N)		205	299.5		3.05	2.95		145
	(C)		226.5	308		1.19	2.93	exists as	145
4-NH_2	(N)			276			1.90		121
	(C)			268			1.61	62% H_2SO_4 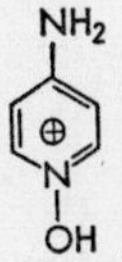	121
	(C)			270			1.81	50% H_2SO_4	58*a*
	(C)			270			1.55	0.1N HCl	58*a*
	(N)			276			1.89		58*a*
(di-cation)				260			0.35	95% H_2SO_4 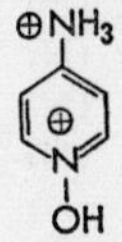	58*a*
4-amino-1-methoxy-pyridinium perchlorate	0.1N HCl			274			1.93		60
	0.1N NaOH			273			1.99		60
4-NHMe	(N)			285			2.28		60
	(C)			279			1.64		60
4-NMe_2	(N)			289			2.38		60
	(C)			288			1.92		60
4-Dimethyl-amino-1-methoxy-pyridinium perchlorate	0.1N HCl			290			2.35		60
	0.1N NaOH			291			2.09		60
2-NH_2	(C)		231	301		0.808	0.523	exists as 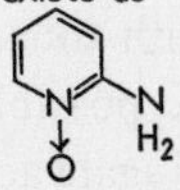	146
	(N)	221	(239)	310	2.33	(0.69)	0.39		146
	(N)	227	251	321	2.50	0.57	0.461	EtOH	146
	(N)	226	(248)	319	2.10	0.40	0.50	EtOH	147
2-NHMe	(N)	226	(246)	324	1.87	(0.55)	0.364		146
	(C)	236	314		0.795	0.425		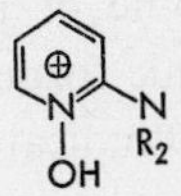	146
2-NMe_2	(N)	236	(261)	319	0.65	(0.54)	0.269		146
	(C)	243	320		0.737	0.447			146
2-Amino-1-methoxy-pyridinium perchlorate	0.1N HCl	230	299		2.77	0.59			146
2-Amino-1-methoxy-pyridinium	0.1N HCl		235	314		1.09	0.66		146
p-toluene-sulfonate	0.1N NaOH	237	297	302	0.94	0.339	0.337		146

(Continued overleaf)

TABLE 4-28 *(continued)*

Substituent[a]		λ(mμ)	ε × 10^{-4}	Remarks	Ref.
$3\text{-}NH_2$	(N)	234 (252) 314	2.28 (1.15) 0.27		121
	(C) (di-cation)	222 261	0.51 0.26	62% H_2SO_4	121
3-NHAc	(N)	240 (254) 295	2.53 (1.50) 0.17		121
$4\text{-}NO_2$	(N)	226 313	0.80 1.25		121
	(C)	244 (280)	0.82 (0.38)		121
			log ε		
	(N)	314	4.08	(H_2O)	116
	(N)	314.5	4.09	(0.1N HCl)	116
	(N)	332.5	4.15	(EtOH)	116
	(N)	347	4.20	(Et_2O)	116
	(N)	348	4.20	(dioxan)	116
4-Cl	(N)	255	1.54		133
	(N)	265	1.34	EtOH	133
	(N)	289	1.48	CCl_4	133
4-Br	(N)	266	1.64		133
	(N)	278	1.48	EtOH	133
	(N)	291	1.44	CCl_4	133

large a change in the electronic configuration of the parent base, and it becomes quite inadequate to analyze the spectra of its *N*-oxide as a simple perturbation of the electronic configuration of the parent heterocycle. This is also one of the reasons why Kubota interpreted and assigned these bands as charge transfer bands (*v.s.*). However, it seems possible to analyze the spectra of *N*-oxides of azanaphthalene or its higher benzenoid homologs in comparison with those of their parent heterocycles.

It is also possible, without exception, to compare and analyze the spectra of the conjugate acids of aromatic *N*-oxides with those of their parent heterocycles. Some of the spectra of substituted pyridine 1-oxides (in neutral and cationic forms) and those of the corresponding parent bases shown in Table 4-29 indicate that the spectra of the conjugate acids of both *N*-oxides and their parent heterocycles are quite similar. The data in this Table show that the N^+–OH compound always has its absorption maximum in a slightly longer wave-length region than that of the corresponding N^+–H compound, though the intensity of the absorption band in the former is slightly weaker than that in the latter. One of the reasons for this slight difference in the spectra of these two kinds of conjugate acids has been attributed by Kubota[41] to the fact that the electron migration from oxygen to the ring system is still effective in the N⁺–OH compounds, though to a very small extent.

The behavior of the charge transfer bands of pyridine 1-oxide arising from methyl substitution has been investigated by Ikekawa and Sato[148]. They found that the introduction of a methyl group into the 3- or 4-position causes a shift of the above bands to a longer wavelength, while substitution into the 2-position has the opposite effect. The wavelength shift of the band maximum as a function of the position of methyl substitution is illustrated in Fig. 4-17. It has also been found that the shift is approximately additive for di- and poly-substitution (*cf.* Table 4-30).

Ikekawa and Sato[148] also pointed out that the marked hypsochromic shift (−4.5 mμ in ether and −2 mμ in 10% ethanol) caused by the 2-methyl group might be due to the presence of a weak hydrogen bonding between the oxygen atom of the *N*-oxide group and the hydrogen atom of the 2-methyl group, as illustrated by (4-35).

TABLE 4-29

π–π^* BANDS OF PYRIDINE AND ITS 1-OXIDE DERIVATIVES

		N-Oxide mμ($\varepsilon \times 10^{-4}$)				Parent base[123] mμ ($\varepsilon \times 10^{-4}$)		
Ethyl β-substituted acrylate[123]								
4-pyridyl	(N)[a]	221(1.24)		300(1.89)	(N)	204(1.33)		259(2.19)
	(C)[a]	203(1.51)		278(2.32)	(C)			271(2.34)
3-pyridyl	(N)	206(1.04)		252(2.86)	(N)	206(1.25)	262(1.71)	282(1.35)
	(C)	223(1.74)		261(1.54)	(C)	218(1.30)	255(1.80)	280(0.99)
2-pyridyl	(N)		244(2.67)	286(1.32)	(N)	206(1.45)	249(1.27)	290(1.50)
	(C)	216(1.42)	251(0.84)	294(1.54)	(C)	206(1.44)	241(0.84)	290(1.86)
β-Substituted acrylamide[123]								
4-pyridyl	(N)	221(1.35)		300(2.07)	(N)	205(1.44)		259(2.17)
	(C)	207(1.38)		282(2.46)	(C)			271(2.32)
3-pyridyl	(N)			251(2.89)	(N)	206(1.34)	259(1.90)	284(1.34)
	(C)	222(1.41)		265(1.64)	(C)	224(1.47)		255(2.34)
2-pyridyl	(N)		239(2.87)	281(1.36)	(N)	206(1.60)	249(1.46)	289(1.52)
	(C)	221(1.40)	256(0.93)	296(1.66)	(C)	206(1.42)	241(0.99)	291(1.80)
Methylpyridine[121,141]								
4-Me	(N)	206(1.85)		256(1.45)	(N)			255(0.21)
	(C)	226(0.83)		254(0.27)	(C)			252.5(0.45)
3-Me	(N)	209(2.00)		254(1.17)	(N)			263(0.31)
	(C)	220(0.42)		263(0.35)	(C)			262.5(0.55)
Nil	(N)			254(1.20)	(N)			257(0.275)
	(C)	217(0.50)		257(0.29)	(C)			256(0.53)

[a] (N) and (C) respectively indicate neutral form and cation.

References p. 167

TABLE 4-30

ULTRAVIOLET ABSORPTION OF METHYLPYRIDINE 1-OXIDES[148]

Position of methyl substituent	Solvent							
	Et_2O		95% EtOH		50% EtOH		10% EtOH	
	λ_{max} (mμ)	ε_{max}	λ_{max} (mμ)	ε_{max}	λ_{max} (mμ)	ε_{max}	λ_{max} (mμ)	ε_{max}
Pyridine	280.5	14780	263.5	12440	258.5	10500	255	10900
2-	276	11230	260.5	10360	255.5	9840	253	10240
					~256			
3-	281	14500	264	9920	258	9580	255	9240
	~282				~259		~255.5	
4-	283	12660	265	11240	259.5	10600	257.5	11480
					~260.5			
2,6-	271.5	11090	258.5	9020	254.5	9300	251.5	8620
2,3-	276.5	11750	261.5	9560	257	9160	253.5	9200
	~277							
2,4-	278.5	12190	261	13600	257.5	12940	255	11240
	~279		~262					
2,4,6-	275	11550	259.5	11180	256	10100	253.5	10380
2,3,4-	279	13220	261.5	12920	258	12240	255	13120
			~262.5					
2,4,5-	278	16050	261	12720	257	12360	254	12400

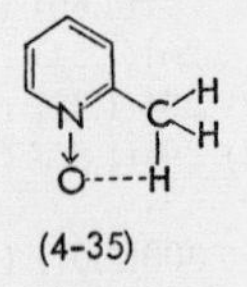

(4-35)

4.8.4 *Ultraviolet Spectra of Azanaphthalene N-Oxides and their Derivatives*

It is well known that the spectra of naphthalene derivatives having a substituent with a comparatively large conjugation effect (such as NR_2, OR, COR, NO_2, etc.) exhibit only one absorption maximum in the case of α-substituted compounds, while two maxima are present in those of β-substituted derivatives, in the region above 250 mμ[150]. This difference was also found to be present in the spectra of substituted azanaphthalenes, such as amino-[151], hydroxy-[152], and nitro-[153] quinolines and -isoquinolines.

The spectra of mononitro-quinolines and -isoquinolines reported by

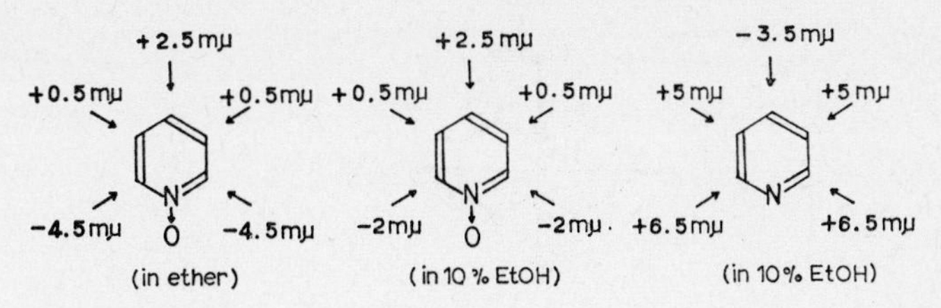

Fig. 4-17. Wavelength length shift of absorption bands with methyl substitution[148,149].

Kaneko[153] are shown as an example in Fig. 4-18. Kubota[118,128,135,141,154] also found this relationship to hold in the spectra of azanaphthalene compounds and their *N*-oxides, and analyzed this relationship in terms of Platt's notation (*cf. Section 4.8.2*).

To illustrate Kubota's point, it would be more convenient to divide the positions of azanaphthalene into two groups; Group I (positions 1, 4, 5, and 8) and Group II (positions 2, 3, 6, and 7), as shown in Fig. 4-19. As the extension of conjugation in a given direction should primarily affect the band polarized in that direction, substitution at the Group II positions extends the conjugation in the longitudinal direction (x-axis), as will be seen from Fig. 4-20, and should therefore cause bathochromic shift predominantly in the longitudinally polarized ^1L_b and ^1B_b bands (*cf.* Table 4-25). As a result, the spectra of azanaphthalenes with substituents in the Group II positions

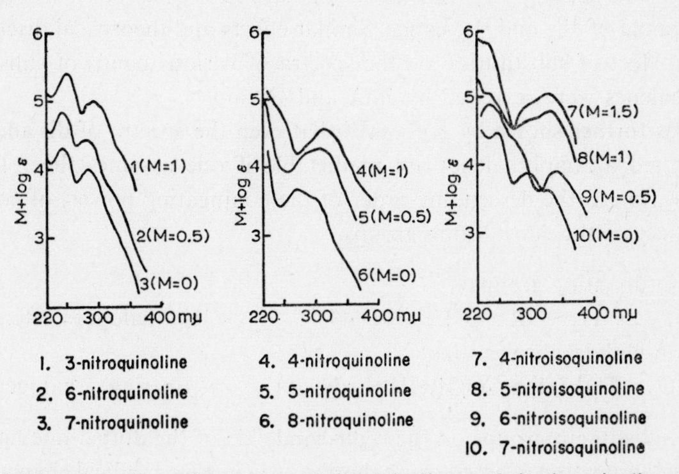

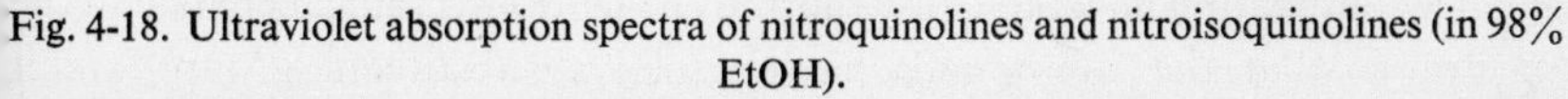

Fig. 4-18. Ultraviolet absorption spectra of nitroquinolines and nitroisoquinolines (in 98% EtOH).

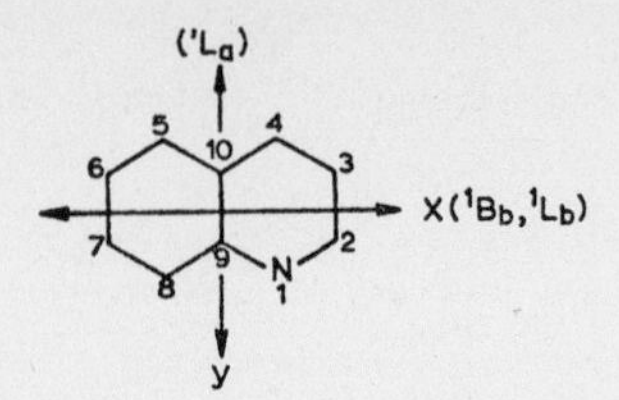

Fig. 4-19. Classification of electronic transitions in azanaphthalene.

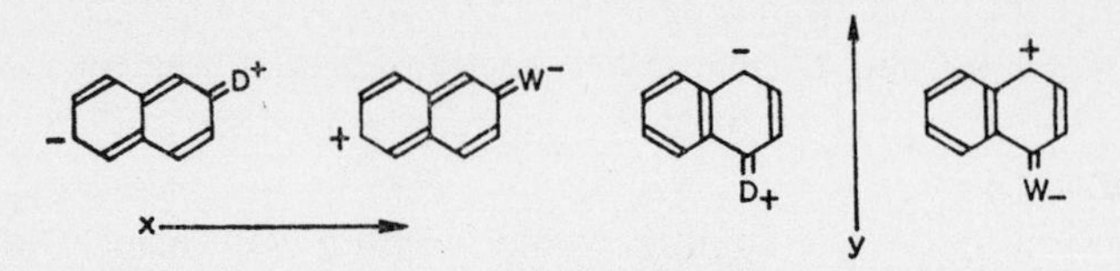

Fig. 4-20. Polarization of naphthalene by substituents.

show three absorption bands ($^1L_b > {}^1L_a > {}^1B_b$, from the longer to shorter wave-length side), as do those of their parent azanaphthalenes. (Quinoline and isoquinoline have three main bands in their spectra which are given in Table 4-26 with their assignments.)

On the other hand, substitution in Group I positions extends the conjugation primarily in the transverse direction (y-axis in Fig. 4-20) and hence causes a red shift of the transversally polarized ^1L_a band, resulting in the overlapping of ^1L_a and ^1L_b bands. Similar effects and theoretical discussions on the effect of substitution on the spectra of various bands of substituted naphthalenes were reviewed by Jaffé and Orchin[155].

It was further shown by Kubota[128] that even the spectra of di- and poly-substituted azanaphthalenes are predictable if one assumes the following order (given in the descending order of the conjugating power) of conjugation effect of the substituting groups.

Electron-donating groups:

$NR_2 \simeq NH_2 > OR \simeq OH \simeq \geqslant N \rightarrow O$ ⋮ $\gg CH_3$, halogens, etc.

Electron-withdrawing groups:

$NO_2 > COR(H) \simeq COOR(H) > \geqslant N \rightarrow O$ ⋮ $\gg N$ (in ring) $>$ halogens, etc.

The substituent groups on the right-hand side of the dotted line have too small a conjugation effect to cause the above-mentioned radical change in the spectra, so that the compounds having such a substituent group exhibit

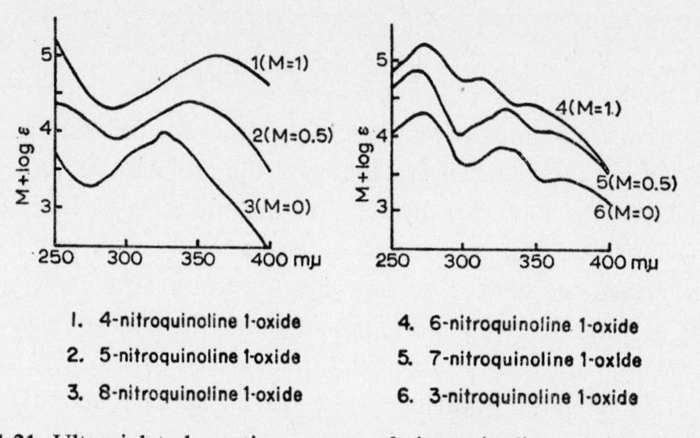

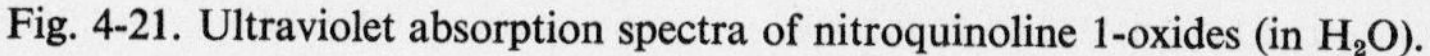
Fig. 4-21. Ultraviolet absorption spectra of nitroquinoline 1-oxides (in H_2O).

TABLE 4-31

ULTRAVIOLET ABSORPTION SPECTRA OF NITROQUINOLINE 1-OXIDE[153,156]
($\lambda_{max}^{H_2O}$ mμ (log ε))

Position of nitro group	^1B_b	^1L_a	^1L_b
3-nitro	272.5(4.33)	327 (3.86)	365.5(3.44)
4-nitro	<250	364 (3.97)	
5-nitro	<250	343.5(3.87)	
6-nitro	273 (4.26)	309 (3.79)	346 (3.43)
7-nitro	268 (4.39)	326.5(3.87)	355 (3.57)
8-nitro	<250	326.5(3.99)	

spectra similar to those of the corresponding non-substituted compounds (but with the expected bathochromic displacement)[128,153].

Figure 4-21 and Table 4-31 show the spectra of nitroquinoline 1-oxides measured in water as a solvent by Ochiai and Kaneko[156]. These data agree with the prediction made above.

4.9 Molecular Orbital Theory

To understand the electronic structure of a molecule, quantum mechanical calculations based on molecular orbital theory were carried out by many workers and their result gave authoritative answers for describing the electronic state of small molecules, conjugate hydrocarbons, etc. However, when such a theory is to be applied for explaining chemical reactivities, there are still some ambiguous points, because complete understanding of the transition state of various kinds of reactions is at present still lacking.

In addition, an extension of this theory to the problems of large molecules having a hetero atom necessarily requires introduction of many theoretical approximations and empirical parameters, and a result derived from such theoretical treatment becomes somewhat uncertain. Fortunately, however, the concepts of inductive and conjugative interactions, originally introduced into organic electron theory (classical electron theory) and which have been very useful in explaining various kinds of chemical reactions, are easily translated into quantum mechanical terms. Since there are good books[155, 157–159] that explain the principle of and practical methods for carrying out the molecular orbital calculations, the discussion of the theoretical background of this subject is kept to a minimum. Rather, this section will be devoted to giving results obtained by various workers from the foregoing theoretical calculations. However, as the evaluation of the results of any calculation requires some understanding of the theory itself, a brief description of the methods and parameters used is included in the present section as a guide to a proper appreciation of the results. For example, many of the workers employed various assumptions and parameters to do this type of work and the important thing is to know the meaning of these parameters, assumptions, and concepts, and what kind of methods have been used to derive such parameters. The discussion here is limited to the problems of substitution reactions of *N*-heteroaromatic compounds and their *N*-oxides as treated by the simplest type of molecular orbital approximation.

In a simple theoretical treatment, which is frequently used for most calculations of chemical reactivity, the principal procedure is to assume that π- and σ-electrons in an aromatic system behave independently of each other and that σ-electrons form a core field in which π-electrons move independently along the π-molecular orbitals that can be constructed as the linear combination of $2p\pi$ atomic orbitals. Chemical reactivity is then assumed to be determined by the distribution and polarizability of the π-electrons.

Consequently, the energy and electron distribution determined by this procedure are concerned with the π-electrons in the system. The energy of molecular orbitals is given by the energy units, α and β, of standard compounds such as benzene, ethylene, etc. The integral α is generally called the coulomb integral and has the physical meaning of π-electron energy pertinent to a π-atomic orbital χ subjected to the whole core potential field; *i.e.*, the expression α_i means that the main term is the energy of an i-th electron moving on the i-th core in a whole core field. Then, a_i would correspond to the electronegativity of an i-th atom. The integral β_{ij} is called the resonance integral and corresponds to the π-bonding energy between the i-th and j-th atomic orbitals. Therefore, β_{ij} would be the measure of strength of the bond concerned. It should also be noted that the integrals α and β are both negative quantities and have the dimensions of energy. For aromatic hydrocarbons such as benzene and naphthalene, only one coulomb integral α_c and one resonance integral β_{cc} are needed, and these are usually abbreviated as α and β. The energy ε_i of molecular orbitals is then given by the equation $\varepsilon_i = \alpha + a_i \cdot \beta$. If α is arbitrarily taken as zero, the energy ε_i will be function only of β. In the case of heteroaromatics like pyridine and pyridine 1-oxide, the calculation becomes more complicated because there appear coulomb (α_N and α_O) and resonance (β_{NO} and β_{NC}) integrals concerning the hetero atoms, which are different from those of simple hydrocarbons. If a comparison is to be made for neutral oxygen and nitrogen atoms, which enter into the conjugate system under the condition that the π-electron number from these atoms is the same†, the following consideration will be reasonable. Since electronegativity of neutral atoms decreases in the order of oxygen, nitrogen, and carbon, it would be reasonable to expect the values of coulomb integrals to fall in the order of $|\alpha_O = \alpha + h_O\beta| > |\alpha_N = \alpha + h_N\beta| > |\alpha|$*, where h_O and h_N are parameters showing a deviation in the β unit from the coulomb integral of a carbon atom in hydrocarbons like benzene.

On the other hand, the effect of a net charge on the coulomb integral of the same atom X is, of course, in the decreasing order of $\alpha_{X^+} > \alpha_X > \alpha_{X^-}$. Another important factor is the so-called inductive effect. A conjugate system having a $C'{=}N$ bond will be taken as an example. Since a nitrogen atom is more electronegative than a carbon atom, a nitrogen atom exerts an inductive effect on the neighboring carbon atoms through its σ and π bonds and hence enhances the electronegativity of the carbon atom. In other words, the cou-

† The nitrogen atom in pyridine contributes one π-electron to the whole conjugate system but the nitrogen atom in aniline contributes two π-electrons. This means that core potential of these two nitrogen atoms will be different and h_N must also differ in these two nitrogen atoms. In addition, if atoms entering into a conjugate system have a large formal charge, such as $\geqslant N^+{-}O^-$ and $\geqslant C{-}O^-$, a theoretical treatment to determine the coulomb integral becomes complicated, although several methods have been proposed for these cases[155, 157–159]. Therefore, the above-mentioned order of α for neutral atoms $\alpha_O > \alpha_N$, will be changed according to the environment of atoms in question.

lomb integral comes out larger* and is written as $\alpha_{C'} = \alpha + h_{ind}\beta$, where $h_{ind}\beta$ is the correction term in the β unit for this inductive effect. Next, the resonance integral between two different atoms must be evaluated. For pyridine and pyridine 1-oxide, β_{CN} and β_{NO} are in question. Generally, however, β_{CN} is taken to be equal to β for the *N*-heterocyclic system, and β_{NO} depends on the method adopted by the individual worker, as will be explained later (*cf.* Table 4-34). It will then be noted that the π-electron charge distribution and its energy will depend on the selection of values assigned to the coulomb and resonance integrals.

Even if the π-electron densities were obtained accurately through molecular orbital calculation, it will not be easy to infer chemical reactivity from such calculations, because a reactivity is also dependent on the polarizability of each position in a molecule. However, theoretical prediction of orientation for chemical reactions is sometimes studied on the basis of a charge distribution in a molecule in static state (Ref. 160)†. In such a case, if the problem is for heteroaromatic compounds, the σ-electron distribution as well as π-electron distribution must be taken into consideration.

In the case of this model, the basic concept is that nucleophilic reagents will attack the position of less net charge (δ^+ position) but electrophilic reaction will occur in the position having δ^- formal charge. Another simple model for treating the orientation property of a chemical reaction was first presented by Wheland[161], whose model for transition states is as follows. Consider, for example, the transition state of electrophilic reactions. Wheland assumed that an activated complex is formed in the transition state; that is, two π-electrons in the whole conjugate system are localized at the carbon atom attacked by the reagent X^+, and a new σ bond C–X is produced there The carbon atom therefore undergoes an sp^3 hybridization having four single bonds, so that the remaining π-electrons (total π-electrons minus two) can move along the carbon cores except for the atom in question, as shown below.

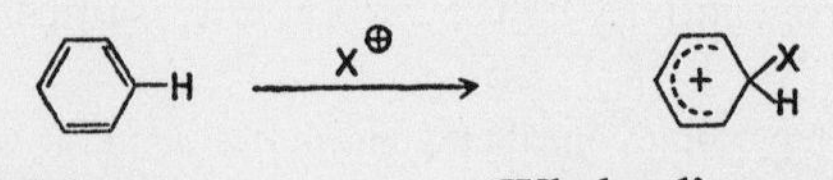

Initial state Wheland's transition state

Consequently, the contribution of the π-electron to the activation energy

* The comparison is made for each absolute value.

† This means that the molecule is not in interaction with reagents.

is calculated as the difference between the total π-electron energy of an initial state and Wheland's transition state. This energy difference is called the localization energy.

Notations L_e, L_n, and L_r are used respectively for electrophilic, nucleophilic, and radical reactions. The two semiquantum mechanical methods described in the foregoing are called static and localization methods[162]. It is now clear that the treatment of transition states is quite different between the above two models; the transition state of a static model assumes that the electronic structure of isolated molecules is very similar to that of a transition state, so that the orientation for chemical reactions is determined mainly by an electrostatic interaction between the electron distributions of a conjugate molecule in an initial state and that of a reagent*. On the other hand, the transition state taken into consideration for the localization model differs chiefly in the electronic state from that of the static model; it is assumed that the origin of cyclic conjugation, which is the main source for aromaticity, is cut off in a transition state, while the loss of stability of cyclic conjugation by the attack of a reagent may be partially compensated by a new σ bond formation between the carbon atom of a reaction center and a reagent†.

The localization energy obtained by this localization model may be considered as a part of the activation energy. If a comparison is made among various reactions similar to one another, it appears that the contribution from the terms, such as entropy change ($\Delta S^{\neq}$) between the ground and transition states, and other miscellaneous factors contributing to the activation energy, is almost the same. Therefore, not only the orientation for chemical reactions but also the relative reactivity at different positions in aromatic molecules should be determined by the magnitude of localization energies at different positions. Fig. 4-22 gives a schematic drawing of the change of free energy ($\Delta F^{\neq}$) with the reaction path as well as the π-electron energy variation (ΔE_{π}) expected from the localization model. The structures in several positions of the reaction coordinate (*a*, *b*, *c*, and *d*) are also assumed to be as in this drawing. It is supposed here that the maximum positions which correspond to the actual transition state and Wheland's σ-complex for the two curves

* The frontier electron theory presented by Fukui *et al.*[163] may be classified, in a broad sense, as a kind of static method although the basic concept of this theory is quite different from that of the static method.

† This discussion was carried out on the assumption that the true transition state is in agreement with that of Wheland's model. This assumption, however, is not always true. As will be explained later, it was recently revealed by theoretical considerations that the cyclic conjugation is not completely cut off at the transition state.

References p. 167

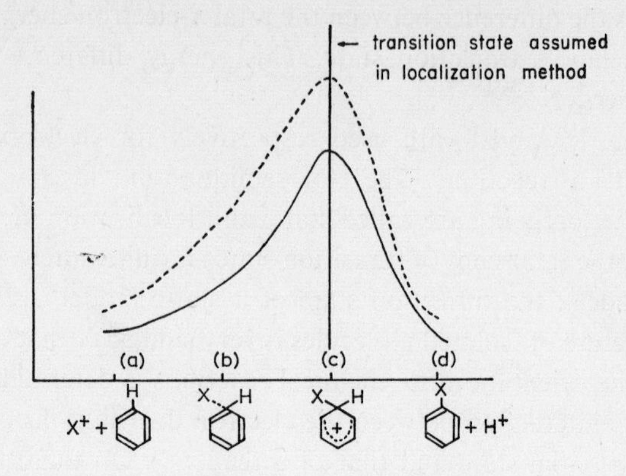

Fig. 4-22. Variation of energy along reaction coordinate (A). - - - -, $\Delta F^{\neq}$; ———, $\Delta E\pi$.

($\Delta F^{\neq}$ and ΔE_{π}) are located at the same position, *c*, in the reaction coordinate.

Although this postulation applies in many cases, it is known in some cases that the rate-determining step is not at the position *c* in Fig. 4-22 but is at the step where the proton is removed from Wheland's σ-complex, as shown in Fig. 4-23 by the dotted line. In the latter case, the reactivity is neither predicted by the localization method nor by the static method.

Even if the reaction, as has been discussed above, is excluded from the consideration, the two methods, static and localization methods, are still not completely suitable for evaluating the true activation energy. In the static method, the contribution of a π-electron energy to the total potential energy is calculated at some positions located before the true transition state shown in Fig. 4-25 (*e.g.*, position *a*). On the other hand, the reactivity calculated by the localization method corresponds to the region beyond the true transition state (*e.g.*, position *d* in Fig. 4-25), as will be discussed later.

Fortunately, however, predictions for the orientation of a reaction by these two methods, static and localization methods, generally give the same conclusion consistent with experimental results, and Brown[164] called this fact the "chemical non-crossing rule". As was already pointed out, electrophilic and nucleophilic reactions occur frequently at the same position in the case of heterocyclic *N*-oxides, such as 2 or 4 in pyridine 1-oxide. This might mean that the static method is not suitable for predicting the chemical reactivity of heterocyclic *N*-oxides* and, in this sense, pyridine 1-oxide may be said to violate Brown's "chemical non-crossing rule".

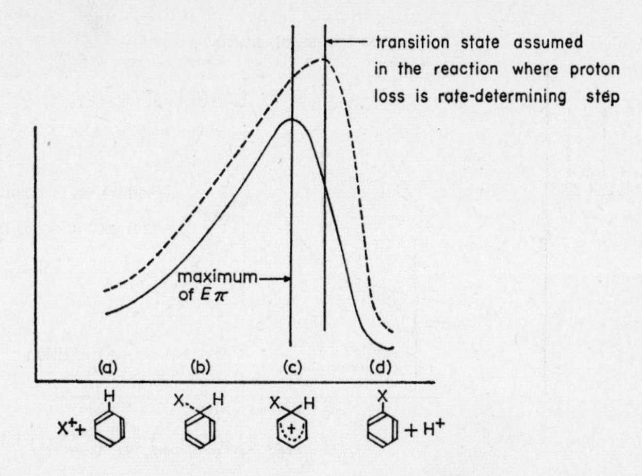

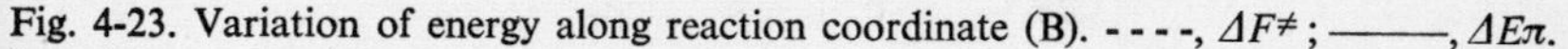
Fig. 4-23. Variation of energy along reaction coordinate (B). - - - -, $\Delta F^{\neq}$; ———, $\Delta E\pi$.

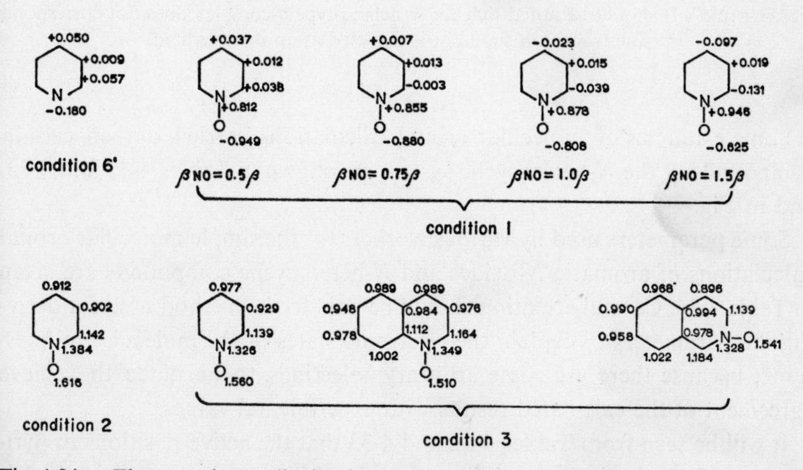

Fig. 4-24. π-Electron charge distribution and π-electron distribution of some aromatic amine oxides and pyridine. (Conditions 1, 2, 3, and 6′ refer to the ones shown in Table 4-34.)

Therefore, it may be reasonable to conclude that the scale giving the chemical reactivity of heterocyclic *N*-oxides should be made by the localization method. This situation in pyridine 1-oxide was already pointed out by Jaffé[56].

* The frontier electron theory was also applied by Kubota[135,154] to predict the chemical reactivity of heterocyclic *N*-oxides and the procedure gave results that agreed well with experimental results in both nucleophilic and electrophilic reactions.

References p. 167

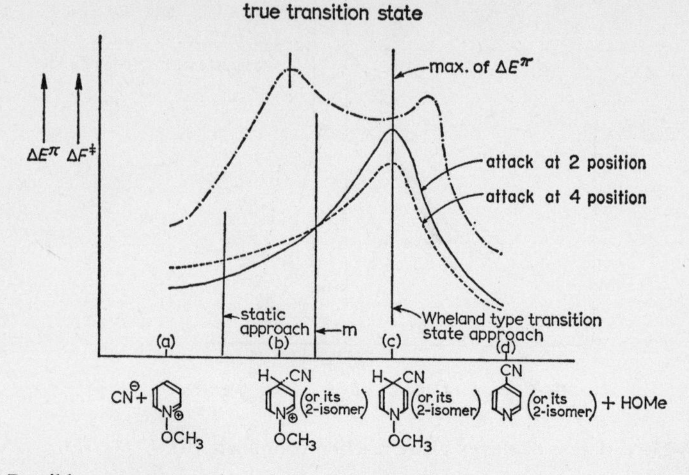

Fig. 4-25. Possible energy curves along the reaction coordinate in the reaction of 1-methoxypyridinium iodide with potassium cyanide. This example violates the "chemical non-crossing rule". It should be noted that the Wheland-type σ-complex does not correspond to the transition state but is depicted as an intermediate.

Some examples of molecular orbital calculations carried out on certain compounds of the *N*-heterocyclic system are shown in Tables 4-32 and 4-33, and in Fig. 4-24.

Some parameters used by various workers for the simple molecular orbital calculations of aromatic *N*-oxides and *N*-heterocyclic compounds are given in Table 4-34. Careful attention should be paid to the method used for deriving the parameters to explain the electronic states of the molecules in question*, because there are some arbitrary selections to be made to achieve agreement of the calculated results with experimental values.

It will be seen from Tables 4-32 and 4-33 that the active positions in pyridine 1-oxide for both electrophilic and nucleophilic reactions are the 2- and 4-positions by the localization method (see footnote on p. 153) and that position 4 is more active than 2. Experimental results from nucleophilic reactions support these calculated results. For example, it has been proved by several kinetic studies[160] that the reactivity of 4-halogenopyridine 1-oxide to nucleophilic reagents is larger than that of 2-halogeno compounds. On the other hand, as will be seen from Fig. 4-24, π-electron distribution falls in the

* This is an important matter and the reader is referred to References 155, 157–159 for more strict discussions on the evaluation of these parameters.

TABLE 4-32

LOCALIZATION ENERGIES[a] (UNIT IN β)[165]

Compound	Position	β_{NO}	L_e	L_r	L_n
Pyridine	2		2.62	2.51	2.40
	3		2.57	2.54	2.51
	4		2.69	2.53	2.36
Pyridinium salt	2		2.85	2.53	2.21
	3		2.62	2.55	2.48
	4		2.84	2.55	2.25
Pyridine 1-oxide	2	0.5	2.51	2.47	2.42
	3	0.5	2.58	2.54	2.50
	4	0.5	2.60	2.50	2.39
Pyridine 1-oxide	2	0.75	2.36	2.42	2.48
	3	0.75	2.58	2.53	2.50
	4	0.75	2.44	2.47	2.46
Pyridine 1-oxide	2	1.0	2.22	2.38	2.54
	3	1.0	2.59	2.53	2.49
	4	1.0	2.34	2.44	2.54
Pyridine 1-oxide	2	1.5	1.92	2.28	2.58
	3	1.5	2.60	2.56	2.52
	4	1.5	2.09	2.42	2.75
Pyridine 1-oxide salt	2	0.5	2.70	2.56	2.39
	3	0.5	2.58	2.54	2.50
	4	0.5	2.67	2.51	2.36

[a] Calculations were made under the condition 1 shown in Table 4-34, as described by Barnes[165]. Barnes' procedure is as follows: First, the coulomb integrals for atoms in pyridine 1-oxide are assumed as $\alpha_O = \alpha + 1\beta$, $\alpha_N = \alpha + 0.6\beta$, and $\alpha_{C_2} = \alpha_{C_6} = \alpha + 006\beta$ and calculations were made using several β_{NO} ($0.5-1.5\beta$). To explain the reactivity, the best agreement between experimental results and this procedure was derived by employing the value of $\beta_N = 0.75\beta$. The parameters adopted by Barnes for pyridine were $\alpha_N = \alpha + 1\beta$ and $\alpha_{C_2} = \alpha_{C_6} = \alpha + 0.1\beta$, and those for pyridine 1-oxide salts, $\alpha_O = \alpha + 1.5\beta$, $\alpha_N = \alpha + 0.65\beta$, $\alpha_{C_2} = \alpha_{C_6} = \alpha + 0.065\beta$, and $\beta_{NO} = 0.5\beta$.

order of 2>4>3 but, if the contribution from σ-electrons is taken into account, the net electron density will perhaps decrease in the order of 4>2>3, because the inductive effect of the $N\delta^+$ atom on position 2 seems to be of considerable importance*.

* *Cf. Section 4.6.* For the case of the 1-hydroxypyridinium ion, the total net electron density may be in the order of 3>4>2. The same would be true for pyridine and its conjugate acid.

References p. 167

TABLE 4-33

LOCALIZATION ENERGIES (Ref. 56)[a,b] (UNIT IN β)

Compound	Position	*Er*	Electrophilic substitution		Nucleophilic substitution	
			E	*Le*	*E*	*Ln*
Benzene		8.000	5.464	2.536	5.464	2.536
Pyridine	2	8.651	5.939	2.712	6.339	2.312
	3	8.651	6.164	2.487	6.264	2.378
	4	8.651	6.014	2.637	6.414	2.228
Pyridine 1-oxide	2	12.894	10.260	2.634	10.205	2.689
	3	12.894	9.676	3.218	10.537	2.357
	4	12.894	10.420	2.474	11.180	1.714

[a] *Er*, *E*, and *L* designate, respectively, the π-electron energies for the starting compound, transition state, and the localization energy. Calculations were made using condition 2 in Table 4-34; *i.e.*, $\alpha_N = \alpha + 2.004\beta$, $\alpha_O = \alpha + 1.016\beta$, and $\alpha_{C_2} = \alpha_{C_6} = \alpha + 0.250\beta$ for pyridine 1-oxide. These parameters were derived on the basis of the relationship between π-electron densities and Hammett's substituent constant (σ), as developed by Jaffé[110a]. The values of σs ($\sigma_m = 1.47\sim1.50$, $\sigma_p = 1.34\sim1.37$) used here were obtained from the measurement of pKa's of mono-substituted pyridine 1-oxides in an aqueous solvent. Therefore, the results obtained by Jaffé should be considered to correspond to their state in aqueous solvents because the electronic structure of heterocyclic *N*-oxides is strongly dependent on the solvent, as was discussed in the foregoing sections.

[b] After Jaffé reported the σ_m and σ_p of pyridine 1-oxide, where the $\geqslant$N$\rightarrow$O group was considered as a substituent, Shindo derived another σ_m and σ_p ($\sigma_m = +1.18$, $\sigma_p = +0.25$) for the same compound from the data obtained from infrared study (*cf. Section* 4.7) with non-polar solvents. Using these new σ values, Kubota also determined the coulomb integrals, $\alpha_N = \alpha + 1.6\beta$ and $\alpha_O = \alpha + 0.595\beta$, by a treatment similar to Jaffé's method[110a], and calculated the molecular orbitals of various heterocyclic *N*-oxides (*cf.* footnote to Table 4-34). The result of Kubota's calculation should therefore be taken as describing the electronic state of the *N*-oxide molecules in a non-polar solvent. As will be understood from the molecular diagrams in Fig. 4-24, both the π-electron transfer from the oxygen atom to the heterocyclic ring and the double-bond character of N=O (bond order) are larger in non-polar solvents than in water, as would reasonably be expected. (see Footnote *a* in Table 4-34 and References 135 and 154). Unfortunately, Kubota did not undertake the calculation of a localization energy.

TABLE 4-34

PARAMETERS USED FOR SIMPLE MOLECULAR ORBITAL CALCULATIONS OF AROMATIC *N*-OXIDES AND PYRIDINES, AND THEIR CONJUGATE ACIDS

i- Solvent	α_N	α_O	Inductive effect	β_{NO} etc.	Ref.
atic heterocyclic *N*-oxides					
	$\alpha + 0.6\beta$	$\alpha + 1\beta$	$\alpha_{C_{2\cdot6}} = \alpha + 0.06\beta$	$\beta_{NO} = 0.75$ etc.	12
	$\alpha + 1\beta$	$\alpha + 1\beta$	$\alpha_{C_{2\cdot6}} = \alpha - 0.1\beta$	$\beta_{NO} = 0.6$	166
\q. soln.	$\alpha + 2.004\beta$	$\alpha + 1.016\beta$	$\alpha_{C_{2\cdot6}} = 0.2505\beta$	$\beta_{NO} = 1$	56
q. soln.	$\alpha + 2\beta$	$\alpha + 1\beta$	$N_{ind} = (1/3)^{n}\cdot 2\beta$	$\beta_{NO} = 1$	135,154,12,13
Ionpolar olvent	$\alpha + 1.6\beta$	$\alpha + 0.595\beta$	$N_{ind} = (1/3)^{n}\cdot 1.6\beta$	$\beta_{NO} = 1$	135,154,12,13,[a]
cid soln.	$\alpha + 2.4\beta$	$\alpha + 2\beta$	$\alpha_{C_{2\cdot6}} = \alpha + 0.4\beta$	$\beta_{NO} = 0.6$	135,154,12,13
cid soln.	$\alpha + 1.5\beta$	$\alpha + 0.65\beta$	$\alpha_{C_{2\cdot6}} = \alpha + 0.065\beta$	$\beta_{NO} = 0.5$	165
	$\alpha + 3.5\beta$	$\alpha + 0.6\beta$			167
atic heterocycles					
	$\alpha + 0.6\beta$		$\alpha_{C_{2\cdot6}} = \alpha + 0.6/8\beta$		56
	$\alpha + 0.5\beta$		$\alpha_{C_{2\cdot6}} = \alpha + 0.05\beta$		165
	$\alpha + 1.6\beta$		$N_{ind} = (1/3)^{n}\cdot 1.6\beta$		135,154
yridinium on	$\alpha + 1\beta$		$\alpha_{C_{2\cdot6}} = \alpha + 0.1\beta$		165
,,	$\alpha + 2.4\beta$		$\alpha_{C_{2\cdot6}} = \alpha + 0.4\beta$		168[b]

[a] Later, these parameters were re-examined by comparison of various experimental results from dipole moments, infrared spectra, N–O bond length, electronic spectra, etc., with theoretical results and the best parameter for the oxygen atom in the N→O bond was found to be $\alpha_O = \alpha + 0.8\beta$ (other parameters being the same as given in References 12 and 13) in nonpolar solvents. The calculation of electronic spectra was based on the ASMO–CI method[155,157–159,12,13].

[b] Calculations for condition 7′ were made with the use of the semi-empirical SCF method[155,157–159].

As mentioned earlier, the electronic structure of aromatic *N*-oxides depends on the solvent. The same seems to be true in the case of chemical reactions involving aromatic *N*-oxides. As an example of such electrophilic and nucleophilic reactions of *N*-oxide compounds, the reaction of an alkoxypyridinium ion with a CN anion giving 2- and 4-cyanopyridines was reported by Okamoto and Tani[169], and the rate of this reaction is highly dependent on the solvent, as shown in Table 4-35.

If it is assumed that the rate-determining step is not at the place of deprotonation from reaction intermediates like a dihydropyridine-type compound, but is at the position before such a reaction intermediate is produced,

TABLE 4-35

REACTION OF 1-METHOXYPYRIDINIUM IODIDE WITH POTASSIUM CYANIDE

Solvent	Ratio of 4/2	pH[a]	Ratio of 4/2
H_2O	0.45	11.91	0.49
H_2O : EtOH (1 : 1)	0.26	11.58	1.31
H_2O : EtOH (3 : 7)	0.17	10.99	1.59
H_2O : EtOH (1 : 4)	0.11	10.07	1.67
EtOH	0.08	9.03	1.19
Dioxan	0.11		

[a] Britton–Robinson buffer (H. T. S. Britton and R. A. Robinson, *J. Chem. Soc.*, (1931) 1456) was used, and the pH value was determined immediately after the addition of potassium cyanide.

the diagrams of $\Delta F^{\neq}$ (free energy change) and ΔE^{π} against the reaction coordinate can be drawn as shown in Fig. 4-25*.

If the true transition state is located on the left-hand side (a) of m in Fig. 4-25 (which is that shown here), formation of the 2-cyano derivative would be favorable, but for the inverse case (the transition state would be on the right-hand side of m), a 4-cyano compound will be the main product. At any rate, that the true transition state is located on the reaction coordinate between that of isolated molecules and a normal Wheland's model was already pointed out by Dewar[170] and has been accepted as an established theory (*cf.* Fig. 4-25). Dewar's concept is based on the fact that, if Wheland's σ-complex is adopted as a true transition state, there will be two difficulties; (*i*) the calculated rate ratio, such as for the nitration of benzene and naphthalene, becomes larger than that determined experimentally, with any reasonable choice of resonance integrals, β†, and (*ii*) the rate ratio observed experimentally is, in general, different for different reactions such as nitration and halogenation. For example, the rate-ratios for chlorination are

* This assumption is clearly not applicable to reactions as in the case of the 3-ethoxycarbonyl-1-hydroxypyridinium ion because ultraviolet spectral studies have proved that proton loss from the Wheland-type σ-complex (reaction intermediate) is the rate-determining step[169]. There is a possibility, therefore, that this assumption is not correct even for reactions involving the unsubstituted 1-alkoxypyridinium ion itself. The discussion here is stressed as an illustration that the position of the true transition state is nearer the starting compounds than that of Wheland's σ-type complex in the reaction coordinate.

† This means the values of β used for calculating the L_e and L_N shown in Table 4-33. The normal value of β is -20 kcal/mole.

much larger than that for nitration in the case of an electrophilic reaction.

The first difficulty may be solved by supposing that the true transition state lies in the position between a and c in Fig. 4-25, so that the cyclic conjugation is still more or less held there because the effective value of β comes out smaller as the true transition state moves from c to a, as will be understood from Fig. 4-25. The second difficulty would be eliminated by assuming that the place of the true transition state depends on the reagent used for the reaction, that is, the effective value of β will also depend on the reagent[170,171].

Thus, the relative rate ratio should be determined by both effects, β-effect and ΔL; the former is the function of various kinds of reagents and solvents, and the latter is a value controlled by substrates. The validity of the discussion given here seems to become clearer if these two effects are compared with the constants ϱ and σ appearing in Hammett's equation[11,172]. With the said Dewar's theory in mind, the solvent effect on the reaction of CN^- with heterocyclic *N*-oxides will now be considered. If solvents are protic or acid, CN^- will be stabilized by solvation, resulting in a weaker nucleophilic reagent*. The position of the true transition state in the reaction coordinate turns out to be nearer the Wheland's σ-complex than that expected in aprotic solvents or alkali solutions, so that the yield of the 4-isomer becomes larger than that of the 2-isomer. On the other hand, the position of the true transition state in non-polar solvents is expected to move to the position closer to the starting system in the reaction coordinate and the main product would be the 2-isomer. The latter conclusion is consistent with that derived from the static method. As will be mentioned later (*cf. Section 6.2.4*), Ochiai and Kaneko assumed the formation of dihydro compounds as the intermediate for the nitration of a compound of (4-36) type with benzoyl nitrate. When X is H

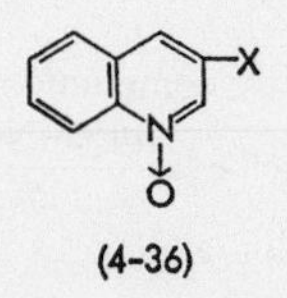

(4-36)

(quinoline 1-oxide), the assumed intermediate would be a 1,4-dihydro compound, while a 1,2-dihydro compound would be formed when X is an electron-withdrawing group. This consideration is based on the fact that, when X is an electron-withdrawing group, activity of the substrate itself (electrophilic

* Strength or weakness of the nucleophilic reagent (CN^-) may also be expressed as the degree of stabilization of a transition state by the formation of a new σ bond, C–CN.

activity of the ring) is higher than when X is H and the true transition state will shift toward the side nearer the starting system, where the static approach becomes quite useful.

The temperature effect[173] and obvious solvent effect[174] were also observed by Ochiai and others in the case of electrophilic reactions such as nitration of quinoline 1-oxide and quinaldine 1-oxide with a mixture of sulfuric and nitric acids (*cf. Section 6.2.1.b*). To explain these phenomena, they concluded that the 4-nitro compound and 5- or 8-nitro compound were formed respectively by the nitration of free *N*-oxide compounds and their conjugate acids (N^+–OH). If this consideration were correct, the kinetic treatment applicable to the usual nitration reaction would give equations (1) and (2).

$$\frac{d[\text{4-NO}_2]}{dt} = k_4(\text{N}^+\text{–O}^-)(\text{NO}_2{}^+) \tag{1}**$$

$$\frac{d[\text{5,8-NO}_2]^*}{dt} = k_{5,8}(\text{N}^+\text{–OH})(\text{NO}_2{}^+) \tag{2}**$$

Since $[(\text{N}^+\text{–O}^-)(\text{H}^+)/(\text{N}^+\text{–OH})] = K$, equation (3) should be derived from equations (1) and (2).

$$\frac{[\text{4-NO}_2]}{[\text{5,8-NO}_2]} = \frac{k_4(\text{N}^+\text{–O}^-)}{k_{5,8}(\text{N}^+\text{–OH})} = \frac{k_4}{k_{5,8}} \cdot \frac{K}{(\text{H}^+)} \tag{3}**$$

$$\therefore \quad \log \frac{[\text{4-NO}_2]}{[\text{5,8-NO}_2]} = \text{const.} - \log(\text{H}^+) \tag{4}**$$

* There is a possibility that a 5- or 8-nitro compound may also be formed by the nitration of the free *N*-oxide compound, in which case $\frac{d[\text{5,8-NO}_2]}{dt} = k'_{5,8}(\text{N}^+\text{–O}^-)(\text{NO}_2{}^+)$. If it is assumed that $k'_{5,8} = K_{5,8}$, the discussion analogous to that mentioned in the text would still be correct.

** The concentration of the reagents such as NO_2^+, etc., in parentheses, appearing on the right-hand side of equations (1) and (2), is not a stoichiometric concentration but is that based on activity. If the stoichiometric concentration is adopted here, equation (3) should be written as

$$\frac{[\text{4-NO}_2]}{[\text{5,8-NO}_2]} = \frac{K_4}{K_{5,8}} \left(\frac{f^{\neq}{}_{5,8}}{f^{\neq}{}_4} \cdot \frac{f_{\text{N}^+\text{–O}^-}}{f_{\text{N}^+\text{–OH}}} \right) \frac{K}{h_0},$$

where $f^{\neq}$ and f are activity coefficients at transition and ground states, respectively, and also $\{[\text{N}^+\text{–O}^-]h_0/[\text{N}^+\text{–OH}]\} = K$ is satisfied.

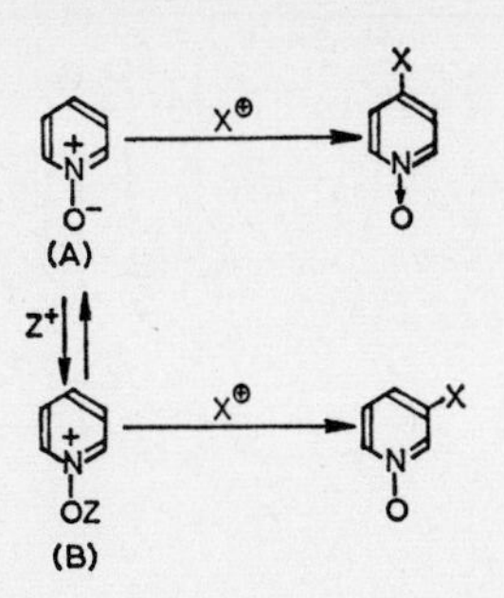

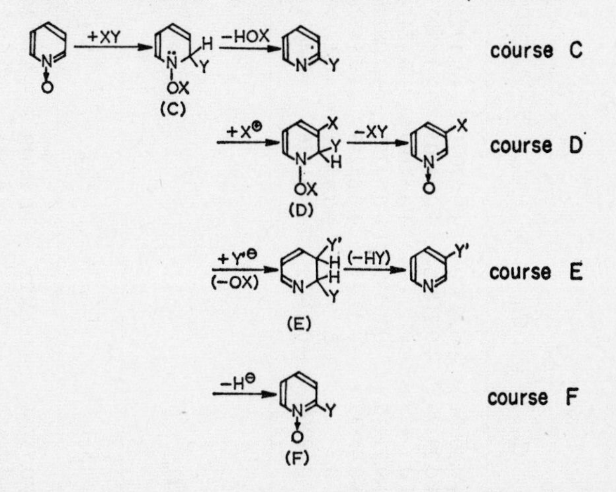

Fig. 4-26. Ionic reaction patterns of pyridine 1-oxide. In the reaction course to C, D, E, and F, examples are given for the 2-isomer. Analogous patterns are also expected for the 4-isomer in certain cases. X^+, electrophile; Y^-, nucleophile.

The experimental facts reported by Ochiai and his associates, and kinetic studies on the nitration of 7-chloroquinoline 1-oxide[175] and pyridine 1-oxide[176] indicate the above assumption to be correct. As may be understood from the above examples on the nitration reaction, reaction of *N*-oxides with electrophilic reagents proceeds through the free base or its conjugate acid under suitable experimental conditions (the conjugate acid is defined as having the general formula $\geqslant N^+–OZ$ where Z designates a cationic species). This is the same as in the case of *N*-heteroaromatic compounds. Consequently, sulfonation and bromination of pyridine 1-oxide occur at the 3-position through its

References p. 167

TABLE 4-36

SIMPLE MOLECULAR ORBITALS

(a) Pyridine 1-oxide (Condition 3) C_{2v}

ψ_i	Eigenvalues (ε_i)	Symmetry	C_{i7}	C_{i1}	C_{i2}	C_{i3}	C_{i4}	Note
ψ_1	+3.00881	b_2	+0.30901	+0.74590	+0.37091	+0.17227	+0.11680	$C_{i5} = \pm C_{i3}$
ψ_2	+1.68693	b_2	+0.29179	+0.31862	−0.13205	−0.47095	−0.57868	$C_{i6} = \pm C_{i2}$
ψ_3	+1.37123	a_2	0	0	+0.54199	+0.45414	0	b_2 orbital : +
ψ_4	+0.501999	b_2	+0.77415	−0.071996	−0.34755	+0.08288	+0.37442	a_2 orbital : −
ψ_5	−0.57249	b_2	+0.45839	−0.53517	+0.35213	+0.14578	−0.46149	
ψ_6	−0.66012	a_2	0	0	+0.45414	−0.54199	0	
ψ_7	−1.65986	b_2	+0.09969	−0.22479	+0.31656	−0.46947	+0.54617	

(b) Pyridine (Condition 6″) C_{2v}

ψ_i	Eigenvalues (ε_i)	Symmetry	C_{i1}	C_{i2}	C_{i3}	C_{i4}	Note
ψ_1	+2.77425	b_2	+0.71744	+0.42122	+0.22649	+0.16684	b_2 orbital { $C_{i5} = C_{i3}$
ψ_2	+1.54738	b_2	+0.44661	−0.01175	−0.45853	−0.61625	$C_{i6} = C_{i2}$
ψ_3	+1.37123	a_2	0	+0.54199	+0.45414	0	a_2 orbital { $C_{i5} = C_{i3}$
ψ_4	−0.31040	b_2	+0.49984	−0.47745	−0.096999	+0.52480	$C_{i6} = C_{i2}$
ψ_5	−0.66012	a_2	0	+0.45414	−0.54199	0	$\varepsilon_6 = -1.6408$ and symmetry is b_2

(c) Pyridine 1-Oxide (Condition 2′)

ψ_i	Eigenvalues (ε_i)	Symmetry	C_{i7}	C_{i1}	C_{i2}	C_{i3}	C_{i4}	Note
ψ_1	+3.31967	b_2	+0.33485	+0.77682	+0.34522	+0.13913	+0.8575	$C_{i5} = C_{i3}$
ψ_2	+1.79055	b_2	+0.32756	+0.25895	−0.19089	−0.47349	−0.55170	$C_{i6} = C_{i2}$
ψ_3	+1.46884	a_2	0	0	+0.55157	+0.44246	0	b_2 orbital: +
ψ_4	+0.78834	b_2	+0.79211	−0.16766	−0.29448	+0.13183	+0.36911	a_2 orbital: −
ψ_5	−0.3448	b_2	+0.38489	−0.51760	+0.41439	+0.09846	−0.47022	

$\varepsilon_6(a)_2 = -0.57996; \varepsilon_7(b_2) = 1.5910$

(d) Pyridine 1-oxide hydrochloride (Condition 4)

ψ_i	Eigenvalues (ε_i)	Symmetry	C_{i1}	C_{i2}	C_{i3}	C_{i4}	C_{i7}	Note
ψ_1	+3.40892	b_2	+0.81461	+0.30687	+0.10873	+0.06379	+0.34691	$C_{i5} = C_{i3}$
ψ_2	+1.93174	b_2	−0.09454	−0.22717	−0.25342	−0.26237	+0.83100	$C_{i6} = C_{i2}$
ψ_3	+1.47039	b_2	−0.37105	+0.046357	+0.42067	+0.57219	+0.42036	b_2 orbital: +
ψ_4	+1.21980	a_2	0	+0.54684	+0.44829	0	0	a_2 orbital: −
ψ_5	−0.26275	b_2	−0.41018	+0.51347	+0.06987	−0.53183	+0.10876	
ψ_6	−0.81980	a_2	0	+0.44829	−0.54683	0	0	

(Continued overleaf)

TABLE 4-36 *(Continued)*

(e) Pyrazine 1,4-dioxide (Condition 3) V_h

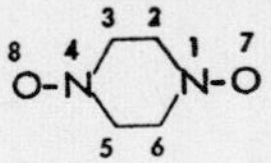

ψ_i	Eigenvalues (ε_i)	Symmetry	C_{i7}	C_{i1}	C_{i2}	Note
ψ_1	+3.26329	b_{2u}	+0.18878	+0.50372	+0.32452	$C_2 = C_3 = C_5 = C_6; C_1 = C_4; C_7 = C_8$
ψ_2	+2.73061	b_{3g}	+0.27605	+0.58954	+0.19524	$C_2 = C_6 = -C_3 = -C_5; C_1 = -C_4; C_7 = -C_8$
ψ_3	+1.71111	b_{1g}	0	0	+ 1/2	$C_7 = C_1 = C_4 = C_8 = 0; C_2 = C_3 = -C_6 = C_5$
ψ_4	+1.02365	b_{2u}	+0.50491	+0.21643	−0.31482	$C_2 = C_3 = C_5 = C_6; C_1 = C_4; C_7 = C_8$
ψ_5	+0.36278	b_{3g}	+0.61829	−0.14358	−0.22033	$C_2 = C_6 = -C_3 = -C_5; C_1 = -C_4; C_7 = -C_8$
ψ_6	−0.28889	a_u	0	0	+ 1/2	$C_7 = C_1 = C_4 = C_8 = 0; C_2 = C_5 = -C_3 = C_6$
ψ_7	−0.38083	b_{2v}	+0.45764	−0.44657	+0.21347	$C_2 = C_3 = C_5 = C_6; C_1 = C_4; C_7 = C_8$
ψ_8	(b_{3g}) = −1.18728					

(f) Quinoline 1-oxide (Condition 3) C_s

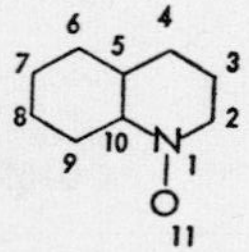

ψ_i	Eigenvalues (ε_i)	C_{i1}	C_{i2}	C_{i3}	C_{i4}	C_{i5}	C_{i6}	C_{i7}	C_{i8}	C_{i9}	C_{i11}	C_{i11}
ψ_1	+3.09063	+0.70014	+0.33562	+0.15813	+0.12498	+0.22074	+0.09051	+0.05363	+0.07458	+0.17239	+0.42749	+0.28055
ψ_2	+2.10244	+0.34186	+0.21189	−0.00938	−0.22994	−0.46043	−0.38932	−0.33503	−0.31091	−0.30022	−0.26691	+0.22678
ψ_3	+1.67910	+0.16011	−0.34416	−0.55443	−0.48822	−0.23641	−0.07585	+0.11355	+0.26511	+0.31589	+0.20914	+0.14769
ψ_4	+1.23341	+0.00178	+0.37674	+0.26197	−0.1002	−0.37962	+0.07964	+0.47313	+0.49808	+0.11169	−0.38018	+0.00278
ψ_5	+1.06768	+0.16534	−0.15012	−0.24556	−0.06840	+0.17658	+0.51322	+0.34096	−0.15339	−0.49565	−0.28768	+0.34979
ψ_6	+0.42113	−0.12055	−0.37598	+0.16274	+0.41558	−0.01235	−0.24333	−0.07571	+0.21238	+0.15256	−0.17525	+0.69335

(g) Isoquinoline N-oxide (Condition 3) C_s

ψ_i	Eigen-values (ε_i)	C_{i1}	C_{i2}	C_{i3}	C_{i4}	C_{i5}	C_{i6}	C_{i7}	C_{i8}	C_{i9}	C_{i10}	C_{i11}
ψ_1	+3.03113	+0.72279	+0.36183	+0.18099	+0.15460	+0.06238	+0.03371	+0.03967	−0.08605	+0.21607	+0.37587	+0.29669
ψ_2	+2.18861	+0.30840	+0.06799	−0.19587	−0.46184	−0.34676	−0.29280	−0.29286	−0.34455	−0.44080	−0.07998	+0.19353
ψ_3	+1.55933	+0.09375	−0.44456	−0.54986	−0.31511	−0.18154	+0.03426	+0.23483	+0.32902	+0.25872	+0.34353	+0.09721
ψ_4	+1.32388	+0.15812	+0.05882	−0.11162	−0.18675	+0.28609	+0.56197	+0.45558	+0.03554	−0.41064	−0.31942	+0.21694
ψ_5	+0.88856	+0.07049	−0.44659	−0.22913	+0.28373	+0.47937	+0.13630	−0.35882	−0.45071	−0.01495	+0.15634	+0.24010
ψ_6	+0.47457	−0.08776	−0.18298	+0.09851	+0.21221	−0.12440	−0.26971	−0.00249	+0.26856	+0.11402	−0.44694	+0.72868
ψ_7	−0.37522	−0.31905	−0.16383	+0.46789	−0.09492	−0.32307	+0.22013	+0.23957	−0.31298	−0.10359	+0.46518	+0.32884

$\varepsilon_8 = -0.67$; $\varepsilon_9 = -1.135$; $\varepsilon_{11} = -2.15$

(h) Acridine N-oxide (Condition 3) C_{2v}

ψ_i	Eigenvalues (ε_i)	Symmetry	C_{i1}	C_{i2}	C_{i3}	C_{i4}	C_{i5}	C_{i6}	C_{i7}	C_{i8}	C_{i9}
ψ_1	+3.15620	b_2	+0.67255	+0.39201	+0.15321	+0.06432	+0.04597	+0.08021	+0.20244	+0.13074	+0.26259
ψ_2	+2.19741	a_2	0	+0.36769	+0.30593	+0.25018	+0.22899	+0.25018	+0.30593	0	0
ψ_3	+2.07512	b_2	+0.41223	−0.04133	−0.10756	−0.16275	−0.22053	−0.29214	−0.36839	−0.36549	+0.27851
ψ_4	+1.50721	b_2	+0.02851	−0.01695	−0.32246	−0.41173	−0.27371	+0.00257	+0.27743	+0.38321	+0.03126

(Continued overleaf)

References p. 167

TABLE 4-36 *(Continued)*

ψ_i	Eigenvalues (ε_i)	Symmetry	C_{i1}	C_{i2}	C_{i3}	C_{i4}	C_{i5}	C_{i6}	C_{i7}	C_{i8}	C_{i9}
ψ_5	+1.20982	a_2	0	+0.39174	+0.13250	−0.25499	−0.42589	−0.25499	+0.13250	0	0
ψ_6	+1.12027	a_2	0	0	+0.36386	+0.34294	0	−0.34294	−0.36386	0	0
ψ_7	+1.02660	b_2	+0.19434	−0.28085	−0.30762	+0.01974	+0.32671	+0.31163	−0.02526	−0.05222	+0.45027
ψ_8	+0.30475	b_2	−0.17995	−0.19344	+0.16901	+0.21490	−0.11626	−0.24890	+0.05515	+0.44934	+0.61997
ψ_9	−0.25300	b_2	−0.39519	+0.13313	+0.22312	−0.22925	−0.15153	+0.26946	+0.06739	−0.43162	+0.46603

$C_{i2} = \pm C_{i2}'$; $C_{i3} = \pm C_{i3}'$; $C_{i4} = \pm C_{i4}'$; $C_{i5} = \pm C_{i5}'$; $C_{i6} = \pm C_{i6}'$; $C_{i7} = \pm C_{i7}'$
(+ and − represent b_2 and a_2 orbitals respectively)
$\varepsilon_{10} = -0.77215$ (a_2); $\varepsilon_{11} = -0.88324$ (a_2); $\varepsilon_{12} = -1.07$ (b_2); ε_{13}, ε_{14} and ε_{15} were not calculated.

(i) Acridine N-oxide (Condition 2′)

ψ_i	Eigenvalues (ε_i)	Symmetry	C_{i1}	C_{i2}	C_{i3}	C_{i4}	C_{i5}	C_{i6}	C_{i7}	C_{i8}	C_{i15}
ψ_1	+3.42879	b_2	+0.72628	+0.36934	+0.30855	+0.04802	+0.03092	+0.05802	+0.16372	+0.09761	+0.29903
ψ_2	+2.25473	a_2	0	+0.38858	+0.30855	+0.23855	+0.21159	+0.23855	+0.30855	0	0
ψ_3	+2.18711	b_2	+0.35232	−0.11543	−0.14127	−0.16215	−0.20135	−0.27823	−0.38656	−0.36588	+0.29679
ψ_4	+1.53061	b_2	+0.02910	−0.03425	−0.33757	−0.40472	−0.25585	+0.01581	+0.27888	+0.38332	+0.05484
ψ_5	+1.25217	a_2	0	+0.38395	+0.11240	−0.26818	−0.42834	−0.26818	+0.11240	0	0
ψ_6	+1.16300	b_2	+0.08527	−0.29724	−0.25895	+0.05363	+0.31734	+0.31544	+0.02615	+0.02722	+0.52311
ψ_7	+1.15088	a_2	0	0	+0.36638	+0.34024	0	−0.34024	−0.36638	0	0
ψ_8	+0.50662	b_2	−0.31099	−0.08295	+0.24854	+0.15363	−0.18208	−0.24588	+0.07573	+0.35037	+0.63033
ψ_9	−0.10861	b_2	−0.37019	+0.22333	+0.14933	−0.27274	−0.09951	+0.28354	+0.04771	−0.52266	+0.33392

$\varepsilon_{10}(a_2) = -0.71877$; $\varepsilon_{11}(a_2) = -0.85459$; $\varepsilon_{12}(b_2) = -0.96488$; $\varepsilon_{13}(b_2) = -1.30892$; $\varepsilon_{14}(a_2) = -1.82514$; $\varepsilon_{15}(b_2) = -2.10040$

conjugate acid, but the bromination may also occur in 2- and 4-positions[177] through the free base under suitable conditions, as will be explained in detail in a separate chapter (*cf. Section 6.3*).

With the exception of a few examples, a nucleophilic reaction of *N*-oxides seems to proceed through the conjugate acid and an addition compound with the nucleophile in 2- or 4-position is formed as the intermediate, as shown in Fig. 4-26 (C). This behavior is similar to that of the corresponding N^+-substituted *N*-heterocycles. Since these two kinds of addition compounds, which are further susceptible to both electrophilic and nucleophilic reactions, may in a sense be taken as a kind of enamines[178], various reaction courses giving rise to C, D, E, F, etc., indicated in Fig. 4-26 may be anticipated for the *N*-oxide system. The course of the reaction giving rise to E is specific for the salt of an *N*-oxide system[178,179] and should be excluded for the reaction with the salts of *N*-heterocyclic compounds. Various reaction patterns applicable to ionic reactions, taking pyridine 1-oxide as an example, are shown in Fig. 4-26.

To close this section, simple molecular orbitals of typical aromatic *N*-oxide compounds calculated by Kubota[135,154,12,13] have been listed in Table 4-36, in which the conditions used for the calculations correspond to those given in Table 4-34.

REFERENCES

1 J. Meisenheimer, *Ber.*, 41 (1908) 3966; *Ann.*, 385 (1911) 117; *Ann.*, 428 (1922) 252; J. Meisenheimer, H. Glawe, H. Greeske, A. Schorning, and E. Vieweg, *Ann.*, 449 (1926) 188.
2 E. P. Linton, *J. Am. Chem. Soc.*, 62 (1940) 1945.
3 E. Ochiai, M. O. Ishikawa, and K. Arima, *Yakugaku Zasshi*, 63 (1943) 140; E. Ochiai and M. O. Ishikawa, *Proc. Imp. Acad. Tokyo*, 18 (1942) 56; E. Ochiai, E. Hayashi, and M. Katada, *Yakugaku Zasshi*, 67 (1947) 33.
4 *cf.* E. Ochiai, *J. Org. Chem.*, 18 (1953) 534.
5 T. Okamoto, H. Hayatsu, and Y. Baba, *Chem. Pharm. Bull. (Tokyo)*, 8 (1960) 892; R. F. Evans and H. C. Brown, *J. Org. Chem.*, 27 (1962) 1329; M. Liveris and J. Miller, *J. Chem. Soc.*, (1963) 3486; G. Coppens, F. Declerck, C. Gillet, and J. Nasielski, *Bull. Soc. Chim. Belges*, 72 (1963) 572. *Cf.* also N. B. Chapmann, D. K. Chaudhury, and J. Shorter, *J. Chem. Soc.*, (1962) 1975; G. Illuminati and G. Marino, *Chem. Ind. (London)*, (1963) 1287; G. Illuminati and G. Marino, *Tetrahedron Letters*, (1963) 1055.
6 A. R. Katritzky, E. W. Randall, and L. E. Sutton, *J. Chem. Soc.*, (1957) 1769.
7 C. M. Bax, A. R. Katritzky, and L. E. Sutton, *J. Chem. Soc.*, (1958) 1254.
8 K. B. Everard, L. Kumar, and L. E. Sutton, *J. Chem. Soc.*, (1951) 2807. See also L. E. Sutton, *Proc. Roy. Soc. (London)*, *Ser. A*, 133 (1931) 668.

9 C. M. BAX, A. R. KATRITZKY, AND L. E. SUTTON, *J. Chem. Soc.*, (1958) 1258.
10 A. N. SHARPE AND S. WALKER, *J. Chem. Soc.*, (1961) 4522.
11 H. H. JAFFÉ, *Chem. Revs.*, 53 (1953) 191.
12 T. KUBOTA AND H. WATANABE, *Bull. Chem. Soc. Japan*, 36 (1963) 1093.
13 T. KUBOTA, *Bull. Chem. Soc. Japan.*, 35 (1962) 946.
14 M. OGATA, H. WATANABE, K. TÔRI, AND H. KANÔ, *Tetrahedron Letters*, 1 (1964) 19
15 H. WATANABE, M. OGATA, AND H. KANÔ, *Chem. Pharm. Bull. (Tokyo)*, 11 (1963) 29 35, 39. *Cf.* also H. KANÔ, M. OGATA, H. WATANABE, AND I. ISHIZUKA, *Chem. Pharm Bull. (Tokyo)*, 9 (1961) 1017.
16 H. OTOMASU, R. YAMAGUCHI, K. ISHIGÔ-OKA, AND H. TAKAHASHI, *Yakugaku Zasshi* 82 (1962) 1434.
17 H. OTOMASU, H. TAKAHASHI, AND M. OGATA, *Chem. Pharm. Bull. (Tokyo)*, 12 (1964) 714.
18 M. OGATA, *Chem. Pharm. Bull. (Tokyo)*, 11 (1963) 1522.
19 T. SASAKI AND K. MINAMOTO, *Chem. Pharm. Bull. (Tokyo)*, 12 (1964) 1329.
20 K. E. BAUGE AND J. W. SMITH, *J. Chem. Soc.*, (1962) 5292.
21 R. M. ACHESON, B. ADCOCK, G. M. GLOVER, AND L. E. SUTTON, *J. Chem. Soc.*, (1960) 3367.
22 Z. V. PUSHKAREVA, L. V. VARYUKHINA, AND Z. YU. KOKOSHKO, *Dokl. Akad. Nauk SSSR*, 108 (1956) 1098 [*Chem. Abstr.*, 51 (1957) 21].
23 Z. V. PUSHKAREVA AND O. N. NECHAEVA, *Zh. Obshch. Khim.*, 28 (1958) 2702 [*Chem Abstr.*, 53 (1959) 9229].
24 M. W. LISTER AND L. E. SUTTON, *Trans. Faraday Soc.*, 35 (1939) 495.
25 C. RÉRAT, *Acta Cryst.*, 13 (1960) 63.
26 A. C. CARON, G. J. PALENIK, E. GOLDISH, AND J. DONOHUE, *Acta Cryst.*, 17 (1964) 102
27 P. W. ALLEN AND L. E. SUTTON, *Acta Cryst.*, 3 (1950) 46.
28 *(a)* E. L. EICHHORN, *Acta Cryst.*, 9 (1956) 787. *(b)* E. L. EICHHORN, *Acta Cryst.*, 1 (1959) 746. *Cf.* also E. L. EICHHORN, Doctoral thesis, Amsterdam, 1956.
29 R. CURTI, V. RIGANTI, AND S. LOCCHI, *Acta Cryst.*, 14 (1961) 133.
30 *(a)* Y. NAMBU, T. ODA, H. ITO, AND T. WATANABE, *Bull. Chem. Soc. Japan*, 33 (1960) 1618. *(b)* Y. NAMBU, T. ODA, AND T. WATANABE, *Bull. Chem. Soc. Japan*, 34 (1961) 889.
31 P. G. TSOUCARIS, *Acta Cryst.*, 14 (1961) 909.
32 E. OCHIAI, *Proc. Imp. Acad. Tokyo*, 19 (1943) 307; E. OCHIAI, T. NAITO, AND M. KATADA, *Proc. Imp. Acad. Tokyo*, 19 (1943) 574; E. OCHIAI, *Yakugaku Zasshi*, 69 (1949) 1.
33 L. V. VARYUKHINA AND Z. V. PUSHKAREVA, *Zh. Obshch. Khim.*, 26 (1956) 1740 [*Chem Abstr.*, 51 (1957) 1960]. *Cf.* D. V. IOFFE AND L. S. EFROS, *Usp. Khim.*, 30 (1961) 132 [*Chem. Abstr.*, 57 (1962) 11175].
34 O. N. NECHAEVA AND Z. V. PUSHKAREVA, *Zh. Obshch. Khim.*, 28 (1958) 2693 [*Chem Abstr.*, 53 (1959) 9229].
35 M. COLONNA, *Boll. Sci. Fac. Chim. Ind. Bologna*, 15 (1957) 1.
36 T. R. EMERSON AND C. W. REES, *J. Chem. Soc.*, (1962) 1923.
37 A. FOFFANI AND E. FORNASARI, *Gazz. Chim. Ital.*, 83 (1953) 1051, 1059 [*Chem. Abstr.* 49 (1955) 2901].
38 M. P. STRIER AND J. C. CAVAGNOL, *J. Am. Chem. Soc.*, 79 (1957) 4331.
39 T. KUBOTA AND H. MIYAZAKI, *Bull. Chem. Soc. Japan*, 35 (1962) 1549.
40 T. KUBOTA AND H. MIYAZAKI, *Bull. Chem. Soc. Japan*, 39 (1966) 2057.
41 T. KUBOTA, *Bunkô Kenkyû*, 10 (1962) 83. *Cf.* also T. KUBOTA, *Yakugaku Zasshi*, 7 (1955) 1546; T. KUBOTA AND H. MIYAZAKI, *Nippon Kagaku Zasshi*, 79 (1958) 924.
42 T. S. LEE, I. M. KOLTHOFF, AND D. L. LEUSSING, *J. Am. Chem. Soc.*, 70 (1948) 2348
43 E. J. COREY, A. L. BORROR, AND T. FOGLIA, *J. Org. Chem.*, 30 (1965) 288.

44 C. Golumbic and M. Orchin, *J. Am .Chem. Soc.*, 72 (1950) 4145.
45 A. R. Osborn, K. Schofield, and L. Short, *J. Chem. Soc.*, (1956) 4191.
46 A. Albert, R. Goldacre, and J. Phillips, *J. Chem. Soc.*, (1948) 2240.
47 H. H. Jaffé and G. O. Doak, *J. Am. Chem. Soc.*, 77 (1955) 4441.
48 P. Nylén, *Tidsskr. Kjemi, Bergvesen Met.*, 18 (1938) 48 [*Chem. Abstr.*, 32 (1938) 8888].
49 T. D. Stewart and S. Maeser, *J. Am. Chem. Soc.*, 46 (1924) 2583.
50 M. M. Davis and H. B. Hetzer, *J. Am. Chem. Soc.*, 76 (1954) 4247.
51 R. Näsänen and E. Uusitalo, *Suomen Kemistilehti*, 29B (1956) 11 (in English) [*Chem. Abstr.*, 51 (1957) 1698].
52 Y. Okamoto and H. C. Brown, *J. Am. Chem. Soc.*, 79 (1957) 1913.
53 N. C. Deno and W. L. Evans, *J. Am. Chem. Soc.*, 79 (1957) 5804.
54 H. H. Jaffé, *J. Org. Chem.*, 23 (1958) 1790. (In this report, Jaffé proposed that the σ^* value should be designated as σ– value, and calculated the reaction constant ϱ as 1.893).
55 J. S. Driscoll, W. Pfleiderer, and E. C. Taylor, *J. Org. Chem.*, 26 (1961) 5230.
56 H. H. Jaffé, *J. Am. Chem. Soc.*, 76 (1954) 3527.
57 *(a)* H. Shindo, *Pharm. Bull. (Japan)*, 4 (1956) 460. *(b) H.* Shindo, *Chem. Pharm. Bull. (Tokyo)*, 6 (1958) 171.
58 *(a)* H. Hirayama and T. Kubota, *Yakugaku Zasshi*, 73 (1953) 140. *(b)* T. Kubota, *Yakugaku Zasshi*, 75 (1955) 1546.
59 E. Shaw, *J. Am. Chem. Soc.*, 71 (1949) 67.
60 J. N. Gardner and A. R. Katritzky, *J. Chem. Soc.*, (1957) 4375.
61 R. A. Jones and A. R. Katritzky, *J. Chem. Soc.*, (1960) 2937.
62 S. Furukawa, *Yakugaku Zasshi*, 79 (1959) 492.
63 A. R. Katritzky and P. Simmons, *J. Chem. Soc.*, (1960) 1511.
64 A. Albert, *Brit. J. Exptl. Pathol.*, 37 (1956) 500.
65 P. L. Corio and B. P. Dailey, *J. Am. Chem. Soc.*, 78 (1956) 3043; A. A. Bothner-By and R. E. Glick, *J. Am. Chem. Soc.*, 78 (1956) 1071; R. W. Taft, Jr., S. Ehrenson, I. C. Leurs, and R. E. Glick, *J. Am. Chem. Soc.*, 81 (1959) 5352; I. Yamaguchi and N. Hayakawa, *Bull. Chem. Soc. Japan*, 33 (1960) 1128; R. R. Fraser, *Can. J. Chem.*, 38 (1960) 2226; H. Spiesecke and W. G. Schneider, *J. Chem. Phys.*, 35 (1961) 731; P. Diehl, *Helv. Chim. Acta*, 44 (1961) 829; H. Suhr, *Z. Electrochem.*, 66 (1962) 466; G. W. Smith, *Monthly Ecumenical Letters from Laboratories of NMR (Mellon Inst.)*, No. 48 (1962) 1.
66 G. Fraenkel, R. E. Carter, A. McLachlan, and J. H. Richards, *J. Am. Chem. Soc*, 82 (1960) 5846.
67 H. Spiesecke and W. G. Schneider, *Tetrahedron Letters*, No. 14 (1961) 468.
68 B. P. Dailey, A. Gawer, and W. C. Neikam, *Discussions Faraday Soc.*, 34 (1962) 18.
69 T. Schaefer and W. G. Schneider, *Can. J. Chem.*, 41 (1963) 966.
70 J. A. Pople, *J. Chem. Phys.*, 24 (1956) 1111.
71 C. E. Johnson, Jr., and F. A. Bovey, *J. Chem. Phys.*, 26 (1958) 1012.
72 L. M. Jackman and J. A. Elvidge, *J. Chem. Soc.*, (1961) 859.
73 I. Yamaguchi, *Mol. Phys.*, 6 (1963) 105.
74 M. W. Hanna and A. L. Ashbaugh, *J. Phys. Chem.*, 68 (1964) 811.
75 M. Ohnishi and Y. Kawazoe, *Chem. Pharm. Bull. (Tokyo)*, 12 (1964) 938; 11 (1963) 243.
76 S. Matsuoka, S. Hattori, N. Nakagawa, and N. Suzuki, *Proc. Symposium on Electronic State of Molecules*, Tokyo, November, 1961.
77 J. D. Baldeschwieler and E. W. Randall, *Proc. Chem. Soc.*, (1961) 303; *Chem. Revs.*, 63 (1963) 81.
78 V. M. S. Gil and J. N. Marrell, *Trans. Faraday Soc.*, 60 (1964) 248.
Cf. R. D. Brown and M. L. Heffernan, *Australian J. Chem.*, 12 (1959) 554.
79 G. Kotowycz, T. Schaefer, and E. Bock, *Can. J. Chem.*, 42 (1964) 2541.

80 T. OKAMATO, Y. KAWAZOE, H. HOTTA, AND M. ITOH, *Abstr. Papers, Symposium on N-Oxide Chemistry at the Institute of Physical and Chemical Research*, Tokyo, December, 1962.
81 A. R. KATRITZKY AND J. M. LAGOWSKI, *J. Chem. Soc.*, (1961) 43.
82 Y. KAWAZOE AND S. NATSUME, *Yakugaku Zasshi*, 83 (1963) 523.
83 K. TÔRI, M. OGATA, AND H. KANÔ, *Chem. Pharm. Bull. (Tokyo)*, 11 (1963) 235.
84 M. OGATA, H. KANÔ, AND K. TÔRI, *Chem. Pharm. Bull. (Tokyo)*, 10 (1962) 1123.
85 M. OGATA, H. KANÔ, AND K. TÔRI, *Chem. Pharm. Bull. (Tokyo)*, 11 (1963) 1527.
86 K. TÔRI, M. OGATA, AND H. KANÔ, *Chem. Pharm. Bull. (Tokyo)*, 11 (1963) 681.
87 M. ITO AND N. HATA, *Bull. Chem. Soc. Japan*, 28 (1955) 353.
88 G. COSTA AND P. BLASINA, *Z. Physik. Chem. (Frankfurt)*, 4 (1955) 24.
G. SARTORI, G. COSTA, AND P. BLASINA, *Gazz. Chim. Ital.*, 85 (1955) 1085.
89 R. H. WILEY AND S. C. SLAYMAKER, *J. Am. Chem. Soc.*, 79 (1957) 2233.
90 H. SHINDO, *Chem. Pharm. Bull. (Tokyo)*, 7 (1959) 791.
91 A. R. KATRITZKY, J. A. T. BEARD, AND N. A. COATS, *J. Chem. Soc.*, (1959) 3680, and references cited therein.
92 H. SHINDO, *Infrared Absorption Spectra*, Vol. 3 (in Japanese), Nankodo, Tokyo, 1958.
93 A. PALM AND H. WERBIN, *Can. J. Chem.*, 32 (1954) 858.
94 R. MATHIS-NOEL, R. WOLF, AND F. GALLAIS, *Compt. Rend.*, 242 (1956) 1873.
95 D. HADŽI, *J. Chem. Soc.*, (1956) 2725.
96 D. G. O'SULLIVAN AND P. W. SADLER, *J. Org. Chem.*, 22 (1957) 283.
97 K. NAKAMOTO AND R. E. RUNDLE, *J. Am. Chem. Soc.*, 78 (1956) 1113.
98 *e.g.* R. D. KROSS AND V. A. FASSEL, *J. Am. Chem. Soc.*, 78 (1956) 4225.
99 A. R. KATRITZKY AND J. N. GARDNER, *J. Chem. Soc.*, (1958) 2192.
100 A. R. KATRITZKY AND A. R. HANDS, *J. Chem. Soc.*, (1958) 2195.
101 H. SHINDO, *Chem. Pharm. Bull. (Tokyo)*, 8 (1960) 33.
102 B. KLEIN AND J. BERKOWITZ, *J. Am. Chem. Soc.*, 81 (1959) 5160.
103 G. F. KOELSCH AND W. H. GUMPRECHT, *J. Org. Chem.*, 23 (1958) 1603.
104 H. SHINDO, *Chem. Pharm. Bull. (Tokyo)*, 8 (1960) 845.
105 G. R. CLEMO AND A. F. DAGLISH, *J. Chem. Soc.*, (1950) 1481.
106 W. LÜTTKE, *Z. Elektrochem.*, 61 (1957) 9761.
107 Y. MATSUI, H. IWATANI, H. MIYAZAKI, M. TAKASUKA, AND T. KUBOTA, *Paper presented at the Symposidium on Infrared and Raman Spectra held under the auspices of the Chemical Society of Japan.* Tokyo, November 19, 1964.
108 H. SHINDO, *Pharm. Bull. (Japan)*, 5 (1957) 472.
109 H. H. JAFFÉ, *J. Chem. Phys.*, 20 (1952) 1554.
110 H. H. JAFFÉ, *J. Chem. Phys.*, 20 (1952) 279, 778.
111 A. R. KATRITZKY, *Rec. Trav. Chim.*, 78 (1959) 995.
112 H. TANI AND K. FUKUSHIMA, *Yakugaku Zasshi*, 81 (1961) 27.
113 A. R. KATRITZKY AND R. A. JONES, *J. Chem. Soc.*, (1959) 3674.
114 J. D. S. GOULDEN, *J. Chem. Soc.*, (1952) 2939.
115 A. R. KATRITZKY AND A. P. AMBLER, *Physical Methods in Heterocyclic Chemistry* Vol. II, Academic Press, Inc., New York, 1963.
116 H. HIRAYAMA AND T. KUBOTA, *Shionogi Kenkyusho Nempo*, 2 (1952) 121.
117 T. KUBOTA, *Shionogi Kenkyusho Nempo*, 6 (1956) 31.
118 H. HIRAYAMA AND T. KUBOTA, *Yakugaku Zasshi*, 72 (1952) 1025.
119 T. KUBOTA, *Yakugaku Zasshi*, 74 (1954) 831.
120 T. KUBOTA, *Yakugaku Zasshi*, 77 (1957) 785.
121 H. H. JAFFÉ, *J. Am. Chem. Soc.*, 77 (1955) 4451.
122 *Cf.* M. COLONNA AND A. RISALITI, *Atti Accad. Naz. Lincei*, [8], 14 (1953) 809.
123 A. R. KATRITZKY, A. M. MONRO, AND J. A. T. BEARD, *J. Chem. Soc.*, (1958) 3721.

124 V. I. Bliznynkov and V. M. Reznikov, *Zh. Obshch. Khim.*, 25 (1955) 401.
125 T. Kubota and M. Yamakawa, *Bull. Chem. Soc. Japan*, 35 (1962) 555.
126 T. Kubota, *Yakugaku Zasshi*, 75 (1955) 1540.
127 T. Kubota, *Nippon Kagaku Zasshi*, 79 (1958) 916.
128 T. Kubota and H. Miyazaki, *Chem. Pharm. Bull. (Tokyo)*, 9 (1961) 948.
129 M. Ito and N. Hata, *Bull. Chem. Soc. Japan*, 28 (1955) 260.
130 N. Hata and I. Tanaka, *J. Chem. Phys.*, 36 (1962) 2072.
131 N. Hata, *Bull. Chem. Soc. Japan*, 34 (1961) 1440, 1444.
132 M. Ito and W. Mizushima, *J. Chem. Phys.*, 24 (1956) 495.
133 N. Hata, *Bull. Chem. Soc. Japan*, 29 (1956) 495.
134 *(a)* H. P. Stephenson, *J. Chem. Phys.*, 22 (1954) 1077. *(b)* H. L. McMurray, *J. Chem. Phys.*, 9 (1941) 241.
135 T. Kubota, *Nippon Kagaku Zasshi*, 80 (1959) 578.
136 *(a)* J. R. Platt, *J. Chem. Phys.*, 17 (1949) 484. *(b)* G. N. Lewis and M. Calvin, *Chem. Rev.*, 25 (1939) 273. G. N. Lewis and J. Bigeleisen, *J. Am. Chem. Soc.*, 65 (1943) 520, 2102, 2107.
137 E. Clar, *Aromatische Kohlenwasserstoffe*, Springer Verlag, Berlin, 1952, 2nd Ed.
138 L. Doub and J. M. Vandenbelt, *J. Am. Chem. Soc.*, 69 (1947) 2714; 71 (1949) 2414.
139 S. F. Mason, in A. R. Katritzky (Editor), *Physical Methods in Heterocyclic Chemistry*, Academic Press, Inc., New York, 1963, Vol. II.
140 H. C. Brown and X. R. Mihm, *J. Am. Chem. Soc.*, 77 (1955) 1723.
141 T. Kubota, *Nippon Kagaku Zasshi*, 79 (1957) 930.
142 E. Ochiai and M. Y. Ishikawa, *Chem. Pharm. Bull. (Tokyo)*, 6 (1958) 208.
143 K. G. Cunningham, G. T. Newbold, F. S. Spring, and J. Stark, *J. Chem. Soc.*, (1949) 2091.
144 D. A. Prins, *Rec. Trav. Chim.*, 76 (1957) 58.
145 R. A. Jones and A. R. Katritzky, *J. Chem. Soc.*, (1960) 2937.
146 A. R. Katritzky, *J. Chem. Soc.*, (1957) 191.
147 G. T. Newbold and F. S. Spring, *J. Chem. Soc.*, (1949) S 133.
148 N. Ikekawa and Y. Sato, *Pharm. Bull. (Japan)*, 2 (1954) 400.
149 N. Ikekawa and Y. Sato, *Pharm. Bull. (Japan)*, 2 (1954) 209.
150 *e.g.* H. H. Hodgson and D. E. Hathway, *Trans. Faraday Soc.*, 41 (1945) 115.
151 E. A. Steck and G. W. Ewing, *J. Am. Chem. Soc.*, 70 (1948) 3397.
152 E. A. Steck and G. W. Ewing, *J. Am. Chem. Soc.*, 68 (1946) 2181.
153 Ch. Kaneko, *Yakugaku Zasshi*, 79 (1959) 433.
154 T. Kubota, *Yakugaku Zasshi*, 79 (1959) 388.
155 H. H. Jaffé and M. Orchin, *Theory and Applications of Ultraviolet Spectroscopy*, John Wiley & Sons, New York, 1962.
156 E. Ochiai and Ch. Kaneko, *Chem. Pharm. Bull. (Tokyo)*, 5 (1957) 56.
157 A. Streitwieser, Jr., *Molecular Orbital Theory for Organic Chemists*, John Wiley & Sons, New York, 1961.
158 C. A. Coulson, *Valence*, Oxford University Press, Oxford, 1961.
159 R. Daudel, R. Lefebre, and C. Moser, *Quantum Chemistry*; *Method and Applications*, Interscience Publishers, New York, 1959.
160 G. W. Wheland and L. Pauling, *J. Am. Chem. Soc.*, 57 (1935) 2086; C. A. Coulson and H. C. Longuet-Higgins, *Proc. Roy. Soc.*, A 191 (1947) 39; A 192 (1947) 16, A 193 (1948) 447.
161 G. W. Wheland, *J. Am. Chem. Soc.*, 64 (1942) 900.
162 *e.g.* H. H. Greenwood, *Trans. Faraday Soc.*, 48 (1952) 585.
163 K. Fukui, T. Yonezawa, Ch. Nagata, and H. Shingu, *J. Chem. Phys.*, 22 (1954) 1433; K. Fukui, T. Yonezawa, and Ch. Nagata, *Bull. Chem. Soc. Japan*, 27 (1954)

423; K. Fukui, T. Yonezawa, and Ch. Nagata, *J. Chem. Phys.*, 27 (1957) 1247, 31 (1959) 550, 32 (1960) 1743.
164 R. D. Brown, *Quart. Rev.*, 6 (1952) 63.
165 R. A. Barnes, *J. Am. Chem. Soc.*, 81 (1959) 1935.
166 S. Basu and K. L. Saha, *Naturwiss.*, 24 (1957) 633.
167 K. Fukui, A. Imamura, and Ch. Nagata, *GANN*, 51 (1960) 119.
168 S. Mataga and N. Mataga, *Z. Physik. Chem. N.F.*, 19 (1959) 231.
169 T. Okamoto and H. Tani, *Chem. Pharm. Bull. (Tokyo)*, 7 (1959) 925; H. Tani, *Chem. Pharm. Bull .(Tokyo)*, 7 (1959) 930.
170 M. J. S. Dewar and P. M. Maitlis, *J. Chem. Soc.*, (1957) 2521.
Cf. also P. B. D. de LaMare and J. H. Ridd, *Aromatic Substitution, Nitration, and Halogenation*, Academic Press, Inc., New York, 1959.
171 M. J. S. Dewar, T. Mole, and E. W. T. Warford, *J. Chem. Soc.*, (1956) 3581.
172 L. P. Hammett, *J. Am. Chem. Soc.*, 59 (1937) 96; L. P. Hammett, *Physical Organic Chemistry*, McGraw-Hill Book Company, New York. 1940.
173 E. Ochiai and T. Okamoto, *Yakugaku Zasshi*, 70 (1950) 384.
174 E. Ochiai and K. Satake, *Yakugaku Zasshi*, 71 (1951) 1078.
175 H. Hotta, private communication.
176 R. B. Moodie, K. Schofield, and M. J. Williams, *Chem. Ind. (London)*, (1964) 1577.
177 M. van Ammers, H. J. den Hertog, and B. Haase, *Tetrahedron*, 18 (1962) 227. *Cf.* A. R. Katritzky, B. J. Ridgewell, and A. M. White, *Chem. Ind. (London)*, (1964) 1576.
178 Ch. Kaneko and T. Miyasaka, *Tokyo Ika-Shika-Daigaku Shikazairyo Kenkyu Nempo*, 2 (1963) 466.
179 L. Bauer and L. A. Gardella, *J. Org. Chem.*, 28 (1963) 1320.

Chapter 5

Reactivity of the Aromatic N-Oxide Group

5.1 Electrophilic Addition to the Oxygen Atom in the Aromatic N-Oxide Group

5.1.1 General

The oxygen atom in the aromatic *N*-oxide group is basic and undergoes addition with a proton, metal ions, Lewis acids, alkyl halides, alkyl sulfonates, and acyl halides to form complex compounds. The aromatic *N*-oxide group also forms a 1,3-dipole with a methine group adjacent to nitrogen, and undergoes 1,3-dipolar cyclic addition with compounds with unsaturated bonds (active double bond, triple bond, isocyanates, etc.)[1]. The hetero ring thereby formed is generally unstable and undergoes decomposition, which in effect means a nucleophilic substitution at the α-position. Therefore, details of these reactions will be given under the section on S_N reactions (*cf. Section 7.3*).

Similarly, the position α to the nitrogen in adducts of *N*-oxides with alkyl halides, alkyl sulfonates, and acyl halides, or the position conjugated with it, is highly reactive to nucleophilic reagents and becomes more susceptible to an S_N reaction. Details of these reactions will also be given under the respective sections (*cf. Section 7.4*).

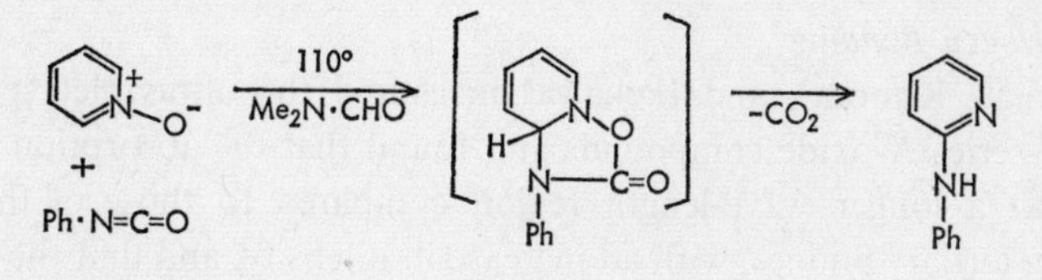

5.1.2. Salt Formation

The aromatic *N*-oxide forms a stable salt with a strong acid. As was described in the preceding Chapter (*cf. Section 4.5*) the basicity of *N*-oxides is far smaller than that of the corresponding tertiary amines and agrees with the decrease of the negativity of oxygen indicated by the resonance,

$\geqslant N^+–O^- \longleftrightarrow \geqslant N^+{=}O$. Therefore, its basicity is dependent on the substituents in the aromatic ring, especially on those in the *ortho* and *para* positions. Jaffé[2] proved that the pKa values of pyridine 1-oxide derivatives are in approximately linear relation with the Hammett's substituent constants.

These salts are generally crystalline and are utilized for the purification and characterization of aromatic *N*-oxide compounds. For these purposes, hydrochlorides, hydrobromides, perchlorates, and picrates are mostly employed.

Aromatic *N*-oxide compounds also combine with Lewis acids, as with a proton, and form complex compounds, as shown below.

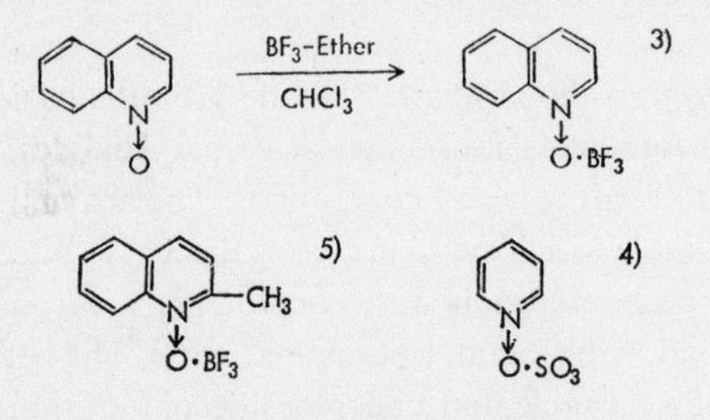

Example. Complex of Quinoline 1-Oxide with Boron Trifluoride[3]

Ether solution of 2 g of boron trifluoride-etherate was added to a solution of 1.95 g of quinoline 1-oxide (anhydrous) dissolved in chloroform. The reaction occurred immediately with evolution of heat, and crystals separated out which were collected by filtration and washed thoroughly with chloroform. Crude crystals melted at 154–155°. Yield, quantitative. This was recrystallized from anhydrous acetone to colorless needles, m.p. 157.5–158.5°

5.1.3 Complex Formation

(a) Hydrogen Bonding

Colonna[6], Kubota[7], and Ikekawa[8] measured the ultraviolet spectra of pyridine series *N*-oxide compounds and found that the absorption maxima shifted to a longer wave-length region compared to those of the corresponding tertiary amines, with an increase in intensity, and that the maxima shifted to the shorter wave-length region with increasing polarity of the solvent, approaching the absorption of the corresponding tertiary amines. Table 4-23 shows this relationship. This fact may be utilized for the identification of the formation of *N*-oxides. Fig. 4-16 illustrates the use of ultraviolet spectra for the identification of the *N*-oxide group during the synthesis of the *N*-oxide of dihydrocinchonine by Ochiai and Ishikawa.

Kubota[7c,d,e] carried out a quantitative examination of the shift of absorptions in the equilibrium systems of these aromatic *N*-oxides with phenol or methanol in carbon tetrachloride solutions of various concentrations, calculated their equilibrium constants, and computed the energy of hydrogen bonding from their temperature coefficients. He showed that the shift of absorptions due to the solvent was caused by the solvation effect through hydrogen bonding. Such a fact indicates the bathochromic shift of the absorption due to resonance of a free electron pair of the oxygen atom in the *N*-oxide group with a π-electron pair in the aromatic ring, and interference of this resonance by hydrogen bonding.

The following is an example of intramolecular hydrogen bonding. Quinaldoin *N,N'*-dioxide (5-1) forms stable, brick red crystals and its enediol type is believed to form a hydrogen bond with the oxygen atom of the *N*-oxide group[9].

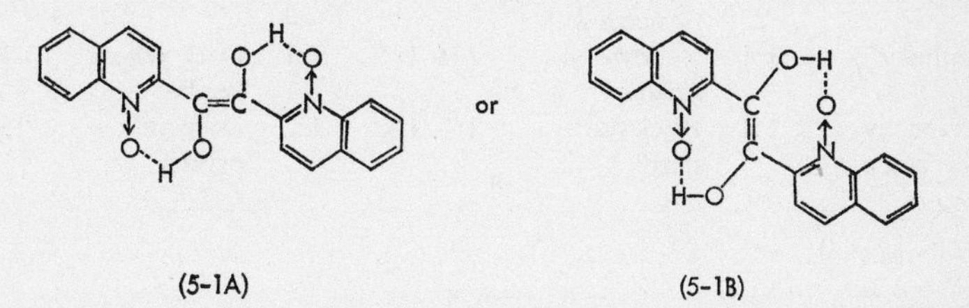

(5-1A) (5-1B)

Vozza[10] showed that the application of hydrogen halide to 2-picoline 1-oxide in ethanol gave a normal type hydrohalide (5-2) of 1:1 ratio, while the same application in anhydrous benzene gave an abnormal salt (5-3), composed of 1 mole of hydrogen halide (chloride, bromide, or iodide) and 2 moles of the *N*-oxide compound. Szafran[11] recognized the same fact in the salts of 2-, 3-, and 4-picoline 1-oxide, and 2,6-lutidine 1-oxide with hydrogen iodide, perchloric acid, hexachloroantimonic acid, and tetrachlorostannic acid, and pointed out that the infrared spectra of these salts lacked the absorption of a free hydroxyl band.

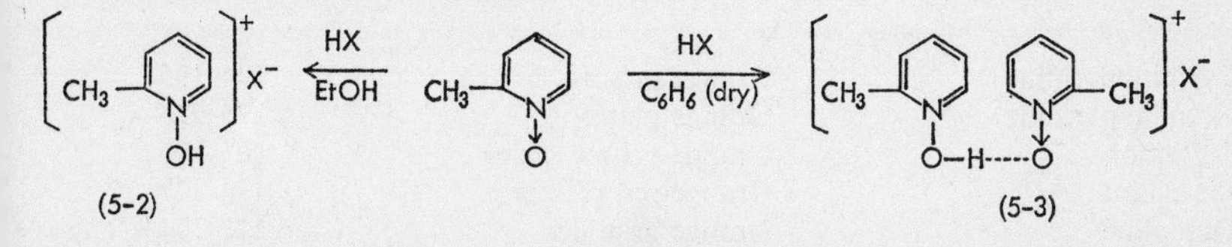

(5-2) (5-3)

This intermolecular hydrogen bonding would cause the formation of a molecular compound. Kröhnke and Schäfer[12] prepared molecular compounds of 4-nitropyridine 1-oxides and phenols in 1:1, 2:1, or 1:2 molar ratio by

References p. 206

TABLE 5-1

MOLECULAR COMPOUNDS OF 4-NITROPYRIDINE 1-OXIDES AND PHENOLS[12]

Phenols	4-Nitropyridine 1-oxide			4-Nitro-3-picoline 1-oxide		
	Molar ratio[a]	Crystal form	m.p. (°C)	Molar ratio[a]	Crystal form	m.p. (°C)
Pyrocatechol	1:1	Golden yellow flat prisms	112–113	1:1	Golden yellow plates	105–107
Resorcinol	2:1	Orange prismatic needles	111–112	1:1	Fine yellow prisms	100–101
Hydroquinone	2:1	Golden yellow prisms or needles	130–131	1:1	Alizarin-red fine prisms	122–124
Pyrogallol	1:1	Dichromate-colored prisms or needles	128–129	1:1	Orange-yellow fine needles	122–123
2-Naphthol	1:1	Yellow-red prisms	134–135	1:1	Dark yellow needles	112–114
2,7-Dihydroxy-naphthalene	1:1	Brick red needles	165–166	1:2	Deep yellow prisms	153–155

[a] *N*-oxide:phenol

TABLE 5-2

AMINE SALTS OF 4-HYDROXY-3,5-DINITROPYRIDINE 1-OXIDE[13]

Amines	Adduct	
	Crystal form	m.p.(decomp.)(°C)
2-Aminopyridine	Orange prisms	194–195
4-Aminopyridine	Orange plates	225–226
4-Pyridone	Orange-yellow scales	205–206
Piperidine	Orange-red prisms	176–177
Quinoline	Orange granules	172–174
4-Quinolone	Orange needles	219–221
Quinine	Orange-red needles	189–190
2-Naphthylamine	Red needles	152–153

recrystallization of the components from acetonitrile, as listed in Table 5-1.

Hayashi[13] found that 4-hydroxy-3,5-dinitropyridine 1-oxide forms a well-crystallizable salt with various amines and that this tendency can be utilized for its purification and characterization. For example, this compound forms an adduct with 0.5 mole of pyridine, crystallizing in orange prisms, decomposing at 207–209°. Some of these adducts are listed in Table 5-2.

(b) Metal Complex

The oxygen atom in aromatic *N*-oxides co-ordinates to metal ions as to a proton and forms a complex salt. For example, Vozza[10] prepared a sodium salt of 2-picoline 1-oxide, corresponding to the "abnormal" salt (5-3) described in the preceding section, by application of sodium iodide to the acetone solution of this *N*-oxide.

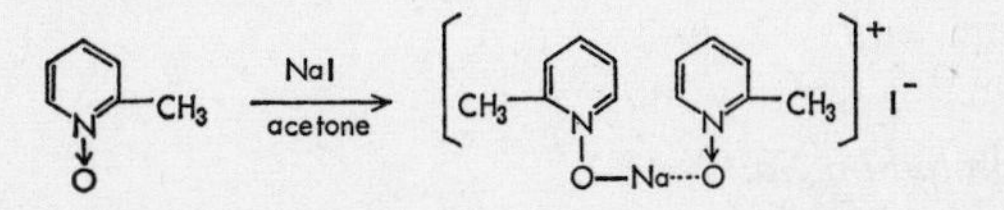

Complex salts reported in past literature are mostly those of pyridine series *N*-oxides, as shown below.

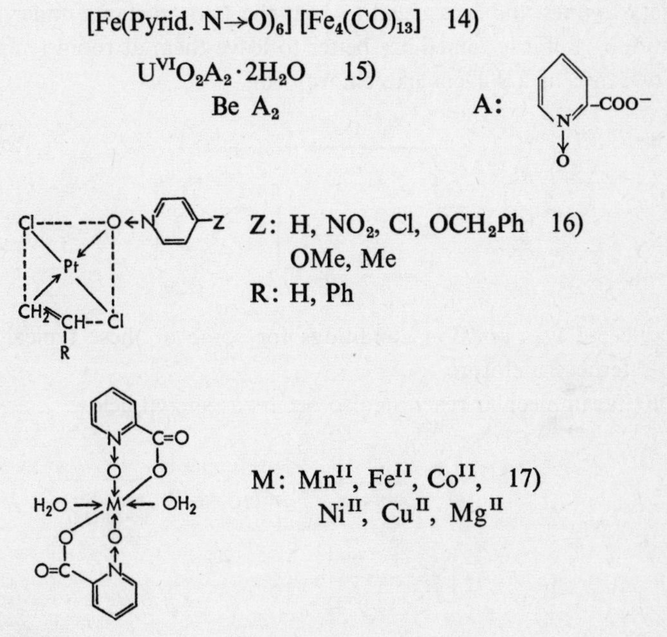

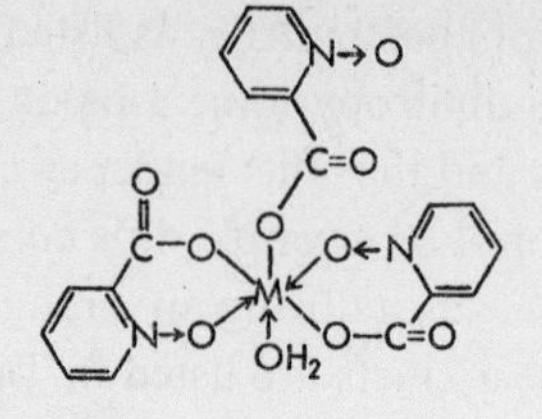

M: Cr^{III}, Co^{III} 17)

$M^{II}(LH)_6(ClO_4)_2$

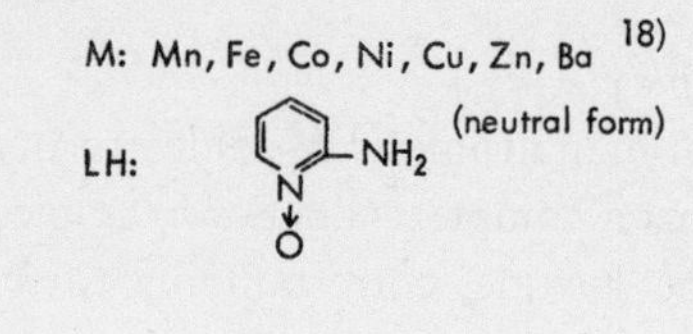

$Cu^{II}(Fe^{III})$ (in higher alkalinity) 19)

5.1.4 Quaternary Salts

(a) Alkoxyammonium Salts

Aromatic *N*-oxide compounds can undergo addition with alkyl halide, dialkyl sulfate, and alkyl sulfonate to form alkoxyammonium salts. In general, *N*-oxide compounds have a lower reactivity than their corresponding tertiary amines and it is usual to heat the two reactants under anhydrous conditions, but it is sometimes better to leave them at room temperature in the presence of a solvent like chloroform.

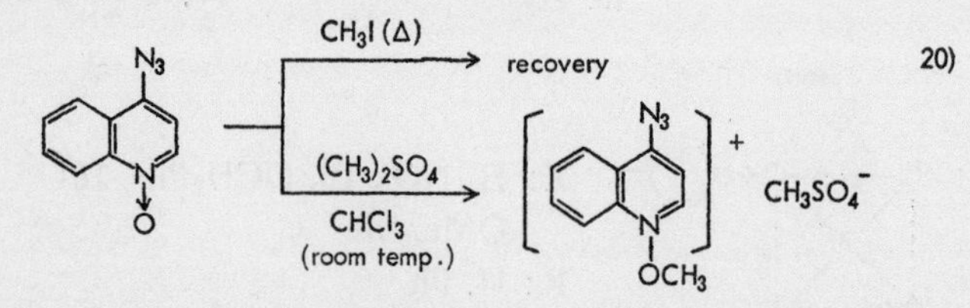

Table 5-3 lists reaction conditions for some of these typical reactions, with reference literature.

The intramolecular reaction also occurs as shown below.

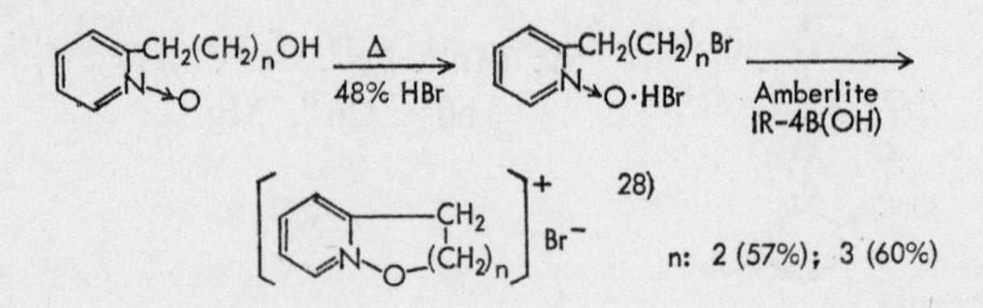

TABLE 5-3

N-ALKYLATION OF AROMATIC AMINE OXIDES

Aromatic *N*-oxide compounds		Reaction conditions			Reference
	R	Reagent	Solvent	Temp.(°C)	
Pyridine N-oxide (R)	H	C_2H_5Br		△	21
	H, 2-CH_3, 4-CH_3	Ts-OCH_3		110	22
	2-CH_3, 4-CH_3, 2,4-$(CH_3)_2$	C_2H_5Br		△	23
	2-CH_3, 4-CH_3	C_2H_5I		△	24
	4-OCH_3, 4-NH_2	CH_3I		△	21
	2-CH_3	$C_6H_5COCH_2Br$		0	25
Quinoline N-oxide (R)	H	CH_3I		room	26
	4-OC_2H_5, 4-NH_2	CH_3I		△	21
	4-N_3	$(CH_3)_2SO_4$	$CHCl_3$	room	20
	2-CH_3, 4-CH_3	C_2H_5I		△	24
Pyrimidine N-oxide (R)	2,4-$(CH_3)_2$	CH_3		△	27
	2,4-$(CH_3)_2$	$C_6H_5CH_2Cl$		△	27

When there is a substituent in the molecule possessing an active hydrogen, there is a tendency for this substituent to be alkylated, but if this substituent is conjugated with the *N*-oxide group, then the *N*-oxide group is alkylated, forming a tautomeric type of *N*-alkoxylated compound or its betain form. In such cases, presence of a base becomes a requisite.

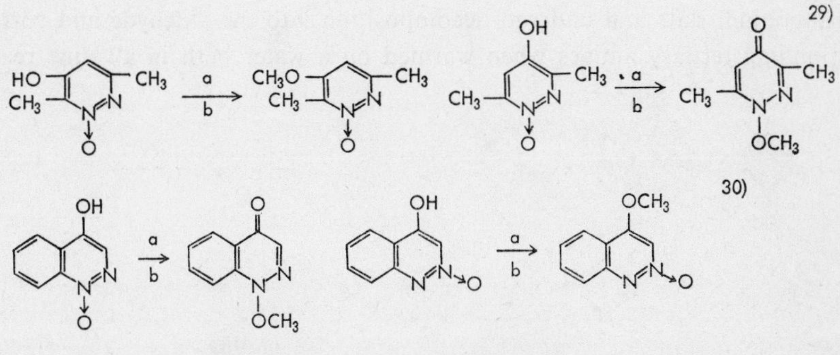

References p. 206

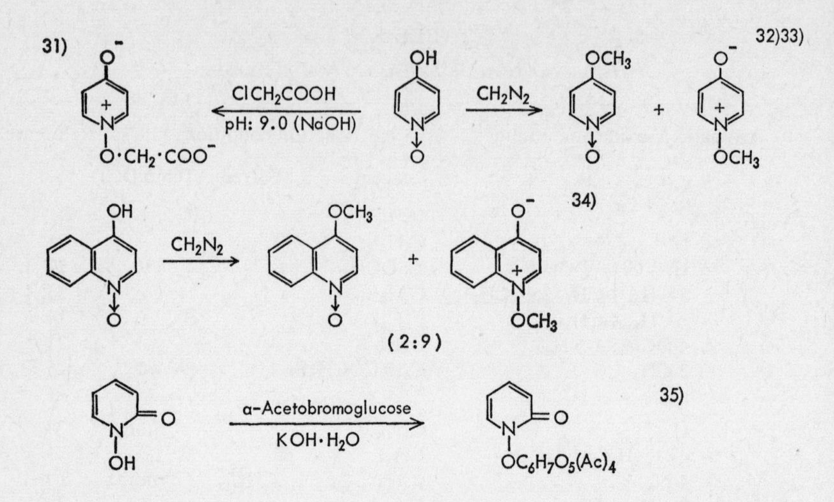

As shown below, both 1-methoxybenzimidazole, obtained by methylation of benzimidazole *N*-oxide, and 1-methylbenzimidazole 3-oxide form an identical methiodide. This fact indicates that π-electrons are delocalized in the methiodide.

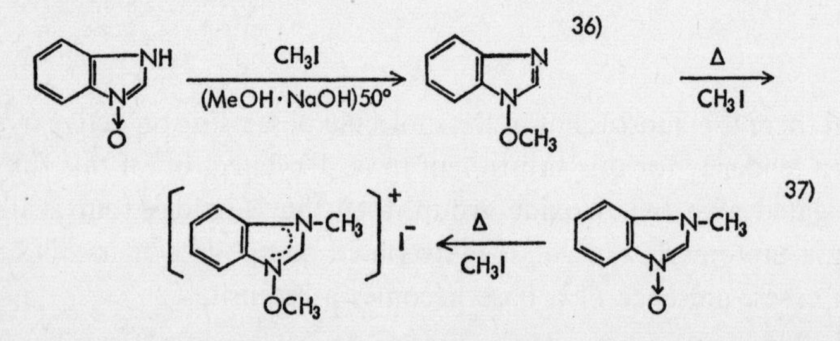

Alkyloxyammonium salts are less stable than their corresponding alkylammonium salts and undergo decomposition into the aldehyde and corresponding tertiary amines when warmed on a water bath in alkaline reac-

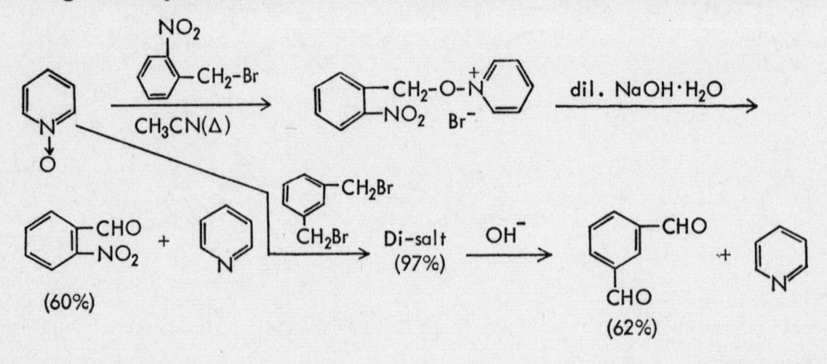

tion[21,26]. Boekelheide and his colleagues[38] successfully used this reaction for the preparation of aromatic aldehydes.

Kato and others[25] reported a convenient method for the preparation of α-ketoaldehydes by the use of the decomposition of the adduct of 2-picoline 1-oxide and α-haloketones.

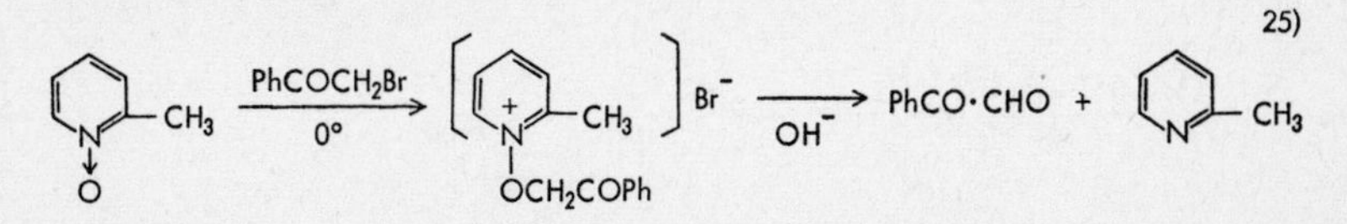

Boekelheide *et al.*[28,39] reported notable decomposition of the following cyclic alkyloxyammonium salts.

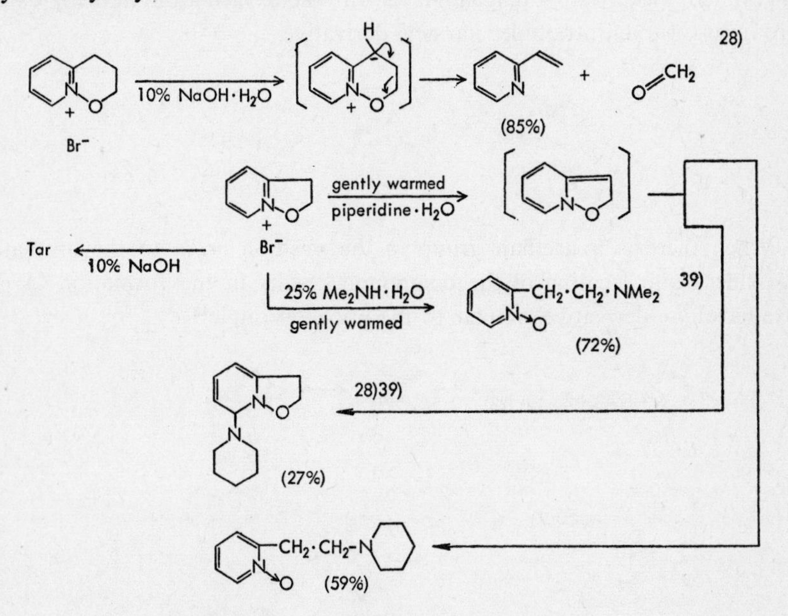

(b) Acyloxyammonium Salts

It is generally difficult to obtain acyl derivatives of aromatic *N*-oxide compounds in a crystalline state, but the perchlorate of the acetyl derivative of pyridine 1-oxide has been obtained as crystals melting at 152–153°[40]. This

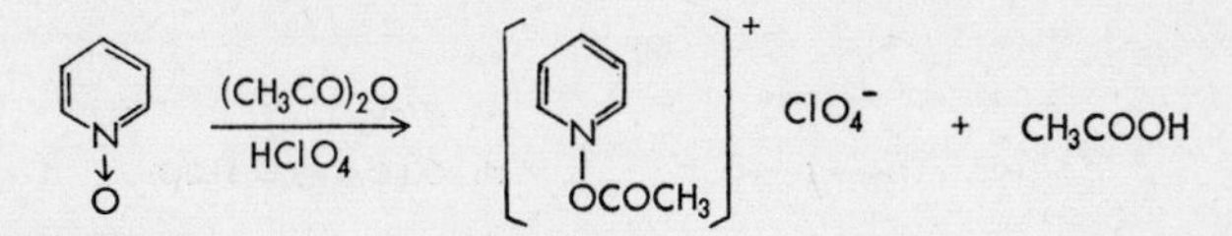

References p. 206

reaction proceeds quantitatively and Muth *et al.*[40] reported quantitative determination of aromatic *N*-oxide compounds by titration in acetic anhydride solution with perchloric acid. Baumgarten and Erbe[4] obtained *N*-sulfoxypyridine betain by the application of sulfur trioxide or chlorosulfonic acid to the hydrochloride of pyridine 1-oxide.

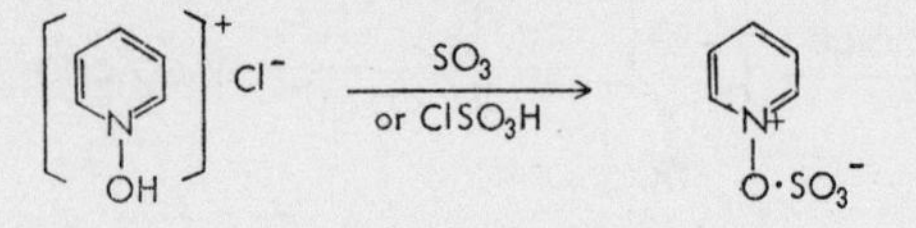

When the amino group in the position *ortho* to an aromatic *N*-oxide group is acylated, the acylating reagent reacts with the oxygen atom in the *N*-oxide group to form an intramolecular acyl derivative.

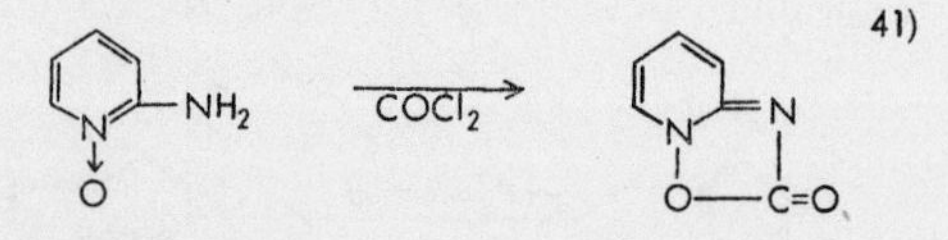

When there is a urethan group in the position *ortho* to the aromatic *N*-oxide group, heating of the compound results in the formation of an oxadiazolone derivative, similar to the above example[42].

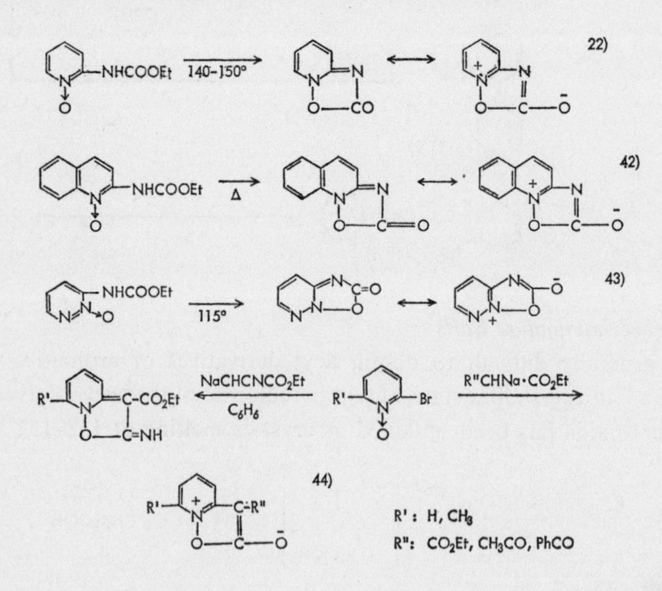

A similar participation of the *N*-oxide group to form isoxazolone derivatives was also observed by the base-catalysed rearrangement of *N*-(acylacetyl) derivatives of 2-pyridinesulfonamide 1-oxide[45] (*cf. Section 8.7*).

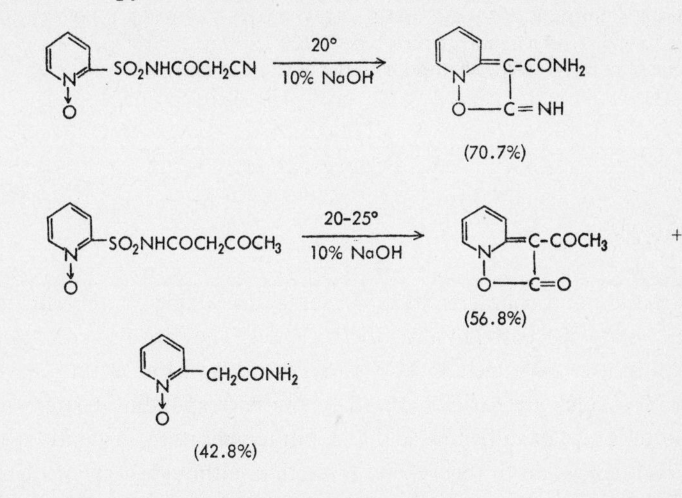

Example 1. Methylation of 4-Hydroxyquinoline 1-Oxide[34,46]

Ether solution of diazomethane (considerably in excess of calculated amount) was added to 1.5 g of 4-hydroxyquinoline 1-oxide dissolved in 300 ml of ethanol. When the reaction has subsided, the solvent was removed by distillation and the oily residue was allowed to stand over night on solid potassium hydroxide in a desiccator, but failed to crystallize. The oily residue was dissolved in acetone and the solution was chromatographed over alumina. The acetone eluate afforded crystals which were recrystallized from a benzene-ether mixture to give colorless needles, m.p. 53–55°, of 4-hydroxy-1-methoxyquinoline betain. Yield, 0.9 g. The methanolic eluate afforded 0.2 g of 4-methoxyquinoline 1-oxide as colorless needles, m.p. *ca.* 50–57°, which formed a picrate as yellow needles (from methanol), m.p. 156–158°.

Example 2. Preparation of 3-Methoxy-1-methylbenzimidazolium Iodide[37]

(i) A mixture of 0.3 g of 1-methylbenzimidazole 3-oxide and 2 ml of methyl iodide was heated under reflux for 10 min. The starting material dissolved and then a crystalline product precipitated out. When cooled, the product (0.46 g) was collected by filtration and recrystallized from ethanol–ethyl acetate to give 3-methoxy-1-methylbenzimidazolium iodide as colorless prisms of m.p. 140° (decomp.).
(ii) A mixture of 0.2 g of 1-methoxybenzimidazole and 1 ml of methyl iodide was heated under reflux for 0.5 h; the product began to separate out within 5 min of heating. By treating as described above, 3-methoxy-1-methylbenzimidazolium iodide was obtained in a quantitative yield.

Example 3. Preparation of Phenylglyoxal from 2-Picoline 1-Oxide and Phenacyl Bromide[25]

While stirring 3 g (0.028 mole) of 2-picoline 1-oxide under ice-cooling, 5.4 g (0.028 mole) of phenacyl bromide was added by which a pale yellow, viscous solution soon turned solid.

References p. 206

This solid mass was washed first with anhydrous ether and then with dehydrated benzene and it afforded 8 g (95%) of hygroscopic white crystals, m.p. 68–70°. To 2 g of the crystals a 10% aqueous solution of sodium hydrogen carbonate was added and the mixture was extracted with ether. The ether layer was washed with 10% hydrochloric acid, dried over anhydrous sodium sulfate, filtered, and the ether evaporated under a reduced pressure affording 0.5 g (50%) of a liquid product, b.p. 90–100°/25 mm (bath temperature). It formed an osazone of m.p. 148–149°, identical with the osazone of phenylglyoxal.

5.2 Deoxygenation

5.2.1 General

One of the main characteristics of aromatic *N*-oxide compounds is their greater stability compared to aliphatic *N*-oxides. They are also comparatively more resistant to deoxygenation[47] (*cf. Section 2.2.1*). For example, whereas aliphatic *N*-oxides are easily reduced to the corresponding tertiary amines by the application of sulfurous acid at room temperature, aromatic *N*-oxides are generally resistant to this type of reduction, although some of them have been found in recent years to be deoxygenated by sulfurous acid at room temperature. Such examples have been found in condensed ring compounds in which more than two benzene rings are condensed to an aromatic hetero ring, and in those possessing an aryl substituent in α- or γ-position of the ring nitrogen in diazine series compounds.

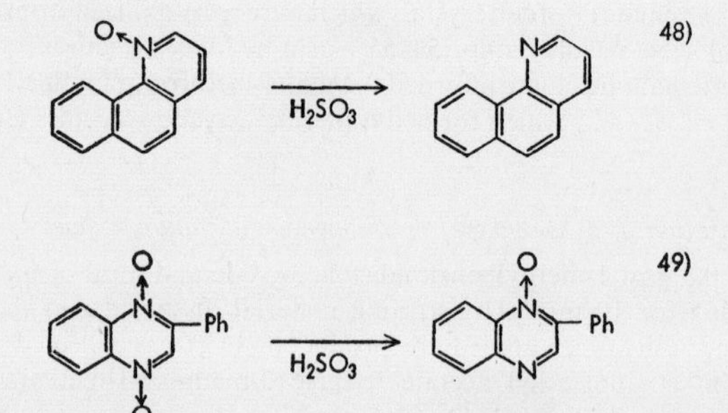

For these deoxygenation reactions, Hayashi assumed a nucleophilic addition of sulfurous acid followed by liberation of sulfuric acid[49].

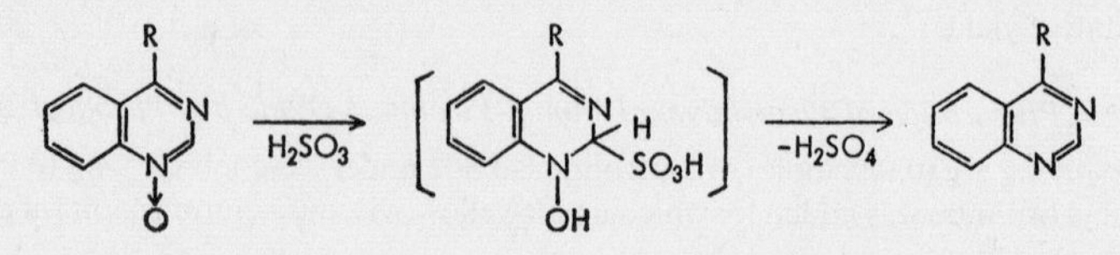

In the early stages of this work, Ochiai and Katada[50] realized that catalytic reduction of 4-nitropyridine 1-oxide over palladium–carbon stopped at the stage of 4-aminopyridine 1-oxide in neutral ethanolic solution, and that 1 mole of hydrogen was further absorbed very gradually in dilute hydrochloric acid solution to give 4-aminopyridine. The same phenomenon was observed in the reduction of 4-nitroquinoline 1-oxide[51], and the fact was proved further by polarography.

As was described in the preceding Chapter (*cf. Section 4.4*) the reduction potential of aromatic *N*-oxide compounds is significantly greater than that of aliphatic *N*-oxides. Moreover, the presence of electron-releasing groups such as a methoxyl or amino group in the position *para* to the *N*-oxide group increases this resistance against reduction, while electron-withdrawing groups like carboxyl and phenyl, as well as a naphthalene-like condensation of benzene rings, and the diazine series compounds, decrease this resistance.

This great resistance of aromatic *N*-oxide compounds to deoxygenation was at first thought to be one of the drawbacks in the application of the nitration reaction of aromatic *N*-oxides to the syntheses of pyridine and quinoline derivatives. However, the work of den Hertog and his collaborators[52] on the development of a method for obtaining 4-aminopyridine by the reduction of 4-nitropyridine 1-oxide with iron powder and acetic acid has led to the use of deoxygenation for other aromatic *N*-oxide compounds.

Hamana[53] found that warming of aromatic *N*-oxide compounds with phosphorus trichloride in an aprotonic neutral solvent resulted in facile deoxygenation of the *N*-oxide group and this method has since often been utilized.

Finally, Hayashi, Yamanaka, and Shimizu[54] found that aromatic *N*-oxides undergo facile reduction, in an excellent yield, when submitted to catalytic reduction over Raney nickel in a neutral solvent.

Deoxygenation of aromatic *N*-oxide compounds can therefore be effected quite smoothly by selecting the various methods described above.

5.2.2 Catalytic Reduction

Yamanaka[55] found that, whereas catalytic reduction of 4-benzyloxy-6-methylpyrimidine 1-oxide (5-4) over palladium–carbon catalyst in methanol results in the formation of 4-hydroxy-6-methylpyrimidine 1-oxide (5-5) and that further 1 mole of hydrogen is absorbed gradually to form 4-hydroxy-6-

References p. 206

methylpyridine (5-6), the same catalytic reduction over Raney nickel results in rapid absorption of 1 mole of hydrogen with selective reduction of the *N*-oxide group alone.

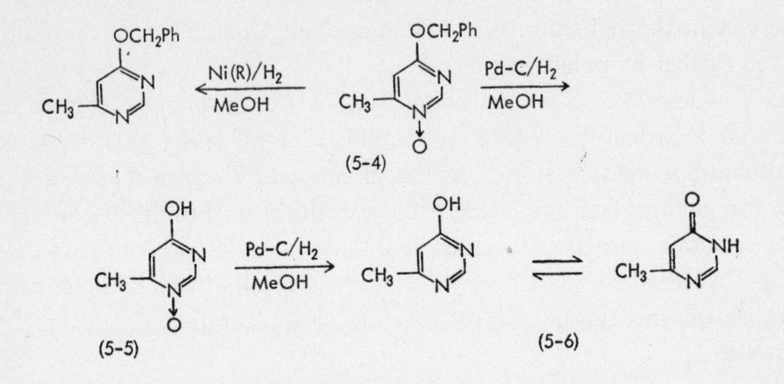

On discovering this fact, Hayashi and his co-workers[54,56] revealed that all pyridine and quinoline series *N*-oxides, without exception, are deoxygenated to the original tertiary amines, generally quantitatively or at least in a good yield, by catalytic reduction over Raney nickel in alcoholic solution. This catalytic reduction over Raney nickel is not accompanied with dehalogenation, although reduction of a nitro group to amino occurs, and sometimes, desulfurization of the mercapto group[56c]. There is one example of the elimination of a benzyl group, in the reduction of 4-benzyloxy-2-hydroxyquinazoline 1-oxide, but even here, the rate of deoxygenation of the *N*-oxide group and that of hydrogenolysis of the benzyloxyl group are clearly different and they can easily be discriminated. Another advantage of this method is that deoxygenation of the *N*-oxide alone can be effected without saturation of the double bond between azo-nitrogen atoms or between carbon atoms by controlling the absorption of hydrogen, and that deoxygenation can also be effected in the presence of a thioether, containing a sulfur atom which is a catalyst poison.

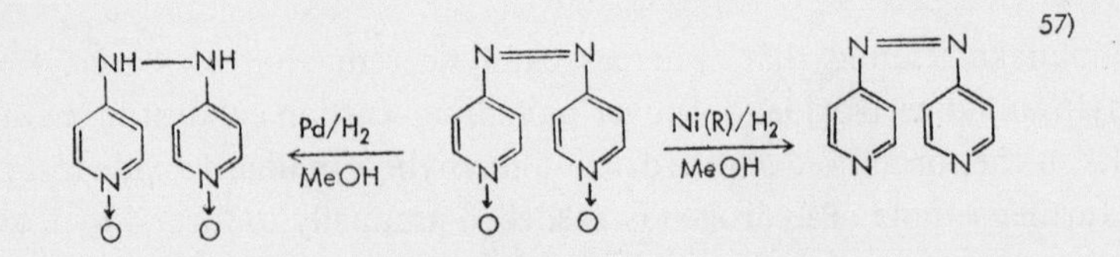

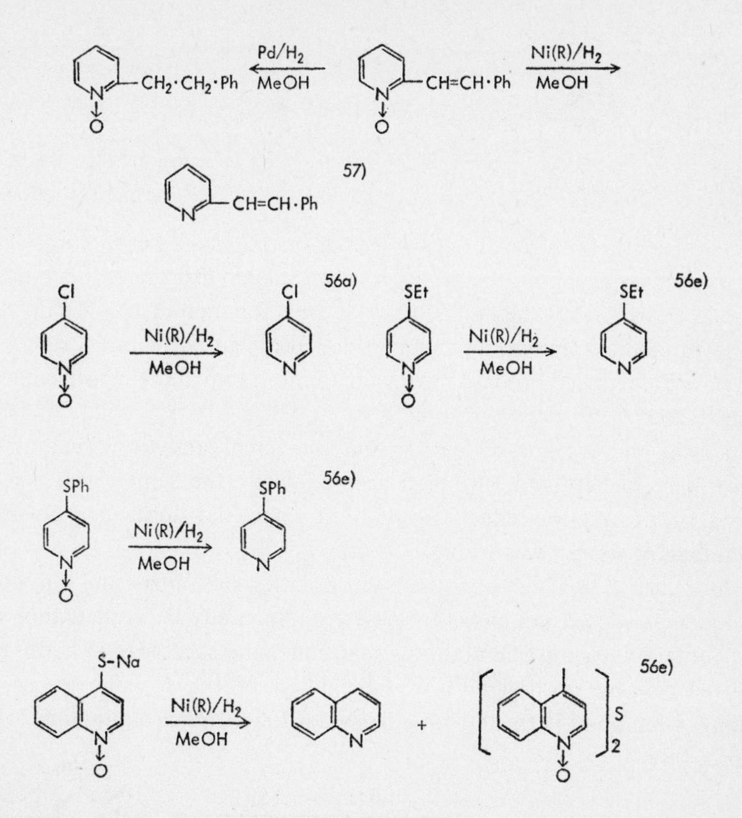

This type of catalytic reduction has been utilized a great deal since then; by Hayashi and others[55,56c,49] for *N*-oxides of the diazine series like pyrimidine and quinazoline, by Itai and others[58,59] for *N*-oxides of the pyridazine series, and by Wagner and others[60] for deoxygenation of 4- and 2-(tetra-*O*-acetyl-β-D-glucopyranosylthio)pyridine 1-oxide (5-7, 5-8) or the corresponding 4-mercaptoquinoline *N*-oxide derivative (5-9).

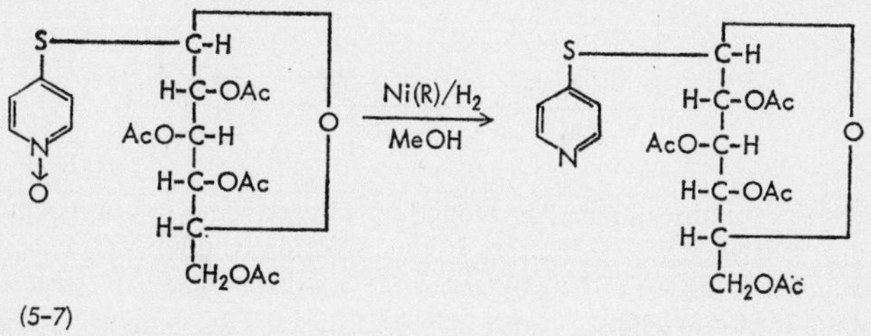

(5-7)

References p. 206

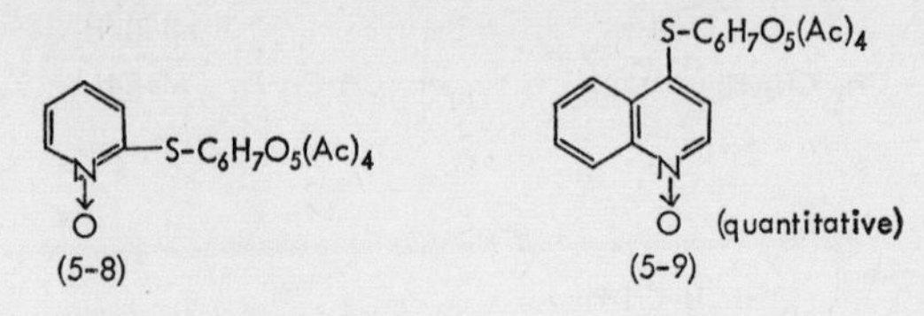

Jerchel and Melloh[61], without knowledge of Hayashi's work, reported that *N*-oxides of pyridine and its derivatives, with some exceptions, were deoxygenated easily when catalytically reduced over Raney nickel in glacial acetic acid solution containing acetic anhydride. However, acetic anhydride might acetylate the reduced product having an amino group and Hayashi's method seems better.

Hayashi and his group[62] carried out the same catalytic reduction with Urushibara nickel* and, although approximately the same result was obtained, they found the reduction rate to be somewhat slower and no special advantage of its use was found.

Ishii[63] carried out high-pressure hydrogenation of 4-nitro- and 4-hydroxyquinoline 1-oxide using nickel formate-paraffin catalyst[64] in methanol solution containing acetic acid and sodium acetate, and succeeded in hydrogenating the benzene portion with deoxygenation of the *N*-oxide group, obtaining 4-amino- (5-10) and 4-hydroxy-5,6,7,8-tetrahydroquinoline (5-11), respectively.

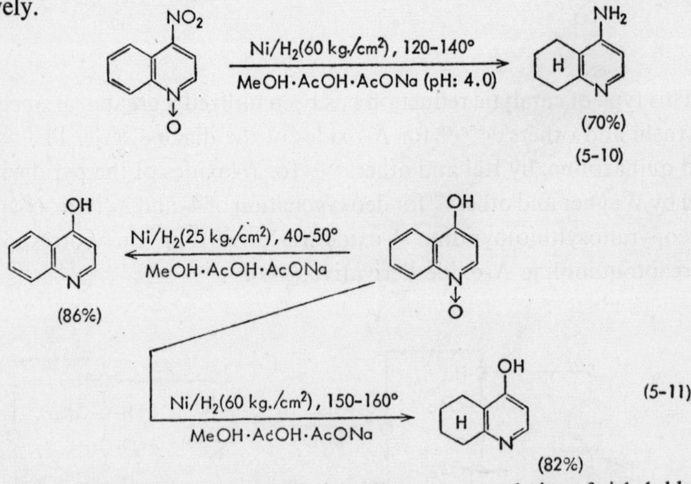

* The nickel–zinc mixture obtained by reduction of hot aqueous solution of nickel chloride with zinc dust is treated with sodium hydroxide solution or dilute acetic acid solution to obtain this nickel catalyst [Y. Urushibara and S. Nishimura, *Bull. Chem. Soc. Japan*, 27 (1954) 480; 28 (1955) 446].

As shown above, nickel catalysts are suitable for facile reduction of the *N*-oxide group, but palladium catalysts are not suited for the deoxygenation of *N*-oxide group, the reaction progressing very gradually only in strong acid solution. Catalytic reduction with palladium catalyst in glacial acetic acid mixed with acetic anhydride results in rapid absorption of hydrogen to effect deoxygenation[65], but the presence of a nitro group will produce an acetamido derivative[66,67].

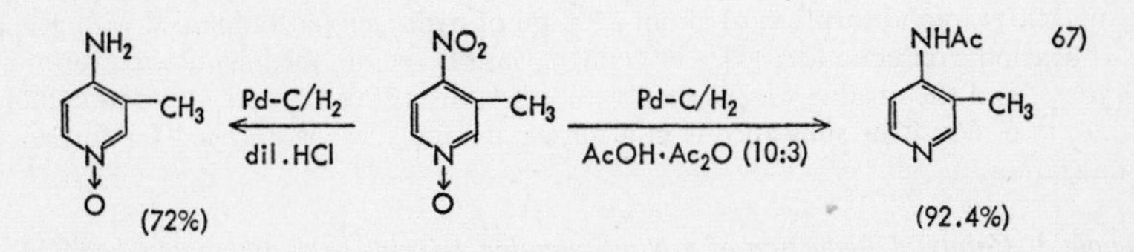

On the other hand, a palladium catalyst has a greater tendency to hydrogenolysis and will selectively hydrogenolyze a benzyloxyl group to a hydroxyl group[55,32,68,69].

Hayashi and his group[56a] recommend the use of Raney nickel and palladium–carbon together in methanolic solution when hydrogenolysis and deoxygenation of the *N*-oxide group are to be effected at the same time.

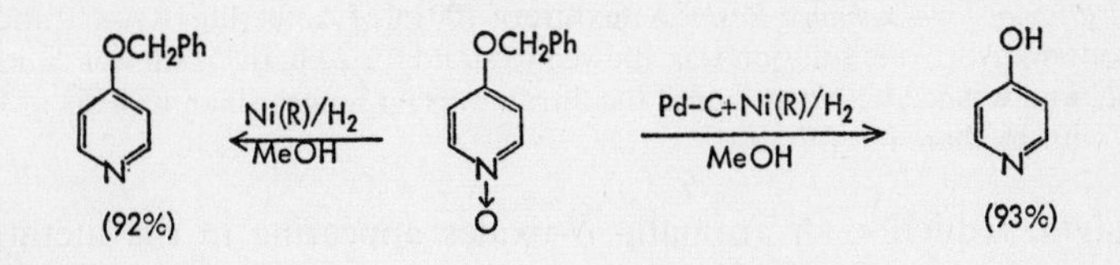

Example 1. Preparation of Raney Nickel for Deoxygenation[56b]

A solution of 10 g of sodium hydroxide (purity, 80%) dissolved in 40 ml of distilled water was placed in a 200-ml beaker, chilled in ice, and 5 g of nickel–aluminium alloy (1:1) was added in small portions. A great generation of heat occurs and an efficient cooling is required. The mixture was then heated at 100° for 2 h with occasional stirring, 20 ml of 20% sodium hydroxide solution was added, and the mixture was further heated at 100° for 1 h. When the evolution of hydrogen was no longer evident, the mixture was cooled and the whole was diluted to 200 ml with water. After the liquid solution was decanted, water was added, the whole was stirred, and the liquid was again decanted. The catalyst thus prepared was washed six times in this manner, finally collected by suctional filtration, and washed with methanol. It was used immediately.

Example 2. Catalytic Reduction of 4-Benzyloxypyridine 1-Oxide[56a]

(*i*) Raney nickel, prepared from 3 g of nickel–aluminium alloy as in Example 1, was added to a solution of 6 g of 4-benzyloxypyridine 1-oxide dissolved in 30 ml of methanol and the mixture was shaken in a hydrogen stream at atmospheric pressure. Hydrogen ab-

References p. 206

sorption was rapid and reduction stopped when about 660 ml of hydrogen had been absorbed. After removal of the catalyst by filtration, methanol was evaporated from the filtrate and the colorless oily residue was taken up in ether. After drying over anhydrous potassium carbonate, ether was distilled off and 5.1 g (92%) of needles were obtained. Recrystallization from petroleum ether afforded crystals of m.p. 57–58°, undepressed on admixture with 4-benzyloxypyridine, m.p. 57–58°.

(ii) A mixture of palladium catalyst, prepared from 10 ml of 1% aqueous palladium dichloride solution and 0.5 g of activated carbon, and Raney nickel, prepared from 5 g of nickel–aluminium alloy (1:1), was added to a solution of 10 g of 4-benzyloxypyridine 1-oxide dissolved in 40 ml of methanol and the whole mixture was shaken in a hydrogen stream. After rapid absorption of about 2200 ml of hydrogen (accompanied with generation of heat), the reduction ceased. The catalyst was filtered off, the filtrate was evaporated to dryness, and the residue was washed with ether, affording 5.2 g of 4-pyridone monohydrate, m.p. 66°. This substance is in a pure state and there is no need for further recrystallization.

Example 3. Catalytic Reduction of 4-Nitroquinoline 1-Oxide to 4-Aminoquinoline[56b]

Raney nickel, prepared from 4 g of nickel–aluminium alloy, was added to a solution of 5 g of 4-nitroquinoline 1-oxide dissolved in 150 ml of methanol, 3 ml of acetic acid was added, and this mixture was reduced at room temperature under atmospheric pressure. The reaction stopped after rapid absorption of about 2350 ml of hydrogen. The catalyst was filtered off and an ion exchanger, prepared as described below, was added to the filtrate. After allowing the mixture to stand overnight at room temperature, the ion exchanger was filtered off, the filtrate was evaporated to dryness, and the residual 4-aminoquinoline monohydrate was dried in a vacuum desiccator. Yield, 3.9 g (79%) of crystals (from ethanol), m.p. 70°.

Preparation of Ion Exchange Resin: A mixture of 100 ml of Amberlite IRA-410 and 300 ml of 4% sodium hydroxide solution was allowed to stand for 24 h, the resin was collected by filtration, and washed with water until the filtrate was no longer alkaline. This was finally washed with methanol before use.

Catalytic reduction of aromatic *N*-oxides appearing in the literature to date is summarized in Table 5-4.

5.2.3 Deoxygenation with Phosphorus Trichloride

In 1951, Hamana[53] found that heating the chloroform solution of the aromatic *N*-oxide of dihydroquinine with phosphorus trichloride resulted in facile deoxygenation to dihydroquinine, and the same reaction of pyridine and quinoline series *N*-oxides also gave the same result[82], indicating that this is a deoxygenation reaction that progresses comparatively easily with the

$$\text{Py}^{\oplus}\text{–O}^{\ominus} \xrightarrow{:PCl_3} \text{Py}^{\oplus}\text{–O–}\overset{\ominus}{P}Cl_3 \longrightarrow \text{Py} + :\ddot{O} \leftarrow PCl_3$$

TABLE 5-4

CATALYTIC REDUCTION OF AROMATIC AMINE OXIDES

N-Oxide	Catalyst	Solvent	H_2 absorbed (mole)	Product (yield, %)	Reference
Pyridine derivatives					
4-OCH_3	Raney Ni	MeOH		4-OCH_3-pyridine (85)	56*a*
4-OCH_3	Ur. Ni	MeOH		4-OCH_3-pyridine	62
4-OC_2H_5	Pd–C	0.1N HCl		4-OC_2H_5-pyridine	70
4-OC_2H_5	Raney Ni	MeOH		4-OC_2H_5-pyridine (86)	56*a*
4-OCH_2Ph	Raney Ni	MeOH		4-OCH_2Ph-pyridine (92)	56*a*
4-OCH_2Ph	Raney Ni + Pd–C	MeOH		4-OH-pyridine (93)	56*a*
4-OCH_2Ph, 3-$CONH_2$	Pd			4-OH-3-$CONH_2$-pyridine	69
4-OH	Raney Ni	MeOH		4-OH-pyridine (98)	56*a*
4-Cl	Raney Ni	MeOH		4-Cl-pyridine (88)	56*a*
4-Cl	Ur. Ni	MeOH		4-Cl-pyridine	62
4-NH_2	Raney Ni	MeOH		4-NH_2-pyridine (98)	56*a*
4-NO_2	Raney Ni	MeOH–AcOH		4-NH_2-pyridine (90)	56*a*
4-NO_2	Ur. Ni	MeOH		4,4′-hydrazopyridine, azopyridine, 4-NH_2-pyridine	62
4-NO_2	Pd–C	MeOH		4-NH_2-pyridine 1-oxide	70
4-NO_2, 2-CH_3	Pd–C	dil.HCl		4-NH_2-2-CH_3-pyridine 1-oxide	66
4-NO_2, 2-CH_3	Pd–C	AcOH–Ac_2O		4-AcNH-2-CH_3-pyridine	66
4-NO_2, 3-CH_3	Pd–C	dil.HCl		4-NH_2-3-CH_3-pyridine 1-oxide (72)	67
4-NO_2, 3-CH_3	Pd–C	AcOH–Ac_2O		4-AcNH-3-CH_3-pyridine (92)	67
4,4′-N=N–	Raney Ni	MeOH	1	4,4′-Azopyridine (14)	57
4,4′-N=N–	Raney Ni	MeOH	3	4,4′-Hydrazopyridine (93)	57
	Pd–C	MeOH	1	4,4′-Hydrazopyridine $N,N' \rightarrow O$ (83)	57
2-Ph–CH=CH–	Raney Ni	MeOH	1	2-Styrylpyridine, 2-Phenylethylpyridine	57
4-SPh–	Raney Ni	MeOH		4-SPh-pyridine (82.5)	56*e*
4-$PhSO_2$–	Raney Ni	MeOH		4-$PhSO_2$-pyridine (98)	56*e*
4-SC_2H_5–	Raney Ni	MeOH		4-C_2H_5S-pyridine (44)	56*e*
4-$SC_6H_7O_5(Ac)_4$	Raney Ni	MeOH		4-(tetra-*O*-acetyl-*β*-D-glucopyranosylthio)-pyridine (100)	60
2-$SC_6H_7O_5(Ac)_4$	Raney Ni	MeOH		2-(tetra-*O*-acetyl-*β*-D-glucopyranosylthio)-pyridine	60
2-CH_2–N · Cl	Raney Ni	MeOH–1%NaOH	4	2-CH_2–N H (91)	72
	Raney Ni	MeOH	1	2-CH_2–N · Cl (43)	72

(*Continued overleaf*)

TABLE 5-4 *(Continued)*

N-Oxide	Catalyst	Solvent	H_2 absorbed (mole)	Product (yield, %)	Reference
Quinoline derivatives					
H	Raney Ni	MeOH		Quinoline (93)	56*b*
4-OCH_3	Raney Ni	MeOH		4-OCH_3-quinoline (90)	56*b*
4-OCH_2Ph	Raney Ni	MeOH	1	4-OCH_2Ph-quinoline (81)	56*b*
4-OCH_2Ph	Raney Ni	MeOH	2	4-OH-quinoline (69)	56*b*
4-OH	Raney Ni	MeOH		4-OH-quinoline (83)	56*b*
2-OH	Raney Ni	MeOH		Carbostyril (83)	56*c*
4-Cl	Raney Ni	MeOH		4-Cl-quinoline (90)	56*b*
4-NH_2	Raney Ni	MeOH		4-NH_2-quinoline (74)	56*b*
4-NO_2	Raney Ni	MeOH		4-NH_2-quinoline (79)	56*b*
4-SPh	Raney Ni	MeOH		4-SPh-quinoline (82.5)	56*e*
4-SO_2Ph	Raney Ni	MeOH		4-SO_2Ph-quinoline (98)	56*e*
4-SC_2H_5	Raney Ni	MeOH		4-SC_2H_5-quinoline (44)	56*e*
4-NaS	Raney Ni	MeOH	1/2	Quinoline (42), Di-(4-quinolyl) sulfide (50)	56*e*
4-$SC_6H_7O_5(Ac)_4$	Raney Ni	MeOH		4-(tetra-*O*-acetyl-*β*-D-glucopyranosylthio)-quinoline	60
Phenanthridine derivatives					
H	Raney Ni	MeOH		Phenanthridine	73
6-CN	Raney Ni	MeOH		6-CN-phenanthridine (60)	74
6-OH	Raney Ni	MeOH–KOH		Phenanthridone (80)	74
Pyridazine derivatives					
3-NO_2-1→O	Pd–C	MeOH–HCl	4	3-NH_2-pyridazine (31)	59
	Pd–C	MeOH	2	3-NHOH-pyridazine 1-oxide	59
	Pd–C	MeOH	3	3-NH_2-pyridazine 1-oxide (32) 3-NH_2-pyridazine (10)	59
4-NO_2-1→O	Pd–C	MeOH–HCl		4-NH_2-pyridazine	75
	Pd–C	MeOH		4-NH_2-pyridazine 1-oxide	75
3-OCH_3-4-NO_2-1→O	Raney Ni	MeOH		3-OCH_3-4-NH_2-pyridazine	58
5-NO_2-1→O	Pd–C	MeOH–HCl		5-NH_2-pyridazine	58
Pyrimidine derivatives					
4-OCH_2Ph-6-CH_3 1→O	Pd–C	MeOH	1	4-OH-6-CH_3-pyrimidine 1-oxide	55*a*
4-OCH_2Ph-6-CH_3 1→O	Raney Ni	MeOH	1	4-OCH_2Ph-6-CH_3-pyrimidine	55*a*
6-Piperidino-4-CH_3 1→O	Raney Ni	MeOH	1	6-Piperidino-4-CH_3-pyrimidine	55*b*
Quinazoline derivatives					
4-OCH_2Ph-1→O	Raney Ni	MeOH	1	4-OCH_2Ph-quinazoline (70)	71
4-OCH_2Ph-1→O	Raney Ni	MeOH	2	4-OH-quinazoline (80)	71
4-OCH_2Ph-1→O	Pd–C	MeOH	1	4-OH-quinazoline 1-oxide	71
4-OCH_2Ph-1→O	Pd–C	MeOH	2	4-OH-quinazoline (44)	71

TABLE 5-4 *(Continued)*

N-Oxide	Catalyst	Solvent	H_2 absorbed (mole)	Product (yield, %)	Reference
4-OCH_2Ph-2-OH-1→O	Raney Ni	MeOH	1	2,4-di-OH-quinazoline 1-oxide (81)	56*c*
4-OCH_2Ph-2-OH-1→O	Raney Ni	MeOH	2	2,4-di-OH-quinazoline(70)	56*c*
4-OCH_3-2-OH-1→O	Raney Ni	MeOH		4-OCH_3-2-OH-quinazoline (74)	56*c*
4-OC_2H_5-2-OH-1→O	Raney Ni	MeOH		4-OC_2H_5-2-OH-quinazoline (65)	56*c*
H-3→O	Raney Ni	MeOH	1	Quinazoline	76
H-3→O	Raney Ni	MeOH	2	3,4-dihydroquinazoline	76
2-CH_3-3→O[a]	Raney Ni	MeOH	2	2-CH_3-3,4-dihydro-quinazoline	77,78
2-C_2H_5-3→O[a]	Raney Ni	MeOH	2	2-C_2H_5-3,4-dihydro-quinazoline	77,78
2-Ph-3→O[a]	Raney Ni	AcOH	2	2-Ph-3,4-dihydro-quinazoline	77,78
2-CH_3-4-Ph-3→O[a]	Raney Ni	EtOH	2	2-CH_3-4-Ph-3,4-dihydro-quinazoline	77,78
2-OH-3→O[a]	Raney Ni	EtOH	2	2-OH-3,4-dihydro-quinazoline	77,78
2-CH_2NMe_2-6,7-di-CH_3-4-Ph-3→O	Pd–C or Raney Ni	EtOH	1	2-CH_2NMe_2-6,7-di-CH_3-4-Ph-quinazoline	78
Quinoxaline derivatives					
2-Ph-1→O	Raney Ni	MeOH	1	2-Ph-quinoxaline (90)	49
2-Ph-4→O	Raney Ni	MeOH	1	2-Ph-quinoxaline (89)	49
2,3-di-OH-1,4-di→O	Raney Ni	MeOH–NaOH aq. or EtOH–AcOH		2,3-di-OH-quinoxaline	79
2-NH_2-3-OH-1,4-di→O	Raney Ni	MeOH–NaOH aq. or EtOH–AcOH		2-NH_2-3-OH-quinoxaline 2-NH_2-3-OH-quinoxaline 4-oxide	79
2-NH_2-1,4-di→O	Raney Ni	MeOH–NaOH aq. or EtOH–AcOH		2-NH_2-quinoxaline	79
2-CH_3-1,4-di→O	Raney Ni	MeOH		2-CH_3-quinoxaline 1-oxide (50)	80
2-OH-4→O	Raney Ni	MeOH		2-OH-quinoxaline	80
2-OH-1→O	Raney Ni	MeOH		2-OH-quinoxaline	80
2-$CH_2CH_2CO_2$H-1→O	Raney Ni	MeOH		2-$CH_2CH_2CO_2$H-quinoxaline	81
Benzimidazole derivative					
2-CH_3-3→O	Raney Ni	MeOH		2-CH_3-benzimidazole	36

[a] W. Ried and P. Stahlhofen[77] gave the formula (5-12) for this compound of the quinazoline series but it was corrected to (5-13) by Sternbach and others[78].

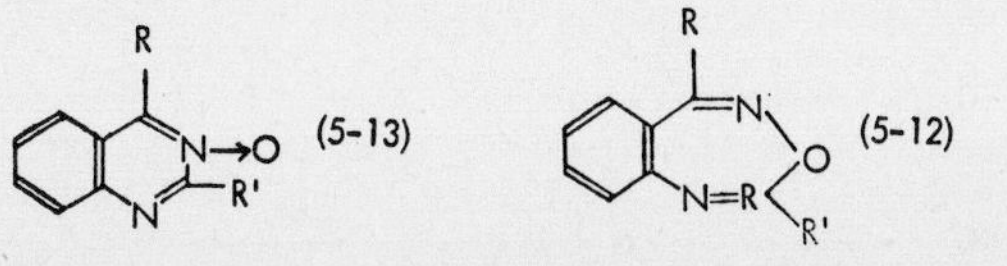

aromatic *N*-oxide group in general. Hamana assumed the foregoing formulae (p. 190) to represent the reaction mechanism[82f].

Chloroform is usually used as the solvent, but non-protonic liquids such as acetic acid ester, dioxan, or dimethylformamide, are sometimes used as a solvent or suspension medium.

Although this reaction gives a good yield, presence of an active hydrogen group, such as hydroxyl and amino, in the position *para* to the *N*-oxide group interferes in this reaction and a considerable amount of the starting material is recovered, with attendant decrease in the yield of deoxygenated product. In the case of quinoline compounds, a part of the hydroxyl is replaced by chlorine.

Addition of acetic anhydride to the reaction system results in increased yield, but the amino group may be acetylated. If the duration of the reaction is controlled, pyridine compounds with a nitro group in the *para*-position give 4-nitropyridine in a good yield, but if the reaction period is prolonged, nucleophilic substitution of the nitro group with chlorine may occur at the same time. This tendency is especially strong in the quinoline series, and 4-chloroquinoline is formed as the main product. Nakayama[83] carried out this reaction on 4-nitroquinoline 1-oxide by adding phosphorus trichloride to the quinoline compound in chloroform at below 10° allowing the reaction to take place below 15°, and obtained 43% of 4-chloroquinoline and 21% of 4-nitroquinoline. He also showed that the reaction of 4-nitroquinoline 1-oxide with about 1.5 equivalents of phosphorus tribromide at below 15° resulted in quantitative formation of 4-nitroquinoline, while reaction above 30° began to cause formation of 4-bromoquinoline. Hamana[82e] observed that heating of this reaction mixture on a boiling water bath resulted in formation of 4-bromoquinoline in 84% yield. These, however, are examples where the nitro group is conjugated to the *N*-oxide group. Ochiai and Kaneko[84] obtained 3-nitroquinoline in a quantitative yield by treatment of 3-nitroquinoline 1-oxide with phosphorus trichloride in chloroform.

Acheson and others[85] found that the same treatment of 9-nitroacridine *N*-oxide with phosphorus trichloride effected concurrent deoxygenation and chlorine substitution to form 9-chloroacridine. This reaction is the same as a reaction with acid chloride, such as phosphoryl chloride and sulfuryl chloride, on aromatic *N*-oxide compounds described elsewhere (*cf. Section 7.4.1*). Kanô and Takahashi[37] showed that the same reaction of 1-methylbenzimidazole 3-oxide (5-14) afforded 1-methylbenzimidazole and 1-methyl-2-chlorobenzimidazole in almost equal quantities, while Itai and Kamiya[86] found that the same treatment of 3-azidopyridazine 1-oxide (5-15) with phosphorus trichloride gave tetrazolo[*b*]pyridazine (5-16). At the same time, it was proved that the reaction of 3,6-dichloropyridazine (5-17) and sodium azide gave 6-azidotetrazolo[*b*]pyridazine (5-18), so that it is certain that azidopyridazine would be formed as an intermediate in the former reaction.

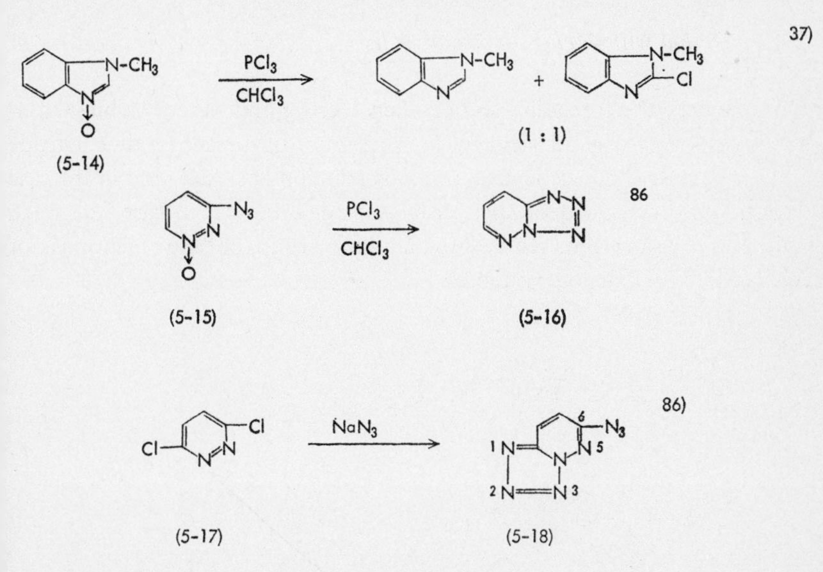

Example 1. Reaction of Pyridine 1-Oxide with Phosphorus Trichloride[82a]

To a solution of 1 g of pyridine 1-oxide dissolved in 10 ml of chloroform, while chilled in ice, 2 ml of phosphorus trichloride was added dropwise, by which the reaction started with evolution of heat. The mixture was heated under reflux on a water bath for 0.5 h, treated with ice water, and the chloroform layer was separated. The chloroform layer failed to yield any substance which might be the product. The acid solution was basified with sodium hydroxide and extracted with ether. The ether extract was dried over potassium carbonate and ether was distilled off, using the Widmer column, by which 0.75 g (90%) of an oil, b.p. 112–114°, was obtained. Picrate, yellow needles, m.p. and mixed m.p. with pyridine picrate, 160°.

Example 2. Preparation of 4-Nitroquinoline from 4-Nitroquinoline 1-Oxide[83]

A solution of 9.5 g of 4-nitroquinoline 1-oxide dissolved in 200 ml of chloroform was concentrated to about one-third the original volume, and ice-cold solution of 25 g of phosphorus tribromide in 20 ml of dehydrated chloroform was added with stirring under ice-cooling at such a rate that the reaction temperature did not rise above 15°. The reaction mixture was allowed to stand at room temperature for 15 min, poured on about 200 g of crushed ice, basified with sodium carbonate, and extracted with chloroform. The chloroform solution was washed with hydrosulfite solution made alkaline with sodium carbonate, dried over anhydrous sodium sulfate, and evaporated. The crystalline residue thereby obtained was distilled in vacuum and a distillate of b.p. 130–135°/5 mm was obtained as the objective product. Yield, 8.7 g. Recrystallization from ether gave yellow needles, m.p. 87–89°.

Results of deoxygenation of aromatic *N*-oxide compounds with phosphorus trihalide are summarized in Table 5-5.

5.2.4 Reduction with Metals or Metal Salts

As stated in the foregoing section, den Hertog and others[52] obtained 4-aminopyridine in a good yield by the reduction of 4-nitropyridine 1-oxide with iron powder and acetic acid, and this reaction has been used in the case of other pyridine and quinoline series *N*-oxides. In such a case, the nitro group is at the same time reduced to an amino group, but the reaction is not accompanied by dehalogenation.

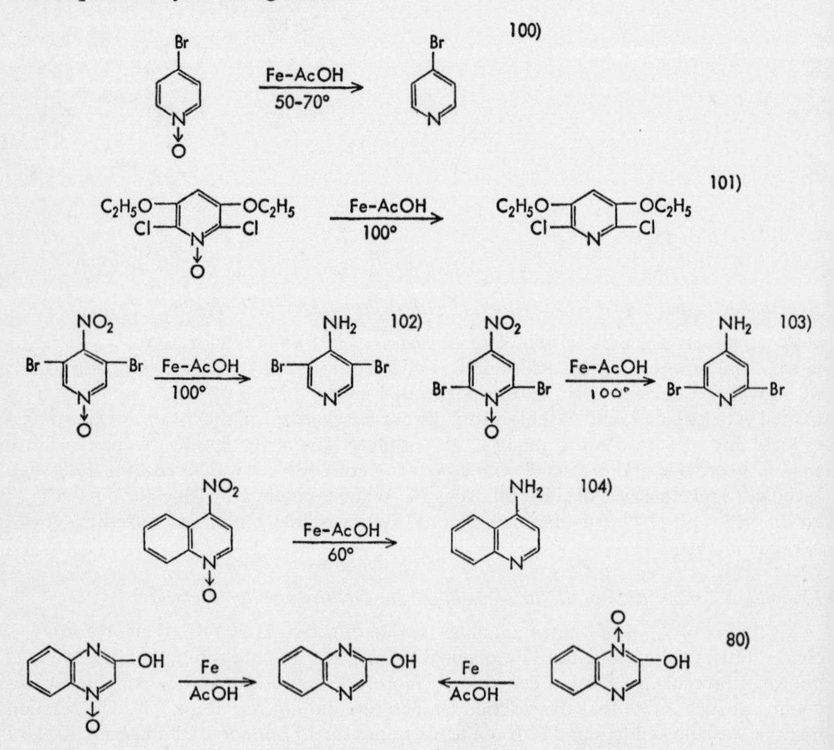

This reaction can also be effected by the use of zinc dust in place of iron powder.

TABLE 5-5

DEOXYGENATION OF AROMATIC AMINE OXIDES WITH PHOSPHORUS TRIHALIDE

Substituent in aromatic *N*-oxides	Solvent	PX_3	Aromatic amines (yield, %)	Reference
Pyridine derivatives				
Nil	$CHCl_3$	PCl_3	Pyridine (90)	53,82*a*
3-NO_2	$CHCl_3$	PCl_3	3-NO_2-pyridine	87
4-NO_2	$CHCl_3$	PCl_3	4-NO_2-pyridine (80)	53
4-NO_2-3-CH_3	$CHCl_3$	PCl_3	4-NO_2-3-CH_3- and 4-Cl-3-CH_3-pyridine	88
4-NO_2-3-CH_3	$CHCl_3$	PBr_3	4-NO_2-3-CH_3-pyridine	67
4-OH	$CHCl_3$	PCl_3	4-OH-pyridine (54)	53
4-NH_2	$CHCl_3$	PCl_3	4-NH_2-pyridine (19.5)	53
4-NH_2	Me_2NCHO	PCl_3	4,4′-Azopyridine 1,1′-dioxide	82*d*
3-C_2H_5-6-CH_3-4-NO_2	$CHCl_3$	PCl_3	3-C_2H_5-6-CH_3-4-NO_2-pyridine	89
2-Cl-4-NO_2	$CHCl_3$	PCl_3	2-Cl-4-NO_2-pyridine	90
2-Br-4-NO_2	$CHCl_3$	PBr_3	2-Br-4-NO_2-pyridine	90
2-I-4-NO_2	$CHCl_3$	PCl_3	2-Cl-4-NO_2- and 2-I-4-NO_2-pyridine	90
2,4-Di-OCH_3	$CHCl_3$	PCl_3	2,4-Di-OCH_3-pyridine	73
4,4′-Di-NO_2-3,3′-dipyridyl di-N→O	$CHCl_3$	PCl_3 (0°)	4,4′-Di-NO_2-3,3′-dipyridyl	91
1,2-Di(4-NO_2-3-pyridyl)-ethane di-N→O	$CHCl_3$	PCl_3	1,2-Di(4-NO_2-3-pyridyl)ethane	92
1,2-Di(4-OC_2H_5-3-pyridyl)-ethane di-N→O	$CHCl_3$	PCl_3	1,2-Di(4-OC_2H_5-3-pyridyl)-ethane	92
Quinoline derivatives				
Nil	$CHCl_3$	PCl_3	Quinoline (92)	53
3-NO_2	$CHCl_3$	PCl_3	3-NO_2-quinoline (100)	84
4-NO_2	$CHCl_3$	PCl_3	4-Cl-quinoline (81.5)	53
4-NO_2	$CHCl_3$	PBr_3 (0–15°)	4-NO_2-quinoline (100)	83
4-OH	$CHCl_3$	PCl_3	4-OH- (36.5) and 4-Cl-quinoline (30)	53
4-NH_2	$CHCl_3$	PCl_3	4-NH_2-quinoline (45)	53
6-Br-5-NO_2	$CHCl_3$	PCl_3	6-Br-5-NO_2-quinoline (96)	53
4-Ph-CH=CH	$CHCl_3$	PCl_3	4-Styrylquinoline	93
Acridine derivatives				
9-NO_2	$CHCl_3$	PCl_3	9-Cl-acridine	85
9-Br	$CHCl_3$	PCl_3	9-Br-acridine	85
Phenanthridine derivative				
Nil	$CHCl_3$	PBr_3	Phenanthridine	73

(Continued overleaf)

TABLE 5-5 *(Continued)*

Substituent in aromatic *N*-oxides	Solvent	PX_3	Aromatic amines (yield, %)	Reference
Pyrimidine derivatives				
4-OCH_3	$CHCl_3$	PCl_3	4-OCH_3-pyrimidine	94
4-OCH_3-6-CH_3	$CHCl_3$	PCl_3	4-OCH_3-6-CH_3-pyrimidine	94
Quinazoline derivatives				
6-Cl-2-CH_3-4-Ph-1→O	$CHCl_3$	PCl_3	6-Cl-2-CH_3-4-Ph-quinazoline	78
6-Cl-2-CH_2Cl-4-Ph-1→O	$CHCl_3$	PCl_3	6-Cl-2-CH_2Cl-4-Ph-quinazoline	78
Nil-3→O	$CHCl_3$	PCl_3	Quinazoline	76
4-CH_3-3→O	$CHCl_3$	PCl_3	4-CH_3-quinazoline	96
4-iC_3H_7-1→O	$CHCl_3$	PBr_3	4-iso-C_3H_7-quinazoline	49
4-OC_2H_5-1→O	$CHCl_3$	PCl_3	4-OC_2H_5-quinazoline	97
Quinoxaline derivatives				
2-Ph-4→O	$CHCl_3$	PCl_3	2-Ph-quinoxaline	49
2-CON(Me)Ph-1→O	$CHCl_3$	PCl_3	recovered	98
2-CON(Me)Ph-4→O	$CHCl_3$	PCl_3	2-CON(Me)Ph-quinoxaline	98
2-CON(Me)Ph-1,4-di→O	$CHCl_3$	PCl_3	2-CON(Me)Ph-quinoxaline 1-oxide	98
Phthalazine derivatives				
Nil-2→O	$CHCl_3$	PCl_3	Phthalazine	99
4-OC_2H_5-2→O	$CHCl_3$	PCl_3	4-OC_2H_5-phthalazine	99
Phenazine derivatives				
3,7-Di-NO_2-5→O	$CHCl_3$	PCl_3	3,7-Di-NO_2-phenazine	95
3,8-Di-NO_2-5→O	$CHCl_3$	PCl_3	3,8-Di-NO_2-phznazine	95
Pyridazine derivatives				
4-OCH_3-3,6-di-CH_3-1→O	$CHCl_3$	PCl_3	4-OCH_3-3,6-di-CH_3-pyridazine (69)	96
3-Azido-1→O	$CHCl_3$	PCl_3	Tetrazolo[*b*]pyridazine	86
Benzimidazole derivative				
1-CH_3-3→O	$CHCl_3$	PCl_3	1-CH_3- and 1-CH_3-2-Cl-benzimidazole	37

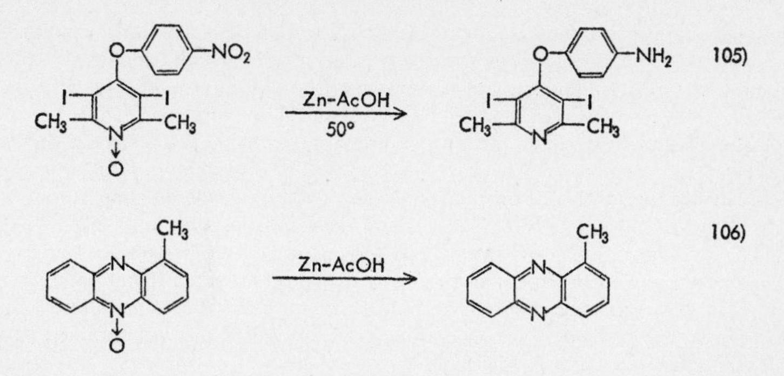

Kröhnke and others [12] used alkali hydroxide instead of acetic acid and obtained the same effect.

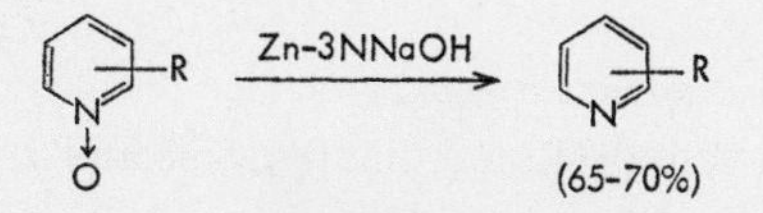

A similar deoxygenation was effected by Talik[107] by the use of ammonia–alkaline ferrous sulfate and by Adachi[108] with iron powder and ferrous sulfate on quinazoline 3-oxide.

Abramovitsch and Adams[109] reported deoxygenation by heating with ferrous oxalate at 300° (see Table 5-6).

Example. Preparation of 4-Aminoquinoline from 4-Nitroquinoline 1-Oxide[104]

After a solution of 10 g of 4-nitroquinoline 1-oxide in a mixture of 100 ml of glacial acetic acid and 20 ml of water had been warmed to 50°, 30 g of iron powder was added slowly in small portions, the reaction temperature being maintained below 60°, with efficient stirring, and the mixture was agitated for 6–7 h thereafter, at around 60°. As the reaction

TABLE 5-6

DEOXYGENATION OF PYRIDINE SERIES OXIDES WITH FERROUS OXALATE

$$\text{Aromatic } N\text{-oxide} \xrightarrow[+\text{Pb granules}]{\text{Ferrous oxalate}\cdot 2H_2O \text{ (4 moles)}} \text{heating (300°)(bath temp.)}$$

N-Oxide compd.	Aromatic tertiary amine (%)	*N*-Oxide compd.	Aromatic tertiary amine (%)
Pyridine	64	2-Aminopyridine	45
2-Picoline	62	2-Phenylpyridine	63
3-Picoline	72	*N,N*-Dimethylaniline	55(crude)

progressed, sludgy precipitate of ferrous acetate separated out in abundance and the stirrin became very difficult. The reaction mixture was then acidified with conc. hydrochloric aci to dissolve the precipitate completely, remaining iron was removed by filtration, and th greenish brown filtrate was evaporated under a reduced pressure. The hydrochloride of th iron complex separated out as greenish cubic crystals which were collected by filtration dissolved in saturated sodium hydroxide solution, and this alkaline solution was filtere while hot from the residue of ferrous hydroxide. Cooling of the alkaline filtrate afforde the monohydrate of 4-aminoquinoline as colorless, long needles, m.p. 67°. The residu obtained on evaporation of its filtrate and ferrous hydroxide precipitate were extracte with methanol and afforded further crops of the monohydrate. Total yield, 76%. Th monohydrate crystals were dissolved in chloroform, the solution was dried over anhydrou potassium carbonate, and then concentrated; 4-aminoquinoline was obtained as slightl yellow, sandy crystals of m.p. 154–155°.

5.2.5 *Other Methods of Reduction*

(a) Sulfur and Its Compounds

Takeda and Tokuyama[110] found that pyridine and quinoline series *N*-oxide compounds could be reduced to the original tertiary amines when heated with sulfur in liquid ammonia at 100–125°, as shown in Table 5-7.

Relyea and others[111] found that mercapto compounds, sulfide, and thiourea could also be used, like sulfur, for such deoxygenation reactions.

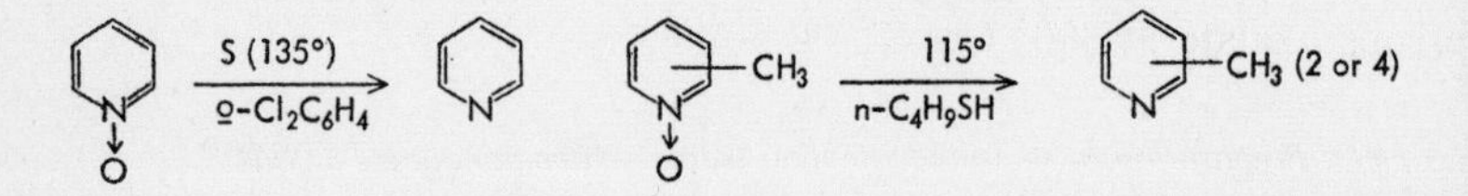

Similarly, the following compounds are also useful as a reducing agent: $(n\text{-}C_4H_9)_2S$ (at 150°), thiourea (at 140°), benzenethiol (at 140°), and 2-mercaptobenzothiazole (at 150°).

TABLE 5-7

DEOXYGENATION OF AROMATIC AMINE OXIDES WITH SULFUR IN LIQUID AMMONIA

N-Oxide of	Product (%)		*N*-Oxide of	Product (%)	
Pyridine	Pyridine	(43.0)	Quinoline	Quinoline	(63.4)
2-Picoline	2-Picoline	(36.8)	Quinaldine	Quinaldine	(56.3)
3-Picoline	3-Picoline	(43.3)	Lepidine	Lepidine	(58.6)
4-Picoline	4-Picoline	(37.5)	8-Nitroquinoline	8-Amino-quinoline	(36.9)
4-Nitropyridine	4-Amino-pyridine	(32.7)			

TABLE 5-8

DEOXYGENATION OF AROMATIC AMINE OXIDES WITH PHENYLSULFENYL CHLORIDE OR SULFUR MONOCHLORIDE

N-Oxide of	Reagent ($CHCl_3$)	Yield of amine (%)	*N*-Oxide of	Reagent ($CHCl_3$)	Yield of amine (%)
Pyridine	PhSCl	80–85	4-Chloropyridine	PhSCl	75–80
Pyridine	S_2Cl_2	80–83	4-Nitropyridine	PhSCl	recovered
2-Picoline	PhSCl	80–85	Quinoline	PhSCl	80–90
2-Picoline	S_2Cl_2	85–90	Quinoline	S_2Cl_2	80–85
4-Picoline	PhSCl	80–82	4-Nitroquinoline	PhSCl	4-Chloro-quinoline (10)
4-Picoline	S_2Cl_2	85–90			

Furukawa[112] further found that deoxygenation can be effected in a good yield by heating the chloroform solution of the *N*-oxide compounds with phenylsulfenyl chloride or sulfur monochloride. This reaction is summarized in Table 5-8.

Lower oxides of sulfur, such as sulfurous acid, do not effect deoxygenation of aromatic *N*-oxides in general, but Hamana and Funakoshi[113] mentioned that the deoxygenation can be effected with this reagent if the reaction is carried out in the presence of an acylation agent as shown below.

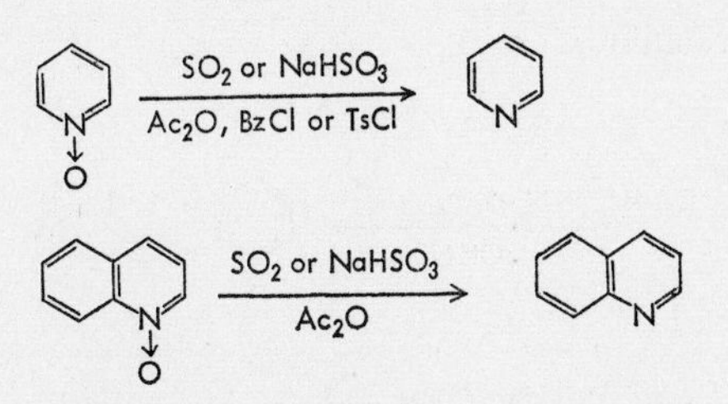

In the case of quinoline 1-oxide, however, the use of benzoyl or tosyl chloride in this reaction results in the formation of carbostyril as the main product, with quinoline as a by-product.

This reaction is somewhat similar to the smooth deoxygenation of aromatic amine oxides by catalytic reduction over palladium catalyst in the presence of acetic anhydride (*cf. Section 5.2.2*).

Analogously, Vozza[10] found that in the treatment of 2-picoline 1-oxide with thionyl chloride in benzene solution, the reaction at 15–20° gives a hydrochloric acid adduct, while that at 50° effects deoxygenation to 2-picoline

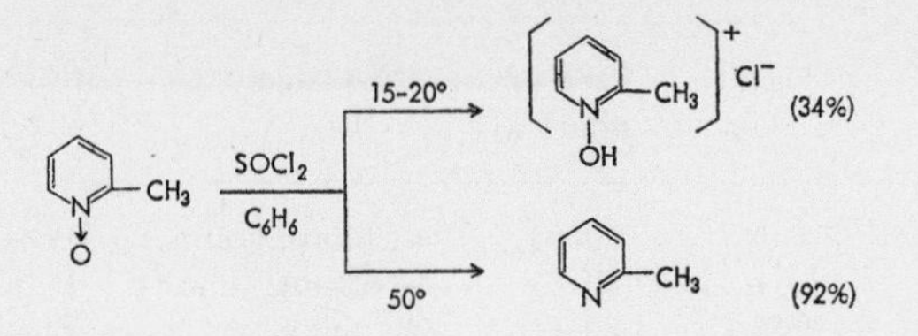

Evans and Brown[114] reported deoxygenation of aromatic *N*-oxide compounds with sodium hydrosulfite, sodium sulfite, or with sodium borohydride and aluminium trichloride.

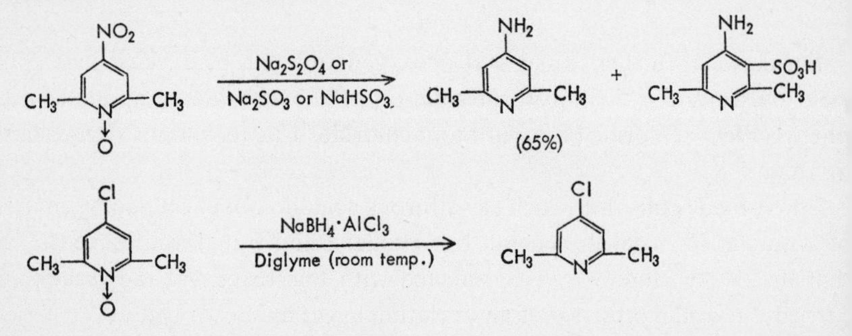

Elina and Magidson[115] also reported deoxygenation of aromatic *N*-oxide with sodium hydrosulfite.

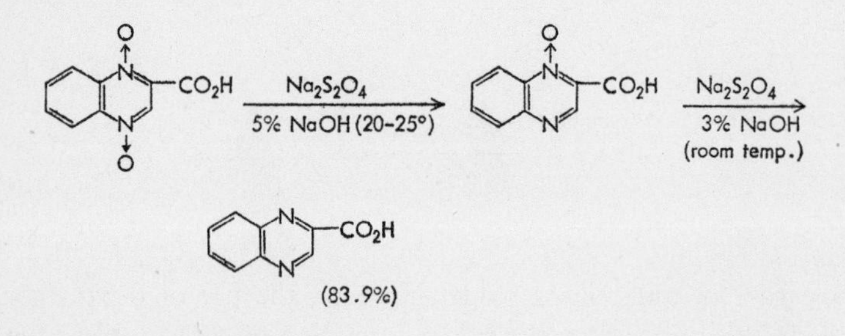

(b) Phosphorus Compounds

In connection with deoxygenation of aromatic *N*-oxide compounds with phosphorus trichloride, Hamana[82f] examined the reaction using triphenyl phosphite, and found that this agent did effect deoxygenation, though less easily than phosphorus trichloride.

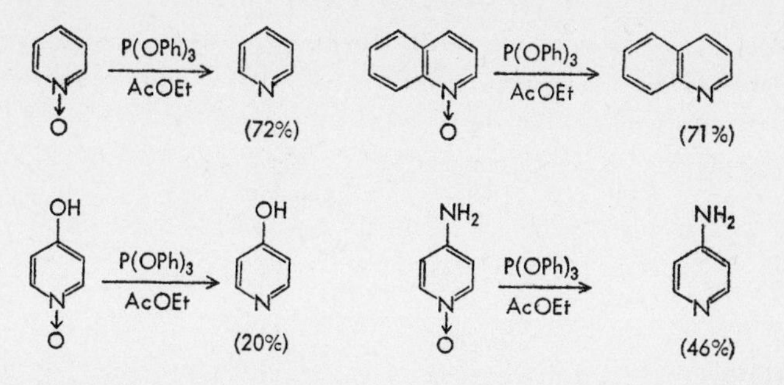

Emerson and Rees[116] examined this reaction with triethyl phosphite and reported that the presence of oxygen and of a peroxide of the solvent is necessary for the progress of the reaction, and that the reaction does not go to completion in their absence, even with triethyl phosphite in a hundred-fold excess.

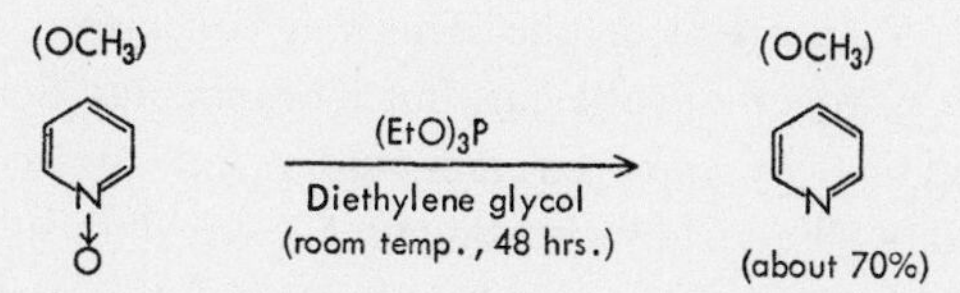

Emerson and Rees considered this to be a kind of radical reaction and assumed the following mechanism for it.

$$RO\text{-}OR \rightleftharpoons 2RO\cdot \qquad (EtO)_3P + RO\cdot \rightleftharpoons (EtO)_3\dot{P}(OR)$$

$$(EtO)_3\dot{P}(OR) + O_2 \rightleftharpoons (EtO)_3P(OR)\text{-}O\text{-}O\cdot$$

$$C_5H_5\overset{+}{N}\text{-}\overset{-}{O}\;\;\dot{O}\text{-}O\text{-}P(OEt)_3(OR) \longrightarrow C_5H_5N: + O_2 + O{=}P(OEt)_3 + \dot{O}R$$

Howard and Olszewski[117] used triphenylphosphine for the deoxygenation of aromatic *N*-oxides and found that the reaction can be effected by heating the components directly at about 230°. The results of this reaction are summarized in Table 5-9.

(c) Other Reducing Agents

Ochiai and Sai[118] found that 5-nitroisoquinoline 2-oxide is deoxygenated to 5-nitroisoquinoline when heated with potassium nitrate in conc. sulfuric acid at 170–180°, and Katada[119] showed that 4-nitropyridine 1-oxide is deoxygenated to 4-nitropyridine when heated with potassium nitrate in conc.

References p. 206

TABLE 5-9

DEOXYGENATION OF AROMATIC AMINE OXIDES WITH TRIPHENYLPHOSPHINE

N-Oxide of	Reaction temperature (°C)	Yield of amine (%)	Note
Pyridine	230–233	89.8	
4-Picoline	275–280	93	
2-Picoline	275–280	51.6	
4-Methoxypyridine	275–280	61.6	
Quinoline	230–233	89.2	
4-Nitropyridine	230–233	?	53.7% of triphenylphosphine oxide. No 4-nitropyridine

sulfuric acid to 165° in a sealed tube. This was considered to be oxidative deoxygenation by nitrogen dioxide gas, but Kröhnke and Schäfer[12] thought that it is a reduction by nitrogen monoxide, and showed that 4-nitropyridine is formed in 93% yield by heating its *N*-oxide with sulfuric and nitric acid mixture at 240° or with nitrosyl sulfate in conc. sulfuric acid at 170–200°, and that the same result can be obtained by passing nitrogen monoxide gas while heating in conc. sulfuric acid at 200°. In a similar manner, deoxygenation of 3-bromo- and 3-methyl-pyridine 1-oxide, and 4-nitroquinoline 1-oxide was effected in a good yield.

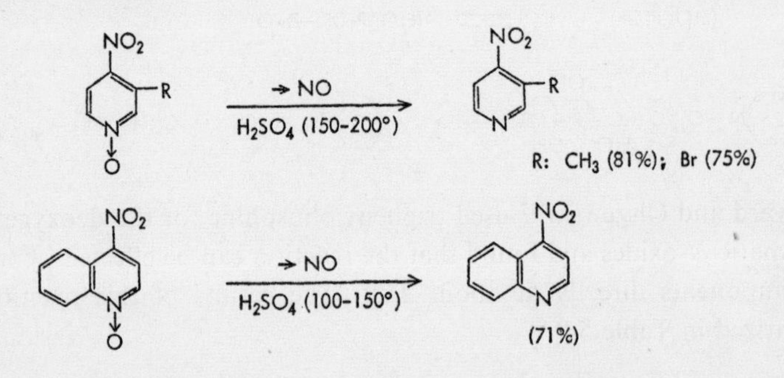

Kröhnke and his associate further reported a simpler method for the preparation of 4-nitropyridine in one step, with 71% yield on pyridine, by *N*-oxidation of pyridine with glacial acetic acid and hydrogen peroxide, followed by nitration, and heating this reaction mixture gradually to 200°, while passing nitrogen monoxide gas through the mixture.

Deoxygenation with other oxidation reagents includes the following.

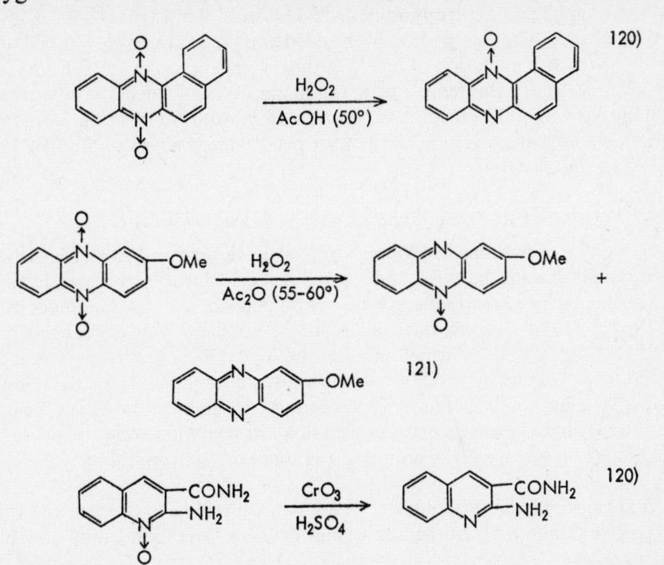

Brooks and Sternglanz[122] utilized the fact that aromatic *N*-oxide compounds are reduced quantitatively to the corresponding tertiary amines by titanium trichloride for the quantitative determination of amine oxides, but this reaction has not been used for syntheses of amines. Kubota and Akita[123] carried out deoxygenation of pyridine series *N*-oxides by reduction with hydrazine, using copper powder as a catalyst.

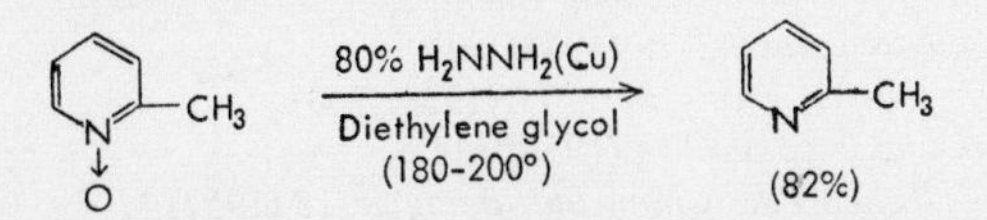

Taylor and his associates[124] showed that deoxygenation can be effected by heating with liquid ammonia in a sealed tube at 150°.

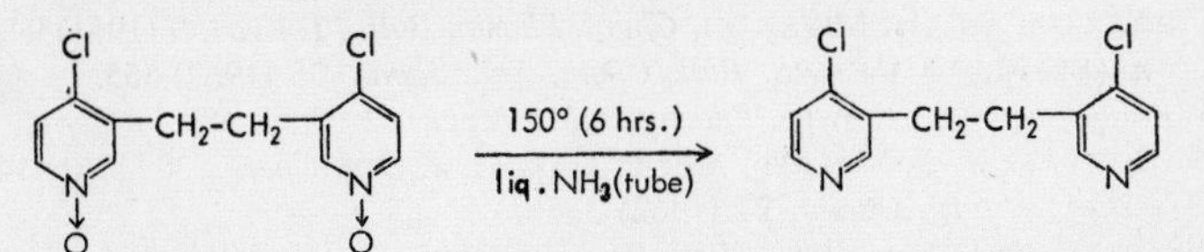

Aromatic *N*-oxide compounds are deoxygenated when converted into an adduct with alkyl halide, alkyl sulfonate, or dialkyl sulfate, and its alkaline solution warmed, resulting in decomposition into tertiary amines and an aldehyde. This reaction has already been described (*cf. Section 5.1.4.a*).

References p. 206

Example 1. Reduction of 4-Nitropyridine 1-Oxide to 4-Nitropyridine*[12]

To a solution of 4.2 g of 4-nitropyridine 1-oxide dissolved in 18 ml of conc. sulfuric acid, 10.5 ml of conc. nitric acid (sp. gr. 1.52) was added and the mixture was warmed on a water bath by which vigorous evolution of nitrogen dioxide gas began at 150°. The mixture was heated to 240° and held at this temperature for 20 min. When cooled, the mixture was basified and extracted with chloroform. Evaporation of chloroform from the extract afforded 4-nitropyridine which was recrystallized from petroleum ether (b.p. 50–70°) to leaflets, m.p. 50°. Yield, 3.3 g (90%).

Example 2. Preparation of 4-Nitropyridine from Pyridine in One Step*[12]

A mixture of 26 g of pyridine in 100 ml of glacial acetic acid and 50 ml of perhydrol was warmed on a boiling water bath for 24 h. After acetic acid was evaporated in vacuum as much as possible, the residue was dissolved in 70 ml of conc. sulfuric acid under ice-cooling and a mixture of 110 ml of conc. nitric acid (sp. gr. 1.52) and 70 ml of conc. sulfuric acid was added. This mixture was warmed in an oil bath, at 130° inner temperature, for 2.5 h and nitrogen monoxide was passed through this hot solution. While the temperature rose gradually to 200°, vigorous evolution of nitrogen dioxide took place. While the brown gas was still a little visible, the mixture was cooled in a nitrogen monoxide stream and worked up as usual, affording 26.4 g (71% on initial pyridine) of 4-nitropyridine.

i*CAUTION*: One should be careful not to bring 4-nitropyridine in contact with the skin as t is likely to give disagreeable burns, sometimes causing severe reddening and blistering.

* Translated from *Chemische Berichte*, by permission of Verlag Chemie G.m.b.H.

REFERENCES

1 R. Huisgen, *Angew. Chem.*, 75 (1963) 604.
2 H. H. Jaffé and G. O. Doak, *J. Am. Chem. Soc.*, 77 (1955) 4441.
3 H. Tanida, *Yakugaku Zasshi*, 78 (1958) 1079.
4 P. Baumgarten and H. Erbe, *Ber.*, 71 (1938) 2603.
5 H. Tanida, *Yakugaku Zasshi*, 78 (1958) 611.
6 M. Colonna, *Boll. Sci. Fac. Chim. Ind. Bologna*, 15 (1957) 1.
7 *(a)* H. Hirayama and T. Kubota, *Yakugaku Zasshi*, 72 (1952) 1025. *(b)* T. Kubota, *Yakugaku Zasshi*, 74 (1954) 831. *(c)* H. Hirayama and T. Kubota, *Yakugaku Zasshi*, 75 (1955) 1540. *(d)* H. Hirayama and T. Kubota, *Yakugaku Zasshi*, 77 (1957) 785. *(e)* T. Kubota and H. Miyazaki, *Chem. Pharm. Bull. (Tokyo)*, 9 (1961) 948. *(f)* T. Kubota and M. Yamakawa, *Bull. Chem. Soc. Japan*, 35 (1962) 555.
8 N. Ikekawa and Y. H. Sato, *Pharm. Bull. (Japan)*, 2 (1954) 400.
9 C. A. Buehler, L. A. Walker, and P. Garcia, *J. Org. Chem.*, 26 (1961) 1410.
10 J. F. Vozza, *J. Org. Chem.*, 27 (1962) 3856.
11 M. Szafran, *Bull. Acad. Polon. Sci., Ser. Sci. Chim.*, 11 (1963) 111 [*Chem. Abstr.*, 59 (1963) 11216].
12 F. Kröhnke and H. Schäfer, *Ber.*, 95 (1962) 1098.
13 E. Hayashi, *Yakugaku Zasshi*, 70 (1950) 142.
14 W. Hieber and A. Lipp, *Ber.*, 92 (1959) 2085.
15 A. Heller, R. Elson, and Y. Marcus, *J. Chem. Soc.*, (1962) 4738.

16 L. Garcia, S. I. Shupack, and M. Orchin, *Inorg. Chem.*, 1 (1962) 893; S. I. Shupack and M. Orchin, *J. Am. Chem. Soc.*, 85 (1963) 902.
17 A. B. P. Lever, J. Lewis, and R. S. Nyholm, *J. Chem. Soc.*, (1962) 6252.
18 D. D. Perrin, *J. Am. Chem. Soc.*, 82 (1960) 5642.
19 H. Sigel and H. Brintzinger, *Helv. Chim. Acta*, 46 (1963) 701.
20 S. Kamiya, *Chem. Pharm. Bull. (Tokyo)*, 10 (1962) 669.
21 E. Ochiai, T. Naito, and M. Katada, *Yakugaku Zasshi*, 64 (1944) 210.
22 A. R. Katritzky, *J. Chem. Soc.*, (1956) 2404.
23 O. Cerrinka, *Collection Czech. Chem. Commun.*, 27 (1962) 567.
24 T. M. Mishina and L. S. Efros, *Zh. Obshch. Khim.*, 32 (1962) 2217 [*Chem. Abstr.*, 58 (1963) 9023].
25 T. Kato, Y. Goto, and Y. Yamamoto, *Yakugaku Zasshi*, 84 (1964) 287.
26 M. Henze, *Ber.*, 70 (1937) 1270.
27 C. F. Koelsch and W. H. Gumprecht, *J. Org. Chem.*, 23 (1958) 1603.
28 V. Boekelheide and W. Feely, *J. Am. Chem. Soc.*, 80 (1958) 2217.
29 S. Sako, *Chem. Pharm. Bull (Tokyo)*, 11 (1963) 337.
30 I. Suzuki and T. Nakajima, *Chem. Pharm. Bull. (Tokyo)*, 12 (1964) 619.
31 H. J. den Hertog and H. N. Bojarska-Dahlig, *Rec. Trav. Chim.*, 77 (1958) 331.
32 E. Ochiai and T. Teshigawara, *Yakugaku Zasshi*, 65 (1945) 435.
33 E. Ochiai and E. Hayashi, *Yakugaku Zasshi*, 67 (1947) 151.
34 E. Ochiai and E. Hayashi, *Yakugaku Zasshi*, 67 (1947) 154.
35 G. Wagner and H. Pischel, *Arch. Pharm.*, 295 (1962) 897.
36 S. Takahashi and H. Kanô, *Chem. Pharm. Bull. (Tokyo)*, 11 (1963) 1380.
37 S. Takahashi and H. Kanô, *Chem. Pharm. Bull. (Tokyo)*, 12 (1964) 783.
38 W. Feely, W. L. Lehn, and V. Boekelheide, *J. Org. Chem.*, 22 (1957) 1135.
39 V. Boekelheide and R. Scharrer, *J. Org. Chem.*, 26 (1961) 3802.
40 C. W. Muth, R. S. Darlak, W. H. English, and A. T. Hamner, *Anal. Chem.*, 34 (1962) 1163.
41 J. H. Boyer, R. Borgers, and L. T. Wolford, *J. Am. Chem. Soc.*, 79 (1957) 678.
42 H. Tanida, *Yakugaku Zasshi*, 79 (1959) 1063.
43 T. Itai and T. Nakajima, *Chem. Pharm. Bull. (Tokyo)*, 10 (1962) 936.
44 R. Adams and W. Reifschneider. *J. Am. Chem. Soc.*, 79 (1957) 2236.
45 R. Dohmori, *Chem. Pharm. Bull. (Tokyo)*, 12 (1964) 595.
46 E. Hayashi, private communication.
47 E. Ochiai, *Proc. Imp. Acad. (Tokyo)*, 19 (1943) 307.
48 I. Iwai, *Yakugaku Zasshi*, 71 (1951) 1288.
49 E. Hayashi and Ch. Iijima, *Yakugaku Zasshi*, 82 (1962) 1093.
50 E. Ochiai and M. Katada, *Yakugaku Zasshi*, 63 (1943) 186.
51 E. Ochiai and T. Naito, *Yakugaku Zasshi*, 64 (1944) 206.
52 H. J. den Hertog and J. Overhoff, *Rec. Trav. Chim.*, 69 (1950) 468.
53 M. Hamana, *Yakugaku Zasshi*, 71 (1951) 263.
54 E. Hayashi, H. Yamanaka, and K. Shimizu, *Chem. Pharm. Bull. (Tokyo)*, 6 (1958) 323.
55 *(a)* H. Yamanaka, *Chem. Pharm. Bull. (Tokyo)*, 7 (1959) 158. *(b)* H. Yamanaka, *Chem. Pharm. Bull. (Tokyo)*, 7 (1959) 505.
56 *(a)* E. Hayashi, H. Yamanaka, and K. Shimizu, *Chem. Pharm. Bull. (Tokyo)*, 7 (1959) 141. *(b)* E. Hayashi, H. Yamanaka, and K. Shimizu, *Chem. Pharm. Bull. (Tokyo)*, 7 (1959) 146. *(c)* H. Yamanaka, T. Higashino, and E. Hayashi, *Chem. Pharm. Bull. (Tokyo)*, 7 (1959) 149. *(d)* H. Yamanaka, *Chem. Pharm. Bull. (Tokyo)*, 7 (1959) 505. *(e)* E. Hayashi, H. Yamanaka, and Ch. Iijima, *Yakugakau Zasshi*, 80 (1960) 1145.

57 E. Hayashi, H. Yamanaka, Ch. Iijima, and S. Matsushita, *Chem. Pharm. Bull. (Tokyo)*, 8 (1960) 649.
58 H. Igeta, *Chem. Pharm. Bull. (Tokyo)*, 8 (1960) 550.
59 T. Itai and S. Natsume (née Suzuki), *Chem. Pharm. Bull. (Tokyo)*, 11 (1963) 342.
60 G. Wagner, H. Pischel and R. Schmidt, *Z. Chem.*, 2 (1962) 86.
61 D. Jerchel and W. Melloh, *Ann.*, 613 (1958) 144.
62 E. Hayashi, H. Yamanaka, and Ch. Iijima, *Yakugaku Zasshi*, 80 (1960) 839.
63 T. Ishii, *Yakugaku Zasshi*, 72 (1952) 1317.
64 F. Allison, J. L. Comte, and H. E. Fierz-David, *Helv. Chim. Acta*, 34 (1951) 818.
65 E. Ochiai, *J. Org. Chem.*, 18 (1953) 539.
66 E. Ochiai and I. Suzuki, *Yakugaku Zasshi*, 67 (1947) 158.
67 T. Itai and H. Ogura, *Yakugaku Zasshi*, 75 (1955) 292.
68 E. Ochiai and T. Naito, *Yakugaku Zasshi*, 65 (1945) 441.
69 T. Wieland and H. Biener, *Ber.*, 95 (1962) 277.
70 E. Ochiai and M. Katada, *Yakugaku Zasshi*, 63 (1943) 265.
71 T. Higashino, *Yakugaku Zasshi*, 79 (1959) 831.
72 M. Hamana, B. Umezawa, and K. Noda, *Chem. Pharm. Bull. (Tokyo)*, 11 (1963) 696.
73 E. Hayashi and Y. Hotta, *Yakugaku Zasshi*, 80 (1960) 834.
74 E. Hayashi and H. Ohki, *Yakugaku Zasshi*, 81 (1961) 1033.
75 T. Itai and S. Natsume (née Suzuki), *Chem. Pharm. Bull. (Tokyo)*, 11 (1963) 83.
76 T. Higashino, *Chem. Pharm. Bull. (Tokyo)*, 9 (1961) 635.
77 W. Ried and P. Stahlhofen, *Ber.*, 87 (1954) 1814.
78 L. H. Sternbach, S. Kaiser, and E. Reeder, *J. Am. Chem. Soc.*, 82 (1960) 475.
79 A. S. Elina and L. G. Tsirulnikova, *Zh. Obshch. Khim.*, 33 (1963) 1544 [*Chem. Abstr.*, 59 (1963) 12807].
80 A. S. Elina, *Zh. Obshch. Khim.*, 32 (1962) 2967 [*Chem. Abstr.*, 58 (1963) 9059].
81 A. S. Elina, *Zh. Obshch. Khim.*, 28 (1958) 1378 [*Chem. Abstr.*, 52 (1958) 20186].
82 *(a)* M. Hamana, *Yakugaku Zasshi*, 75 (1955) 121. *(b)* M. Hamana, *Yakugaku Zasshi*, 75 (1955) 123. *(c)* M. Hamana, *Yakugaku Zasshi*, 75 (1955) 127. *(d)* M. Hamana, *Yakugaku Zasshi*, 75 (1955) 130. *(e)* M. Hamana, *Yakugaku Zasshi*, 75 (1955) 135. *(f)* M. Hamana, *Yakugaku Zasshi*, 75 (1955) 139.
83 I. Nakayama, *Yakugaku Zasshi*, 71 (1951) 1088.
84 E. Ochiai and Ch. Kaneko, *Pharm. Bull. (Japan)*, 5 (1957) 59.
85 R. M. Acheson, B. Adcock, G. M. Glover, and L. E. Sutton, *J. Chem. Soc.*, (1960)3367.
86 T. Itai and S. Kamiya, *Chem. Pharm., Bull. (Tokyo)*, 11 (1963) 348.
87 E. C. Taylor and J. S. Driscoll, *J. Org. Chem.*, 25 (1960) 1716.
88 W. Herz and D. R. K. Murty, *J. Org. Chem.*, 26 (1961) 122.
89 C. Hansch and W. Carpenter, *J. Org. Chem.*, 22 (1957) 936.
90 Z. Talik and T. Talik, *Roczniki Chem.*, 36 (1962) 417 [*Chem. Abstr.*, 58 (1963) 5627]
91 J. Haginiwa, *Yakugaku Zasshi.*, 75 (1955) 731.
92 E. C. Taylor and J. S. Driscoll, *J. Org. Chem.*, 26 (1961) 3796.
93 M. Hamana and H. Noda, *Yakugaku Zasshi*, 83 (1963) 342.
94 E. Ochiai and H. Yamanaka, *Pharm. Bull. (Japan)*, 3 (1955) 175.
95 H. Otomasu, *Chem. Pharm. Bull. (Tokyo)*, 6 (1958) 77.
96 S. Sako, *Chem. Pharm. Bull. (Tokyo)*, 11 (1963) 340.
97 H. Yamanaka, *Chem. Pharm. Bull. (Tokyo)*, 7 (1959) 152.
98 M. S. Habib and C. W. Rees, *J. Chem. Soc.*, (1960) 3386.
99 E. Hayashi, T. Higashino, Ch. Iijima, Y. Kôno, and T. Doihara, *Yakugaku Zasshi*, 82 (1962) 584.
100 H. J. den Hertog and W. P. Combé, *Rec. Trav. Chim.*, 70 (1951) 581.

101 H. J. den Hertog and C. Hoogzand, *Rec. Trav. Chim.*, 76 (1957) 261.
102 H. J. den Hertog, C. H. Henkens, and K. Dilz, *Rec. Trav. Chim.*, 72 (1953) 299.
103 H. J. den Hertog and M. van Ammers, *Rec. Trav. Chim.*, 77 (1958) 344.
104 D. U. Mizoguchi, *Pharm. Bull. (Japan)*, 3 (1955) 227.
105 M. Fujimoto, *Pharm. Bull. (Japan)*, 4 (1956) 1.
106 Y. Kidani and H. Otomasu, *Pharm. Bull. (Japan)*, 4 (1956) 391.
107 Z. Talik, *Roczniki Chem.*, 35 (1961) 475 [*Chem. Abstr.*, 57 (1962) 15065].
108 K. Adachi, *Yakugaku Zasshi*, 77 (1957) 514.
109 R. A. Abramovitsch and K. A. H. Adams, *Can. J. Chem.*, 39 (1961) 2134.
110 K. Takeda and K. Tokuyama, *Yakugaku Zasshi*, 75 (1955) 620.
111 D. I. Relyea, P. O. Tawney, and A. R. Williams, *J. Org. Chem.*, 27 (1962) 477.
112 S. Furukawa, *Pharm. Bull. (Japan)*, 3 (1955) 230.
113 M. Hamana and K. Funakoshi, *Yakugaku Zasshi*, 80 (1960) 1027.
114 R. F. Evans and H. C. Brown, *J. Org. Chem.*, 27 (1962) 1665.
115 A. S. Elina and O. Yu. Magidson, *Zh. Obshch. Khim.*, 25 (1955) 149 [*Chem. Abstr.*, 50 (1956) 1839].
116 T. R. Emerson and C. W. Rees, *J. Chem. Soc.*, (1962) 1917; *Proc. Chem. Soc.*, (1960) 418.
117 E. Howard and W. F. Olszewski, *J. Am. Chem. Soc.*, 81 (1959) 1483.
118 E. Ochiai and Z. Sai, *Yakugaku Zasshi*, 65 (1945) 418.
119 M. Katada, *Yakugaku Zasshi*, 67 (1947) 59.
120 I. J. Pachter and M. C. Kloetzel, *J. Am. Chem. Soc.*, 73 (1951) 4958.
121 H. Otomasu, H. Takahashi, and K. Yoshida, *Yakugaku Zasshi*, 84 (1964) 1080.
122 R. T. Brooks and P. D. Sternglanz, *Anal. Chem.*, 31 (1959) 561 [*Chem. Abstr.*, 53 (1959) 13894].
123 S. Kubota and T. Akita, *Yakugaku Zasshi*, 78 (1958) 248.
124 E. C. Taylor, A. J. Crovetti, and N. E. Boyer, *J. Am. Chem. Soc.*, 79 (1957) 3549.

Chapter 6

Electrophilic Substitution

6.1 General

Electrophilic substitution reactions of aromatic *N*-oxides have developed around nitration of pyridine series *N*-oxides. As has been stated in a section on amine oxides of the pyridine series (*cf. Section 2.1.2*), Ochiai and his collaborators[1] reported in 1942 that heating of pyridine 1-oxide with potassium nitrate in a mixture of fuming sulfuric acid and sulfuric acid on a water-bath resulted in formation of 4-nitropyridine 1-oxide in 94% yield. With this as the beginning, nitration of pyridine and quinoline series *N*-oxides in their 4-position was developed. In 1950, without knowledge of this work, den Hertog and his co-workers[2] obtained 4-nitropyridine 1-oxide by the nitration of pyridine l-oxide by a similar method, and reported the reactivity of the position *para* to the nitrogen towards nitration.

Later, Otomasu[3] found that whereas nitration of phenazine with sulfuric and nitric acids did not occur at a low temperature and produced 1,3-dinitrophenazine only at 60°, nitration of phenazine 5-oxide (6-1) occurred readily at 0°, giving its 3-nitro compound (6-2) mainly, with a small amount of l-nitro derivative (6-3), and pointed out that the polar effect of the *N*-oxide function was transmitted through the *para*-nitrogen atom to the *para* and *ortho* positions of one of the benzene portions, that is, the 3 or 1 position.

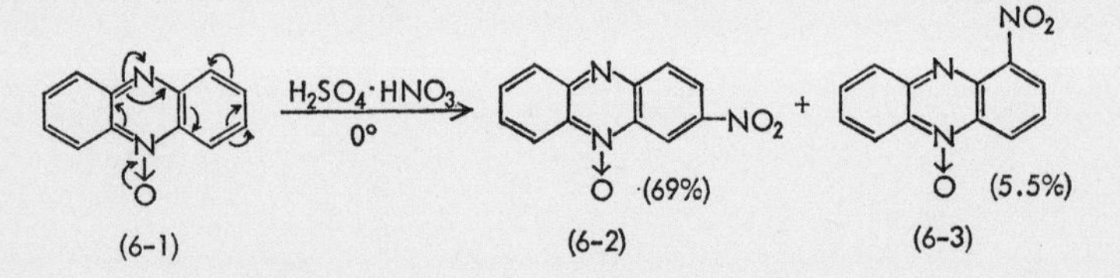

Itai and Igeta[4], Nakagome[5], and Kanô and his co-workers[6] proved that nitration of *N*-oxides of the pyridazine and cinnoline series similarly resulted in the nitration of the position *para* to the *N*-oxide group.

Thus, the polar effect of the aromatic *N*-oxide function to nitration appears most markedly in the position *para* to the *N*-oxide group or to the positions corresponding to it, but other kinds of electrophilic substitution reactions do not progress so smoothly.

Ochiai and Okamoto[7] obtained 4-bromoquinoline 1-oxide in 30% yield by shaking the perbromide of quinoline 1-oxide with bromine water in a sealed tube and noted the activation of the 4-position to bromination, but Mosher and Welch[8] failed to obtain any successful result from sulfonation, bromination, or Friedel–Crafts reaction of pyridine 1-oxide. They therefore reported that they could not recognize any polar effect of aromatic *N*-oxide groups to these reactions. Further examination of this sulfonation[9] and bromination[10] by den Hertog and others proved a slight activation of the 4-position to bromination and sulfonation. Acheson[11] also showed the activation of the position *para* to the *N*-oxide group when he obtained 9-bromoacridine 10-oxide by the bromination of acridine *N*-oxide in glacial acetic acid.

Ukai and his school[12] reported that mercuration of pyridine and quinoline 1-oxides with mercuric acetate resulted in the substitution in 4- and 8-positions, respectively, although later examination by den Hertog and van Ammers[13] proved that 2- and 2,6-positions were substituted in pyridine 1-oxide. This fact suggests that the mercuration of quinoline l-oxide should be re-examined.

Ochiai and Okamoto[14,15] pointed out that there is a marked temperature effect in the nitration of quinoline l-oxide. At 0° to 10°, a polar effect of the *N*-oxide function is not apparent and nitration of quinoline 1-oxide gives 5- and 8-nitro derivatives as in the case of nitration of quinoline. Activity of the 4-position becomes apparent from about 40° upward, and the 4-nitro compound is formed in the main by nitration at 65–70°. Ochiai and Kaneko[16] found, during examination of the mechanism of this reaction, that the application of benzoyl nitrate or of benzoyl chloride and silver nitrate to quinoline l-oxide dissolved in an aprotic solvent like chloroform or dioxan results in the formation of 3-nitro compound in 40% yield, with by-product formation of *N*-benzoyloxy-3,6-dinitrocarbostyril. This has resulted in the development of nitration of the 3-position of quinoline 1-oxide. This reaction occurs also with pyridine 1-oxide, but the yield is very poor. If, however, *p*-nitrobenzoyl chloride, having a stronger electrophilic reactivity than benzoyl chloride, is used as the reagent, 3-nitro- and 3,5-dinitropyridine 1-oxides are obtained in yields of 9 and 10% respectively[17].

Hamana and Yamazaki[18] applied this reaction to the bromination of pyridine and quinoline 1-oxides, and revealed that the application of bromine

References p. 243

in chloroform solution, in the presence of an acylation agent, gave 3,6-di bromoquinoline 1-oxide or 3,5-dibromopyridine 1-oxide. Itai and Natsume[1 used this reaction for the nitration of pyridazine 1-oxide and obtained main ly 3-nitropyridazine 1-oxide, with a small amount of 5-nitro compound the similar nitration of cinnoline 1-oxide by Itai, Suzuki, and Nakashima[4 gave only the 3-nitro compound.

Ochiai and Ohta[20] found that heating of the methosulfate or boron tri fluoride complex of quinoline 1-oxide with a nitrate in dimethyl sulfoxid solution at 140° results in the formation of 3-nitroquinoline 1-oxide i 30–50% yield. This reaction does not occur with pyridine 1-oxide, but it ma be expected to open a way for the development of substitution reactions i position *meta* to the *N*-oxide group.

6.2 Nitration

Nitration reactions of aromatic *N*-oxides utilizing the polar effect of thei *N*-oxide function have developed around the *N*-oxides of the pyridine an benzopyridine series, and these reactions are being applied to the *N*-oxide in the phenazine, pyridazine, and cinnoline series. Activation to nitration b the *N*-oxide group has not been found at all in the *N*-oxides of pyrimidine an pyrazine series.

6.2.1 Nitration of Pyridine Series N-Oxides

(a) Pyridine Derivatives

Pyridine 1-oxide (6-4) gives 4-nitropyridine 1-oxide (6-5) in over 90% yiel when warmed on a water bath with potassium nitrate in conc. sulfuric aci solution containing fuming sulfuric acid. The structure of the product wa determined by its reduction to 4-aminopyridine[1]. Ochiai and his collaborator[2 re-examined this reaction and found that a small amount of 2-nitropyridin (6-6) was formed as a by-product. Formation of the 2-nitro compound in- creased slightly when the reaction temperature was raised above 110°, with a corresponding decrease of 4-nitropyridine 1-oxide and formation of 4-nitro- pyridine. This fact suggests deoxygenation of primarily formed 2-nitro- pyridine 1-oxide and shows that this nitration results in overwhelming sub- stitution of the position *para* to the ring nitrogen but that at the same time, substitution also occurs, though only slightly, in the position *ortho* to the nitrogen.

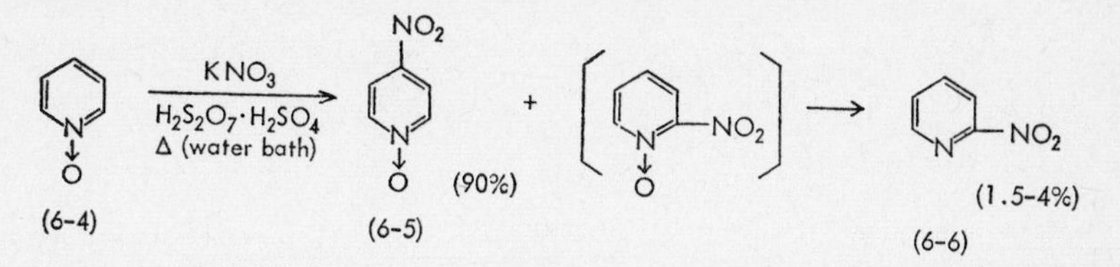

In this case, the use of fuming sulfuric acid increases the yield of the nitrated derivative and the yield decreases by the use of a mixture of sulfuric and nitric acids. The yield of 4-nitropyridine 1-oxide increases to about 70% by raising the reaction temperature to around 130° (Ref. 22). Den Hertog and others[2a] obtained 4-nitropyridine 1-oxide in 80–90% yield by heating pyridine 1-oxide with fuming nitric acid in conc. sulfuric acid at 90°. Ochiai and Sai[23] reported a simplified method for the synthesis of 4-nitropyridine 1-oxide by direct nitration of the crude pyridine 1-oxide obtained by heating pyridine with 30% hydrogen peroxide in glacial acetic acid solution. Den Hertog and others[2b] also used a similar simplified method and obtained 4-nitropyridine 1-oxide in 90–95% yield.

The same nitration reaction of the *N*-oxides of pyridine homologs also gives 4-nitro derivatives in a good yield. These are summarized in Table 6-1.

2,4-Lutidine 1-oxide and 2,4,6-collidine 1-oxide, in which the 4-position is occupied by an alkyl group, resist nitration under the same conditions[28], but the alkyl group in 2-position increases the electron-releasing effect of the *N*-oxide group through its inductive effect and promotes nitration of the 4-position, and the presence of an alkyl group in the 3-position seems to cause a slight steric hindrance. This is especially noted in the case of 3-tert-butylpyridine 1-oxide in which the 2-position and not the 4-position is nitrated, and the deoxygenated 3-tert-butyl-2- or -6-nitropyridine is known to be formed[26].

Compared to alkyl groups, the steric hindrance of halogen groups like chlorine and bromine is not so remarkable and, as shown in Table 6-2, both the 2,6- and 3,5-substituted compounds give 4-nitro compounds in a satisfactory yield, although the results reported differ according to workers.

Krause and Langenbeck[35] obtained a 4-nitro derivative (6-8) by the nitration of picolinic acid 1-oxide (6-7). There is no report on the nitration of nicotinic acid 1-oxide, and Taylor and Crovetti[36] obtained it through the oxidation of 4-nitro-3-picoline 1-oxide.

References p. 243

TABLE 6-1

NITRATION OF ALKYL DERIVATIVES OF PYRIDINE 1-OXIDE

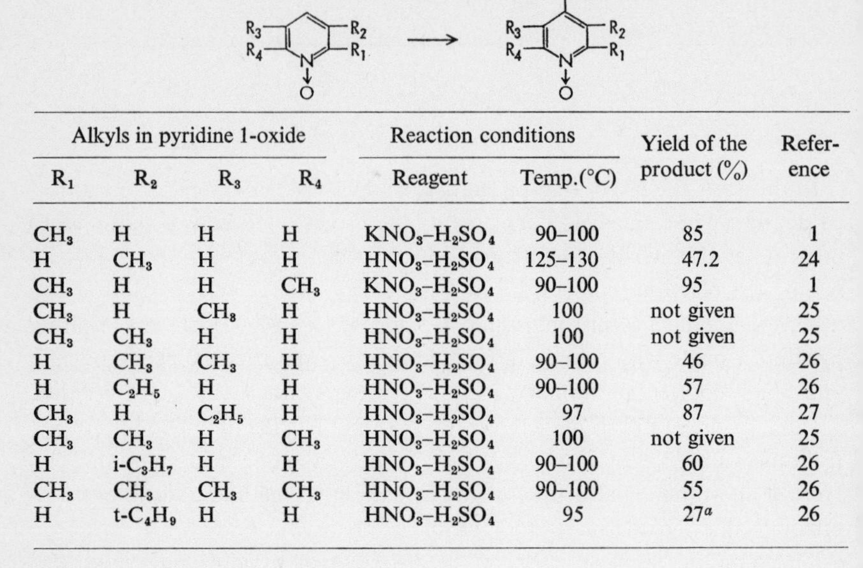

Alkyls in pyridine 1-oxide				Reaction conditions		Yield of the product (%)	Reference
R_1	R_2	R_3	R_4	Reagent	Temp.(°C)		
CH_3	H	H	H	KNO_3–H_2SO_4	90–100	85	1
H	CH_3	H	H	HNO_3–H_2SO_4	125–130	47.2	24
CH_3	H	H	CH_3	KNO_3–H_2SO_4	90–100	95	1
CH_3	H	CH_3	H	HNO_3–H_2SO_4	100	not given	25
CH_3	CH_3	H	H	HNO_3–H_2SO_4	100	not given	25
H	CH_3	CH_3	H	HNO_3–H_2SO_4	90–100	46	26
H	C_2H_5	H	H	HNO_3–H_2SO_4	90–100	57	26
CH_3	H	C_2H_5	H	HNO_3–H_2SO_4	97	87	27
CH_3	CH_3	H	CH_3	HNO_3–H_2SO_4	100	not given	25
H	i-C_3H_7	H	H	HNO_3–H_2SO_4	90–100	60	26
CH_3	CH_3	CH_3	CH_3	HNO_3–H_2SO_4	90–100	55	26
H	t-C_4H_9	H	H	HNO_3–H_2SO_4	95	27[a]	26

[a] 2- or 6-nitro deoxygenated derivative.

TABLE 6-2

NITRATION OF HALOGENOPYRIDINE 1-OXIDES

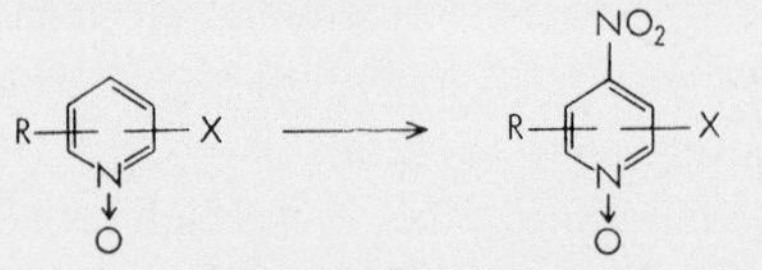

Halogenopyridine 1-oxide		Reaction conditions		Yield of the product (%)	Reference
R	X	Reagent	Temp.(°C)		
H	2-Cl	fum. HNO_3–H_2SO_4	100	not given	29, 30
3-CH_3	2-Cl	fum. HNO_3–H_2SO_4	100	not given	29
5-CH_3	2-Cl	fum. HNO_3–H_2SO_4	100	not given	29
6-CH_3	2-Cl	fum. HNO_3–H_2SO_4	100	not given	29
H	2-Br	fum. HNO_3–H_2SO_4	90	not given	31
H	3-Br	KNO_3–$H_2SO_4(SO_3)$	90–100	91	32
H	2,6-Br_2	fum. HNO_3–H_2SO_4	90	65	33
H	3,5-Br_2	fum. HNO_3–H_2SO_4	90	90–95	34

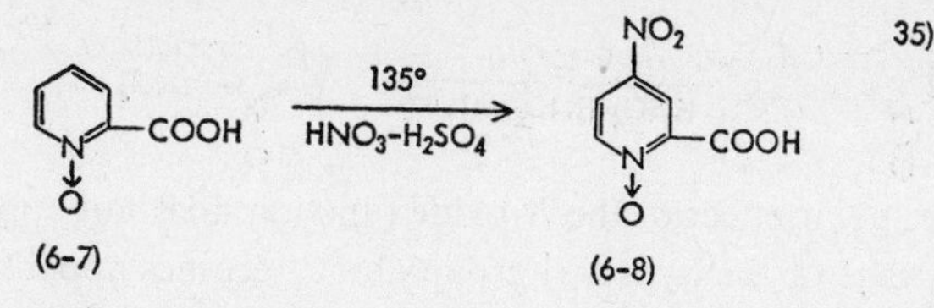

The most interesting fact is that the orienting effect of the *N*-oxide group is larger than that of the alkoxyl group towards nitration. Den Hertog and others[31] showed that whereas nitration of 2-ethoxy-[37–39] (6-9) and 3-ethoxy-pyridine[40–42] (6-10) resulted in the nitration of 5- (6-11) and 2-positions (6-12), respectively, nitration after *N*-oxidation invariably gave 4-nitro compounds (6-13, 6-14).

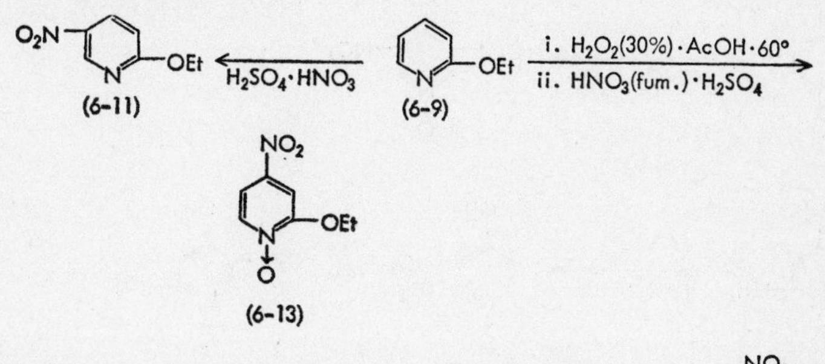

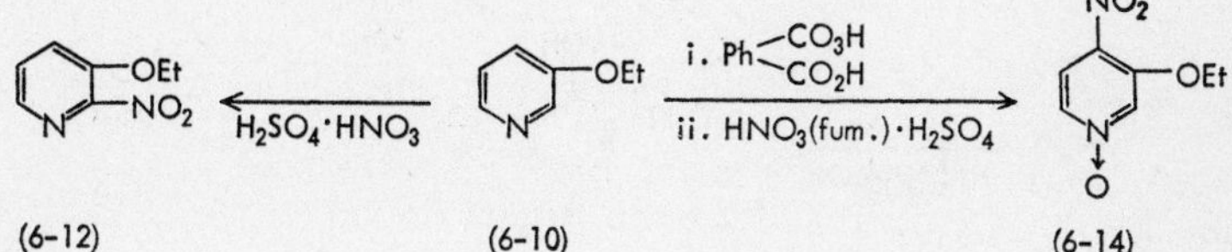

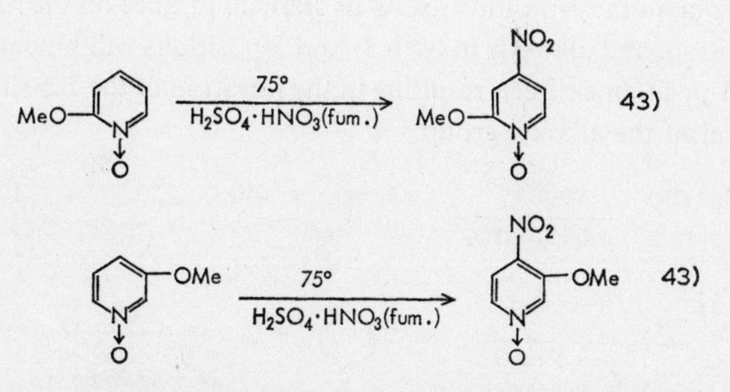

If the alkoxyl is changed to a hydroxyl, the orienting effect of the hydroxyl group predominates over that of the *N*-oxide group, and the positions *ortho* and *para* to the hydroxyl group are nitrated.

References p. 243

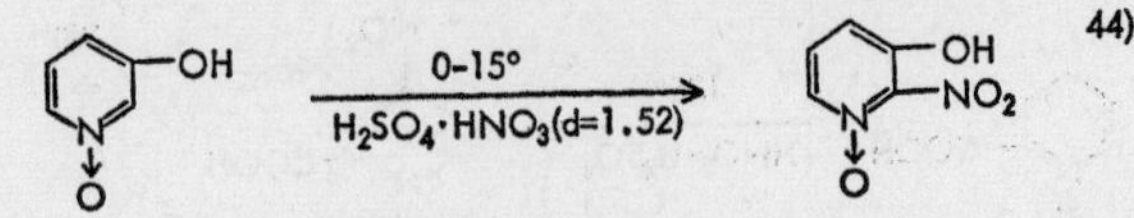

In this case, polar effect of the *N*-oxide function does not appear at a low temperature; that of the hydroxyl group alone becomes apparent, and only the 2-position is nitrated, as in the case of 3-hydroxypyridine[45].

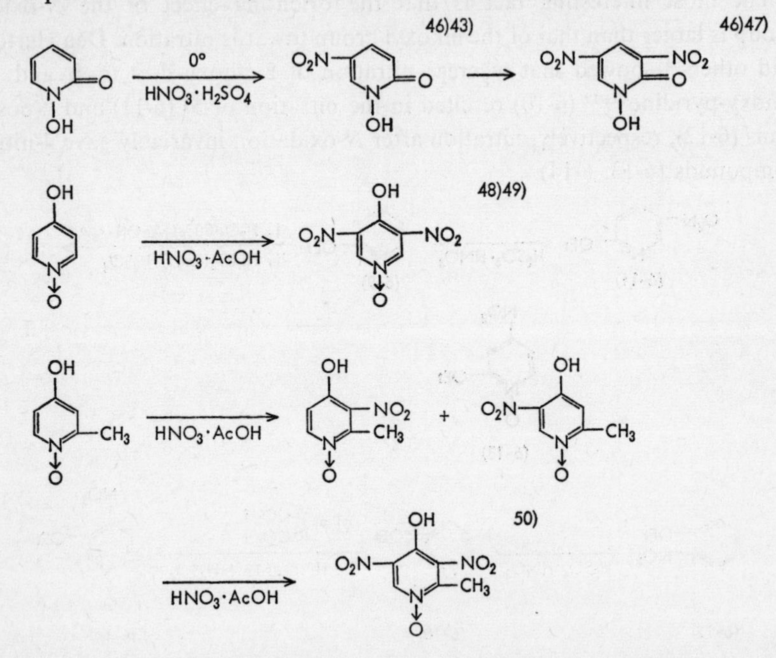

The alkoxyl group in the 3-position exerts its steric hindrance on the nitration of the 4-position, and alkoxyls in both 3- and 5-positions will hinder the nitration of the 4-position entirely, resulting in the nitration of the 2-position due to orientation of the alkoxyl group.

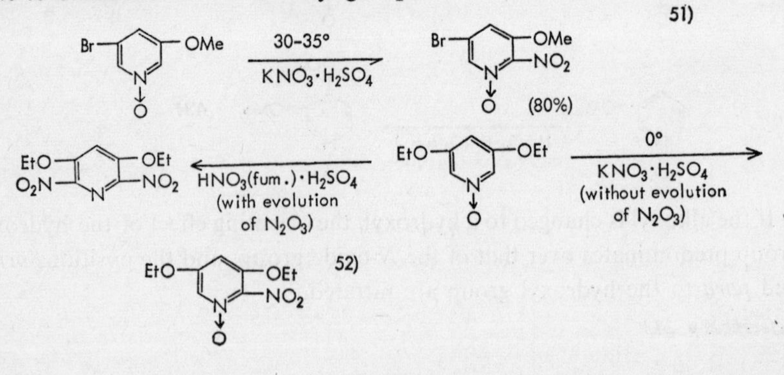

TABLE 6-3

NITRATION OF ALKOXY- AND HYDROXY-PYRIDINE 1-OXIDES

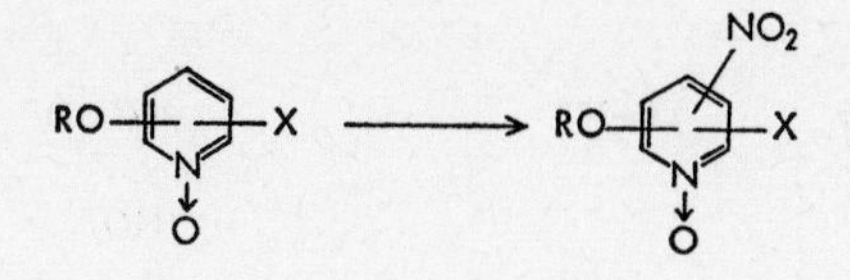

Pyridine 1-oxide		Reaction conditions		Product		
RO	X	Reagent	Temp.(°C)	Position of NO_2	Yield (%)	Reference
2-EtO	H	fum. HNO_3–H_2SO_4	85	4	10–15	31
2-MeO	H	fum. HNO_3–H_2SO_4	75	4	55–60	43
3-EtO	H	fum. HNO_3–H_2SO_4	85	4	70–80	31
3-MeO	H	fum. HNO_3–H_2SO_4	75	4	70–75	43
3-OH	H	HNO_3(1.52)–H_2SO_4	0–15	2	60.5	44
2-OH	H	HNO_3–AcOH	0	5	67	46
2-OH	H	HNO_3–AcOH	40	3,5	not given	47
4-OH	H	HNO_3–AcOH	90–100	3, 5	80	48, 49
4-OH	2-CH_3	HNO_3–AcOH	65–75	3, 5	15	50
4-OH	3- or 5-NO_2	HNO_3–AcOH	65–75	3, 5	not given	50
3-OEt	5-OEt	KNO_3–H_2SO_4	0	2	100	52
3-OEt	5-OEt	fum. HNO_3–H_2SO_4	85	2, 6	90–95[a]	52
3-MeO	5-Br	KNO_3–H_2SO_4	30–35	2	80	51
3-MeO	5-MeO	KNO_3–H_2SO_4	0	2	not given	51

[a] Deoxygenated.

Results of nitration of these alkoxy- and hydroxy-pyridine 1-oxides are summarized in Table 6-3.

(b) Quinoline Derivatives

The 4-position in quinoline 1-oxide (6-15) is more active than that in pyridine 1-oxide, and 4-nitroquinoline 1-oxide (6-16) is formed in a good yield of 60–70% under milder conditions (such as heating with nitric and sulfuric acids at 70°)[53]. Compared to the formation of 5- and 8-nitro derivatives (6-18, 6-19) by a similar nitration of quinoline (6-17)[54-57], the powerful polar effect of the *N*-oxide group becomes apparent.

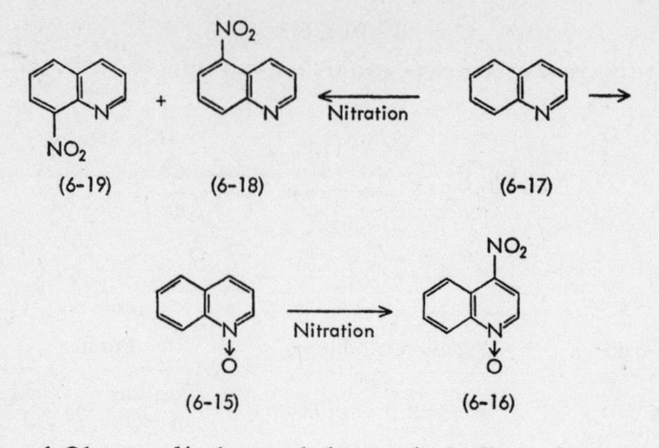

Ochiai and Okamoto[14] observed the marked effect of temperature in th
nitration of quinoline 1-oxide with potassium nitrate and sulfuric acid. Th
effect of the *N*-oxide group is hardly observed at a low temperature, and .
and 8-nitro derivatives are formed as in the case of quinoline, but above 40
the yield of the 4-nitro compound increases and 4-nitroquinoline l-oxide
mainly formed. At above 100°, deoxygenation of the *N*-oxide group occu
and the yield of the 4-nitro compound decreases. A similar temperatu
effect was observed in the nitration of 3- (Ref. 15) and 6-bromoquinolir
l-oxides[58]. The effect of various factors influencing the nitration of quinaldir
1-oxide (6-23) was examined in detail by Ochiai and Satake[59] who found
similar temperature effect also to exist in this case and the activity of the
position in quinaldine l-oxide to be greater than that in quinoline l-oxid
The yield of the nitrated products given in Table 6-4 and Fig. 6-1 is tha
obtained on nitration of 3 g of quinaldine 1-oxide dissolved in 10 g of 96
sulfuric acid and treated with 2 g (1.1 equiv.) of potassium nitrate at −1
to 160°. Table 6-5 shows the change in the amount of the products on varyin
the reaction time during a nitration reaction carried out at 18–24°.

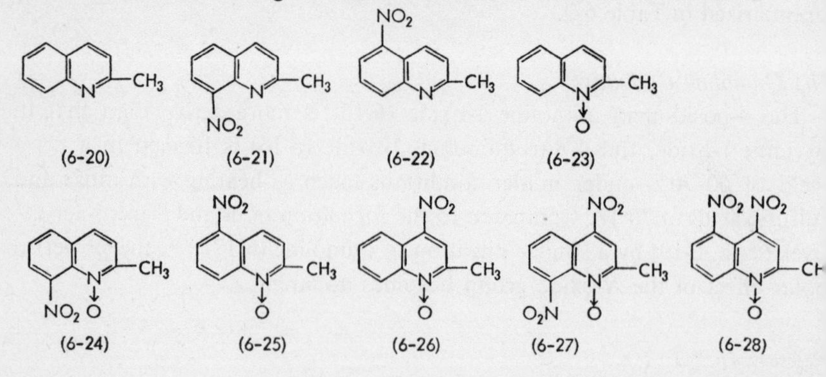

TABLE 6-4

NITRATION OF QUINALDINE 1-OXIDE AT VARIOUS TEMPERATURES

Reaction conditions		Yield of reaction products (%)									
		Quinaldine			Quinaldine 1-oxide						
Temp.(°C)	Time(h)	6-20	6-21	6-22	6-23	6-24	6-25	6-26	6-27	6-28	Total
−5 to −10	8				0.5	29	27	1.5	6.3	5.1	67
5–10	5				0.7	33	31	2.9	2.8	trace	70
35–40	2				trace	35	29	3.8	trace		67
55–60	2				0	16	10.7	58			85
85–90	2	trace				9	7.4	56			73
110–120	1	2.8	1.0	1.3		7.1	11.5	51			75
150–160	0.5	7.3	3.1	5.2		4	7	38			65

TABLE 6-5

NITRATION OF QUINALDINE 1-OXIDE WITH VARIOUS REACTION DURATIONS

Reaction duration (days)	Yield of reaction products (%)					
	6-26	6-24	6-25	6-27	6-28	Total
0.4	6	15	29	19	0	69
2	3.3	8.2	20	28	2.4	64
5	trace	4	11	34	11	59
10	0	trace	7	41	23	71

TABLE 6-6

EFFECT OF THE CONCENTRATION OF SULFURIC ACID ON THE NITRATION OF QUINALDINE 1-OXIDE

Reaction conditions		Yield of reaction products (%)					
Temp.(°C)	Concn. of H_2SO_4 (%)	6-23	6-24	6-25	6-26	6-27	Total
−5 to 10	96	0.7	33	31	2	2.7	69
	91	4.9	31	27	6.2	trace	69
	86	91	0	0	1.5	0	93
	96	0	16	9	58	0	83
55–60	91	trace	2.2	5.8	76	0	84
	86	6	0	0	82	0	88

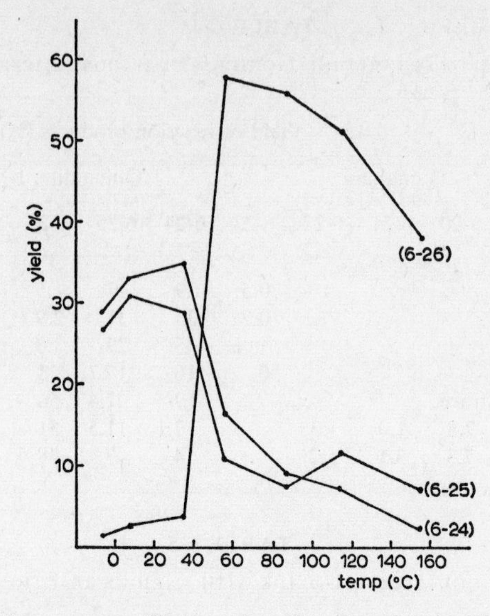

Fig. 6-1. Temperature effect on the nitration of quinaldine 1-oxide.

Table 6-6 is the result obtained by examination of the effect of the concer tration of sulfuric acid on the amount of the products formed when 3 g c quinaldine 1-oxide (6-23) was treated with 2 g (1.1 equiv.) of potassium nitrat in 10 g of sulfuric acid at −5 to 10° and at 55–60°, both for 2 h. It may b seen from this table that 86% sulfuric acid has a better effect on the yield c 4-nitroquinaldine 1-oxide (6-26) than 96% acid.

This effect of the *N*-oxide group for orientation of the 4-position to nitra tion is also observed in the methyl homologs[58,60,61], halogen derivatives[58,6 and in nitro derivatives[63]. Ishikawa[60] reported that in lepidine 1-oxide (6-29 in which the 4-position is already occupied, nitration chiefly took place a the 8-position, while Ames and others[64] reported the difficulty of nitrating 2 heptylquinoline 1-oxide (6-30).

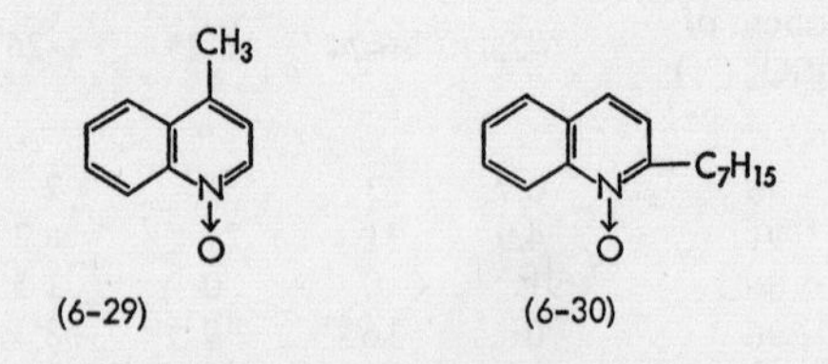

A bromine atom in 3-position exerts a slight hindrance on the nitration c the 4-position. A nitro group in the benzene portion of the quinoline rin inhibits nitration of that ring due to its polar effect, so that nitration of the 4-position progresses markedly even at a low temperature. At the same time, a nitro group in the 4-position also interferes in the nitration of the pyridine portion, and nitration occurs only in the 6- and 8-positions, 4,6,8-trinitroquinoline 1-oxide being formed under drastic conditions[63].

Ishikawa[65] obtained the 5-nitro compound in 56% yield by nitration of 6-methoxyquinoline 1-oxide with potassium nitrate in conc. sulfuric acid at 40°, and did not obtain any 4-nitro compound. This fact indicates the greater orienting effect of the methoxyl function than of the *N*-oxide function, which differs from the case of pyridine series *N*-oxides. When the 4-position is occupied by a methoxyl group, quinoline *N*-oxides are comparatively resistant to nitration, and 6- and 7-nitro compounds are mainly formed[65,66].

In contrast to the above, the orienting effect of the hydroxyl group is still stronger and positions *ortho* and *para* to the hydroxyl group are nitrated under milder conditions (nitric acid and glacial acetic acid).

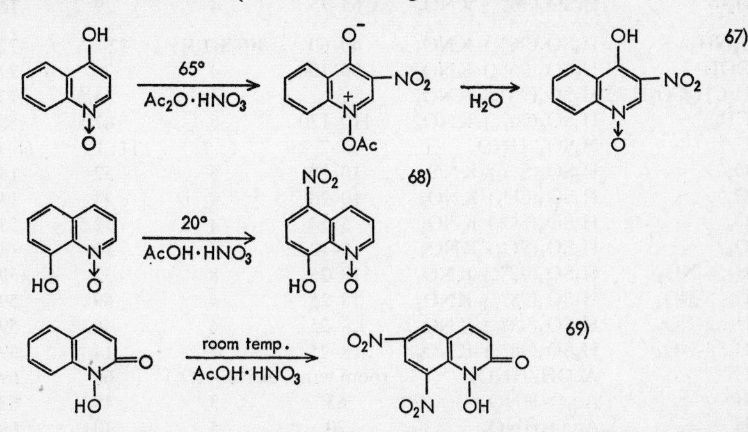

An analogous case is the orientation of the amino group, as for example:

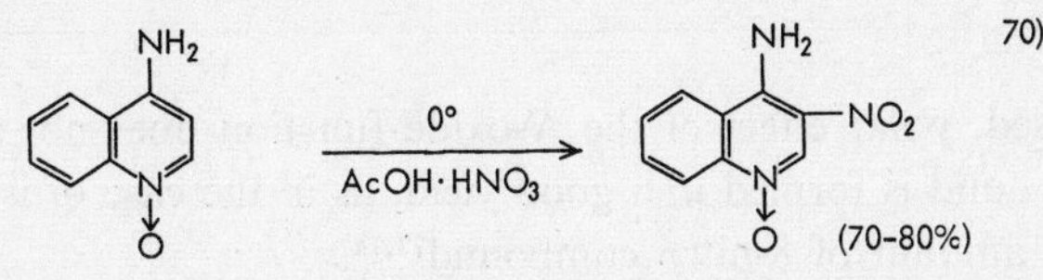

Results of the nitration of these quinoline series *N*-oxides are summarized in Table 6-7.

(c) Other Benzopyridines

In isoquinoline 2-oxide (6-31), in which the position *para* to the *N*-oxide

TABLE 6-7

NITRATION OF QUINOLINE SERIES *N*-OXIDES

Subst. in quinoline 1-oxide	Reaction conditions		Nitroquinoline 1-oxide		Reference
	Reagent	Temp.(°C)	Position of NO_2	Yield (%)	
Nil	H_2SO_4–KNO_3	70	4	60–70	53, 14, 71
2-CH_3	H_2SO_4(86%)–KNO_3	55–60	4	82	59, 60
4-CH_3	H_2SO_4–HNO_3(d=1.5)	5–7	8	39	60
6-CH_3	H_2SO_4–KNO_3	60–80	4	80	61, 60
6-Cl	H_2SO_4(95%)–KNO_3	110–120	4; 5	16; 32	58
7-Cl	H_2SO_4–KNO_3	60	4	90	62
3-Br	H_2SO_4(98%)–KNO_3	60–70	4; 5; 8	6.1; 50; 4.4	15
6-Br	H_2SO_4(95%)–KNO_3	60–70	4; 5	8.5; 33	58
		110–120	4; 5	25; 25	58
2-CH_2OH	H_2SO_4(86%)–KNO_3	55	4	35.1	72
2-CH_2-N(piperidino)	H_2SO_4(96%)–KNO_3	65–75	4	29	72
6-SO_2NH_2	H_2SO_4(96%)–KNO_3	40–60	4(6-SO_3H)	32–36	72
6-COOH	H_2SO_4(96%)–KNO_3	90–100	4	28	72
2-$CH_2CH_2CO_2H$	H_2SO_4(96%)–KNO_3	55	4	54.6	72
6-OCH_3	H_2SO_4(96%)–KNO_3	110–120	5	47.5	58
4-OCH_3	H_2SO_4–HNO_3	5–7	6; 7	11; 18	66, 65
4-NO_2	H_2SO_4(SO_3)–KNO_3	10–20	5	52	14
5-NO_2	H_2SO_4(SO_3)–KNO_3	10–20	4	24	14
6-NO_2	H_2SO_4(SO_3)–KNO_3	5–10	4	72.5	58
8-NO_2	H_2SO_4(SO_3)–KNO_3	0–10	4	73	14
2-CH_3, 4-NO_2	H_2SO_4(98%)–KNO_3	18–25	8	32	59
2-CH_3, 8-NO_2	H_2SO_4(98%)–KNO_3	18–25	4	69	59
2-CH_3, 5-NO_2	H_2SO_4(98%)–KNO_3	1 8–25	4	46	59
2-CH_3, 4-NO_2	H_2SO_4(98%)–KNO_3	18–25	5	14	59
2-OH	AcOH–HNO_3	room temp.	6,8(di-NO_2)	60	69
4-OH	Ac_2O–HNO_3	65	3	48	67
8-OH	Ac_2O–HNO_3	20	5	70	68
4-NH_2	Ac_2O–HNO_3	0	3	70–80	70

group is closed, polar effect of the *N*-oxide function does not appear and 5-nitro compound is formed in a good yield, as in the case of isoquinoline, with a small amount of 8-nitro compound[73,74].

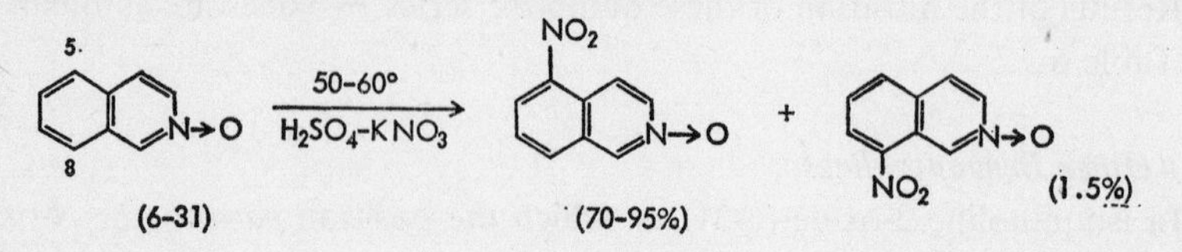

Of the ring systems formed by fusion of a benzene ring to the pyridine portion of the quinoline ring, acridine *N*-oxide is nitrated in the position *para* to the ring nitrogen, that is, position 9 (Ref. 11). Considering the fact that electrophilic substitution in acridines takes place first at 2- and 7-positions, and then at 4- and 5-positions[75], nitration of acridine *N*-oxide (6-32) indicates the marked polar effect of the *N*-oxide function.

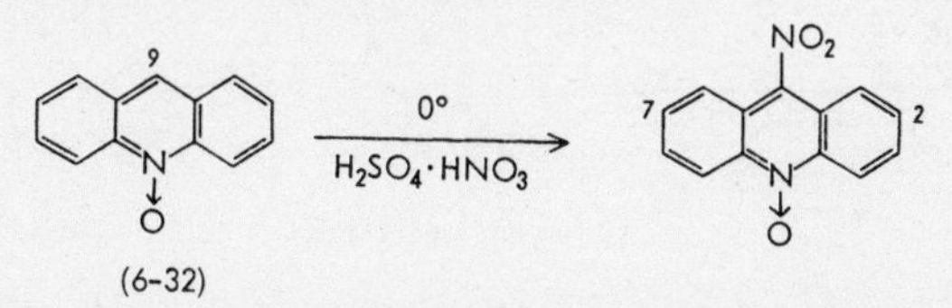

In the ring systems formed by fusion of a benzene ring to the benzene portion of the quinoline ring at 7-8 or 5-6 positions, the mode of nitration of benzo[*h*]quinoline 1-oxide (6-34)[76,77] and benzo[*f*]quinoline 4-oxide (6-33)[78] is not so greatly different from that of benzoquinolines corresponding to the condensed benzene portion, in spite of the vacancy of the position *para* to the *N*-oxide function.

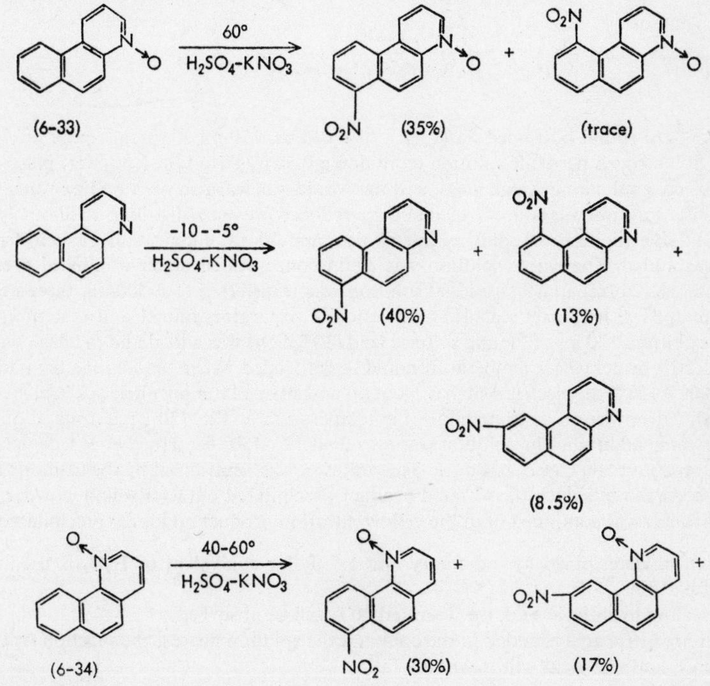

References p. 243

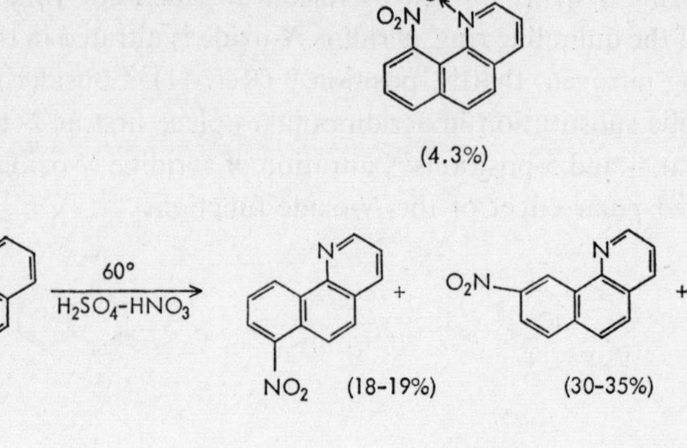

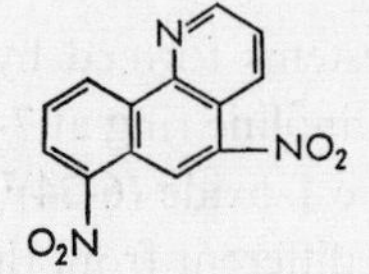

Example 1. Preparation of 4-Nitropyridine 1-Oxide from Pyridine
(from a note of E. Hayashi)

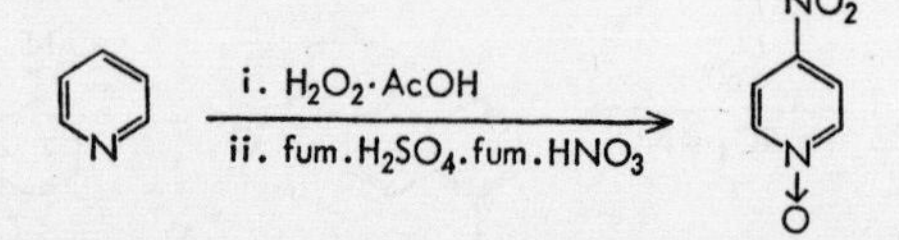

In a 500-ml round-bottomed flask, 40 g of pyridine, 100 ml of glacial acetic acid, and 68 ml* of hydrogen peroxide solution (containing 0.30 g of H_2O_2 in 1 ml) were placed, the flask was covered with a watch glass, and the whole was warmed on a boiling water bath for 3 h. The solution was concentrated under a reduced pressure (distilling off about 135 ml of a distillate), the residual solution was cooled, and 30 ml of conc. sulfuric acid (sp. gr. 1.84) was added. The whole solution was again concentrated under a reduced pressure (distillate, *ca.* 20 ml) and the residual solution was transferred to a 500-ml, three-necked flask equipped with a stirrer sealed in sulfuric acid**, separatory funnel, and a thermometer. To this solution, 120 ml of fuming sulfuric acid (30% SO_3) was added and the flask was set on an electric heater whose temperature could be controlled. When the solution temperature had reached 120°† the electric switch was cut off and 40 ml of fuming nitric acid (sp. gr. 1.52) was added dropwise so as to maintain the temperature at 125–130°, requiring about 1 h for the whole addition. The solution was stirred at 125–130° for a further 1 h, cooled, and poured on about 300 g of crushed ice. This solution was neutralized by the cautious addition of soda ash by which the nitrated product precipitated out as a yellow powder. The neutralization was continued until the yellow nitration product no longer precipitated out.

* Content is determined by iodometry and 1.5 molar equivalent of H_2O_2 is used for a mole of pyridine.

** By sealing in sulfuric acid, the liberated NO_2 will be absorbed.

† If fuming nitric acid is added in the cold and the solution heated, the reaction will start all at once and the yield will inevitably fall.

requiring about 230 g of soda ash. The yellow product was collected by suctional filtration*, washed with 5% sodium carbonate solution and water, and dried in air. The dried product was recrystallized from acetone. Yield of 4-nitropyridine 1-oxide, m.p. 159°, 49.5 g (70%).

Example 2. Nitration of Quinoline 1-Oxide
(from a note of M. Horiuchi)

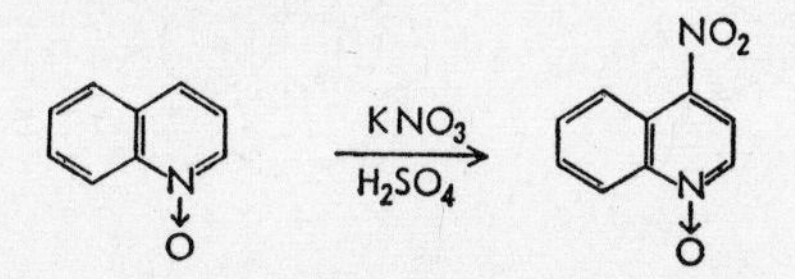

To 30 g of conc. sulfuric acid (98% H_2SO_4 diluted with 0.1 volume of ice water) placed in a three-necked flask provided with a stirrer and a thermometer, 50 g of quinoline 1-oxide dihydrate was added gradually, while cooling the flask with water, and 35 g of potassium nitrate was then added in small portions, by which the solution gradually warmed by the heat of the reaction. When the addition was complete, the flask was warmed on a water bath to 65–70° and maintained there for 1.5 h. The reaction mixture was poured into a large amount of ice water with stirring, by which the nitrated product precipitated out as a yellow to orange-yellow powder. The product was collected by filtration, washed with water until the washings were no longer acidic, and the precipitate was washed with dilute sodium carbonate solution until the washing was almost free from color. The filtered product was again washed with water, dried in air, and recrystallized from acetone to yellow needles, m.p. 153–154°. Yield of 4-nitroquinoline 1-oxide, 45 g (68%).

6.2.2 Nitration of Diazine Series N-Oxides

Nitration of diazine series compounds has been carried out on *N*-oxides in the phenazine, pyridazine, and cinnoline series.

(a) Phenazine Derivatives

Otomasu stated that, whereas nitration of phenazine (6-35) with sulfuric and nitric acids did not occur at a low temperature and 1,6- and 1,9-dinitro derivatives were formed by the same nitration at 100° (Ref. 79), nitration of its 5-oxide (6-36) started already at 0°, forming chiefly the 3-nitro compound, with a small amount of 1-nitro compound[3], and that nitration at 100° afforded 3,7- and 1,7-dinitrophenazine 5-oxides[80].

* Soda ash should be added to the filtrate to see that no further precipitation of the nitrated product will occur. If a precipitation occurs, further neutralization will be necessary. Extraction of the filtrate with chloroform may raise the yield of the product to about 75%. This procedure has been omitted in the present method of preparation.

References p. 243

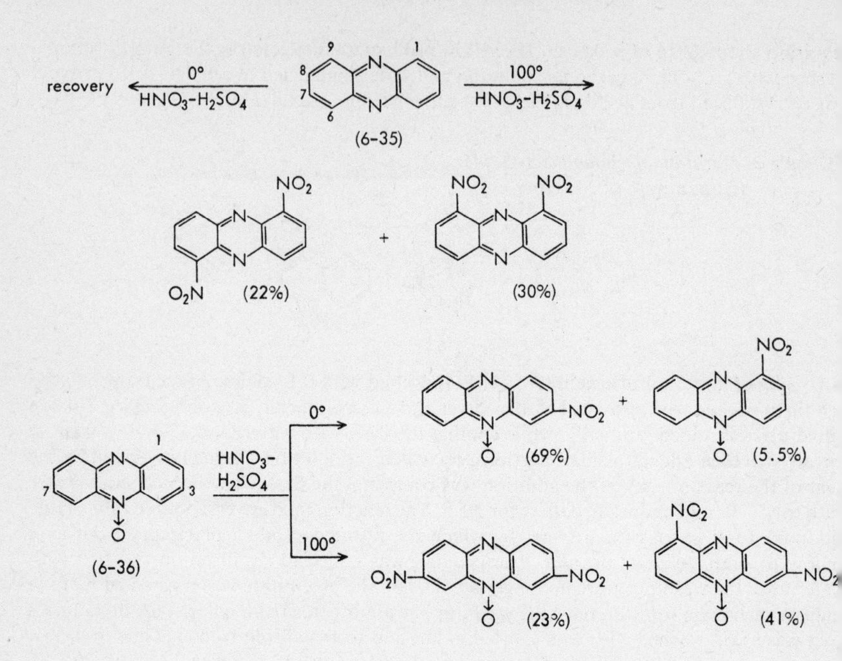

In a similar manner, while 1- and 2-chlorophenazine resist nitration wit sulfuric and nitric acids at 80°, 1-, 2-, and 3-chlorophenazine 5-oxides ar nitrated with sulfuric and nitric acids at 0°, giving the corresponding 7-nitr compounds in a good yield[81].

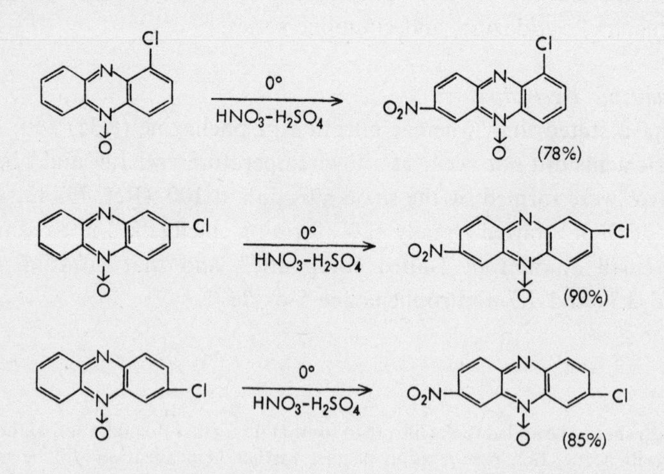

These facts indicate that the *N*-oxide group of phenazine *N*-oxide exert its polar effect through the nitrogen atom to its *para* position by the contribution of the following resonance system.

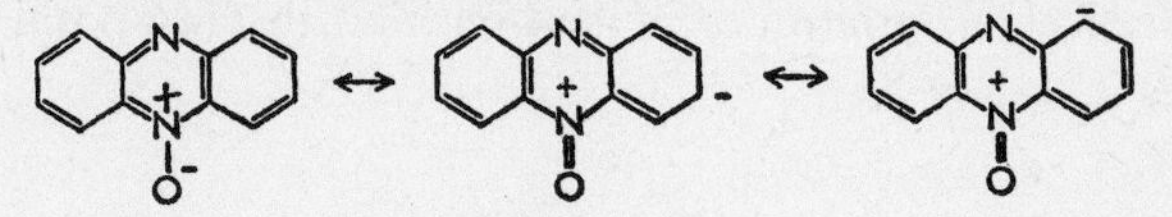

This orienting effect of its *N*-oxide function is not as great as that of the methoxyl function, as in quinoline 1-oxide, and nitration of methoxyphenazine 5-oxides is under the domination of the methoxyl group[3].

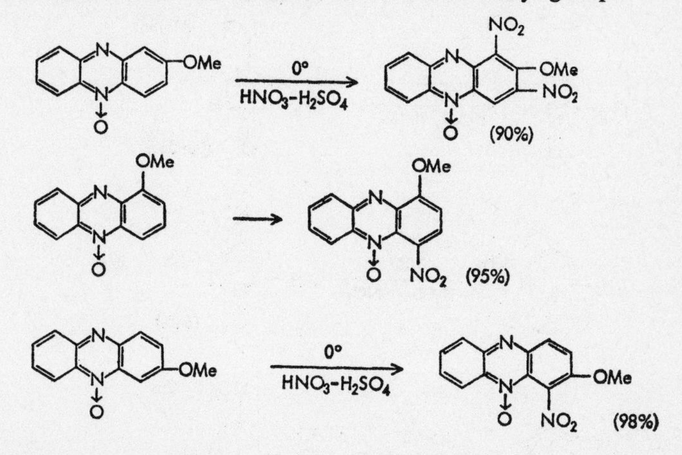

(b) Pyridazine Derivatives

According to a calculation by Orgel, Calvet, and Sandorfy, electrophilic substitution of pyridazine is more difficult than that of pyridines[82] and the compound resists nitration[83], undergoing decomposition under a drastic condition. When the pyridazine is converted into its *N*-oxide, the compound becomes active to nitration. Itai and Natsume[84] showed that, while the nitration of pyridazine 1-oxide (6-37) with sulfuric and nitric acids at 100° ended in recovery of the starting material, the same nitration at 130–140° gave 4-nitropyridazine 1-oxide (6-38) in 22% yield, with recovery of 26% of the starting material. They proved that the position *para* to the *N*-oxide group is activated.

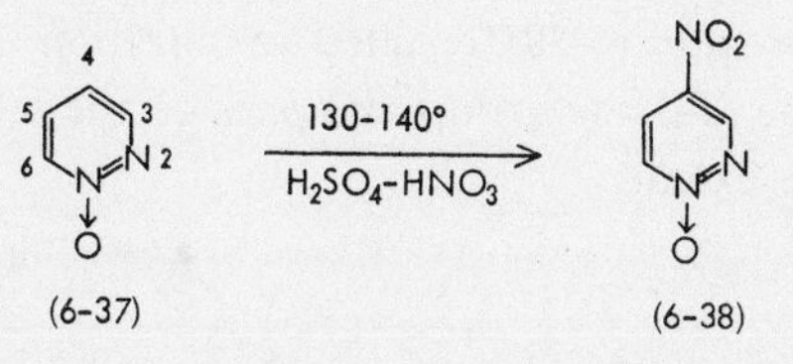

References p. 243

Nitration of the methyl homologs of pyridazine 1-oxide has also been examined and the results indicate that although the methyl group in 3-position has no great effect on the nitration of the 4-position, that in 6-position markedly promotes it.

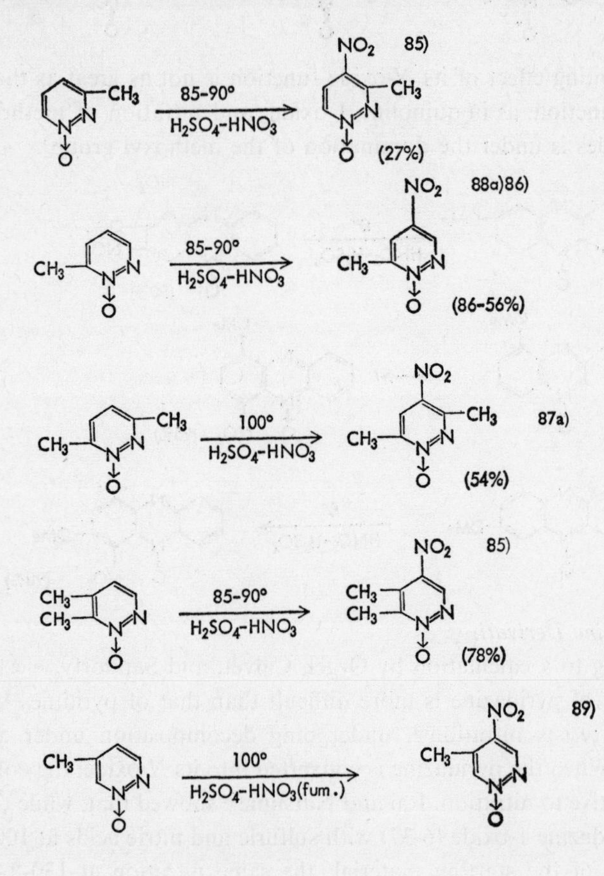

Kanô and Ogata[89] confirmed that 4-methylpyridazine 1-oxide, in which the position *para* to the *N*-oxide group is occupied by a methyl group, strongly resists nitration, while Nakagome[85] found that nitration of 3,4-dimethylpyridazine 1-oxide (6-39) resulted in nitration of the 6-position, a position *ortho* to the *N*-oxide group and *para* to the second methyl group, though to a minute extent.

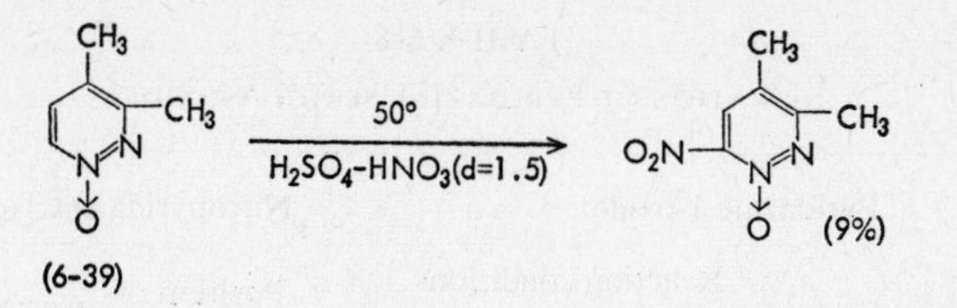

When the methyl group is changed to an alkoxyl or hydroxyl group, its polar effect is added to that of the *N*-oxide function and the positions *ortho* and *para* to both groups are activated. Igeta[4b] observed the formation of 4-nitro and 4,6-dinitro derivatives by the nitration of 3-methoxypyridazine 1-oxide (6-40), while Nakagome[88c] found from the same nitration reaction that a molecular compound of the 4-nitro compound and the starting material is formed in the main, with the 6-nitro compound as a by-product.

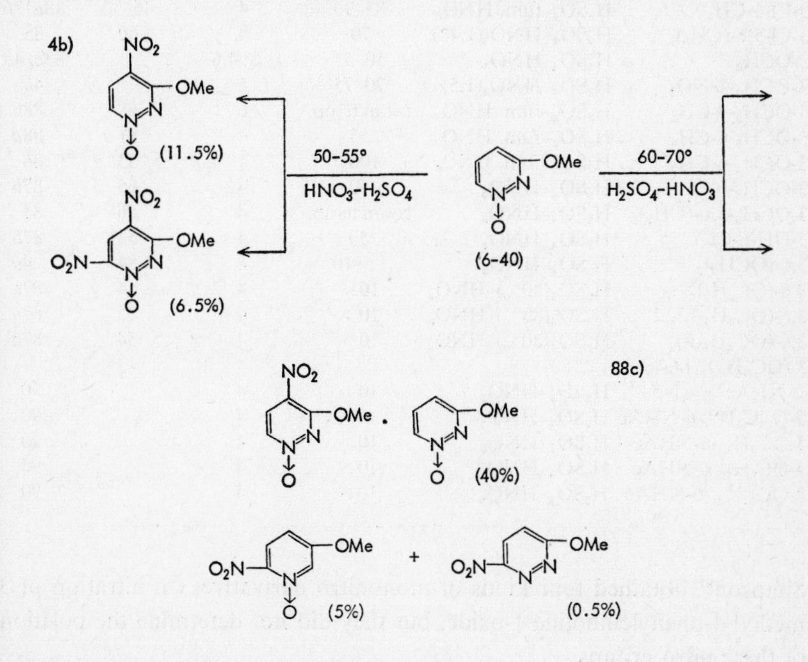

Nitration reactions of pyridazine series *N*-oxides reported to date in literature are summarized in Table 6-8.

(c) Cinnoline Derivatives

Suzuki, Itai, and Nakajima[4c] obtained a 4-nitro derivative on nitration of cinnoline 1-oxide (6-41) with sulfuric and nitric acids, while Atkinson and

TABLE 6-8

NITRATION OF PYRIDAZINE SERIES *N*-OXIDES

Pyridazine 1-oxide			Nitropyridazine 1-oxide		Reference
Substituent	Reaction conditions		Position of NO_2 group	Yield (%)	
	Reagent	Temp.(°C)			
	H_2SO_4–HNO_3	130–140	4	22	84, 88[illegible]
3-CH_3	H_2SO_4–HNO_3	85–90	4	27	85
5-CH_3	H_2SO_4–fum. HNO_3	100	4	18	89
6-CH_3	H_2SO_4–fum. HNO_3	85–90	4	86–56	86, 88[illegible]
3,4-$(CH_3)_2$	H_2SO_4–HNO_3(1.5)	50	6	9	85
3,6-$(CH_3)_2$	H_2SO_4–fum. HNO_3	90–100	4	54	87*a*
5,6-$(CH_3)_2$	H_2SO_4–HNO_3	85–90	4	78	85
3-Cl-6-CH_3	H_2SO_4–fum. HNO_3	85–90	4	46–32	88*a*, 8[illegible]
3-Cl-5,6-$(CH_3)_2$	H_2SO_4–HNO_3(1.48)	70	4	66	85
3-OCH_3	H_2SO_4–HNO_3	50–55	4; 6; 4,6		88*c*, 4[illegible]
3-OCH_3-4-NO_2	H_2SO_4–HNO_3(1.5)	70–75	6		4*b*
3-OCH_3-4-CH_3	H_2SO_4–fum. HNO_3	room temp.	6	68	88*b*
3-OCH_3-6-CH_3	H_2SO_4–fum. HNO_3	55	4	81	88*a*
3-OCH_3-6-CH_3	H_2SO_4–fum. HNO_3	100	4	53	86
3-OCH_3-6-Cl	H_2SO_4–HNO_3	50	4	65	87*b*
3-OCH_3-5,6-$(CH_3)_2$	H_2SO_4–HNO_3	room temp.	4	86	85
3-OH-6-Cl	H_2SO_4–HNO_3	50	4	53	87*b*
3,6-$(OCH_3)_2$	H_2SO_4–HNO_3	5–10	4	84	4*a*
3,6-$(OC_2H_5)_2$	H_2SO_4(80%)–HNO_3	10>	4	44	87*a*
3,6-$(OC_3H_7$-n$)_2$	H_2SO_4(80%)–HNO_3	10>	4	35	87*a*
3,6-$(OC_4H_9$-n$)_2$	H_2SO_4(80%)–HNO_3	10>	4	54	87*a*
3-(OC_nH_{2n+1})-6-NHAc $n=1$–5	H_2SO_4–HNO_3	10>	4		90
3-O-iC_5H_{11}-6-NHAc	H_2SO_4–HNO_3	10>	4		90
3-OC_6H_{13}-6-NHAc	H_2SO_4–HNO_3	10>	4		90
3-OC_8H_{17}-6-NHAc	H_2SO_4–HNO_3	10>	4		90
3-$OC_{11}H_{21}$-6-NHAc	H_2SO_4–HNO_3	10>	4		90

Simpson[91] obtained four kinds of mononitro derivatives on nitration of [illegible]methyl-4-phenylcinnoline 1-oxide, but they did not determine the position of these nitro groups.

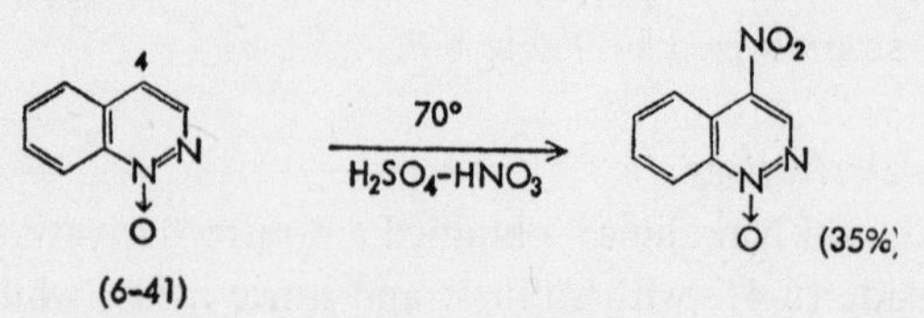

King and King[92] reported that the 9-nitro compound was chiefly obtained on nitration of benzo[c]cinnoline 5-oxide (6-42), with formation of the 8-nitro compound as by-product. Barton *et al.*[93,94] proved the formation of 10- and 7-nitro compounds on nitration of the same benzo[c]cinnoline 5-oxide with sulfuric and nitric acids.

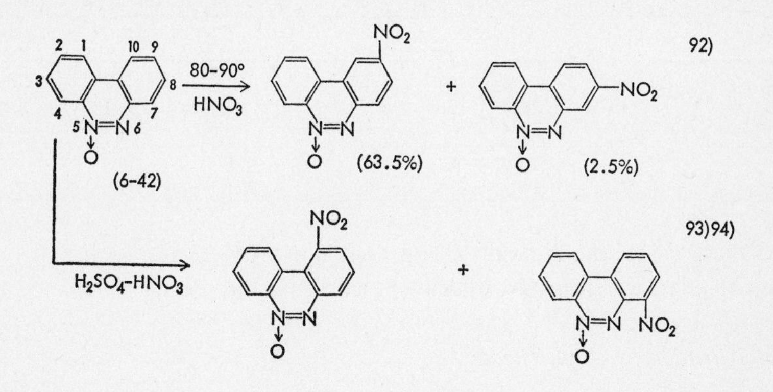

6.2.3 *Phenyl Derivatives of Aromatic N-Oxides*

Katritzky and others[95] examined the nitration of 2-, 3-, and 4-phenylpyridine 1-oxides and observed that only the phenyl group is nitrated instead of the 4-position of the pyridine nucleus, and they compared this result with the nitration of phenylpyridines. This comparison is listed in Table 6-9.

In this reaction the proportion of substitution at the position *meta* to the phenyl group is greatest in 2- and 4-phenylpyridine 1-oxides, while *para* substitution is overwhelming in 3-phenylpyridine 1-oxide. At the same time, it was revealed that *N*-oxidation increased the ratio of *meta*-substitution to the phenyl group and decreased that of *para*-substitution.

In a similar manner, Beech[96] showed that the *meta* position of the phenyl group is nitrated in the nitration of 3,6-diphenyl-2,5-dimethylpyrazine 1,4-dioxide (6-43).

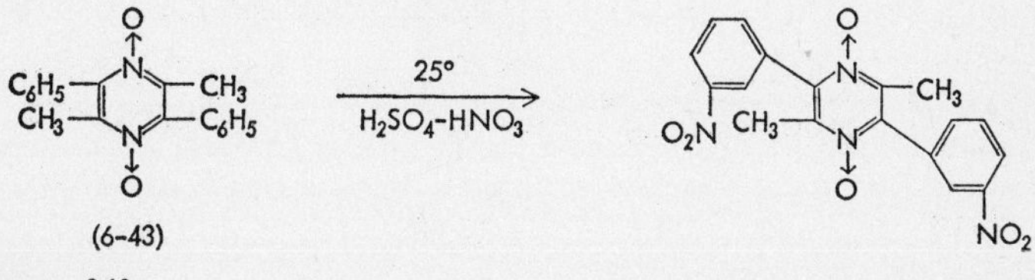

References p. 243

TABLE 6-9

NITRATION OF PHENYLPYRIDINES AND THEIR *N*-OXIDES[95]

Position of phenyl group	Yield of nitropyridine (%)				Yield of nitropyridine 1-oxide (%)			
	o	*m*	*p*	Total	*o*	*m*	*p*	Total
2	5	35	42	82	0–3	58–54	5–7	63–64
3			64	64			38	38
4	13	28	38	79	12	51	13	76

In these cases, the *N*-oxide group does not promote nitration and only seems to exert an electron-withdrawing effect on the phenyl group.

6.2.4 Nitration by Acyl Nitrate

The foregoing sections dealt with nitration of *para* and *ortho* positions in aromatic *N*-oxide compounds. Nitration of a *meta* position was developed by Ochiai and Kaneko, during the examination of the mechanism of the temperature effect in the nitration of quinoline 1-oxide. In order to examine nitration in a neutral solvent, quinoline 1-oxide (6-15) was dissolved in chloroform or dioxan, and benzoyl nitrate was applied at room temperature. It was found that 3-nitroquinoline 1-oxide (6-44) is formed in about 40% yield, with a by-product formation of *N*-benzoyloxy-3,6-dinitrocarbostyril (6-45)[16a,16b]. The same reaction was carried out on pyridine 1-oxide, and formation of 3-nitropyridine 1-oxide, though in a small amount, was observed. Further, these workers showed that quinoline 1-oxides with a substituent in 3-, 5-, 6-, or 7-position were also nitrated in the position β to the *N*-oxide group, or a position corresponding to it[16c], namely the 3- or 6-position.

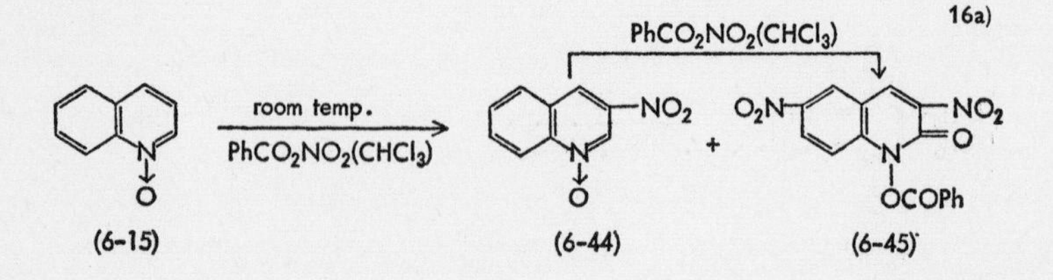

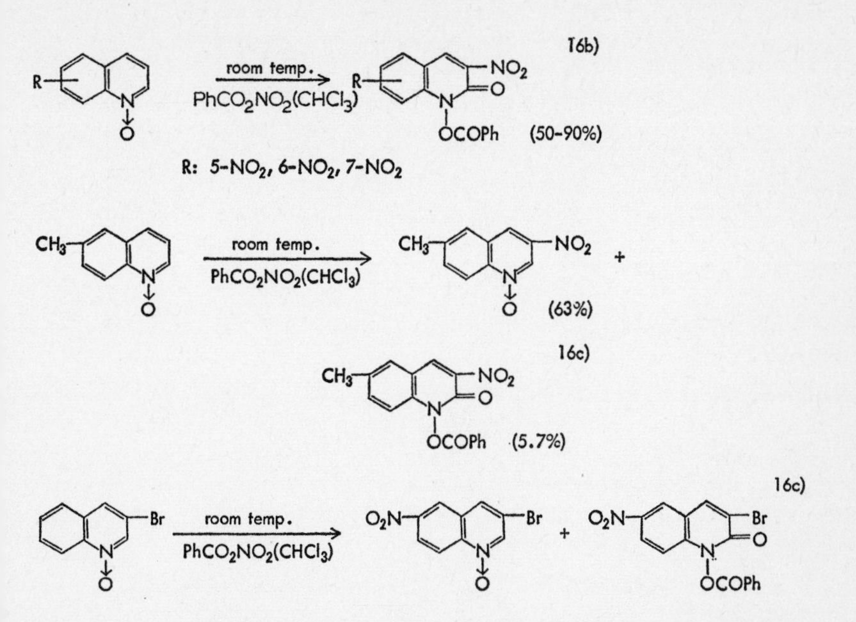

The most troublesome point in this reaction, the synthesis of benzoyl nitrate, was solved by the development of a simple method of first applying benzoyl chloride and subsequent addition of silver nitrate[97]. Later, it was found that the same result is obtained by the use of potassium nitrate in place of silver nitrate[98]. The use of acetyl chloride, instead of benzoyl chloride, in this reaction was found to give 3-nitro-, 6-nitro, and 3,6-dinitroquinoline l-oxides; the reaction did not progress to the formation of carbostyril[97].

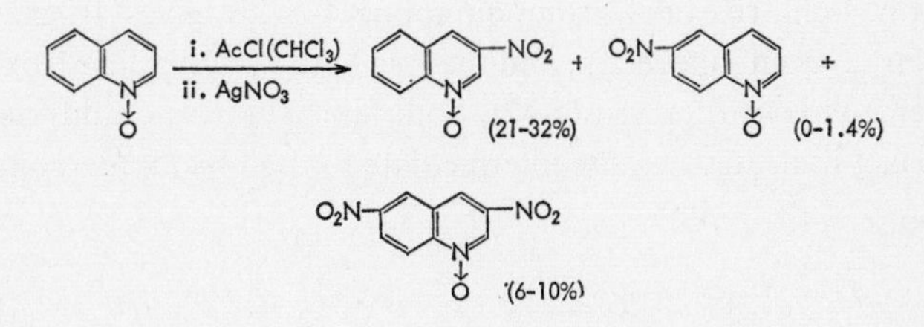

On the basis of these facts, Ochiai and Kaneko[97] assumed the following mechanism for this reaction instead of a traditional mechanism of simple electrophilic substitution.

References p. 243

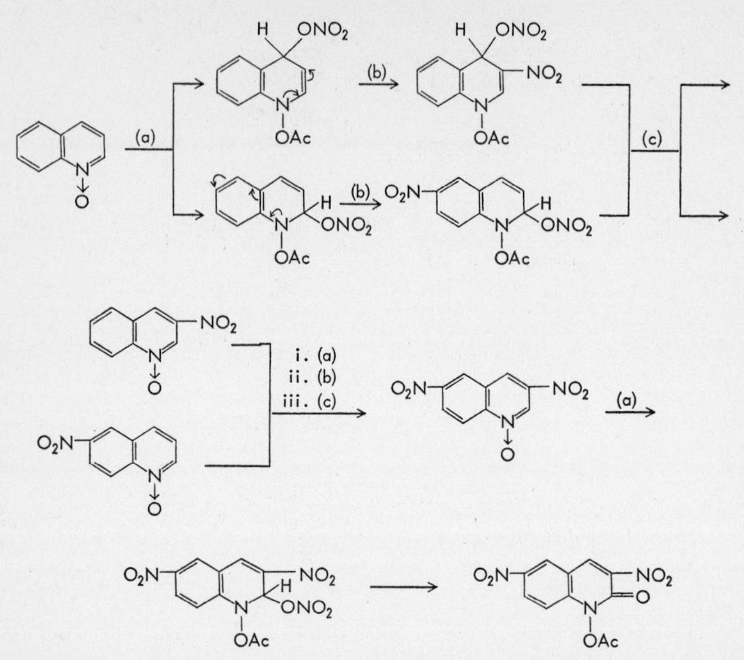

This assumption was endorsed by the fact that the application of *p*-nitrobenzoyl chloride, in which the electrophilic activity of the acyl group is stronger, and silver nitrate to pyridine 1-oxide resulted in some increase in the yield of 3-nitro derivative, with simultaneous formation of 3,5-dinitropyridine 1-oxide[17].

Kimura and his collaborators[99] obtained derivatives of 5-nitropyridine l-oxide in 40-50% yields by the application of this reaction to pyridine 1-oxides having a cyano or ethoxycarbonyl group in the 3-position. The same reaction on 4-chloro- or 4-bromoquinoline 1-oxide gives 1-benzoyloxy-3-halo-4-hydroxyquinolinebetain, and that on 2-ethoxyquinoline 1-oxide (6-46) gives 1-benzoyloxycarbostyril (6-47), both failing to form a nitrated derivative. This fact indicates that the intermediate formed had undergone reaction in a different direction[100].

i. p-NO_2-PhCOCl($CHCl_3$)
ii. $AgNO_3$

(9-10%) + (9-10%) + recovery (40-50%) 17)

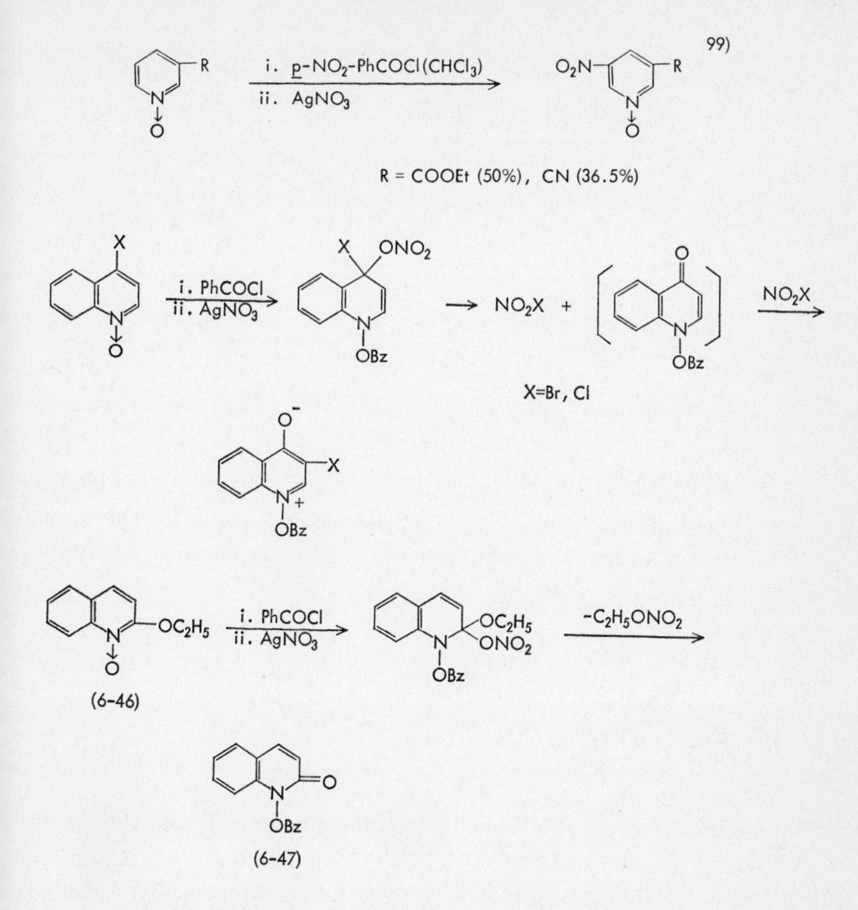

Tanida[101,102] applied this reaction to the methyl homologs of quinoline 1-oxide substituted in the pyridine portion. Application of benzoyl nitrate to quinaldine 1-oxide (6-32) gave, besides 3-nitroquinaldine 1-oxide (6-48), a rearranged product, 2-(benzoyloxymethyl)quinoline (6-49), and a small amount of 2-cyano-3-nitroquinoline 1-oxide (6-50). If this reaction is carried out in chloroform solution with acetyl chloride and silver nitrate, no nitro derivative is formed; the reaction gives 2-(acetyloxymethyl)quinoline and 2-cyanoquinoline l-oxide (6-51) when carried out at the boiling point, and quinaldaldoxime acetate l-oxide (6-52), considered to be an intermediate of the latter, at room temperature.

References p. 243

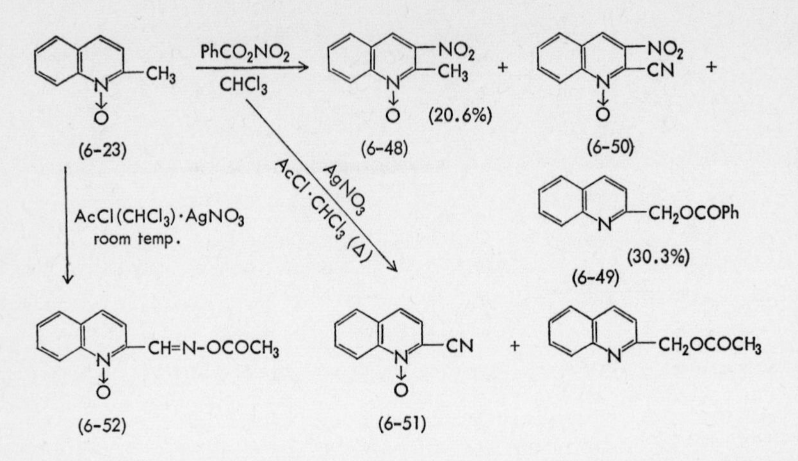

Reaction of lepidine 1-oxide (6-29) with benzoyl nitrate at room temperature results in recovery of 40% of the starting material, with formation of a small amount of 3-benzoyloxylepidine (6-53), and no evidence for the formation of a nitrated compound[103].

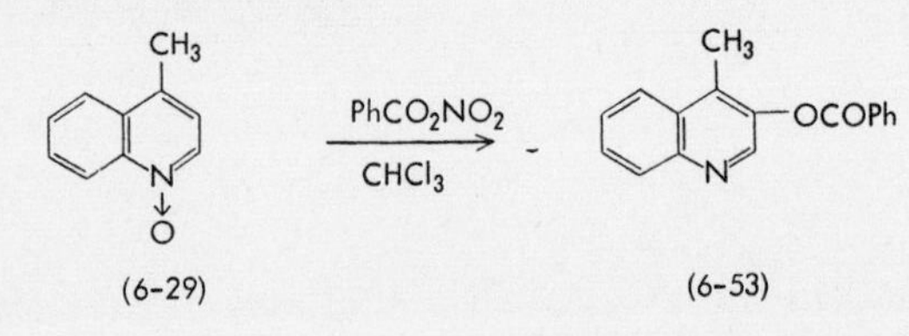

The action of benzoyl nitrate on 2,4-dimethylquinoline 1-oxide (6-54) under exactly the same conditions fails to give a nitrated product, forming 3-benzoyloxy-2,4-dimethylquinoline (6-55), 2-benzoyloxymethyl-4-methylquinoline (6-56), and 2-cyano-4-methylquinoline 1-oxide (6-57)[102,103].

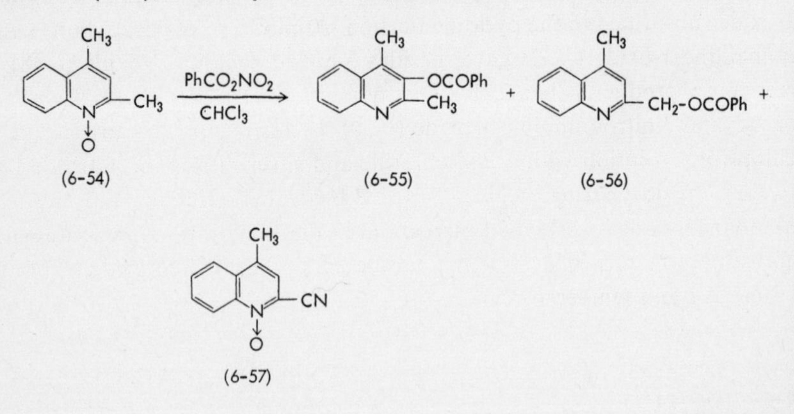

The fact that these quinoline compounds in which a methyl group is present in their 4-position do not form a nitrated product is understandable from the foregoing reaction mechanism.

Mitsuhashi and others[104] examined this reaction with quinoline 1-oxides having a substituent with a great polar or steric effect, such as methoxyl and tosyloxyl, in the benzene portion of the quinoline ring. They obtained the following results.

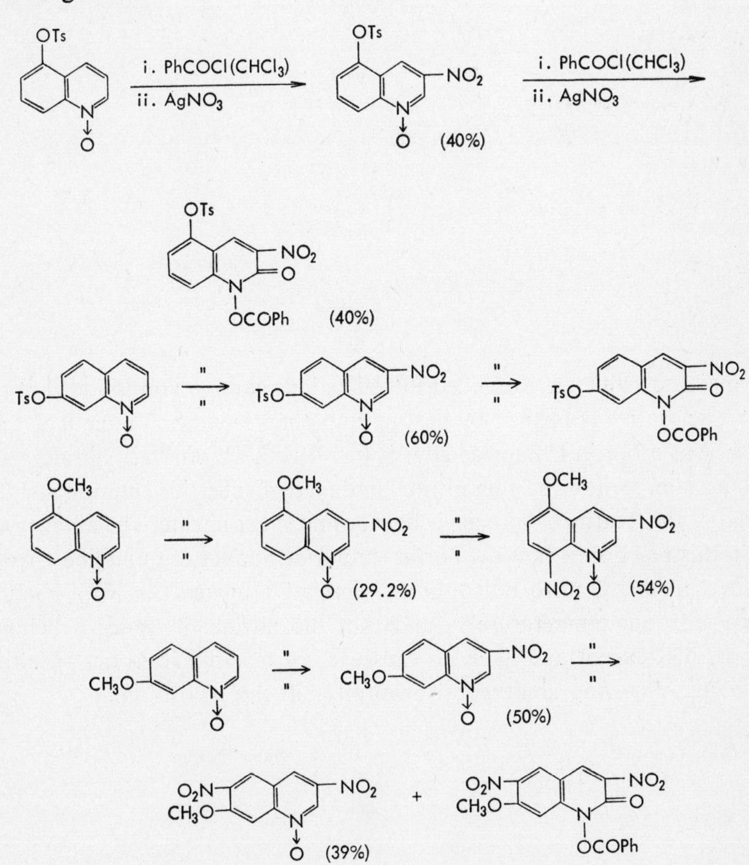

This experimental evidence indicates that the activity of the 3-position towards nitration predominates over orientation of the methoxyl and tosyloxyl groups, and that both groups, especially the tosyloxyl group, interfere with the nitration of the 6-position.

This reaction was applied to the nitration of pyridazine 1-oxide[19] (6-37)

References p. 243

and cinnoline 1-oxide[4c] (6-41) by Itai and others, and nitration of the positio *meta* to the *N*-oxide group was effected. In the case of pyridazine 1-oxid a small amount of 5-nitro derivative was obtained besides the main 3-nitr compound.

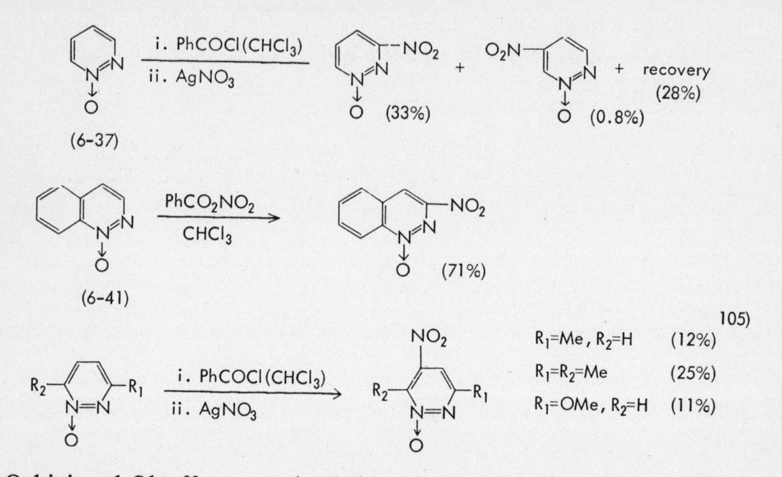

Ochiai and Ohta[20] re-examined this nitration reaction and found that th 3-nitro derivative is formed by heating quinoline 1-oxide as a methosulfat (6-58) or as a boron trifluoride adduct (6-59) with an ordinary nitrate, suc as potassium, lead, or barium nitrate, in dimethyl sulfoxide solution at 140° instead of the treatment with acyl chloride and silver nitrate. The noteworth fact in this reaction is that the boron trifluoride adduct of quinoline 1-oxid does not form a 3-nitro but a 4-nitro derivative by reaction with benzoy nitrate at room temperature[103], and that the nitrate, in spite of being a nucleophilic reagent and likely to cause an S_N reaction in 2- or 4-positio of the quinoline ring, undergoes substitution at the 3-position.

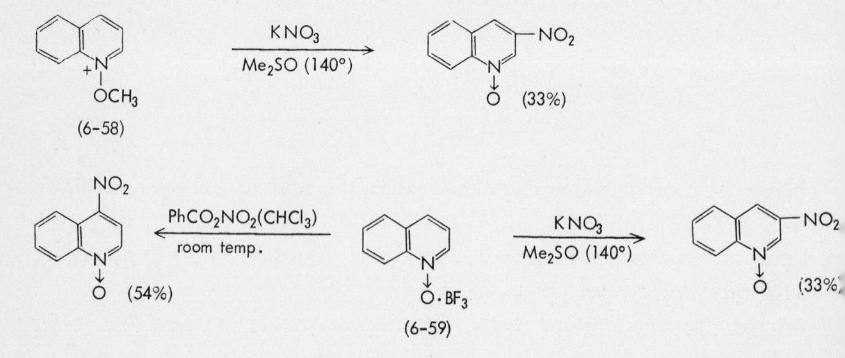

Example. Nitration of Quinoline 1-Oxide with Benzoyl Chloride and Silver Nitrate[97]

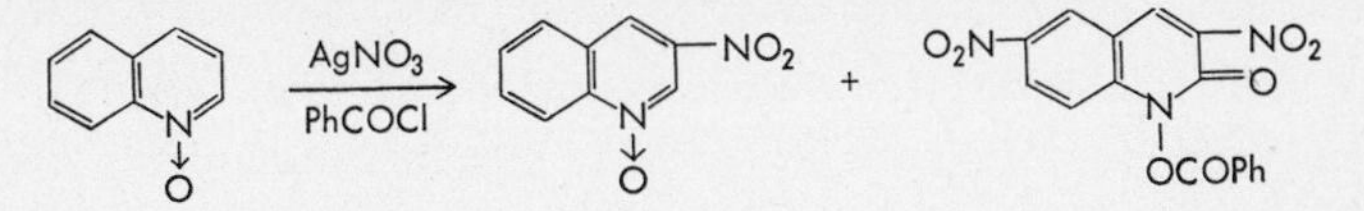

To a solution of 10 g of quinoline 1-oxide dissolved in 200 ml of chloroform, 9.4 g (*ca.* 1 equiv.) of benzoyl chloride was added under ice-cooling. The mixture was chilled to −20 to −15° by external application of a freezing mixture and 13.7 g (*ca.* 1.2 equiv.) of finely pulverized silver nitrate was added to the mixture in small portions under agitation, over a period of about 30 min. The reaction mixture was stirred under chilling for 5 h and further at room temperature for 5 h. The mixture was allowed to stand for 2 days, silver chloride was removed by filtration, and washed with chloroform. The combined filtrate and washings were washed with saturated solution of sodium hydrogen carbonate and then with 5% hydrochloric acid, and 3.82 g of unreacted quinoline 1-oxide was recovered from the hydrochloric acid solution. The chloroform solution was dried over anhydrous sodium sulfate and the solvent was distilled off. The residue (8 g) was recrystallized from acetone and 4.19 g (32.3%) of 3-nitroquinoline 1-oxide, m.p. 189–191°, was obtained from the easily soluble portion, and 1.96 g (7.3%) of 1-benzoyloxy-3,6-dinitrocarbostyril, m.p. 253–255° (decomp.), from the sparingly soluble portion.

6.3 Bromination

Ochiai and Okamoto[7] found that when the perbromide of quinoline 1-oxide[106] is heated, it undergoes decomposition into quinoline and 3-bromoquinoline, but if the perbromide is sealed in a tube with bromine water and shaken at room temperature, 4-bromoquinoline 1-oxide (6-60) is formed in 32% yield, while heating of such a solution to 100° results in the decrease of the 4-bromo compound to 2–3% with concurrent formation of 3,4,6,8-tetrabromocarbostyril (6-61) in about 48% yield. This latter compound is also formed by the same treatment of 3-, 4-, or 6-bromoquinoline 1-oxide or 4-bromocarbostyril with bromine water in the same manner. It follows, therefore, that 3,4,6,8-tetrabromocarbostyril is formed via 4-bromoquinoline 1-oxide.

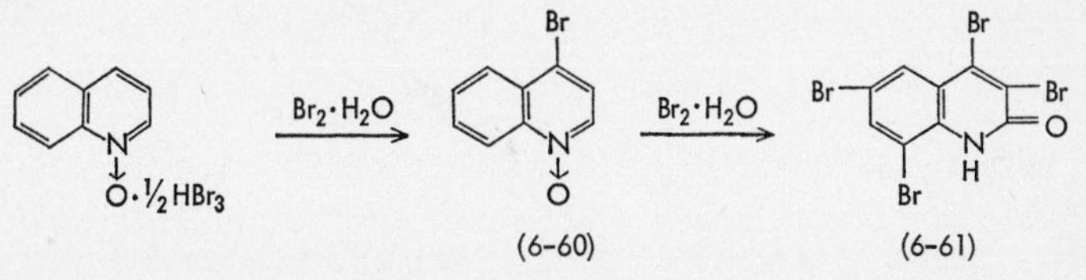

Quinoline 1-oxide is brominated in the position *para* to the *N*-oxide group under the above condition, but the reaction seems to proceed further, although the yield of 4-bromoquinoline l-oxide does not increase. Since the 3-position of quinoline is active to bromination[107], bromination of quinoline 1-oxide in the 4-position clearly indicates the effect of the *N*-oxide function.

Acheson and his group[11] obtained a 9-brominated compound by the application of bromine to acridine *N*-oxide (6-32) in glacial acetic acid solution, with warming on a water bath. Since the S_E reaction on acridine occurs generally at 2,7- and then at 4,5-positions, the fact that the 9-position was substituted indicates the orienting effect of the *N*-oxide function.

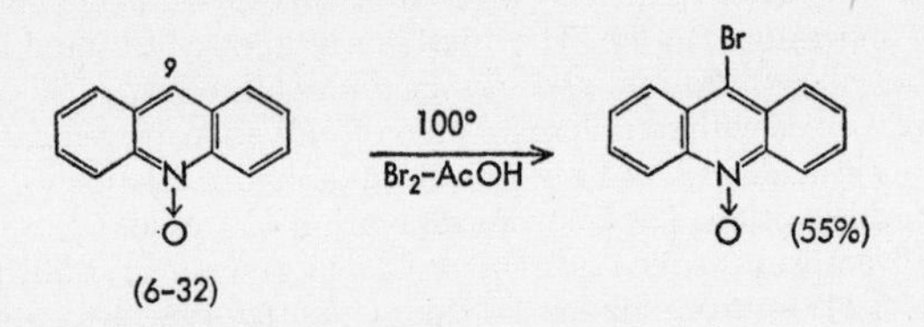

In contrast to such a marked polar effect of the *N*-oxide group towards bromination in quinoline and acridine *N*-oxides, the effect is very small in pyridine 1-oxide. Den Hertog and his collaborators[10a] found that although the reaction did not take place when pyridine 1-oxide was heated with excess of bromine in 90% sulfuric acid at 200°, addition of silver sulfate to this reaction mixture resulted in the formation of 2- and 4-bromopyridine 1-oxide in a total yield of about 10%, and the formation ratio of the 2- and 4-bromo derivatives was found to be 1:2 by gas-chromatographic analysis. These workers also found that the 3-position was chiefly substituted, and that the 2- and 4-positions were inactive towards bromination in 65% fuming sulfuric acid[108], and they assumed that the sulfur trioxide adduct of pyridine l-oxide was comparatively stable in fuming sulfuric acid and bromination might be hindered, as compared to that in 90% sulfuric acid.

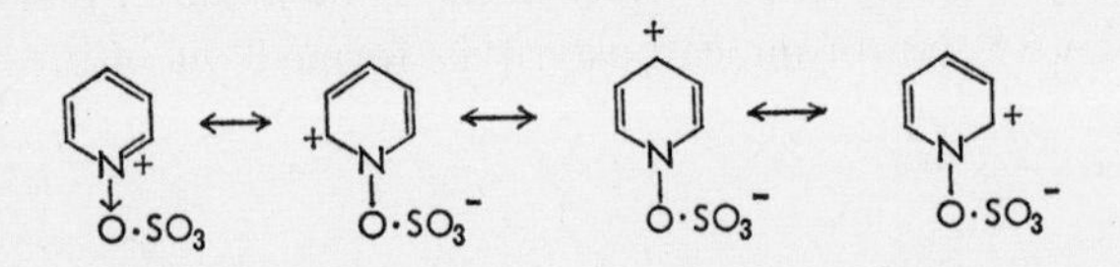

In any case, the yield of brominated product is extremely poor, compared to that in the case of nitration, and the reaction has no practical value.

When a hydroxyl or amino group is present in the pyridine ring, halogenation is affected by the polar effect of these groups and substitution occurs under a milder condition.

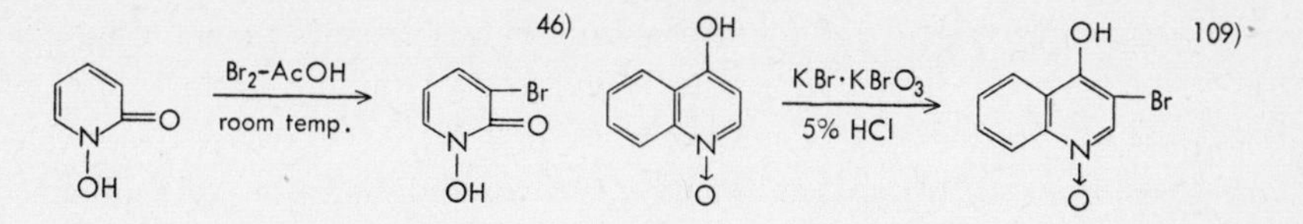

Hamana and others[18] applied the method of the nitration of the 3-position of quinoline and pyridine l-oxides by acyl nitrate to the bromination. Quinoline l-oxide (6-15) was dissolved in acetic anhydride, anhydrous sodium acetate and chloroform solution of bromine were added to this solution, and the solution was boiled; they obtainted 3,6-dibromoquinoline l-oxide (6-62) in 60% yield from it. By the same reaction, 3,6-dibromoquinoline l-oxide was obtained in 40% and 26% yield, respectively, from 6- and 3-bromoquinoline 1-oxides, and 3,5-dibromopyridine 1-oxide (6-63) in 35% yield from pyridine 1-oxide.

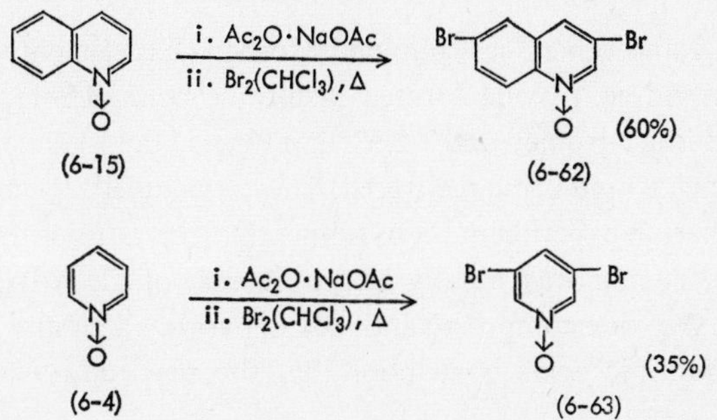

A similar reaction with quinaldine 1-oxide (6-23) gave ω-dibromoquinaldine 1-oxide (6-64).

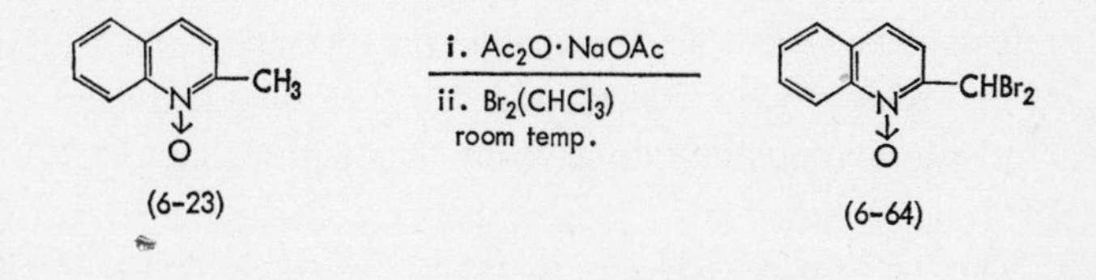

6.4 Other Electrophilic Substitutions

6.4.1 Sulfonation

Mosher and Welch[8] found that pyridine 1-oxide was highly resistant to sulfonation and obtained 3-sulfonic acid in a crude yield of 51% by using con-

References p. 243

ditions similar to that used for sulfonation of pyridine (addition of a catalytic amount of mercuric sulfate to a solution in 20% fuming sulfuric acid and heating at 220–240° for 22 h), indicating that there is no activation by the *N*-oxide function towards sulfonation. Brown and his associate[110] obtained the 3-sulfonic acid of pyridine 1-oxide and of 2,6-lutidine 1-oxide (6-65) by the same reaction.

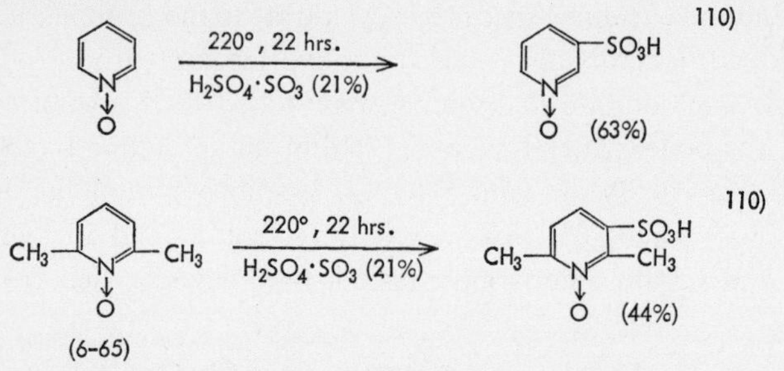

Den Hertog and others[9] re-examined Mosher's experiment and found that 40–45% of pyridine 1-oxide formed 3-sulfonic acid, 0.5–1% of it formed the corresponding 2-isomer, and 2-2.5% of it formed the 4-isomer, about 45% of the starting compound being recovered. They concluded, by comparison of a similar sulfonation experiment on pyridine, that the ratio of the 2- and 4-isomers to the 3-isomer was slightly greater in case of the *N*-oxide compound (up to 0.07 in *N*-oxide compound and 0.01 in pyridine). In any case, the yield of 2- and 4-sulfonic acids is so poor that the reaction is valueless.

6.4.2 *Mercuration*

Ukai and his group[12] reported that heating of pyridine 1-oxide with mercuric acetate in glacial acetic acid resulted in the formation of a 4-substituted compound in a relatively good yield, and that the same treatment of quinoline 1-oxide with mercuric acetate gave an 8-substituted compound.

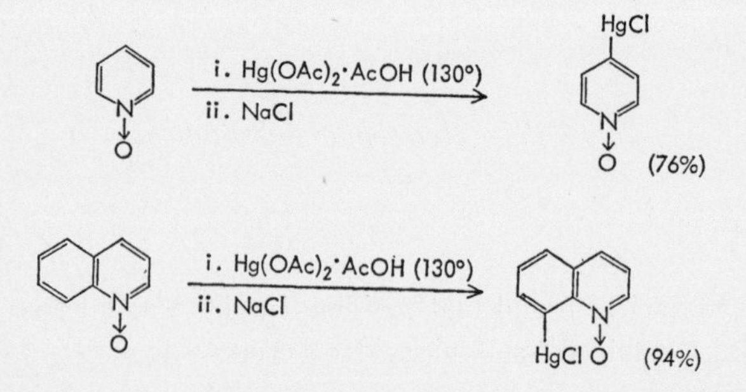

Of these, mercuration of pyridine 1-oxide was re-examined by den Hertog and his co-workers[13], who proved that only 2- and 2,6-substituted compounds were formed and not the 4-substituted compound. It is certain that, at least in the mercuration of quinoline l-oxide, an 8-substituted compound is formed*, but this point should be re-examined.

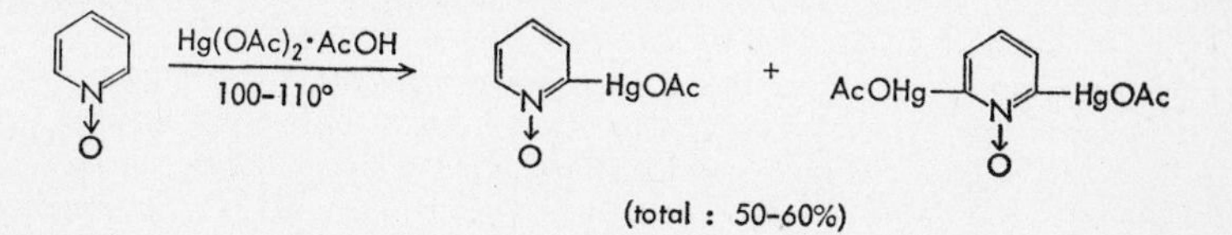

Besides with mercuric acetate, den Hertog and Van Ammers[13] carried out the mercuration of pyridine l-oxide with mercuric sulfate, chloride, or bromide in sulfuric acid, and converted the mercurated compounds to the corresponding bromopyridines in order to examine the ratio of their formation by gas chromatography. They found that invariably the 2-substituted compound was formed in the greatest ratio, and that the formation of 3- and 2,6-substituted compounds was very small. It is assumed that the mechanism of this mercuration is different from that of the foregoing nitration and bromination reactions, but the reaction may be used as an indirect method for the preparation of 2- and 6-halogenated compounds.

* E. Hayashi, private communication.

REFERENCES

1 *(a)* E. Ochiai and M. O. Ishikawa, *Proc. Imp. Acad. (Tokyo)*, 18 (1942) 561.
(b) E. Ochiai, K. Arima, and M. O. Ishikawa, *Yakugaku Zasshi*, 63 (1943) 79.
2 *(a)* H. J. den Hertog and J. Overhoff, *Rec. Trav. Chim.*, 69 (1950) 468.
(b) H. J. den Hertog and W. P. Combé, *Rec. Trav. Chim.*, 70 (1951) 581.
3 H. Otomasu, *Pharm. Bull. (Japan)*, 2 (1954) 283.
4 *(a)* T. Itai and H. Igeta, *Yakugaku Zasshi*, 75 (1955) 966. *(b)* H. Igeta, *Chem. Pharm. Bull. (Tokyo)*, 8 (1960) 550. *(c)* I. Suzuki, T. Nakashima, and T. Itai, *Chem. Pharm. Bull. (Tokyo)*, 11 (1963) 268.
5 T. Nakagome, *Yakugaku Zasshi*, 80 (1960) 712.
6 H. Kanô, M. Ogata, H. Watanabe, and I. Ishizuka, *Chem. Pharm. Bull. (Tokyo)*, 9 (1961) 1017.
7 E. Ochiai and T. Okamoto, *Yakugaku Zasshi*, 67 (1947) 87.
8 H. S. Mosher and F. J. Welch, *J. Am. Chem. Soc.*, 77 (1944) 2902.
9 M. van Ammers and H. J. den Hertog, *Rec. Trav. Chim.*, 78 (1959) 586.
10 *(a)* H. C. van der Plas, H. J. den Hertog, M. van Ammers, and B. Haase, *Tetrahedron Letters*, 1 (1961) 32. *(b)* M. van Ammers, H. J. den Hertog, and B. Haase, *Tetrahedron*, 18 (1962) 227.

11 R. M. Acheson, B. Adcock, G. M. Glover, and L. E. Sutton, *J. Chem. Soc.*, (1960) 3367.
12 T. Ukai, Y. Yamamoto, and S. Hirano, *Yakugaku Zasshi*, 73 (1953) 823.
13 H. J. den Hertog and M. van Ammers, *Rec. Trav. Chim.*, 77 (1958) 340; 81 (1962) 124
14 E. Ochiai and T. Okamoto, *Yakugaku Zasshi*, 70 (1950) 384.
15 T. Okamoto, *Yakugaku Zasshi*, 70 (1950) 376.
16 *(a)* E. Ochiai and Ch. Kaneko, *Chem. Pharm. Bull. (Tokyo)*, 5 (1957) 56. *(b)* E. Ochiai and Ch. Kaneko, *Chem. Pharm. Bull. (Tokyo)*, 7 (1959) 191. *(c)* E. Ochiai and Ch. Kaneko, *Chem. Pharm. Bull. (Tokyo)*, 7 (1959) 195.
17 E. Ochiai and Ch. Kaneko, *Chem. Pharm. Bull. (Tokyo)*, 8 (1960) 28.
18 M. Hamana and M. Yamazaki, *Chem. Pharm. Bull. (Tokyo)*, 9 (1961) 414.
19 T. Itai and S. Natsume (née Suzuki), *Chem. Pharm. Bull. (Tokyo)*, 11 (1963) 342.
20 E. Ochiai and A. Ohta, *Chem. Pharm. Bull. (Tokyo)*, 10 (1962) 349; *Sci. Papers Inst. Phys. Chem. Res. (Tokyo)*, 56 (1962) 189.
21 E. Ochiai, E. Hayashi, and M. Katada, *Yakugaku Zasshi*, 67 (1947) 79.
22 E. Ochiai and E. Hayashi, *Yakugaku Zasshi*, 67 (1947) 157.
23 E. Ochiai and Z. Sai, *Yakugaku Zasshi*, 65 (1945) 18.
24 T. Itai and H. Ogura, *Yakugaku Zasshi*, 75 (1955) 292.
25 R. F. Evans and W. Kynaston, *J. Chem. Soc.*, (1961) 5556.
26 J. M. Essery and K. Schofield, *J. Chem. Soc.*, (1960) 4953.
27 J. A. Berson and T. Cohen, *J. Org. Chem.*, 20 (1955) 1461.
28 M. O. Ishikawa, *Yakugaku Zasshi*, 65B (1945) 105.
29 E. V. Brown, *J. Am. Chem. Soc.*, 79 (1957) 3565.
30 F. Leonard and A. Wajngurt, *J. Org. Chem.*, 21 (1957) 1077.
31 H. J. den Hertog, C. R. Kolder, and W. P. Combé, *Rec. Trav. Chim.*, 70 (1951) 598.
32 R. Jûjo, *Yakugaku Zasshi*, 66B (1946) 49,
33 M. van Ammers and H. J. den Hertog, *Rec. Trav. Chim.*, 77 (1958) 344.
34 H. J. den Hertog, C. H. Henkens, and K. Dilz, *Rec. Trav. Chim.*, 72 (1953) 298.
35 H. W. Krause and W. Langenbeck, *Ber.*, 92 (1959) 155.
36 E. C. Taylor and A. J. Crovetti, *J. Am. Chem. Soc.*, 78 (1956) 215.
37 A. E. Tschitschibabin and I. G. Rylinkin, *J. Russ. Phys. Chem. Ges.*, 50 (1918) 471 [*Chem. Zentr.*, III (1923) 1020].
38 A. Pieroni, *Atti Reale Acad. Lincei*, (6) 5 (1927) 304 [*Chem. Zentr.*, I (1927) 3003].
39 C. Räth, *Ann.*, 484 (1930) 52.
40 E. Königs, H. C. Gerdes, and A. Sirot, *Ber.*, 61 (1928) 1022.
41 O. van Schickh, A. Binz, and A. Schulz, *Ber.*, 69 (1936) 2593.
42 H. J. den Hertog, C. Jouwersma, A. A. van der Wal, and E. C. C. Willebrands-Schogt, *Rec. Trav. Chim.*, 68 (1949) 433.
43 H. J. den Hertog and M. van Ammers, *Rec. Trav. Chim.*, 74 (1955) 1166.
44 K. Lewicka and E. Plazék, *Rec. Trav. Chim.*, 78 (1959) 644.
45 E. Plazék and Z. Rodewald, *Rocznikki Chem.*, 16 (1936) 502 [*Chem. Zentr.*, 1 (1937) 3634].
46 E. Shaw and W. A. Lott, *J. Am. Chem. Soc.*, 71 (1949) 70.
47 H. J. den Hertog and M. van Ammers, *Rec. Trav. Chim.*, 75 (1956) 1260.
48 E. Hayashi, *Yakugaku Zasshi*, 70 (1950) 142.
49 E. Ochiai and K. Futaki, *Yakugaku Zasshi*, 72 (1952) 274.
50 I. Suzuki, *Yakugaku Zasshi*, 71 (1951) 789.
51 H. J. den Hertog, M. van Ammers, and S. Schukking, *Rec. Trav. Chim.*, 74 (1955) 1171.
52 H. J. den Hertog, C. H. Henkens, and K. Dilz, *Rec. Trav. Chim.*, 72 (1953) 296.
53 E. Ochiai, M. O. Ishikawa, and Z. Sai, *Yakugaku Zasshi*, 63 (1943) 280.

54 W. Königs, *Ber.*, 12 (1879) 449.
55 A. Claus and K. Krämer, *Ber.*, 18 (1885) 1243.
56 E. Nölting and E. Trautmann, *Ber.*, 23 (1890) 3655.
57 L. F. Fieser and E. B. Hershberg, *J. Am. Chem. Soc.*, 62 (1940) 1640.
58 T. Okamoto, *Yakugaku Zasshi*, 71 (1951) 727.
59 E. Ochiai and K. Satake, *Yakugaku Zasshi*, 71 (1951) 1078.
60 M. O. Ishikawa, *Yakugaku Zasshi*, 65B (1945) 98.
61 E. Profft, G. Buchmann, and N. Wackrow, *Arzneimittel Forsch.*, 10 (1960) 1029 [*Chem. Abstr.*, 55 (1961) 10442, 1616].
62 S. Yoshida, *Yakugaku Zasshi*, 66B (1946) 158.
63 M. O. Ishikawa, *Proc. Imp. Acad. (Tokyo)*, 20 (1944) 599.
64 D. E. Ames, C. S. Franklin, and T. F. Grey, *J. Chem. Soc.*, (1956) 3079.
65 M. O. Ishikawa, *Yakugaku Zasshi*, 65A (1945) 102.
66 T. Naito, *Yakugaku Zasshi*, 67 (1947) 144.
67 E. Ochiai and H. Ogura, *Yakugaku Zasshi*, 72 (1952) 767.
68 V. Petrow and B. Sturgeon, *J. Chem. Soc.*, (1954) 570.
69 Ch. Kaneko, *Chem. Pharm. Bull. (Tokyo)*, 7 (1959) 274.
70 T. Naito, *Yakugaku Zasshi*, 67 (1947) 141.
71 E. Ochiai, *J. Org. Chem.*, 18 (1953) 534.
72 E. Ochiai, S. Suzuki, Y. Utsunomiya, E. Nagatomo, M. Itoh, and T. Ohmoto, *Yakugaku Zasshi*, 80 (1960) 339.
73 E. Ochiai and Z. Sai, *Yakugaku Zasshi*, 65 (1945) 418.
74 E. Ochiai and M. Ikehara, *Yakugaku Zasshi*, 73 (1953) 666.
75 L. E. Orgel, in R. M. Acheson (Ed.), *Acridines*, Interscience Publishers, Inc., New York, 1956, p. 9.
76 I. Iwai, *Yakugaku Zasshi*, 71 (1951) 1291.
77 I. Iwai, S. Hara, and S. Sayegi, *Yakugaku Zasshi*, 71 (1951) 1152.
78 E. Ochiai and S. Tamura, *Yakugaku Zasshi*, 72 (1952) 985.
79 S. Maffei and M. Aymon, *Gazz. Chim. Ital.*, 84 (1954) 667.
80 H. Otomasu, *Chem. Pharm. Bull. (Tokyo)*, 6 (1958) 77.
81 H. Otomasu, *Pharm. Bull. (Japan)*, 4 (1956) 117.
82 T. L. Jacobs in R. C. Elderfield (Editor), *Heterocyclic Compounds*, John Wiley & Sons, New York, 1957, Vol. 6, p. 111.
83 S. Dixon and L. F. Wiggins, *J. Chem. Soc.*, (1950) 3236.
84 T. Itai and S. Natsume (née Suzuki), *Chem. Pharm. Bull. (Tokyo)*, 11 (1963) 83.
85 T. Nakagome, *Chem. Pharm. Bull. (Tokyo)*, 11 (1963) 726.
86 M. Ogata and H. Kanô, *Chem. Pharm. Bull. (Tokyo)*, 11 (1963) 29.
87 *(a)* T. Itai and S. Sako, *Chem. Pharm. Bull. (Tokyo)*, 9 (1961) 149. *(b)* T. Itai and S. Sako, *Chem. Pharm. Bull. (Tokyo)*, 10 (1962) 934.
88 *(a)* T. Nakagome, *Yakugaku Zasshi*, 82 (1962) 253. *(b)* T. Nakagome, *Yakugaku Zasshi*, 82 (1962) 1009. *(c)* T. Nakagome, *Yakugaku Zasshi*, 81 (1961) 554.
89 M. Ogata and H. Kanô, *Chem. Pharm. Bull. (Tokyo)*, 11 (1963) 35.
90 T. Horie, *Chem. Pharm. Bull. (Tokyo)*, 11 (1963) 1157.
91 C. M. Atkinson and J. C. E. Simpson, *J. Chem. Soc.*, (1947) 1649.
92 F. E. King and T. J. King, *J. Chem. Soc.*, (1945) 824.
93 J. W. Barton and J. F. Thomas, *J. Chem. Soc.*, (1964) 1265.
94 J. W. Barton and J. M. A. Cockett, *J. Chem. Soc.*, (1962) 2454.
95 A. R. Hands and A. R. Katritzky, *J. Chem. Soc.*, (1958) 1754.
96 W. F. Beech, *J. Chem. Soc.*, (1955) 3094.
97 E. Ochiai and Ch. Kaneko, *Chem. Pharm. Bull. (Tokyo)*, 7 (1959) 267.
98 E. Ochiai and A. Ohta, *Sci. Papers, Inst. Phys. Chem. Res.*, 56 (1962) 192.

99 M. Nakadate, Y. Takano, T. Hirayama, S. Sakaizawa, T. Hirano, K. Okamoto, K. Hirao, T. Kawamura, and M. Kimura, *Chem. Pharm. Bull. (Tokyo)*, 13 (1965) 113.
100 E. Ochiai and Ch. Kaneko, *Chem. Pharm. Bull. (Tokyo)*, 8 (1960) 284.
101 E. Ochiai and H. Tanida, *Chem. Pharm. Bull. (Tokyo)*, 5 (1957) 313.
102 H. Tanida, *Chem. Pharm. Bull. (Tokyo)*, 7 (1959) 540.
103 H. Tanida, *Yakugaku Zasshi*, 78 (1958) 1079.
104 K. Mitsuhashi and S. Shiotani, *Yakugaku Zasshi*, 82 (1962) 773.
105 T. Itai and S. Natsume (née Suzuki), *Chem. Pharm. Bull. (Tokyo)*, 12 (1964) 228.
106 E. Ochiai and T. Okamoto, *Yakugaku Zasshi*, 67 (1947) 86.
107 A. Edinger, *J. prakt. Chem.*, [2], 54 (1896) 357.
108 M. van Ammers, H. J. den Hertog, and B. Haase, *Tetrahedron*, 18 (1962) 227.
109 E. Hayashi, *Yakugaku Zasshi*, 71 (1951) 213.
110 R. F. Evans and H. C. Brown, *J. Org. Chem.*, 27 (1962) 1329.

Chapter 7

Nucleophilic Substitution

7.1 General

As has already been stated (*cf. Section 2.2.1*), an aromatic *N*-oxide group has the property of increasing the activity of its α and γ positions to electrophilic and nucleophilic reagents at the same time. Consequently, a large number of nucleophilic substitutions (by S_N reaction) as well as electrophilic substitutions have been elucidated.

In 1926, Meisenheimer[1] found that heating of quinoline 1-oxide with phosphoryl chloride or, more effectively, with sulfuryl chloride, resulted in the formation of 2- and 4-chloroquinoline. Later, Bobranski made a detailed examination of the reaction of quinoline 1-oxide[2] and pyridine 1-oxide[3] with sulfuryl chloride and observed that 4- and 2-chloro-quinolines or -pyridines were formed in 62:38 or 43:57 ratio. In these reactions, deoxygenation of the *N*-oxide group is accompanied by chlorination of its α- and γ-positions.

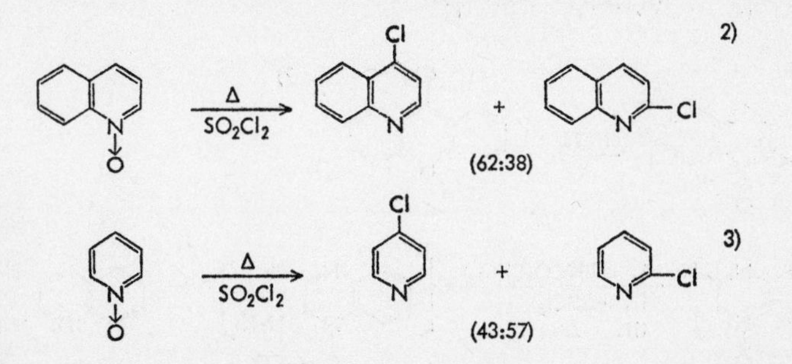

In 1936, Henze[4] found that the treatment of quinoline 1-oxide with sodium hydroxide or potassium cyanide, in the presence of benzoyl chloride, respectively gave carbostyril or 2-cyanoquinoline. In 1947, Katada[5] showed that heating of pyridine 1-oxide with acetic anhydride or benzoic anhydride resulted in the formation of α-pyridone. These reactions are all assumed to begin with acylation of the *N*-oxide group, and to result in the nucleophilic addition in the α- or γ-position and deacyloxylation.

References p. 334

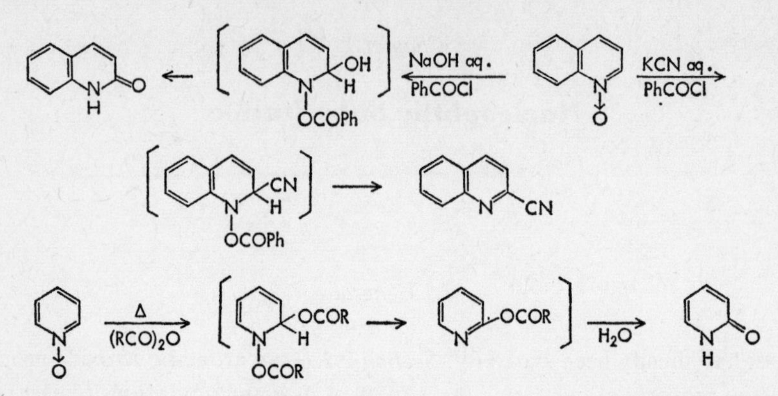

In some cases, 1,3-addition involving oxygen, instead of acylation of the *N*-oxide group, and subsequent decomposition accompanying deoxygenation to give an α- or γ-substituted compound are known to take place. Such examples, shown below, indicate increased polarity of the aromatic ring.

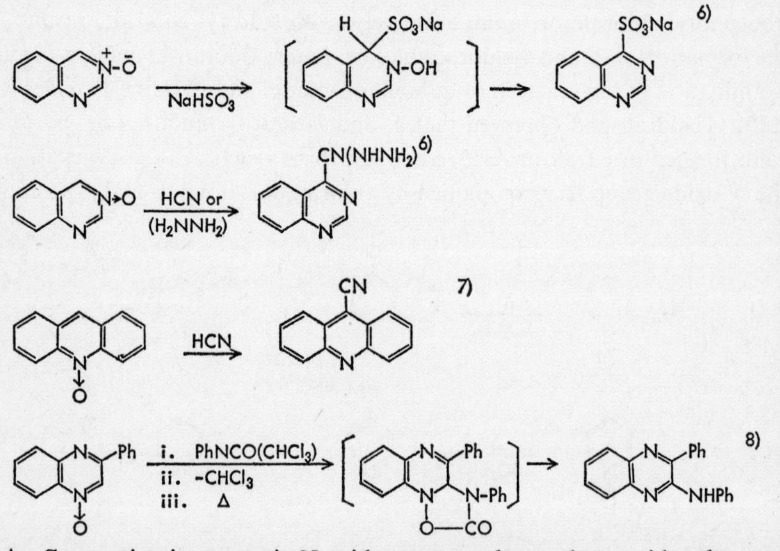

An S_N reaction in aromatic *N*-oxide compounds may be considered to progress, in the majority of cases, through the formation of such intermediate adducts. Consequently, substitution occurs in the position α or γ to the *N*-oxide group, or positions conjugated with such positions. If the bonding between oxygen and the adduct molecule is strong, a deoxygenated compound results, as shown above, and if such a bond is weak, the substitution occurs with retention of the *N*-oxide group, as shown below.

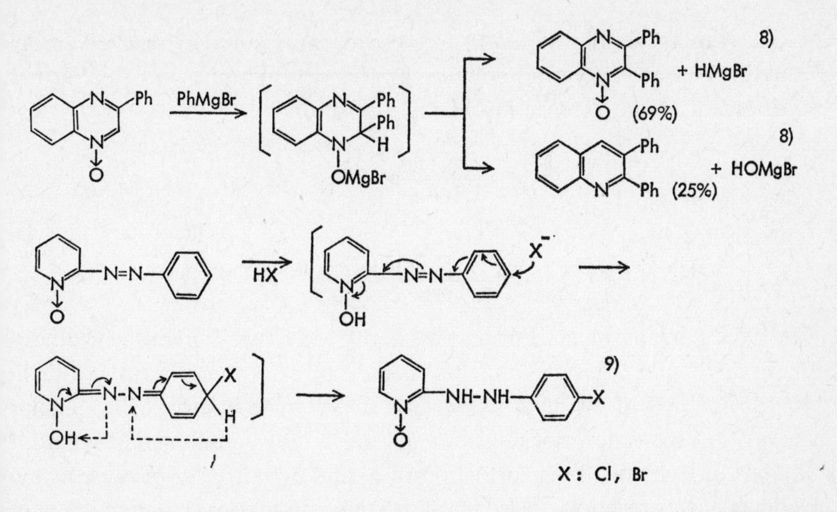

When the nitrogen–oxygen bond of the intermediate adduct is easily split, forming a stable anion with oxygen and a cationic radical, a center active to S_N reaction is formed in the position β to the ring-nitrogen or positions conjugated with it, due to the resonance of the cationic part having nitrogen with an electron-sextet, resulting in the formation of a compound substituted in the position β to the ring-nitrogen or positions corresponding to it, as shown below.

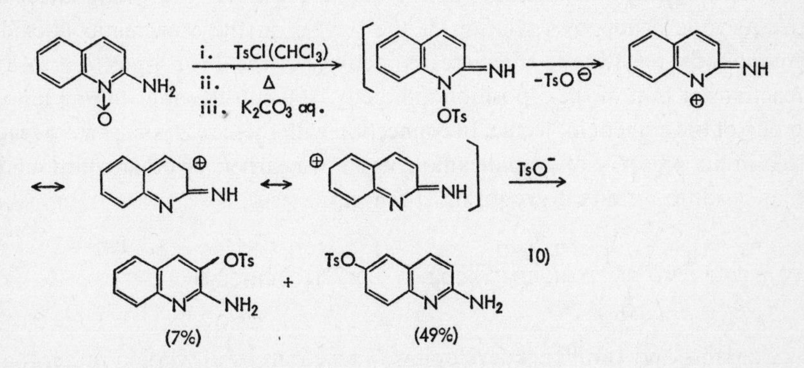

It is assumed that these changes actually progress as a concerted process. Such S_N reaction of the β position was first noted by Murakami and Matsumura[11] in 1949, when heating of pyridine 1-oxide with tosyl chloride at a bath temperature of 205–210° afforded 3-tosyloxypyridine, although in a very poor yield. Later, Ochiai and Ikehara[12] found that treatment of isoquinoline

References p. 334

2-oxide with tosyl chloride and sodium carbonate solution resulted in the formation of 4-tosyloxyisoquinoline in a good yield. This has initiated further development of an S_N reaction of the β position.

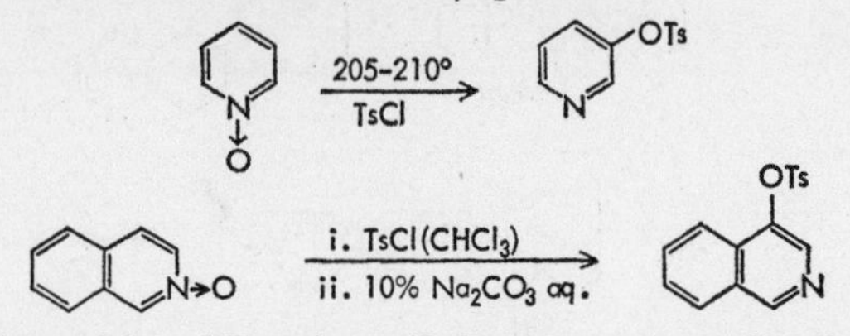

In 1953, Kobayashi and Furukawa[13] developed a reaction for the synthesis of 2-pyridylmethanol by heating 2-picoline 1-oxide with acetic anhydride at 100–140°, followed by hydrolysis with dilute hydrochloric acid. Similar results were reported by Boekelheide and Linn[14], and by Bullit and Maynard.[15] Okuda[16] demonstrated the formation of 3- and 5-hydroxy-2-picoline as by-products of this reaction. Based on these studies, a hydroxylation reaction of active methyl and methylene groups in the 2- and 4-positions, with attendant hydroxylation in the 3-position, has been developed. Discussions about its reaction mechanism are still rife.

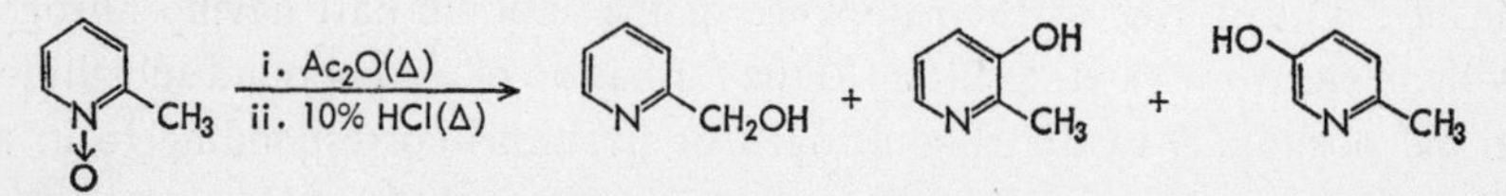

The foregoing reactions are mainly due to the electron-drawing effect of the *N*-oxide group by acylation of the oxygen in these aromatic *N*-oxide compounds, resulting in the increased activity of the 2- or 4-position to S_N reaction, or that of the 3-position indirectly, and substitution with an anion group of the reagent molecule. In connection with these reactions, an S_N reaction in the presence of cyanide anion, as shown earlier, or other compounds that produce an anion group, has been developed.

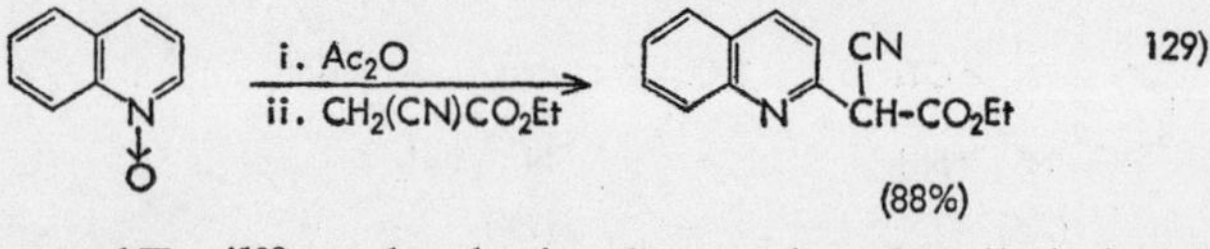

Okamoto and Tani[193] are developing S_N reactions by alkylation rather than acylation of aromatic *N*-oxide compounds.

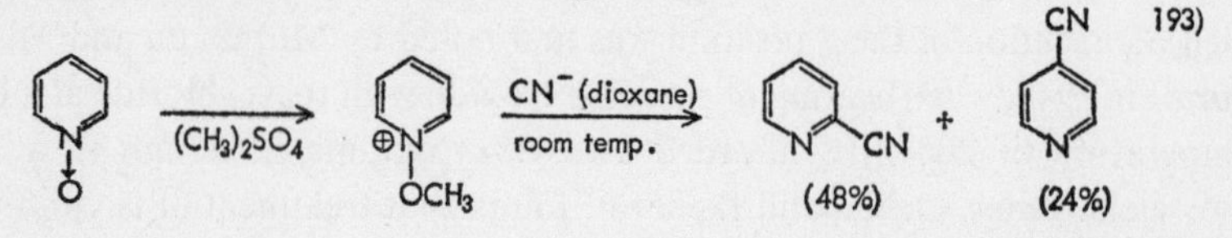

As shown above, nucleophilic substitution in the pyridine and benzopyridine series has made great progress by conversion of these compounds to their *N*-oxides and these experimental results have been utilized for *N*-oxides of diazine and azole systems, providing useful tools in the synthesis of aromatic heterocyclic compounds containing nitrogen.

7.2 *Nucleophilic Substitution by Organometallic Compounds*

In 1940, Colonna showed that the reaction of pyridine and quinoline 1-oxides with phenylmagnesium bromide in ether respectively gives 2-phenyl-pyridine and -quinoline[17a]; he obtained similar results with 6-methoxy and 6-methyl derivatives of quinoline 1-oxide[17b], and with benzo[*h*]- and benzo[*f*]-quinoline 1-oxide[17c]. Ochiai and Arima[18] re-examined this reaction with pyridine 1-oxide, adding benzene to increase its solubility, and found the formation of presumably 2,2′-diphenyl-4,4′-bipyridyl (4%) and biphenyl, besides 2-phenylpyridine (13%). Risaliti[19] carried out this reaction with 2- and 4-methoxyquinoline 1-oxide and found that while the 4-methoxy derivative afforded 4-methoxy-2-phenylquinoline, the 2-methoxy derivative formed 2-phenylquinoline, liberating the methoxyl group.

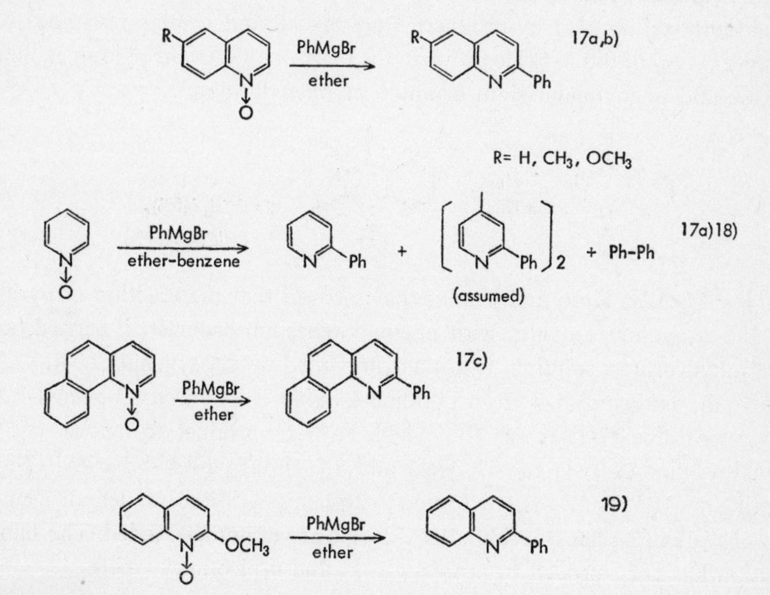

References p. 334

Kobayashi[20] applied this reaction to aromatic *N*-oxides of quinine and dihydroquinine to synthesize their respective 2′-phenyl derivatives.

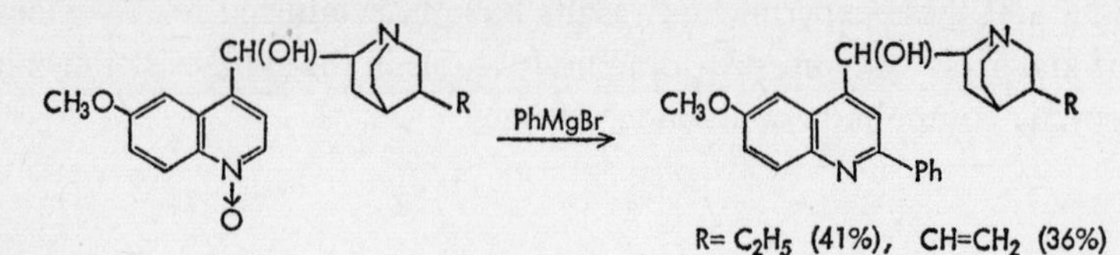

Similarly

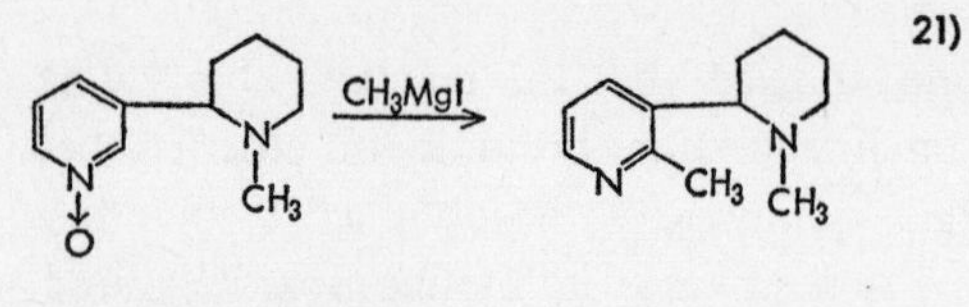

These reactions are all examples of nucleophilic substitution by the Grignard reagent accompanying deoxygenation of the *N*-oxide group, but such a reaction without deoxygenation was found to occur in diazine series *N*-oxides with marked nucleophilic activity. Hayashi and others[8] found that the reaction of 2-phenylquinoxaline 4-oxide with phenylmagnesium bromide or phenyllithium mainly gave 2,3-diphenylquinoxaline 4-oxide, with 2,3-diphenylquinoxaline as a by-product. They also found similar resistance to deoxygenation of the *N*-oxide group in the reaction of 4-isopropylquinazoline 1-oxide and phenylmagnesium bromide or phenyllithium[22].

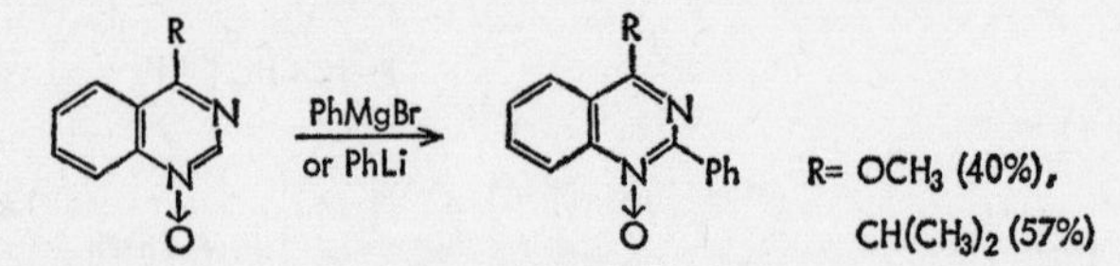

More recently, Kato and Yamanaka[23] proved that the reaction of pyridine and quinoline 1-oxides with phenylmagnesium bromide, if carried out in tetrahydrofuran solution, is not accompanied by deoxygenation. In such a case, the main product from pyridine 1-oxide is 1-hydroxy-2-phenyl-1,2-dihydropyridine (7-1) in 60–80% yield, with by-product formation of 2-phenylpyridine (7-2) in *ca.* 5% yield and 2,2′-diphenyl-4,4′-bipyridyl (7-3). Heating of (7-1) results in dehydration to 2-phenylpyridine, but dehydrogenation of (7-1) to 2-phenylpyridine 1-oxide has as yet not succeeded. The same reaction has been proved to progress in 2- and 4-picoline 1-oxides.

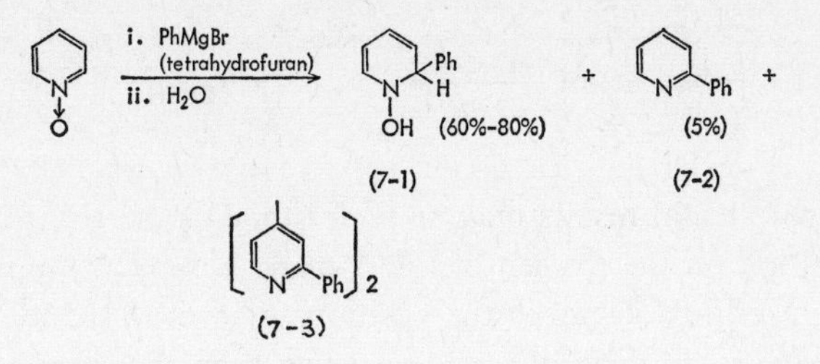

In contrast, the same reaction of quinoline 1-oxide in boiling tetrahydrofuran solution gives 2-phenylquinoline 1-oxide in *ca.* 60% yield as the main product and 2-phenylquinoline in *ca.* 30% yield as a by-product. 1-Hydroxy-2-phenyl-1,2-dihydroquinoline (7-4) was obtained only when the reaction was carried out at room temperature. This dihydroquinoline derivative (7-4) is more labile than the corresponding dihydropyridine derivative and is dehydrated to 2-phenylquinoline under mild conditions. It is dehydrogenated to 2-phenylquinoline 1-oxide by passage of air through its tetrahydrofuran solution.

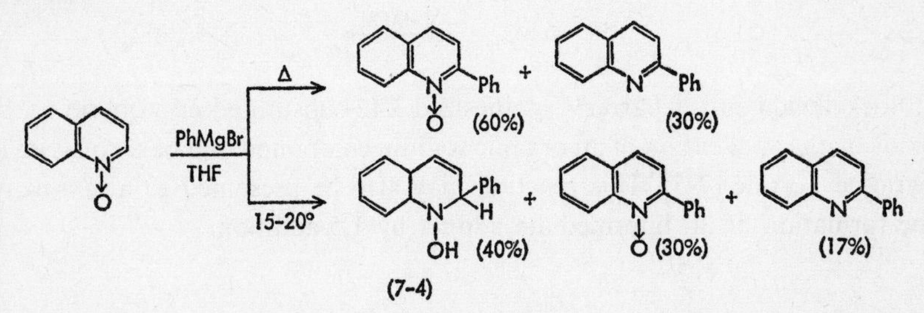

It has been confirmed that a similar reaction takes place with quinoline 1-oxides substituted in 4-position with methoxyl, methyl, chlorine, etc.

Blumenthal and Plainfield[24] took out a patent for the synthesis of 2-ethynyl derivatives by the reaction of pyridine and quinoline l-oxides with ethynylsodium in dimethyl sulfoxide at room temperature, by which the corresponding 2-ethynyl derivatives are obtained in a good yield without deoxygenation of the *N*-oxide group. Kaneko and Miyasaka[25] used the same reaction for derivation of 2-(phenylethynyl)pyridine 1-oxide from pyridine 1-oxide.

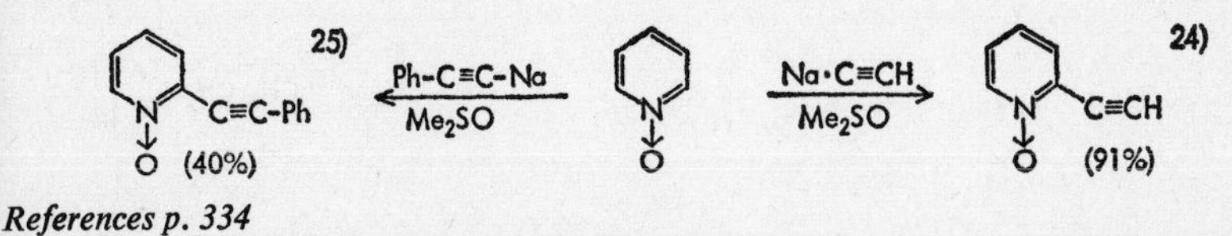

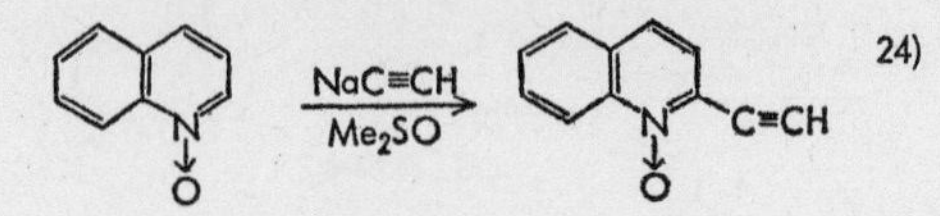

These experimental results may be understood as the electrophilic attack of the metal ion on the oxygen of the *N*-oxide group, accompanied by the bonding of a residual group in the position adjacent to the *N*-oxide group, whose activity to nucleophilic addition has been increased, followed by liberation of a metal hydroxide or metal hydride from the intermediate adduct so formed, and formation of either a deoxygenated derivative or a compound still retaining the *N*-oxide group. Kato[26] presumed that, in the reaction with Grignard reagent in tetrahydrofuran, the fact that tetrahydrofuran more easily forms a peroxide than ether has some bearing on the retention of the *N*-oxide group.

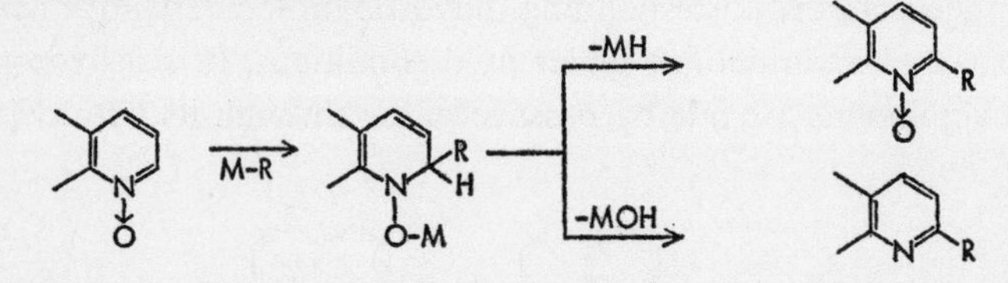

Boekelheide and Scharrer[27] synthesized 2-(2-substituted ethyl)pyridine 1-oxide by the application of an organic sodium compound or a base to 2-vinylpyridine 1-oxide (7-5). This reaction may also be presumed to accompany the formation of an intermediate adduct by 1,5-addition.

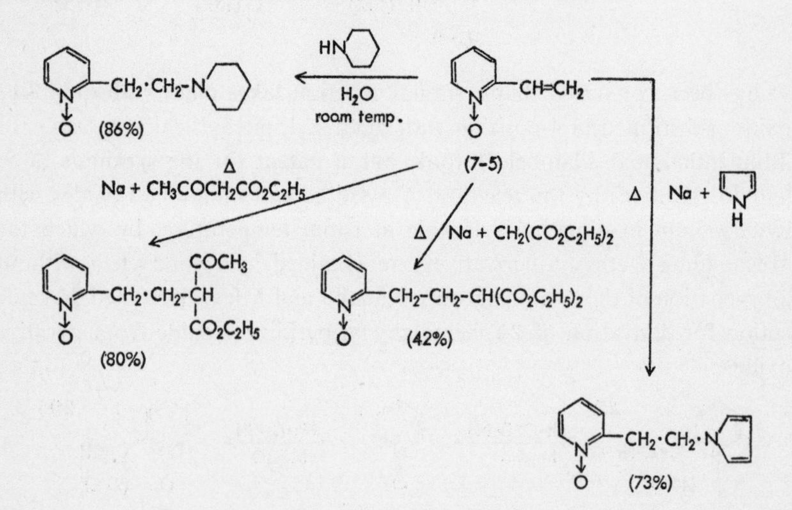

A reaction with the same trend is found in the one already described for quinazoline 3-oxide[6] and acridine *N*-oxide[7], in which the proton, instead of a metal ion, co-ordinates with the oxygen and dehydration from the adduct bonded with the residual group gives a nucleophilic substitution product accompanied by deoxygenation of the *N*-oxide group (*cf. Section 7.1*, p. 248). This reaction does not take place with 4-methoxyquinazoline 1-oxide (7-6), with the exception of a substitution of the methoxyl group with hydrazine.[6]

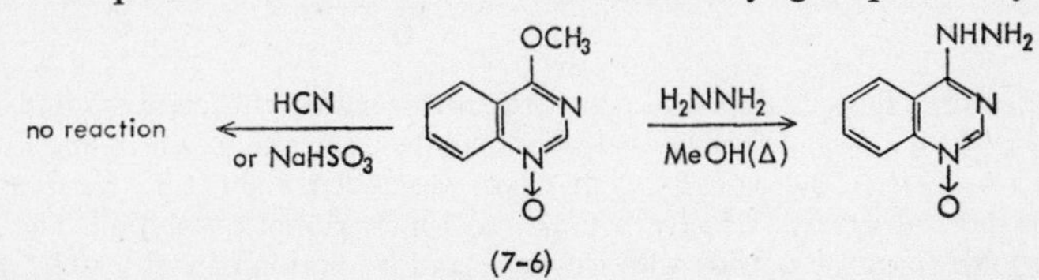

Hayashi and others[8] found that the reaction of 2-phenylquinoxaline 4-oxide (7-7) with hydrogen cyanide does not take place at room temperature, but is effected only when its hydrochloride is heated with potassium cyanide in a sealed tube at 100°.

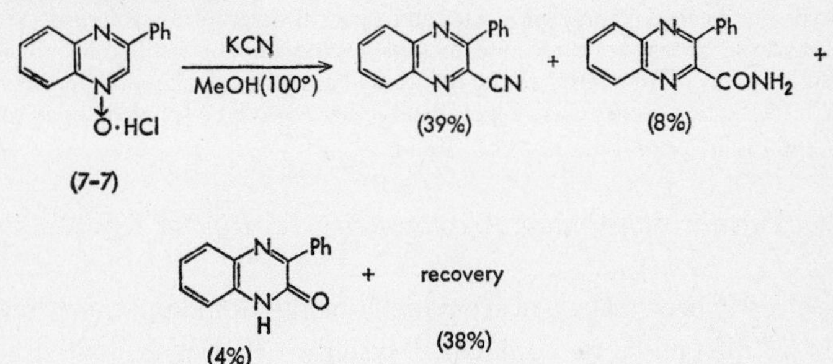

Example 1. Reaction of 2-Phenylquinoxaline 4-Oxide and Phenyllithium[8]

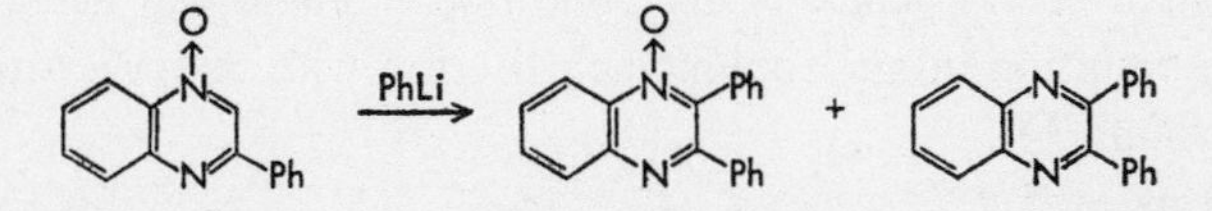

To a solution of 1 g of 2-phenylquinoxaline 4-oxide dissolved in 10 ml of benzene, 20 ml of an ether solution of phenyllithium, prepared from 50 mg of lithium and 1.0 g of bromobenzene, was added slowly with ice cooling, by which the solution became dark red. After mild refluxing for 7.5 h, the cooled solution was poured on ice, 10 ml of 20% sulfuric acid was added, and the mixture was warmed on a water bath for 5 min. When cooled, the reaction mixture was basified with potassium carbonate, separated lithium carbonate was removed by filtration, and the filtrate was extracted with chloroform. After drying over anhydrous sodium sulfate, the solvent was evaporated from the chloroform extract, the residue was dissolved in benzene, and the solution was passed through a column packed with 20 cc. of activated alumina (200 mesh). The column was eluted with benzene and, from the least adsorptive portion, a minute amount of crystals was obtained. Recrystallization

References p. 334

from methanol gave white needle crystals of m.p. 125° which agreed with 2,3-diphenylquinoxaline. Crystals obtained from later eluted fractions were recrystallized from benzene and light petroleum (b.p. 60–80°) to 760 mg (54%) of 2,3-diphenylquinoxaline 1-oxide as white needles, m.p. 197°.

Example 2. 2-(Phenylethynyl)pyridine 1-Oxide from Pyridine 1-Oxide[25]

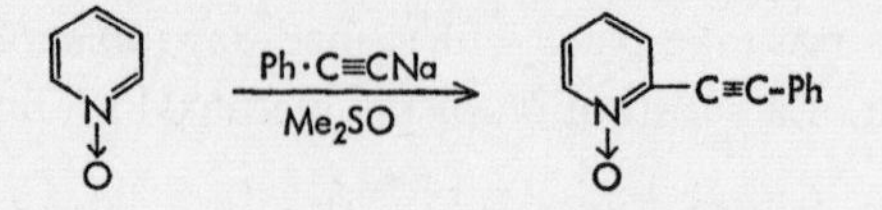

In a three-necked flask equipped with a stirrer, nitrogen inlet tube, and calcium chloride tube, 2.042 g (0.02 mole) of freshly distilled ethynylbenzene and 40 ml of absolute ether were placed, 0.460 g (0.02 mole) of sodium flakes was added, and the mixture was stirred until all the flakes dissolved. Usually, it took 3 to 7 h for complete solution. The mixture was then cooled from the outside with ice-water and a solution of 1.91 g (0.02 mole) of freshly distilled pyridine 1-oxide in 20 ml of dimethyl sulfoxide (freshly distilled in vacuum after drying over calcium hydride) was added dropwise. This mixture was stirred for 2 h in the cold in a nitrogen stream and for 12 h at room temperature, and then placed on a water bath (around 70°) for 1 h. The dark-colored solution was treated with 50 ml of ice-water and extracted three times with 100 ml each of methylene chloride. The combined organic layer was washed several times with ammoniacal silver nitrate solution. After drying over anhydrous potassium carbonate, methylene chloride was distilled off; the residual crude crystals weighed 1.6 g (40% of the theoretical amount calculated from the *N*-oxide used). Recrystallization from benzene–petroleum ether afforded 2-(phenylethynyl)pyridine 1-oxide as pale yellow needles, m.p. 102–103°.

7.3 Nucleophilic Substitution Accompanying 1,3-Dipolar Cycloaddition

As has already been stated, an aromatic *N*-oxide compound can be regarded as a kind of nitrone or its conjugated system[17a,28]. Consequently, as with nitrones, 1,3-dipolar cycloaddition[28] also takes place in these compounds and the adduct formed is generally so labile that it undergoes decomposition, apparently resulting in nucleophilic substitution with deoxygenation of the *N*-oxide group[8,29,30].

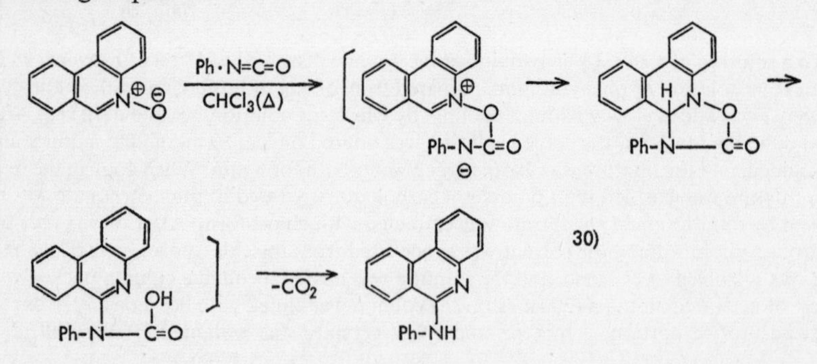

In these reactions, Hayashi and his group were not able to obtain the intermediate adduct, but Huisgen[29] obtained the adduct shown in the following example for nitrone.

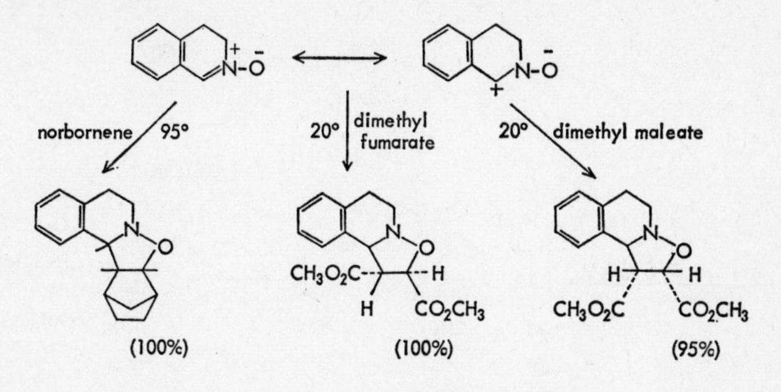

Huisgen also reported similar reactions, shown below.

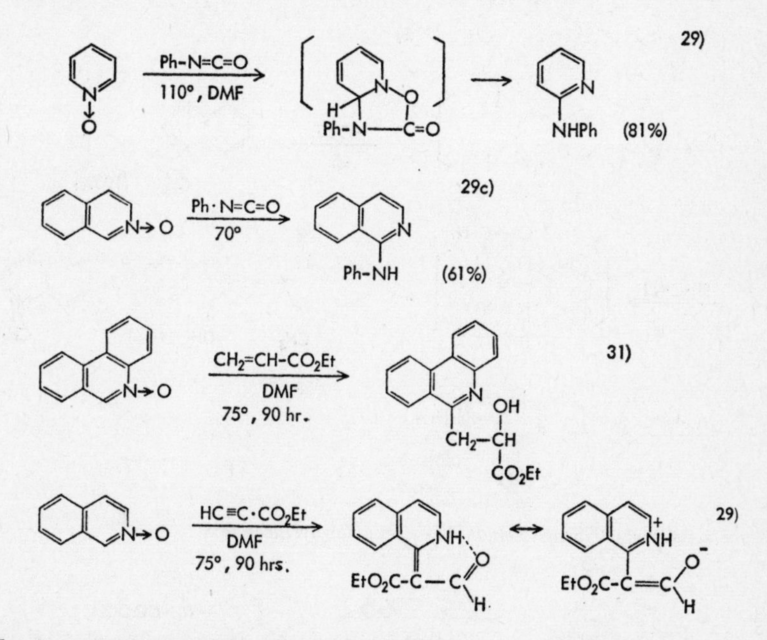

Kanô and Takahashi[32] used this reaction for 1-methylbenzimidazole 3-oxide by treating it in chloroform solution with phenyl isocyanate, phenyl isothiocyanate, carbon disulfide, methyl propiolate, or dimethyl ethynyldicarboxylate as dipolarophiles at room temperature, or with benzonitrile or phenyl isocyanide with heating, and obtained the corresponding 2-substituted 1-methylbenzimidazole derivatives, generally in good yield.

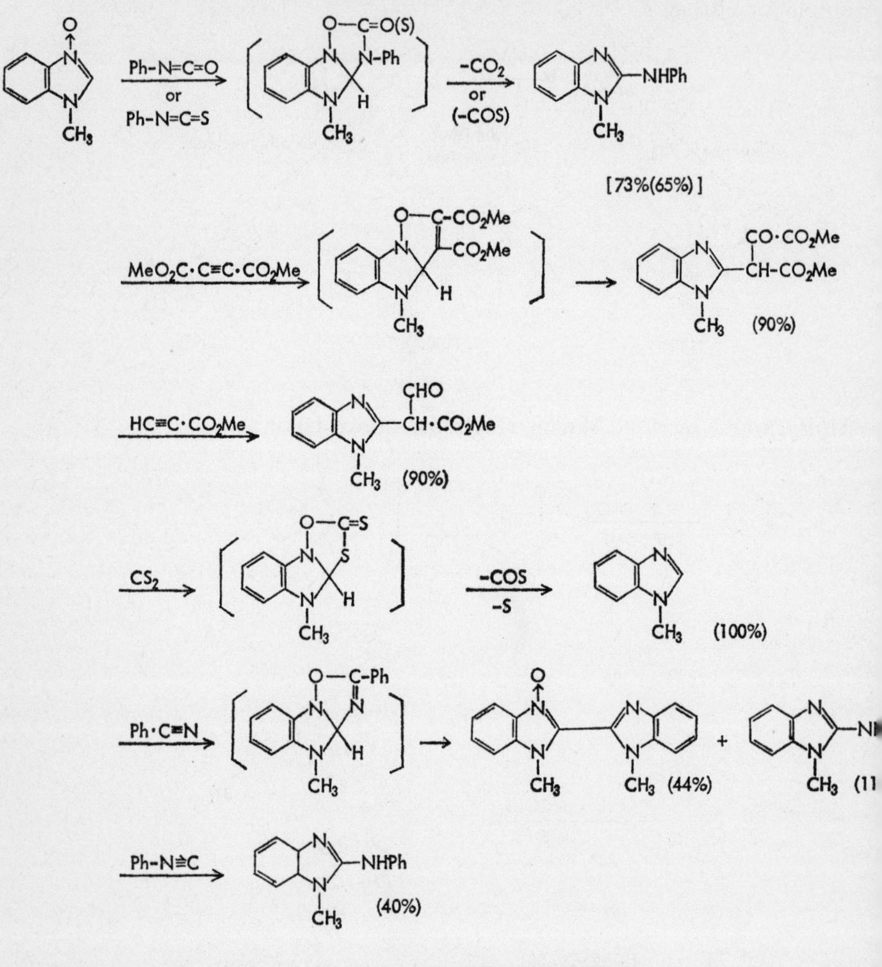

Example. Methyl α-Formyl(1-methyl-2-benzimidazolyl)acetate[32b]

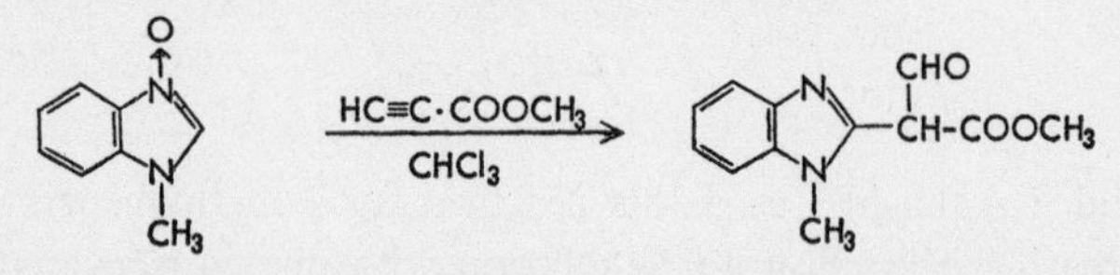

After azeotropic distillation to remove the water of crystallization, using chloroform, from 1.84 g (0.010 mole) of 1-methylbenzimidazole 3-oxide dihydrate, the residue was dissolved in 10 ml of chloroform and 0.84 g (0.010 mole) of methyl propiolate was added dropwise while shaking the chloroform solution with ice-cooling. The mixture was allowed to stand at room temperature for 1 h after completion of the addition and the solvent was evaporated under reduced pressure. The crystals so obtained were recrystallized from methanol to 2.10 g (90%) of colorless short prisms, m.p. 183° (decomp.).

7.4 Nucleophilic Substitution by Reactive Halides and Acid Anhydrides

7.4.1 Reaction of Inorganic Acid Halides

As has already been stated (*cf. Section 7.1*), reactions of pyridine and quinoline 1-oxides with phosphoryl chloride or sulfuryl chloride have been developed by Meisenheimer[1] and by Bobranski[3]. Later, it became evident that this reaction is applicable to aromatic *N*-oxide compounds in general. Some additional points to be noted in this reaction will be described in this section.

When the position α or γ to the *N*-oxide group is occupied, substitution with chlorine takes place in the remaining γ or α' position.

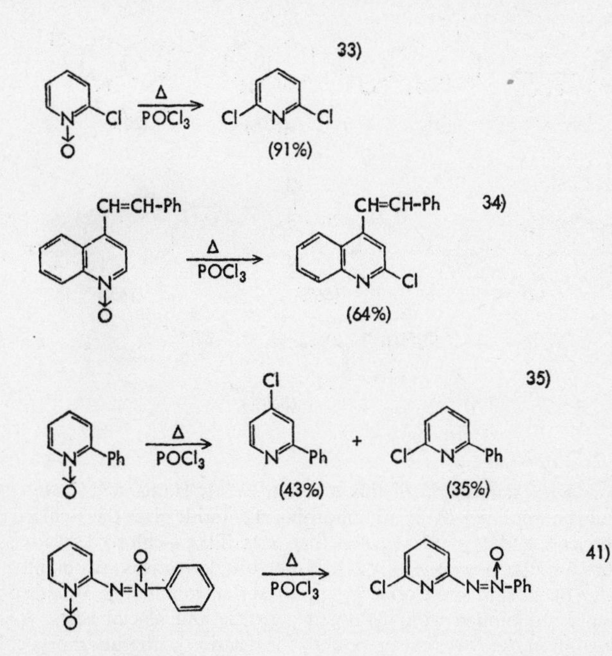

When a methyl group is present in the α-position, the methyl groups are partly substituted by chlorine.

References p. 334

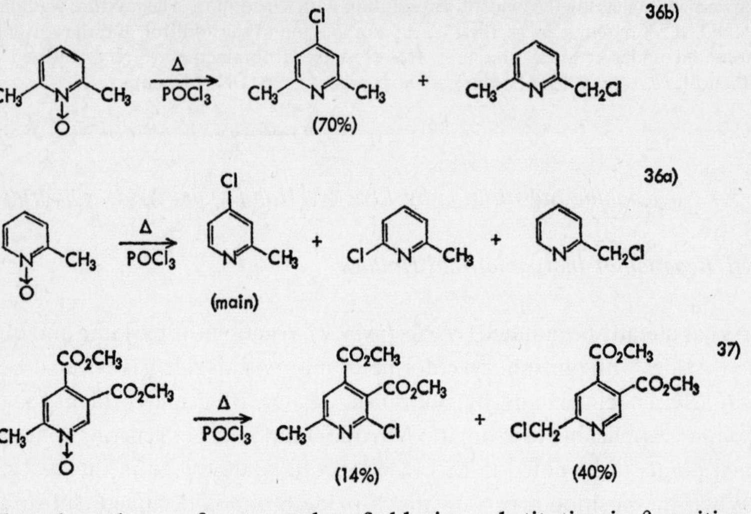

There have been a few examples of chlorine substituting in β-position.

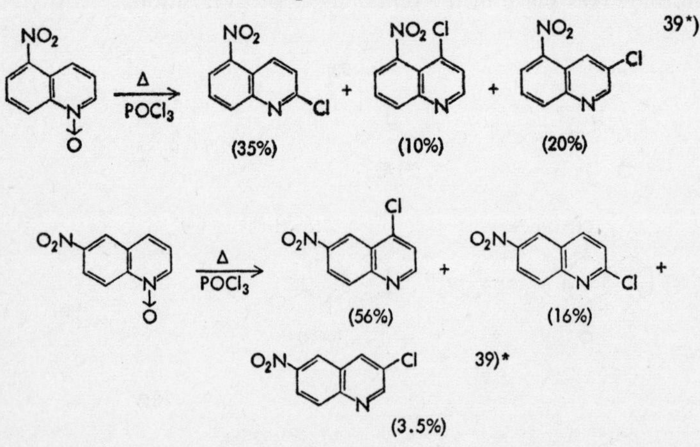

* According to a re-examination of this reaction by M. Hamana (private comm.), the reaction of 5-nitroquinoline 1-oxide and phosphoryl chloride gives the 3-chloro derivative, as in past evidence[38], and its yield is greater than that of the 4-chloro derivative. In case of 6-nitroquinoline 1-oxide, however, only 2- and 4-chloro derivatives are obtained. The reaction reported by Bachmann and Cooper[39] employed 6-nitroquinoline 1-oxide obtained by oxidation of 6-nitroquinoline with hydrogen peroxide and glacial acetic acid, without further purification of the *N*-oxide compound, for reaction with phosphoryl chloride and Hamana presumed that 3-hydroxy-6-nitroquinoline, which is formed as a by-product in a small amount during *N*-oxidation (*cf. Section* 3.1.3.*b*), might have taken part in this reaction.

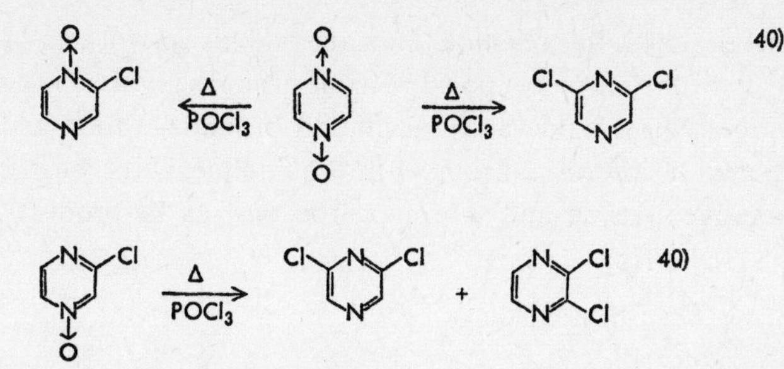

There are also examples of chlorine substitution occurring in other positions conjugated with the *N*-oxide group.

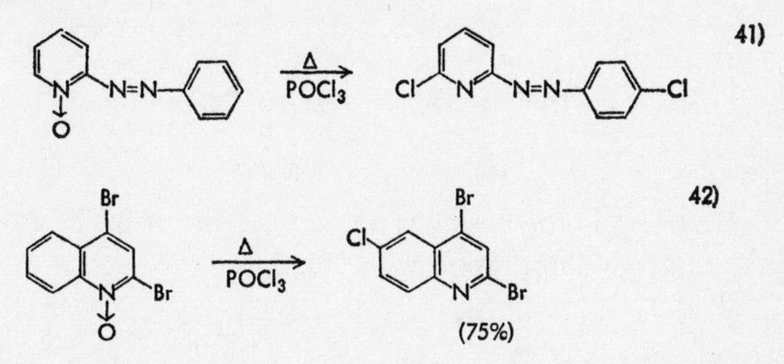

The nitro group in the position γ to the *N*-oxide group also undergoes substitution with chlorine by the same reaction, simultaneously with substitution of chlorine in the α-position. If this reaction is carried out under a mild condition, substitution of the nitro group with chlorine is the main reaction.

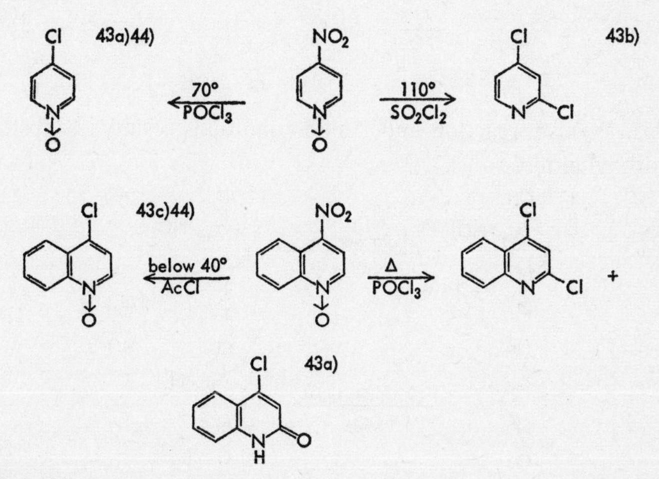

References p. 334

In connection with this reaction, Hamana and his co-workers[45] found that the reactivity of the nitro group and *N*-oxide group is reversed in the reaction of 4-nitroquinoline 1-oxide and phosphoryl bromide, which under a mild condition mainly affords 2-bromo-4-nitroquinoline, with formation of 4-bromoquinoline 1-oxide and 4-bromocarbostyril as by-products.

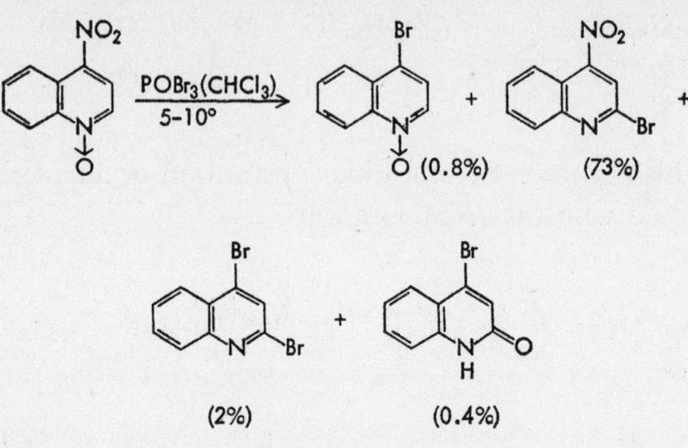

A hydroxyl group in the positions α and γ to the *N*-oxide group has a tendency to undergo substitution with chlorine.

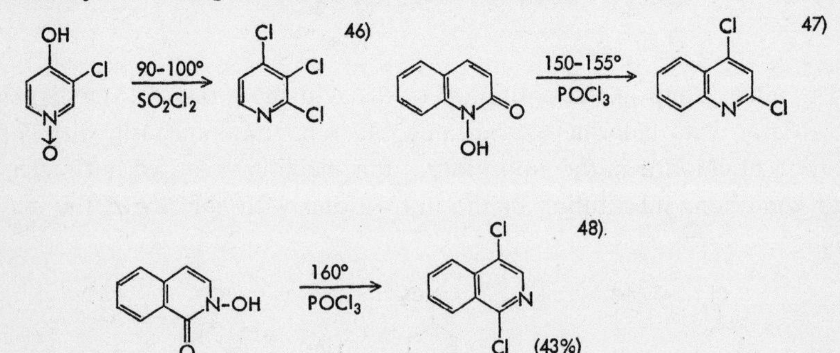

In this case also, reaction under mild conditions results in substitution of the hydroxyl alone.

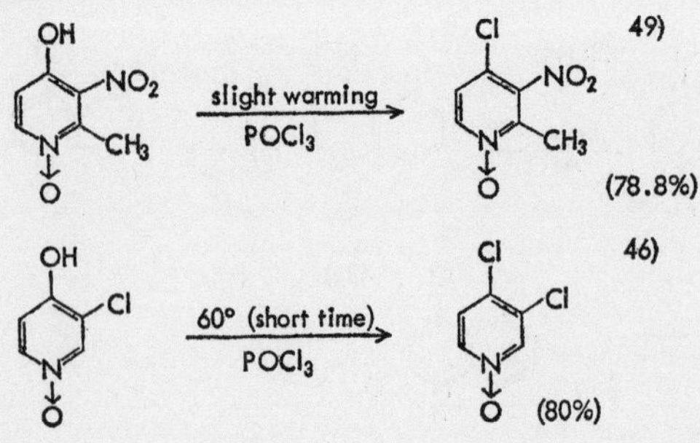

When there is a secondary alcoholic hydroxyl group in the molecule that is easily substituted with chlorine, a reaction carried out in chloroform may prevent substitution of the hydroxyl with chlorine and only cause alteration at the site of the aromatic *N*-oxide group. Such examples have been observed with cinchona bases.

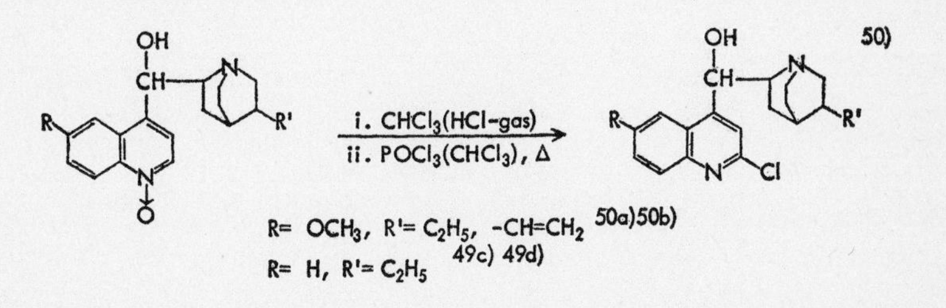

When such a state is not expected, it is usual to protect the alcoholic hydroxyl group by acylation before submitting the compound to reaction with phosphoryl chloride.

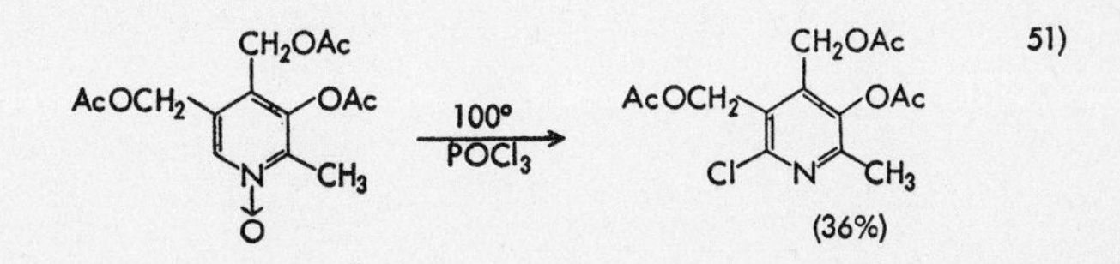

When there are substituents like hydroxyl and alkoxyl in the ring, further chlorination of the ring may occur, besides the foregoing substitutions.

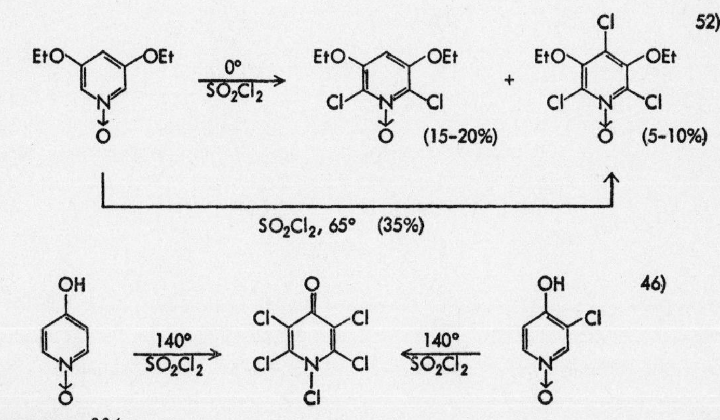

References p. 334

TABLE 7-1

NUCLEOPHILIC REACTIONS OF *N*-OXIDE COMPOUNDS WITH INORGANIC ACID HALIDES

N-Oxide	Reagent	Temp. (°C)	Solvent	Product (tert. amine) Yield (%)	Ref.
Pyridine Series					
Nil	SO_2Cl_2	120		4-Cl/2-Cl = 43:57	3
2-CH_3	$POCl_3$	Δ		4-Cl (main), 6-Cl, 2-CH_2Cl	36*a*
3-CH_3	$POCl_3$	Δ	$CHCl_3$	4-Cl (35.5), 2-Cl (8.3). 6-Cl (20)	53
3-CH_3	SO_2Cl_2	<45	$CHCl_3$	4-Cl	54
2,6-$(CH_3)_2$	$POCl_3$	140–150		4-Cl (70), 2-CH_2Cl (2)	36*b*
2-Ph	SO_2Cl_2	Δ		4-Cl (43), 6-Cl (35)	35
4-NO_2	SO_2Cl_2	110		2,4-Cl_2 (25-30)	43*a*,*b*
4-NO_2	$POCl_3$	70		4-Cl-N→O, 2,4-Cl_2	43*a*
3-NO_2	$POCl_3$	Δ		2-Cl-3-NO_2 (30), 6-Cl-3-NO_2 (8.4)	55
3-NO_2	$POCl_3$	150		2-Cl-3-NO_2 (45)	56
2-Anisyl	$POCl_3$	Δ		2-Anisyl-6-Cl (34), 2-Anisyl-4-Cl (5)	35
2-Cl	$POCl_3$	Δ		2,6-Cl_2 (91)	33
3-Br	SO_2Cl_2	110–120		3-Br-4-Cl>3-Br-2-Cl> 3-Br-6-Cl	57
4-OH	SO_2Cl_2	140		Cl Cl O= N-Cl Cl Cl	46
4-OH	SO_2Cl_2	Δ		2,4-Cl_2	46
3-Cl-4-OH	$POCl_3$	60		3,4-Cl_2-N→O (80)	46
3-Cl-4-OH	SO_2Cl_2	90–100		2,3,4-Cl_3	46
3-Cl-4-OH	SO_2Cl_2	140		Cl Cl O= N-Cl Cl Cl	46
4-C_2H_5O	$POCl_3$	Δ		4-C_2H_5O-2-Cl	43*a*
3-COOH	$POCl_3$–PCl_5	Δ		2-Cl-3-COOH (41), 4-Cl-3-COOH (4)	58
3-$CONH_2$	$POCl_3$–PCl_5	Δ		2-Cl-3-CN (22)	58
4-$COOC_2H_5$	(*i*) $POCl_3$ (*ii*) KOH	130–140		2-Cl-4-COOH (70)	59
3-$CONH_2$-4-CH_3O	$POCl_3$–PCl_5	Δ		3-CN-2,4-Cl_2 (33)	60
2-N=N(O)Ph	$POCl_3$ or SO_2Cl_2	Δ		6-Cl-2-N=N(O)Ph	41
2-N=N–Ph	$POCl_3$	Δ		6-Cl-2-N=N–Ph–Cl (*p*)	41
3-Br-4-NO_2	$POCl_3$	room		4-Cl-3-Br–N→O (quant.)	61

TABLE 7-1 (*continued*)

N-Oxide	Reagent	Temp. (°C)	Solvent	Product (tert. amine) Yield (%)	Ref.
3-NO_2-6-OH	$POBr_3$	150		3-NO_2-6-Br	62
2-CH_3-4-NO_2	$POCl_3$–PCl_5	90–95		6-Cl	63
2,6-$(CH_3)_2$-3-NO_2	$POCl_3$–PCl_5	90–100		4-Cl	63
3,5-Br_2	SO_2Cl_2	110		6-Cl/4-Cl = 60:40	52
3,4-Cl_2	SO_2Cl_2	90–100		2,3,4-Cl_3 (50–60)	46
3,5-$(OC_2H_5)_2$	SO_2Cl_2	0		2,4,6-Cl_3-3,5-$(OC_2H_5)_2$ N→O (5–10) 2,6-Cl_2-3,5-$(OC_2H_5)_2$–N→O (15–20)	52
3,5-$(OC_2H_5)_2$	SO_2Cl_2	65		2,4,6-Cl_3-3,5-$(OC_2H_5)_2$–N→O (35)	52
3,5-Br_2-2-OH	$POBr_3$	160		2,3,5-Br_3	64
2-OH-3,5-$(NO_2)_2$	$POBr_3$	160		2-Br-3,5-$(NO_2)_2$	64
3,5-$(NO_2)_2$	$POCl_3$	140–150		2-Cl-3,5-$(NO_2)_2$	56
2,6-$(CH_3)_2$-4-NO_2	$POCl_3$–PCl_5	90–100		2,6-$(CH_3)_2$-4-Cl	63
2-CH_3-4-OH-5-NO_2	$POCl_3$	slight warming		2-CH_3-4-Cl-5-NO_2 (75.3)	49
3,4-$(COOCH_3)_2$-6-CH_3	$POCl_3$	120		3,4-$(COOCH_3)_2$-6-CH_2Cl (40) 2-Cl-3,4-$(COOCH_3)_2$-6-CH_3 (14)	37
3-C_2H_5-4-CN	$POCl_3$	Δ	$CHCl_3$	3-C_2H_5-4-CN-6-Cl 2-Cl-3-C_2H_5-4-CN	65
2-CH_3-3-CH_3COO-4,5-$(CH_3COOCH_2)_2$	$POCl_3$	100		2-CH_3-3-CH_3COO-4,5-$(CH_3COOCH_2)_2$-6-Cl (36)	51
Quinoline Series					
Nil	SO_2Cl_2	Δ		4-Cl/2-Cl = 62:38	2
2-CH_3	SO_2Cl_2	Δ		2-CH_3-4-Cl	72
2-C_6H_5	$POCl_3$	130–140		2-C_6H_5-4-Cl	66
4-NO_2	$POCl_3$ or SO_2Cl_2	Δ		2,4-Cl_2, 2-OH-4-Cl	43*a,c*
4-NO_2	$POCl_3$, SO_2Cl_2 or AcCl	0–40		4-Cl–N→O (98)	43*a*, 44
4-NO_2	$POBr_3$	5–10	$CHCl_3$	4-Br–N→O, 2-OH-4-Br (0.4) 2-Br-4-NO_2 (73), 2,4-Br_2(2)	45
5-NO_2	SO_2Cl_2	90–100		2-Cl-5-NO_2	67
5-NO_2	$POCl_3$	Δ		3-Cl-5-NO_2 (20), 2-Cl-5-NO_2 (35), 4-Cl-5-NO_2(10)	38
6-NO_2	$POCl_3$	Δ		4-Cl-6-NO_2 (56), 2-Cl-6-NO_2 (16), 3-Cl-6-NO_2 (3.5)	39

 Continued overleaf

TABLE 7-1 (*continued*)

N-Oxide	Reagent	Temp. (°C)	Solvent	Product (tert. amine) Yield (%)	Ref.
8-NO_2	SO_2Cl_2	Δ		2-Cl-8-NO_2	67
3-NO_2	$POCl_3$	Δ		2-Cl-3-NO_2	47
3-Br	$POCl_3$	Δ		2-Cl-3-Br	47
6-Cl	$POCl_3$	Δ		2,6-Cl_2 (49), 4,6-Cl_2 (35)	39
2-OH	$POCl_3$	150–155		2,4-Cl_2	47
6-CH_3O	$POCl_3$	Δ		2-Cl-6-CH_3O/4-Cl-6-CH_3O = 55:35	39, 68
4-C_2H_5O	$POCl_3$	Δ		2-Cl-4-C_2H_5O	43*a*
4-NH_2	$POCl_3$	90–100		2-Cl-4-NH_2	69
4-NHOH	$POCl_3$	90–100		2-Cl-4-NH_2	69
3-$NHCOCH_3$	*(i)* $POCl_3$ *(ii)* H_2O	Δ		3-$NHCOCH_3$–carbostyril	74
4-COOH	$POCl_3$	90–100		2-Cl-4-COCl	70, 71
3,6-$(NO_2)_2$	$POCl_3$	90–100		2-Cl-3,6-$(NO_2)_2$ (100)	73
4-CH=CH-C_6H_5	$POCl_3$	Δ		2-Cl-4-CH=CH–C_6H_5	34
4-NO_2-7-Cl	AcCl	40		2,7-Cl_2–N→O (83)	75
4,6-$(NO_2)_2$	$POCl_3$	90–100		2,4-Cl_2-6-NO_2	76
4,6,8-$(NO_2)_3$	$POCl_3$–PCl_5	90–100		2,4-Cl_2-6,8-$(NO_2)_2$	76
2-CH_3-4-NO_2	$POCl_3$	100–110		2-CH_3-4-Cl–N→O	77*a*, 78*a*,*b*
2-CH_3-4-NO_2	$POCl_3$	below 120		2-CH_3-4,7-Cl_2	77*b*
2,4-Br_2	$POCl_3$	Δ		2,4-Br_2-6-Cl (75)	42
4-CH_3-8-NO_2	$POCl_3$	60–78		2-Cl-4-CH_3-8-NO_2	78
4-NO_2-6-CH_3	$POCl_3$	77–87		2,4-Cl_2-6-CH_3	78
4-CH_3O-7-NO_2	$POCl_3$	140–150		2-Cl-4-CH_3O-7-NO_2	79
Dihydrocinchonine ar. *N*-oxide	$POCl_3$	Δ	$CHCl_3$	2′-Cl-dihydrocinchonine (quant.)	50*c*
Dihydrocinchonidine ar. N-oxide	$POCl_3$	Δ	$CHCl_3$	2′-Cl-dihydrocinchonidine (quant.)	50*d*
Quinine ar. N-oxide	$POCl_3$	Δ	$CHCl_3$	2′-Cl-quinine (50)	50*a*
Dihydroquinine ar. N-oxide	$POCl_3$	Δ	$CHCl_3$	2′-Cl-dihydroquinine (70–90)	50*a*,*b*
Dihydrocinchoninone ar. N-oxide	$POCl_3$	Δ	$CHCl_3$	2′-Cl-dihydrocinchoninone	50*e*
Dihydroquininone ar. N-oxide	$POCl_3$	Δ	$CHCl_3$	2′-Cl-dihydroquininone	50*e*
aci-N-Benzoyl-dihydroniquine ar. N-oxide	$POCl_3$	Δ	$CHCl_3$	2′-Cl-*aci*-N-benzoyl-dihydroniquine	80
3-NO_2-4-OH	$POCl_3$	150–155		2,4-Cl_2-3-NO_2	47
Benzoquinoline Series					
Benzo[*f*]-	$POCl_3$	Δ		2-Cl-benzo[*f*]	17*c*
Benzo[*f*]-	$POCl_3$	Δ		2-Cl-(32), 4-Cl- (21) benzo[*f*]	39

TABLE 7-1 (*continued*)

N-Oxide	Reagent	Temp. (°C)	Solvent	Product (tert. amine) Yield (%)	Ref.
7-NO_2-benzo[*f*]	$POCl_3$	90–100		2-Cl-7-NO_2-benzo[*f*]	81, 82
9-NO_2-benzo[*h*]	$POCl_3$	90–100		2-Cl-9-NO_2-benzo[*h*]	81
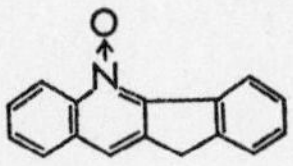	$POCl_3$	Δ		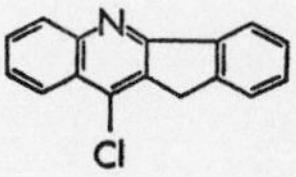 (44.5)	83
Isoquinoline Series					
	$POCl_3$	Δ	$CHCl_3$	1-Cl (62)	84
5,6,7,8-tetrahydro	$POCl_3$	Δ	$CHCl_3$	1-Cl	85
4-Br	$POCl_3$	Δ	$CHCl_3$	1-Cl-4-Br	86
1-OH	$POCl_3$	160		1,4-Cl_2 (43)	48
5-NO_2	$POCl_3$	100		1-Cl-5-NO_2 (77)	87
Naphthyridine Series					
1,5-Naphthyridine 1-oxide	$POCl_3$	Δ		2-Cl	88
1,5-Naphthyridine 1,5-dioxide	$POCl_3$	Δ		2,6-Cl_2	88
Pyridazine Series					
3-C_6H_5	$POCl_3$	Δ		3-C_6H_5-6-Cl	89
3-CH_3O	$POCl_3$	room	$CHCl_3$	3-CH_3O-6-Cl (52)	90
3-N_3	$POCl_3$	Δ	$CHCl_3$	5-Cl-pyridazino(2,3-*d*)-tetrazole (57)	93
3,6-$(CH_3)_2$	$POCl_3$	60–70		3,6-$(CH_3)_2$-4-Cl (28)	92
3-CH_3O-6-CH_3	$POCl_3$	Δ		3-CH_3O-4-Cl-6-CH_3 (58)	94
3,6-$(CH_3O)_2$	$POCl_3$	room		3,6-$(CH_3O)_2$-4-Cl (72)	91
Cinnoline Series					
1-Oxide	$POCl_3$	room	$CHCl_3$	4-Cl (50)	95
3-CH_3O-4-NO_2 1-oxide	conc. HCl	Δ		3-CH_3O-4-Cl 1-oxide	96
Pyrazine Series					
1,4-dioxide	$POCl_3$	Δ		2,6-Cl_2	40
1,4-dioxide	$POCl_3$	slight warming		2-Cl-1$\rightarrow$O	40
3-Cl 1-oxide	$POCl_3$	Δ		2,6-Cl_2 + 2,3-Cl_2 (1:1)	40
3-$CONH_2$ 1-oxide	$POCl_3$	Δ		2-Cl-6-CN	40
2,5-$(CH_3)_2$ 1-oxide	$POCl_3$	Δ		2,5-$(CH_3)_2$-6-Cl	97
2,5-$(CH_3)_2$ 1,4-dioxide	$POCl_3$	Δ		2,5-$(CH_3)_2$-3,6-Cl_2	97

References p. 334

Continued overleaf

TABLE 7-1 (*continued*)

N-Oxide	Reagent	Temp. (°C)	Solvent	Product (tert. amine) Yield (%)	Ref.
2,5-(sec-C_4H_9)$_2$ 1,4-dioxide	$POCl_3$	Δ		2,5-(sec-C_4H_9)$_2$-3,6-Cl_2	97
3-NH_2 1-oxide	$POCl_3$	Δ		2-Cl-3-NH_2	98
2-C_6H_5-3-CH_3O 1-oxide	$POCl_3$	Δ		2-C_6H_5-3-CH_3O-6-Cl	99
Quinoxaline Series					
1-oxide	$POCl_3$	Δ		2-Cl	100
1,4-dioxide	$POCl_3$	Δ		2,3-Cl_2	100
2-C_6H_5 4-oxide	SO_2Cl_2	Δ		2-Cl-3-C_6H_5 (quant.)	8
2-C_6H_5 1,4-dioxide	SO_2Cl_2	Δ		2-Cl-3-C_6H_5-4→O (73)	101
Quinazoline Series					
4-OR 1-oxide	SO_2Cl_2 or $POCl_3$	Δ		2-Cl-4-OR (55–57) R = CH_3, C_2H_5	102
4-i-C_3H_7 1-oxide	SO_2Cl_2 or $POCl_3$	Δ		2-Cl-4-*i*-C_3H_7 (63)	103
Benzimidazole Series					
1-CH_3 3-oxide	SO_2Cl_2 or $POCl_3$	ice-cooling	$CHCl_3$	1-CH_3-2-Cl (70, 90)	111

Example 1. Reaction of 2,6-Lutidine 1-Oxide and Phosphoryl Chloride[36b]

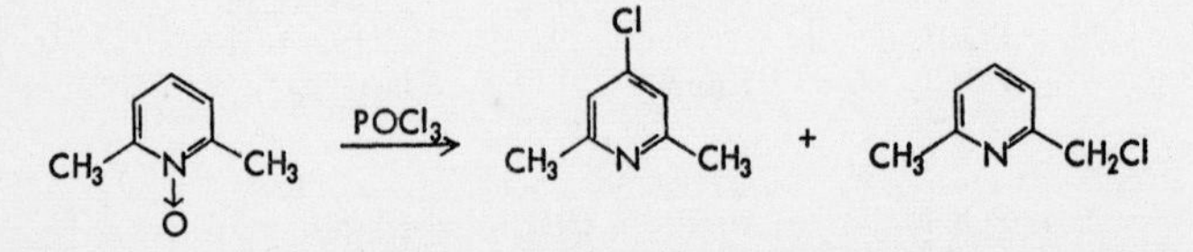

A mixture of 8 g (0.05 mole) of 2,6-lutidine 1-oxide hydrochloride and 23 g (0.15 mole) of phosphoryl chloride, placed in a flask fitted with a calcium chloride tube, was heated on a water bath for 30 min, at 120° in an oil bath for 30 min, then at 140–150° for 5 h. When cooled, the clear solution was poured into 100 ml of ice-water and this was basified with sodium carbonate while cooling from outside. The separated oil was extracted with ether, which was evaporated after drying over anhydrous sodium sulfate. About 7 g of the residue so obtained was fractionated at a reduced pressure into 5 g of b.p. 110–115°/90 mm and 0.5 g of b.p. 120–130°/90 mm. Redistillation of the former fraction gave 4.5 g (64%) of b.p. 112–115°/85–90 mm, whose picrate of yellow needles (from ethanol), m.p. 166–167°, was identical with the picrate of 4-chloro-2,6-lutidine. Redistillation of the latter fraction gave 0.2 g (2%) of b.p. 124–126°/85 mm, whose picrate of yellow plates (from acetone), m.p. 163–164°, agreed with the picrate of 2-chloromethyl-6-methylpyridine.

Example 2. 2′-Chlorodihydrocinchonidine[50d]

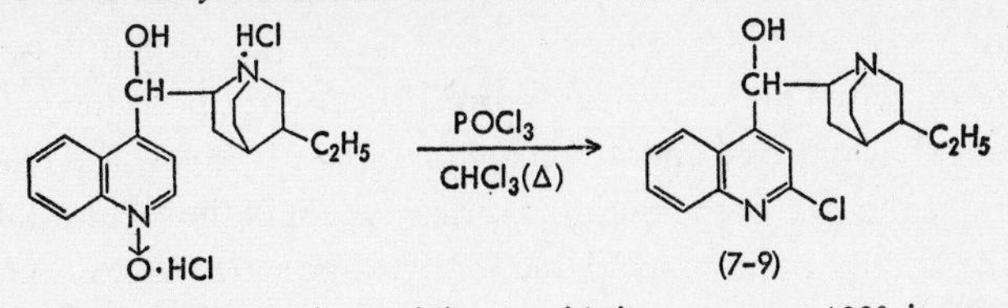

After drying 10 g of dihydrocinchonidine 1-oxide in vacuum at 100°, it was dissolved in 200 ml of chloroform, the solution was saturated with dry hydrogen chloride gas, and 10 g of phosphoryl chloride was added. This mixture was refluxed on a water bath at 75° for 2 h, the pale yellow reaction mixture was poured into ice-water, and this was basified with 10% sodium hydroxide solution. This was extracted with chloroform and the solvent was evaporated from the extract after drying over anhydrous sodium sulfate. The crystalline residue was recrystallized from methanol to 9.5 g of 2′-chlorodihydrocinchonidine as needles, m.p. 221–222.5°.

7.4.2 *Reaction in the Presence of Organic Acid Halides*

The reaction of acyl chloride with aromatic *N*-oxide compounds is milder than that of inorganic chlorides and formation of α- or γ-substituted compounds is rather difficult. If the reaction is carried out in the presence of a nucleophilic reagent, nucleophilic substitution is known to occur in α- or γ-position.

As has already been stated, Henze[4] found that the reaction of quinoline 1-oxide with potassium cyanide or sodium hydroxide in the presence of benzoyl chloride results in the formation of 2-cyanoquinoline or carbostyril, respectively. The former corresponds to the so-called Reissert reaction of quinolines[104] but while the Reissert reaction of quinoline gives 1-benzoyl-2-cyano-1,2-dihydroquinoline (7-8), that of quinoline *N*-oxide results in concurrent deoxygenation of the *N*-oxide group and 2-cyanoquinoline is formed immediately. Consequently, this reaction is very useful for synthetic purpose.

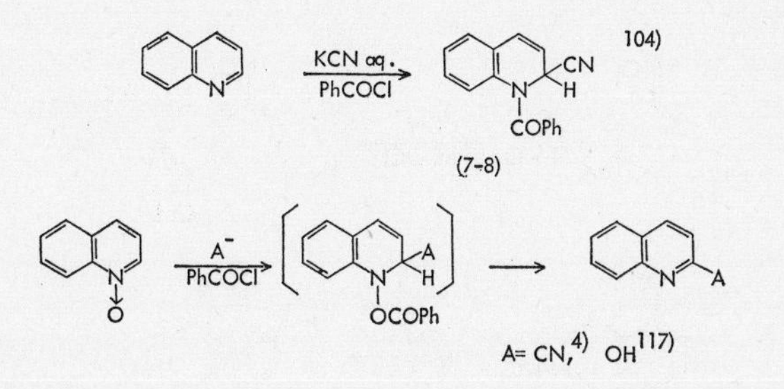

References p. 334

Ochiai and Nakayama[105] examined this Reissert reaction with 4-substituted pyridine and quinoline 1-oxides and found that, while the corresponding 2-cyanoquinoline derivatives are formed from the quinoline system compounds, except 4-hydroxy and 4-benzoylamino derivatives, the reaction was unsuccessful in the case of the pyridine series, except in the 4-chloro derivative. The reaction finally became available for pyridine series *N*-oxide compounds by the application of the cyanide ion to their methosulfate[195], as will be described later (*cf. Section 7.4.5*).

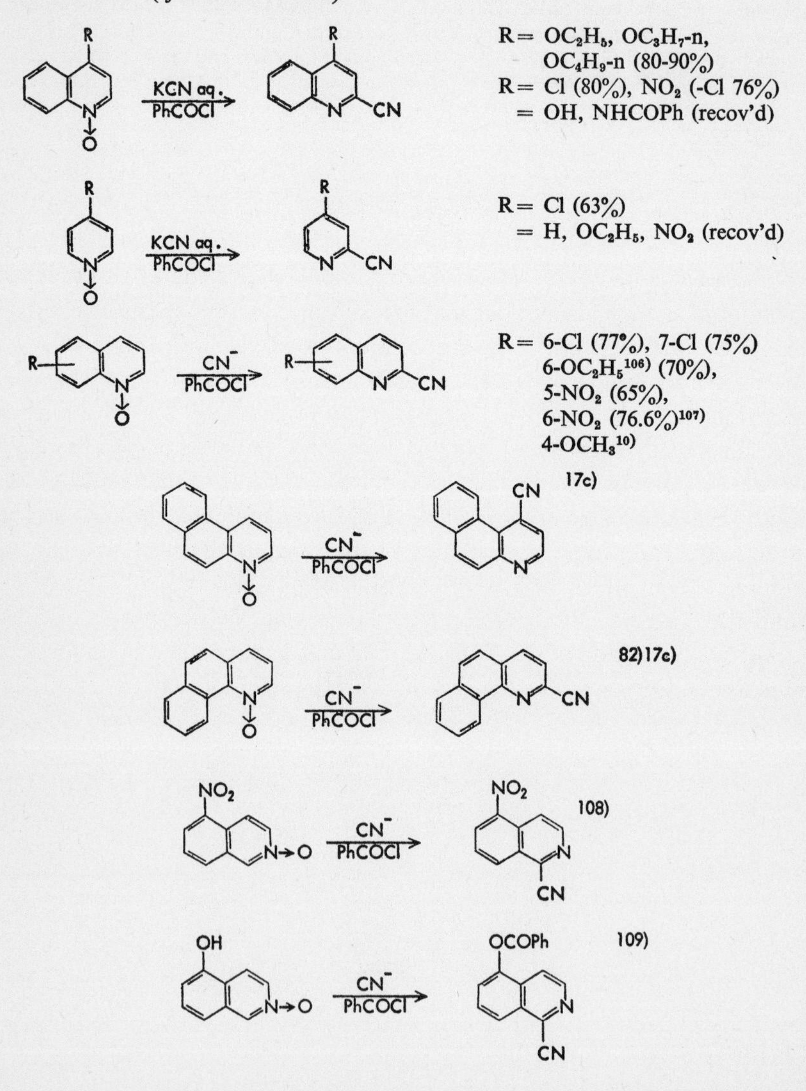

Later, this reaction was found applicable not only to benzopyridine series *N*-oxides but also to aromatic *N*-oxide compounds in general, giving objective products in a good yield. Some such reactions reported to date are quoted below.

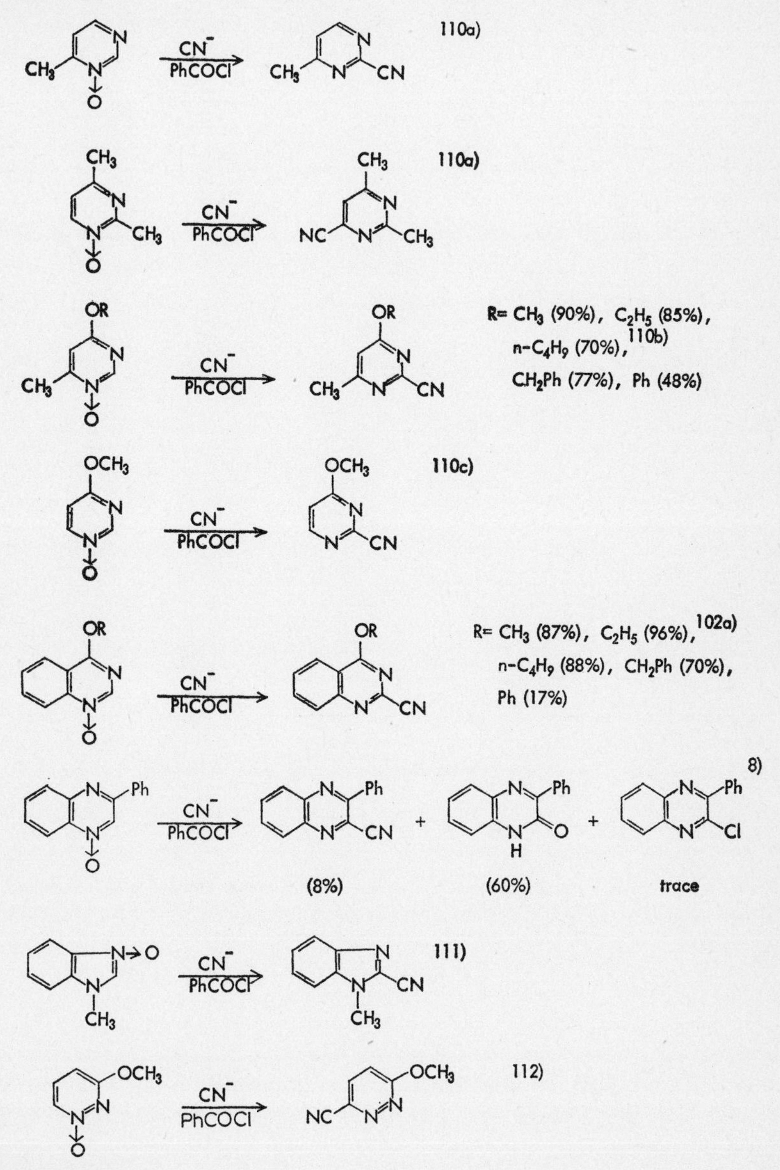

References p. 334

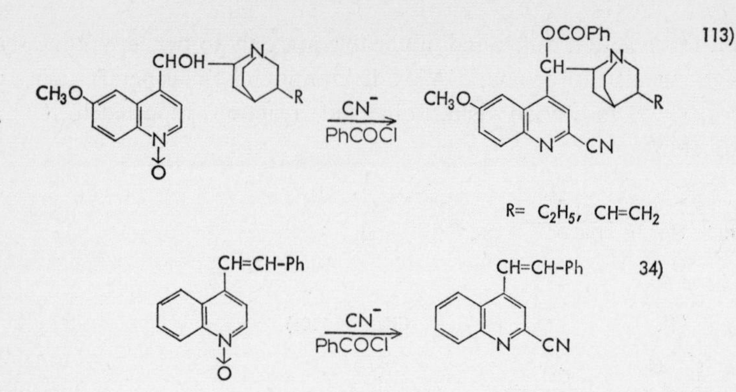

Hayashi and Higashino[22] found that in the Reissert reaction of 4-isopropylquinazoline 1-oxide (7-10), α,α-dimethyl-4-quinazolylacetamide (7-11) (35%) is formed besides the normal product, 2-cyano-4-isopropylquinazoline (33%), and proposed the following mechanism for the formation of (7-11).

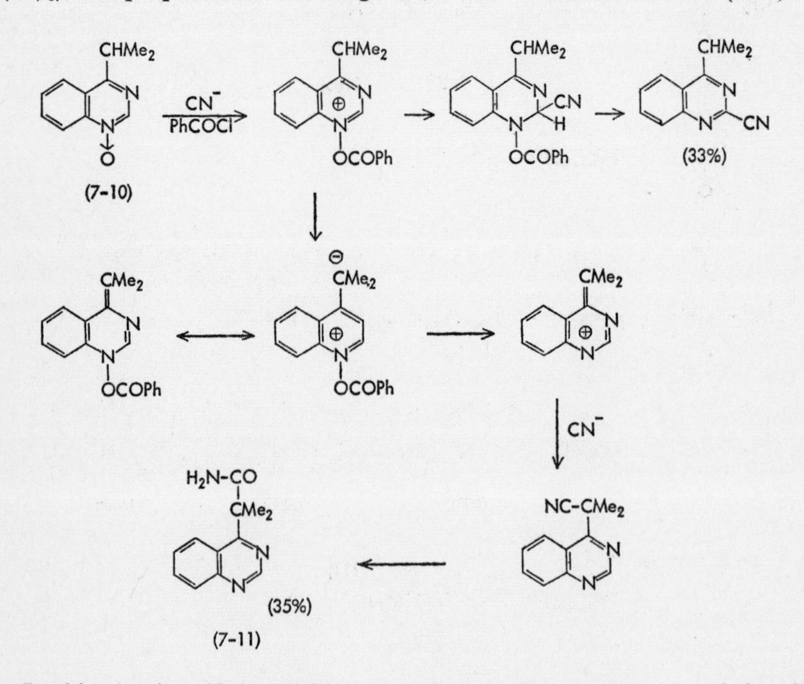

In this reaction, if the position α to the *N*-oxide group is occupied, substitution with the cyanide group is sometimes known to take place in the γ-position.

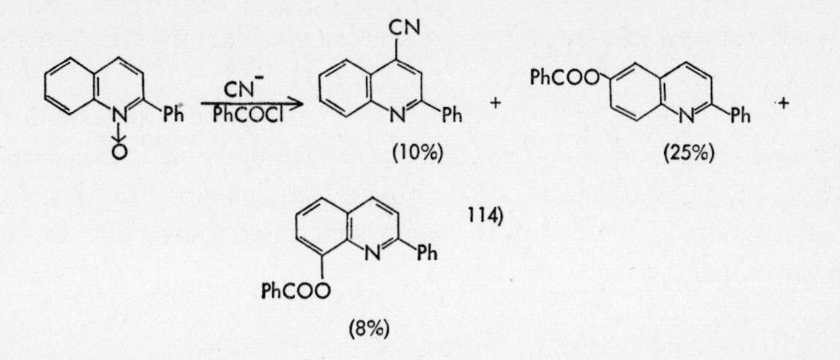

The main reaction in this case is rather the introduction of the benzoyloxyl group into the benzene portion, corresponding to the positions *ortho* and *para* to the ring-nitrogen, but the same substitution with a benzyloxyl group has been observed in 2,4-diphenylquinoline 1-oxide, in which both the 2- and 4-positions are already occupied[114].

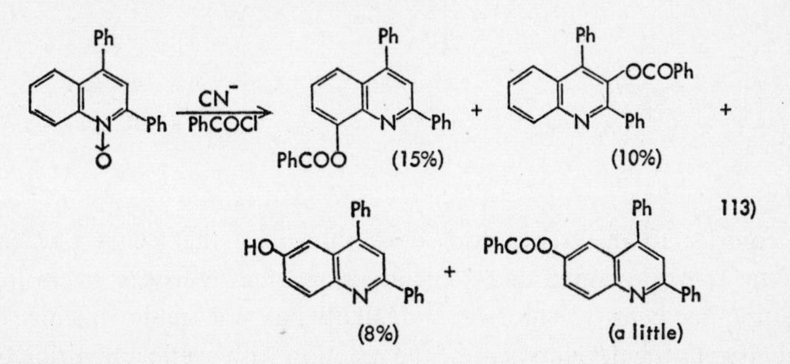

Nucleophilic substitution in the position β to the *N*-oxide group, and into the 8- and 6-positions in the benzene portion, often occurs when there is an electron-donating group in 2-position. As will be described later (*cf. Section 7.4.6*), this phenomenon may be explained as a result of preliminary splitting of the N–OCOPh bond in the intermediate compound.

In contrast to the formation of 4-alkoxy-2-cyanoquinazoline from 4-alkoxyquinazoline 1-oxide in a good yield[102a], this reaction of quinazoline 3-oxide results in ring cleavage, while reaction with hydrogen cyanide, without the use of benzoyl chloride, at room temperature easily gives 4-cyanoquinazoline[102b].

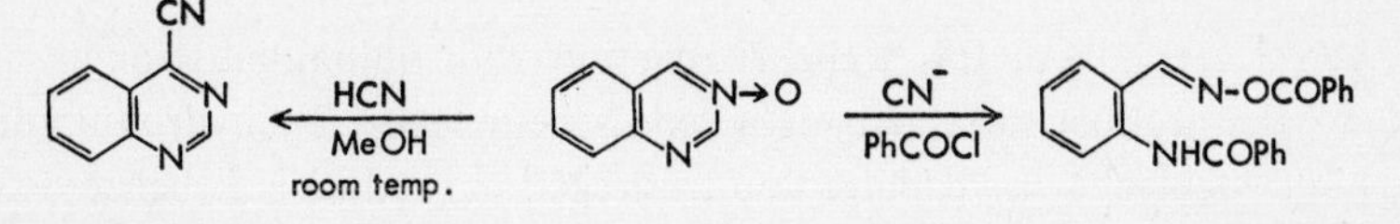

References p. 334

This reaction with hydrogen cyanide does not take place in 4-methoxyquinazoline 1-oxide[102b].

Kaneko[115] found that the use of silver cyanide in place of potassium cyanide in this reaction, carried out in chloroform solution, sometimes gave a good result. For example, the yield of 2-cyano-3-nitroquinoline l-oxide is only 20% with postassium cyanide, while this yield is increased to over 80% by the use of silver cyanide.

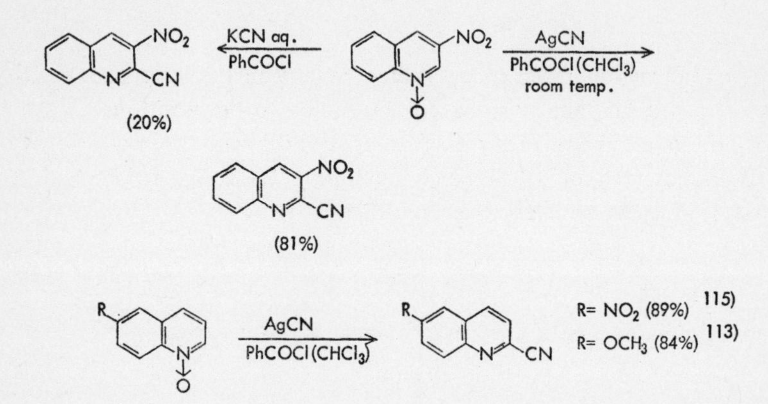

In contrast to the tremendous development of the Reissert reaction of aromatic *N*-oxide compounds, introduction of a hydroxyl group into the α-position by benzoyl chloride and alkali has not made much progress, mainly due to the development of the reaction with acetic anhydride or with tosyl chloride and alkali, as will be described later (*cf. Section 7.4.3* and *.4*). Hamana and Funakoshi[116] succeeded in isolating the reaction intermediate from quinoline 1-oxide to carbostyril and revealed that this intermediate is so labile that it easily undergoes decomposition into benzoic acid and carbostyril; they explained the mechanism of this reaction as follows.

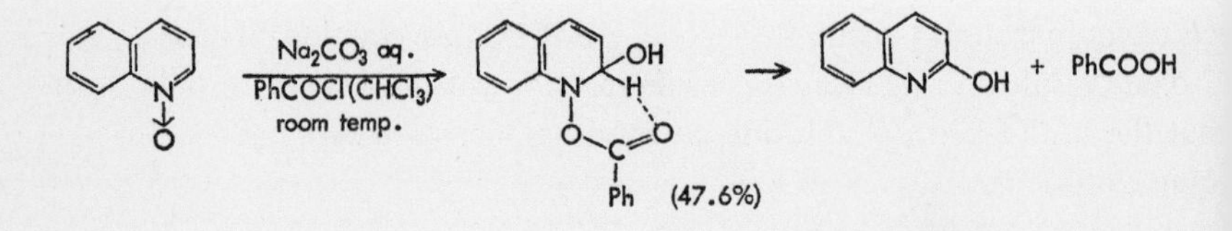

Henze[117] carried out this Reissert reaction with quinaldine 1-oxide and assumed the product to be *N*-benzoyloxy-2-methylene-1,2-dihydroquinoline (7-12) but Pachter[118] proved later that this reaction product is 2-(benzoyloxymethyl)quinoline (7-13) by the identification of its hydrolysis product as quinaldyl alcohol (7-15), obtained by the reduction of methyl quinaldinate (7-14) with lithium aluminium hydride.

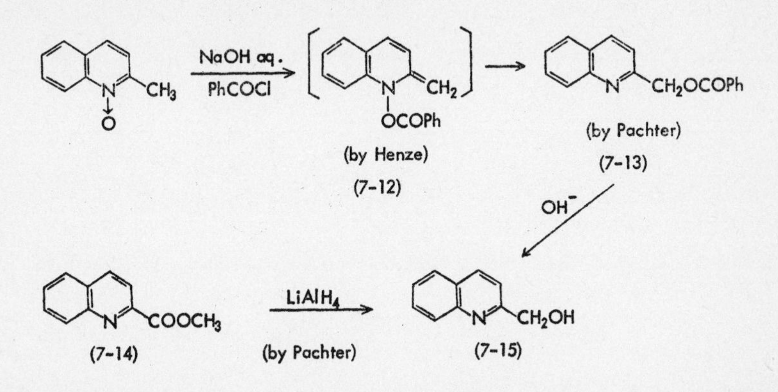

Kaneko[115] found that the Reissert reaction carried out on 2-ethoxyquinoline l-oxide (7-16) results in its smooth conversion into *N*-benzoyloxycarbostyril (7-17).

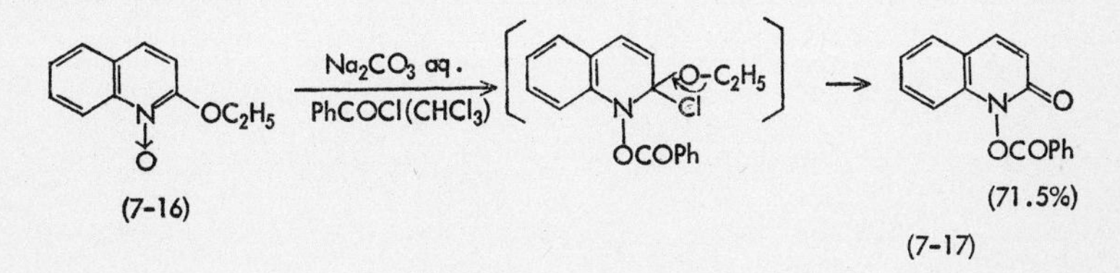

A similar reaction was carried out by Itai and Natsume[119] with pyridazine derivatives.

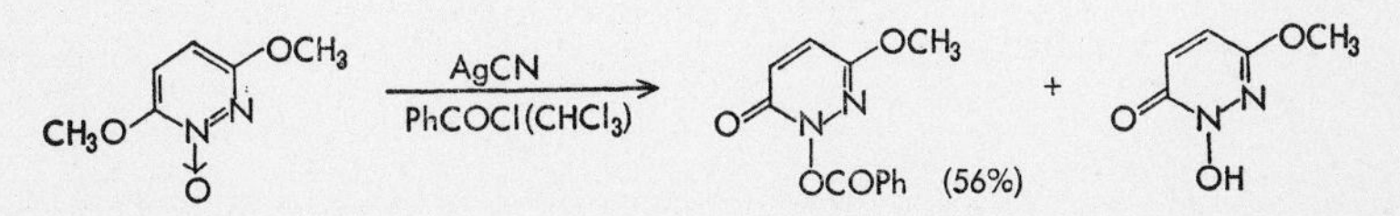

Hamana and Noda[120] carried out the reaction of quinoline 1-oxide and morpholine enamine of cyclohexanone (7-18) (2 equiv.) in chloroform solution with addition of benzoyl chloride (1.2 equiv.). Treatment of the reaction mixture with 20% hydrochloric acid solution gave 2-(2-quinolyl)cyclohexanone (7-19) in 74% yield. This reaction does not take place if benzoyl chloride is not added and the mechanism of the formation of (7-19) was assumed to be as follows:

References p. 334

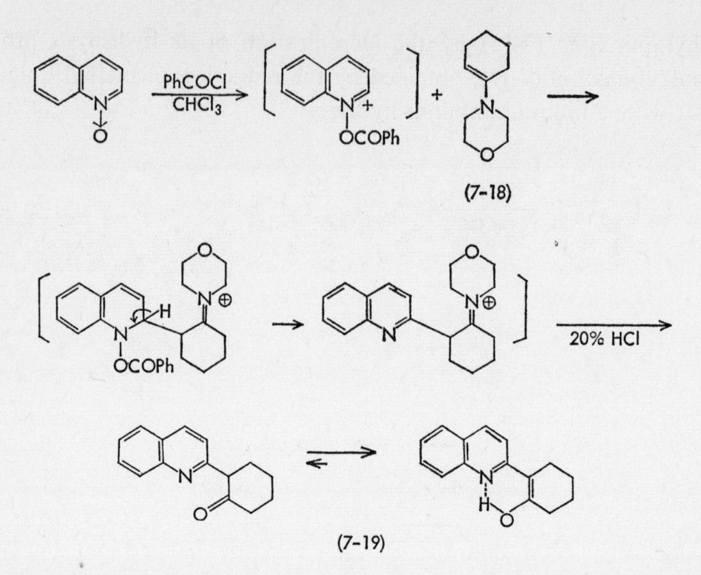

The infrared spectrum of this 2-(2-quinolyl)cyclohexanone lacks the absorption of a ketone group, indicating that the enol type is more stable. Its structure was determined by the sequence of reactions shown below.

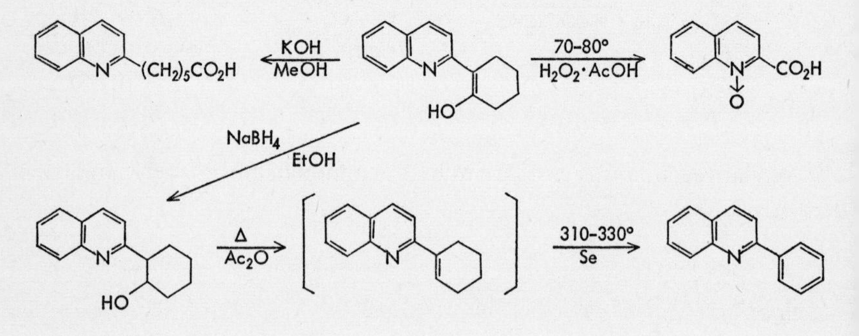

A similar reaction of enamines is known to occur by the use of tosyl chloride or with pyridine 1-oxide.

Example 1. Reissert Reaction of 4-Benzyloxyquinoline 1-Oxide[121]

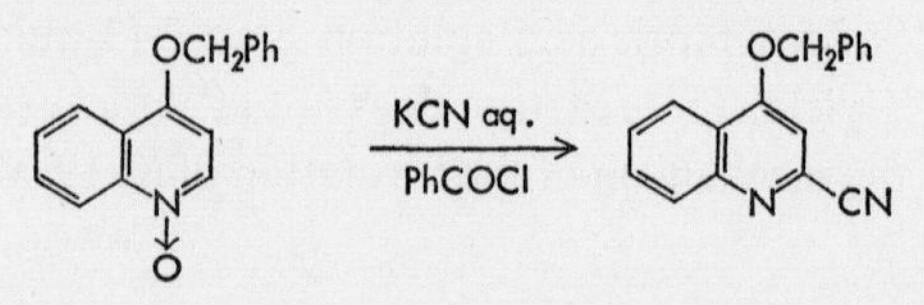

To a solution of 2.6 g of 4-benzyloxyquinoline 1-oxide and 1 g of potassium cyanide dissolved in 100 ml of water, 1.7 g of benzoyl chloride was added dropwise with shaking, and the mixture was allowed to stand for 2 h at room temperature with occasional shaking. The mixture was basified with sodium carbonate and extracted with chloroform. The chloroform extract was dried over anhydrous sodium sulfate, the solvent was evaporated, and the residue was recrystallized from methanol to 2.4 g (93%) of 4-benzyloxyquinoline-2-carbonitrile, m.p. 164°.

Example 2. Modified Reissert Reaction of 6-Methoxyquinoline 1-Oxide[113]

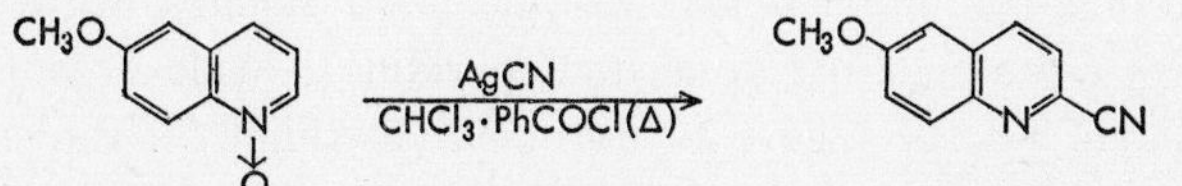

A solution of 1 g of 6-methoxyquinoline 1-oxide hydrate dissolved in 60 ml of chloroform was dried over anhydrous sodium sulfate, the solution was evaporated to one-half its volume, and to the cooled solution was added 1.6 ml of benzoyl chloride and 1.84 g of silver cyanide, with stirring and water-cooling. The reaction mixture was warmed on a water bath for 4 h and filtered from insoluble matter. The filtrate was evaporated and the crystalline residue was washed with ether, giving 0.8 g of crystals melting at 170–175°. Recrystallization from acetone gave 2-cyano-6-methoxyquinoline as needles, m.p. 176–177.5°.

Example 3. 2-(2-Quinolyl)cyclohexanone from Quinoline 1-Oxide[120]

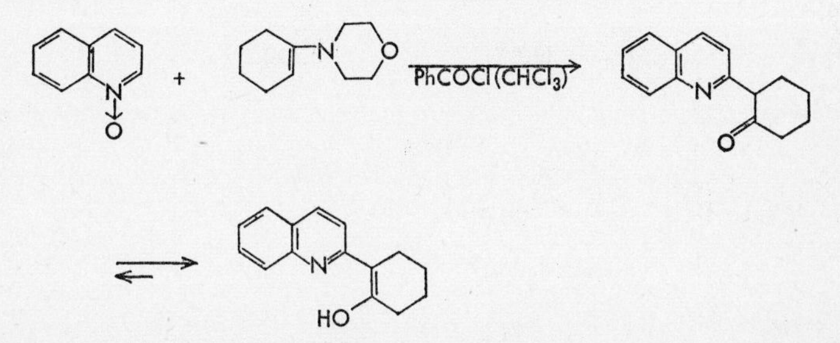

A suspension of 5.43 g (0.03 mole) of quinoline 1-oxide dihydrate in benzene was heated to effect azeotropic dehydration. The compound was dissolved in 30 ml of chloroform, 10 g (0.06 mole) of the morpholine enanime of cyclohexanone was added, and 5.06 g (0.044 mole) of benzoyl chloride was added dropwise with stirring under ice cooling by which some heat evolved and the solution became deep red. After allowing the mixture to stand overnight at room temperature, 60 ml of 20% hydrochloric acid solution was added and the mixture was allowed to stand for 2 h with occasional shaking. The solution was evaporated on a water bath under a reduced pressure, the residue was dissolved in 20% hydrochloric acid, and the solution was washed with benzene–ether mixture. This solution was made alkaline with potassium carbonate and extracted with chloroform. The solid residue obtained from the extract was recrystallized from methanol to 4.59 g of orange-red prisms, m.p. 121–122°. The mother liquor was chromatographed as a benzene solution over an alumina column and 0.36 g of the same substance was obtained. Total yield, 4.95 g (73.9%). Oxime: White scales (from ethanol), m.p. 190–192°.

References p. 334

7.4.3 Reaction in the Presence of Tosyl Chloride

Murakami and Matsumura[11] found that heating of pyridine 1-oxide and tosyl chloride at 200–205° gave 3-tosyloxypyridine. Although its yield was rather poor (*ca.* 6.2%), this reaction was notable in that the position β to the *N*-oxide group had been substituted with a tosyloxyl group. Matsumura further examined this reaction and showed that the yield of 5-tosyloxy-3-picoline from 3-picoline 1-oxide was increased slightly, with concurrent formation of 5-methyl-2-pyridone, that pyridine 1-oxide homologs with a methyl group in 2-position gave 2-chloromethylpyridine derivatives in a good yield, and that derivatives with a carboxyl in the 2-position underwent decarboxylation at the same time.

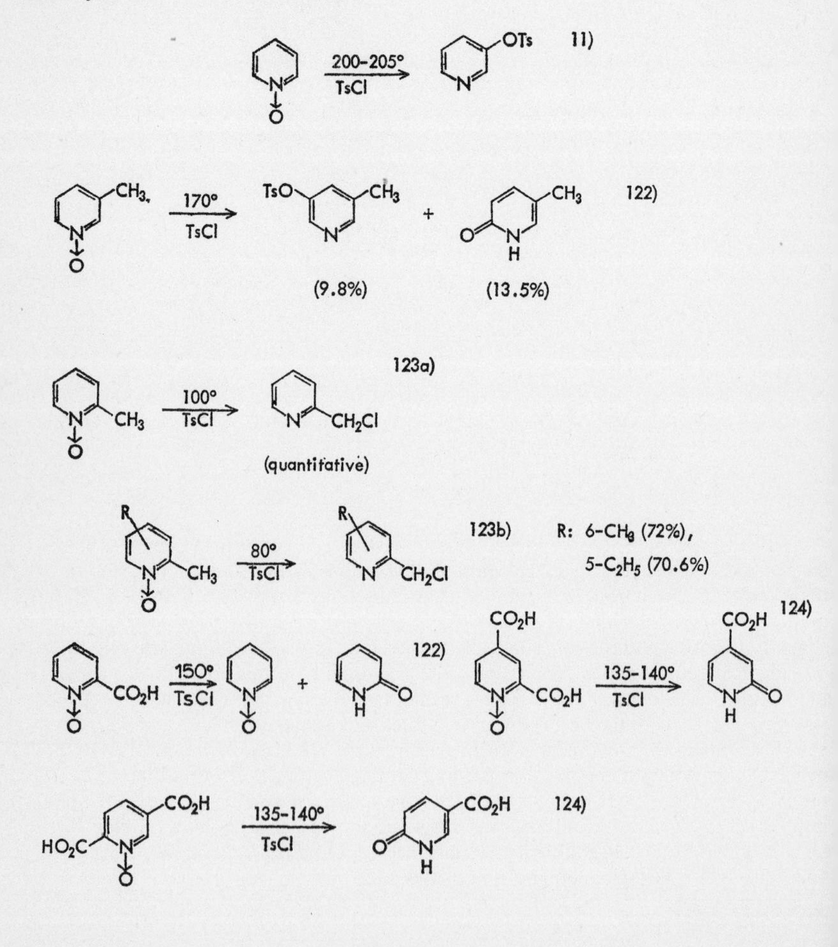

Matsumura and others extended this reaction to 2,4-dimethylthiazole *N*-oxide (7-20) and phenazine *N*-oxide (7-21)[125], and McOmie and others to 4,6-dimethylpyrimidine 1-oxide (7-22)[126].

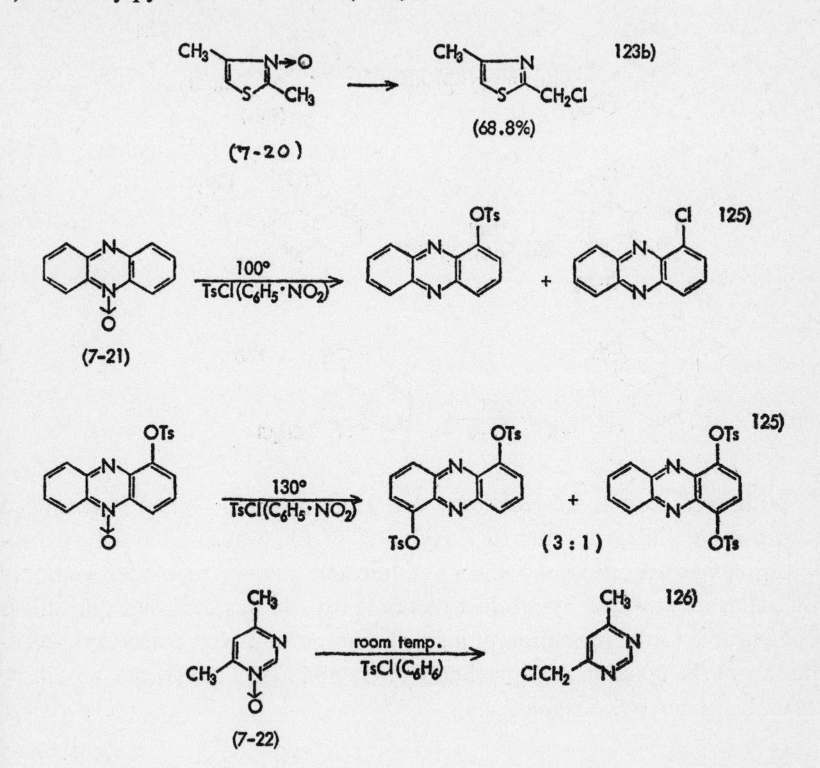

Quinoline 1-oxide does not form its 3-tosyloxy compound by this reaction[127] and Tanida[128a] obtained 3-tosyloxyquinoline by the reaction of tosyl chloride with the boron trifluoride adduct of quinoline 1-oxide at 160–180°.

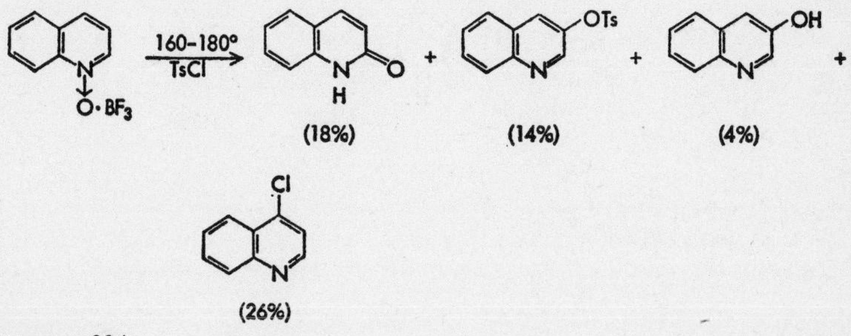

References p. 334

In quinoline compounds, the methyl group in 2- or 4-position is likewise chlorinated and Tanida obtained a good result by the addition of a catalytic quantity of boron trifluoride in this reaction[128b].

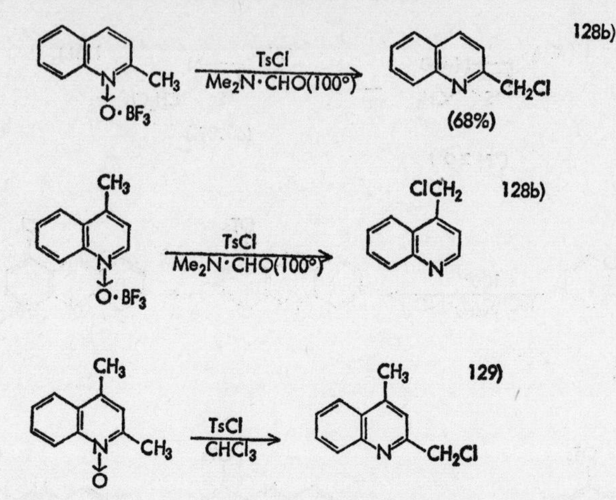

With regard to these reactions, den Hertog[130] and Ochiai[131] examined reaction conditions for the tosyloxylation of the 3-position but such substitution was unsuccessful both in pyridine and quinoline 1-oxides, and only a minute amount of a by-product was obtained. It was assumed that this is comparable to the formation of a small amount of 3- and 5-acetoxy-2-picoline from the reaction of 2-picoline 1-oxide and acetic anhydride, as will be described later (*cf. Section 7.4.4*).

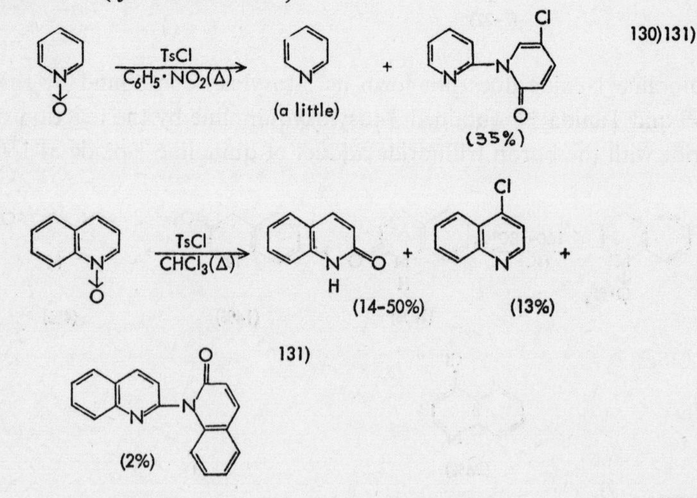

In the case of isoquinoline series *N*-oxides, 4-tosyloxy derivatives are obtained in a good yield.

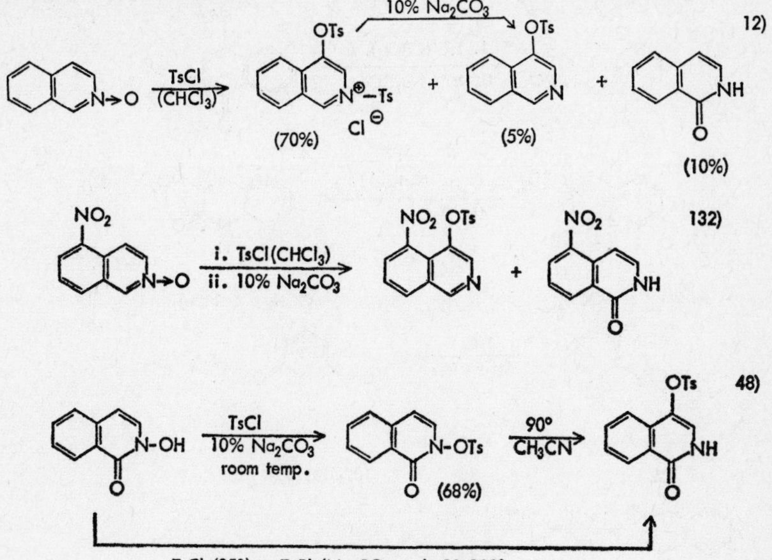

Ochiai and Yokokawa[133] found that the reaction of quinoline 1-oxide and tosyl chloride in chloroform solution and treatment of the reaction mixture with 10% sodium carbonate solution gives carbostyril in a good yield. It is assumed that this reaction, similar to the reaction with acyl chloride (*cf. Section 7.4.2*), results from increased activity of the 2-position to nucleophilic reaction by tosylation of the oxygen in the aromatic *N*-oxide group. Yoshikawa[134] carried out this reaction on 5-methylquinoline 1-oxide and obtained a stable intermediate as needle crystals.

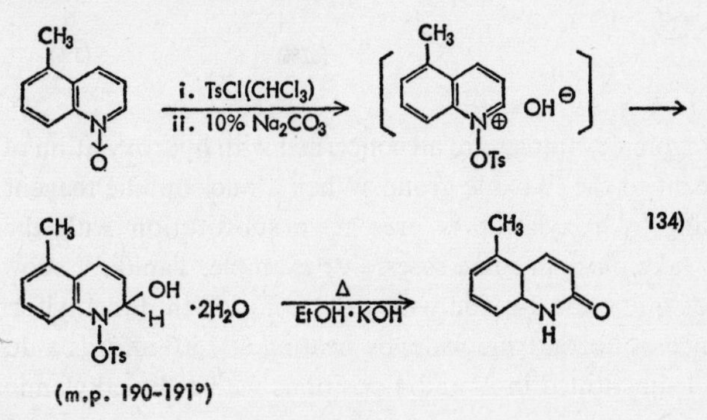

References p. 334

Similar reactions are found in the following examples.

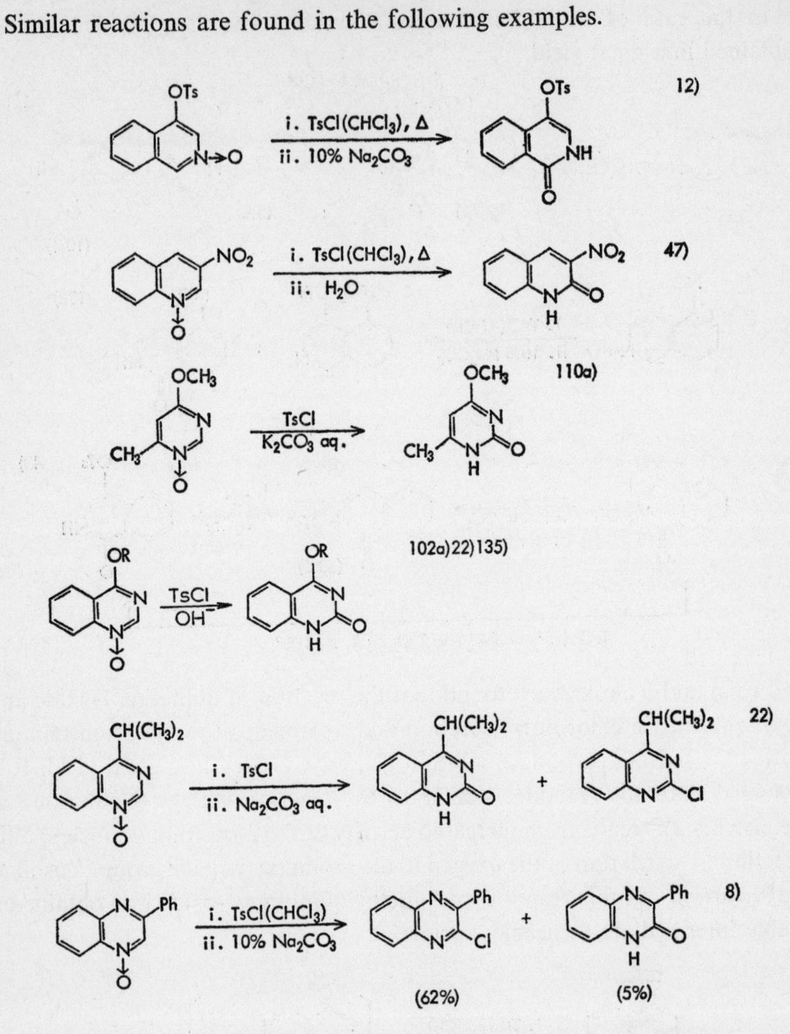

The foregoing examples are all concerned with hydroxylation of the 2-position adjacent to the *N*-oxide group. When a nucleophilic reagent other than that causing hydroxylation is present, a substitution with that group is known to take place in some cases. For example, Tanida[136] showed that the reaction of quinoline 1-oxide with tosyl chloride in dimethylformamide, in the presence of boron trifluoride, by heating at 150° affords a deoxygenated compound substituted in 2- and 4-positions with a dimethylamino group.

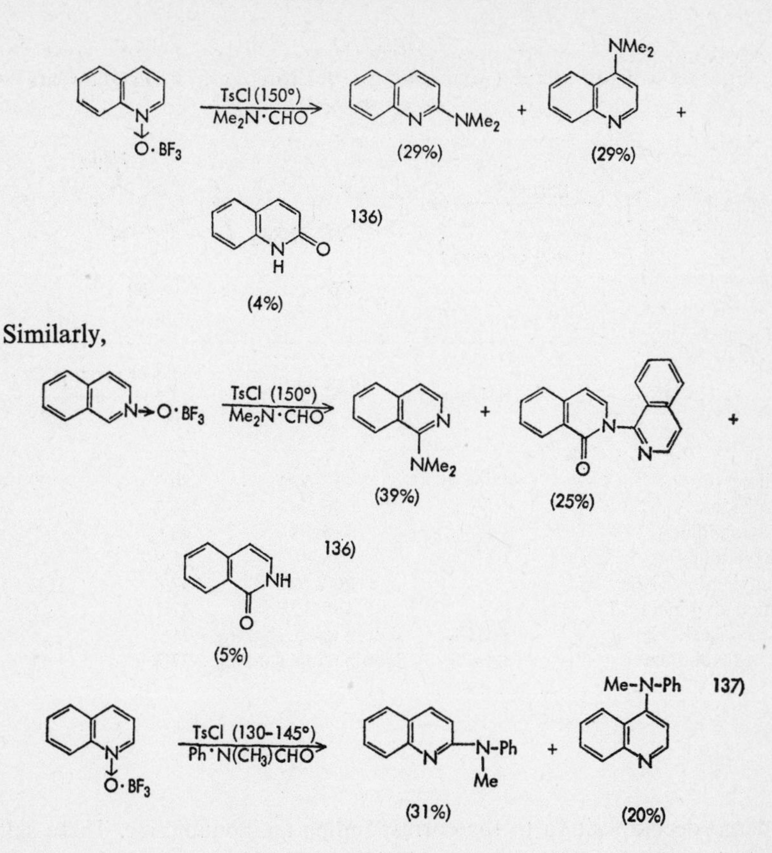

However, the same reaction of quinaldine and lepidine 1-oxides only gave the corresponding ω-(chloromethyl)quinoline[128]. Hamana and others[138] developed this reaction and showed that heating of quinoline 1-oxide with tosyl chloride in the presence of a primary or secondary amine, either directly or in chloroform or dioxan, results in the substitution of its 2- or 4-position with the amino group, with concurrent deoxygenation of the *N*-oxide group. Results of these experiments are summarized in Table 7-2.

Hamana and Funakoshi[139] found that addition of tosyl chloride to the pyridine solution of quinoline l-oxide under ice-cooling, the mixture being allowed to stand at room temperature or warmed on a water bath, resulted in the formation of a 1-(4-quinolyl)pyridinium salt (7-23), and that the same reaction using benzoyl chloride mainly gave 1-(2-quinolyl)pyridinium chloride (7-24). These salts, when heated on a water bath in absolute ethanol solution with the addition of an amine, according to the formula of Zincke *et al.*[140],

TABLE 7-2

REACTION OF QUINOLINE 1-OXIDE WITH TOSYL CHLORIDE IN THE PRESENCE OF AMINE[138]

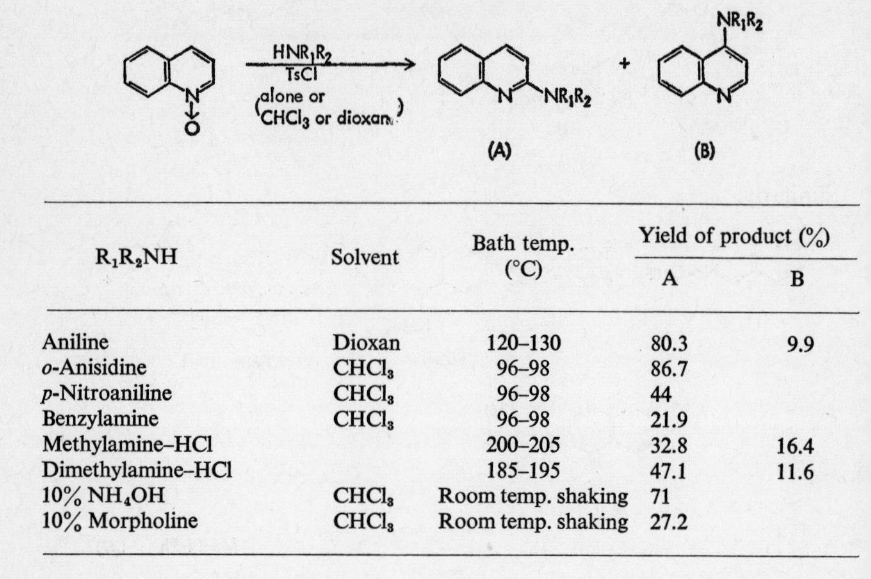

R_1R_2NH	Solvent	Bath temp. (°C)	Yield of product (%)	
			A	B
Aniline	Dioxan	120–130	80.3	9.9
o-Anisidine	$CHCl_3$	96–98	86.7	
p-Nitroaniline	$CHCl_3$	96–98	44	
Benzylamine	$CHCl_3$	96–98	21.9	
Methylamine–HCl		200–205	32.8	16.4
Dimethylamine–HCl		185–195	47.1	11.6
10% NH_4OH	$CHCl_3$	Room temp. shaking	71	
10% Morpholine	$CHCl_3$	Room temp. shaking	27.2	

undergo decomposition to the corresponding aminoquinoline. These salts, when heated with aniline hydrochloride while passing hydrogen chloride at 200°, in accordance with the method of Jerchel *et al.*[141], give the corresponding anilinoquinoline.

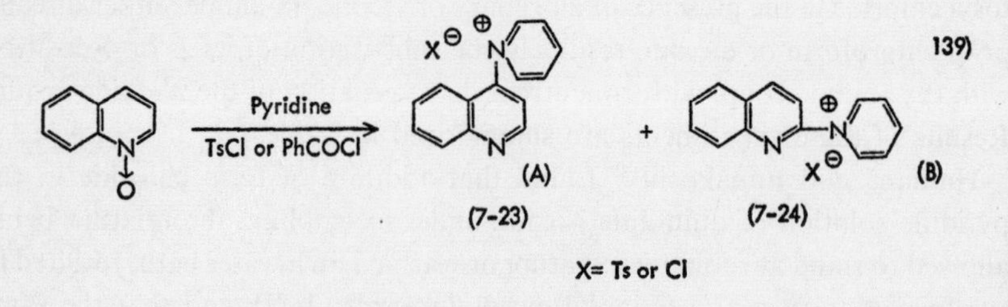

Table 7-3 gives the result of the same experiment carried out on the *N*-

TABLE 7-3

REACTION OF BENZOPYRIDINE *N*-OXIDE WITH ACYL CHLORIDE IN PYRIDINE SOLUTION[139]

N-Oxide of	Acylating agent	Position of pyridinium group (Yield[a], %)	
		As amino deriv.	As anilino deriv.
Quinoline	PhCOCl	2 (50.5)	2 (92.9) 4 (5.3)
Quinoline	TsCl	4 (85.9)	2 (4.5) 4 (95.5)
Lepidine	TsCl	2 (58.9)	2 (74.6)
Quinaldine	TsCl	4 (83.0)	4 (70.5)
Benzo[*f*]quinoline	TsCl	3 (60.0)	3 (96.5)
Isoquinoline	TsCl	1 (41.7)	1 (71.5)

[a] The yield is based on the amine oxide.

oxides of quinoline and its homologs, and isoquinoline, the product being determined by conversion of the pyridinium salts formed to the corresponding anilino or amino substitution compounds.

The same reaction was also found to take place in pyridine series *N*-oxides, but the reaction with benzoyl chloride does not give the pyridinium salt. As opposed to the case of quinaldine and lepidine *N*-oxides, the active methyl group takes part in the reaction, and ω-substituted product is also obtained in the reaction of 2- and 4-picoline 1-oxides. If the latter reaction is carried out in chloroform, only the ω-substituted compound is formed, while the reaction of 2,6-lutidine 1-oxide gives only the ring-substituted derivatives.

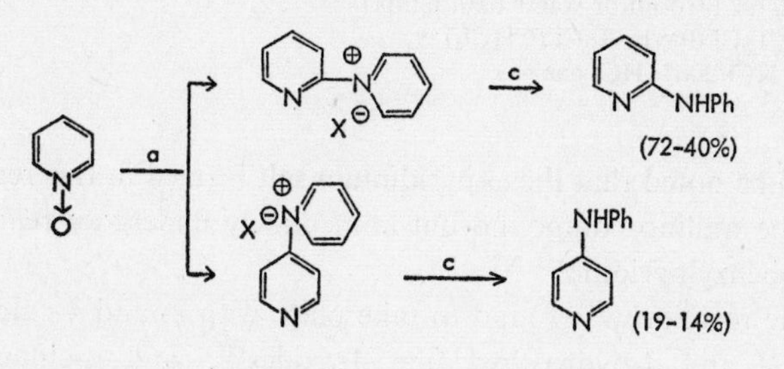

References p. 334

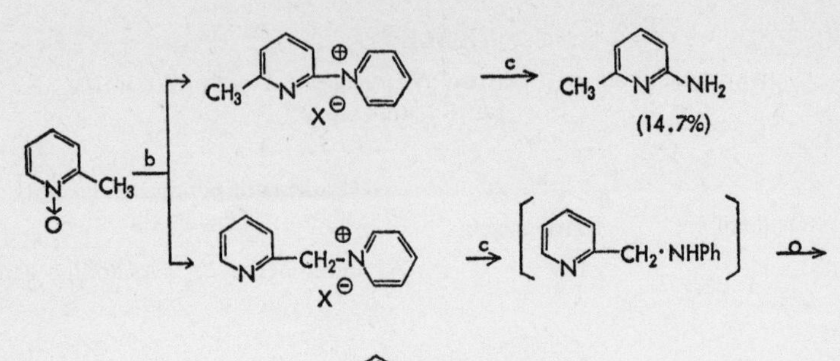

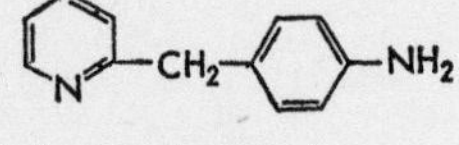

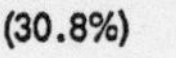

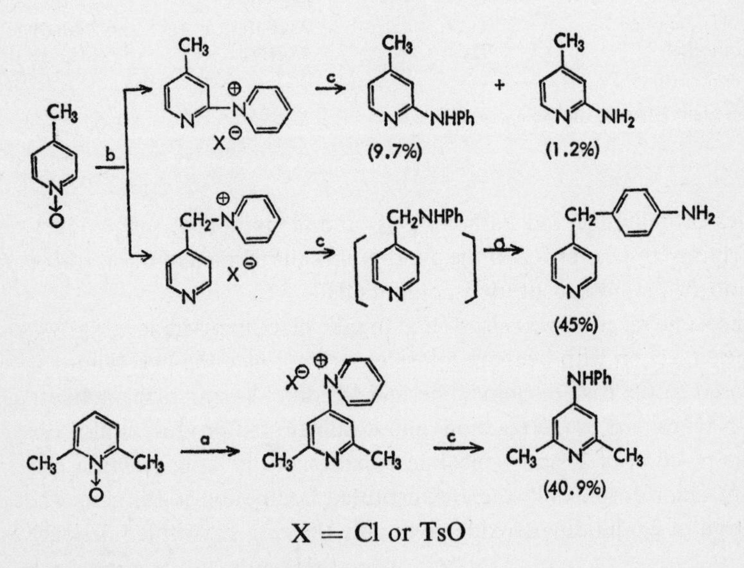

X = Cl or TsO

a = Pyridine/TsCl (room or water bath temp.)
b = Pyridine/TsCl (0°) or TsCl ($CHCl_3$)
c = $PhNH_2 \cdot HCl$/200° (HCl gas →)

It should be noted that the ω-pyridinium salt formed in this reaction does not give the anilino compound but immediately undergoes rearrangement to *p*-aminobenzylpyridine[142,143].

The same reaction was found to take place with 2- and 4-chloropyridine 1-oxides, 2- and 4-hydroxypyridine 1-oxides[145], and 4-chloroquinoline 1-oxide[146], forming the corresponding pyridinium salt. Hamana and Funakoshi[144] examined the exchange decomposition of 2- and 4-quinolylpyridinium salts so obtained, not only with various amines but also with hydroxylamine, phenol, benzenethiol, phosphorus pentachloride, and alkoxides, and their experimental results are summarized below.

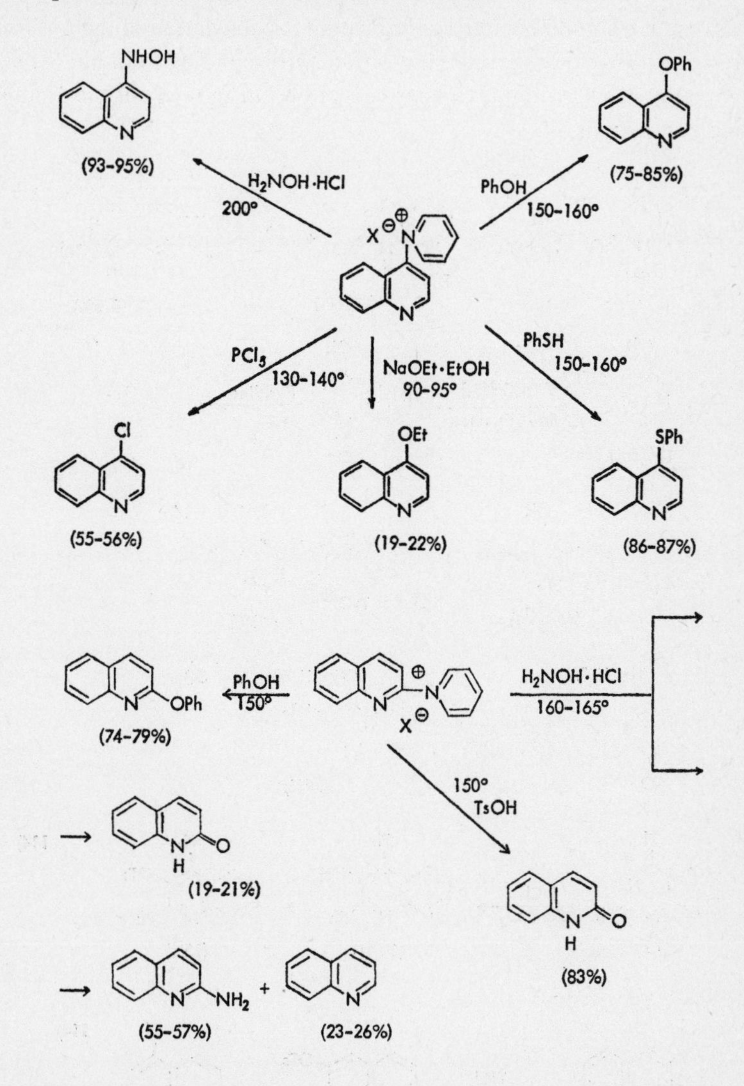

Tanida[10] found that heating of 2-aminoquinoline 1-oxide and tosyl chloride in chloroform results in formation of 2-amino-3- and -6-tosyloxyquinoline.

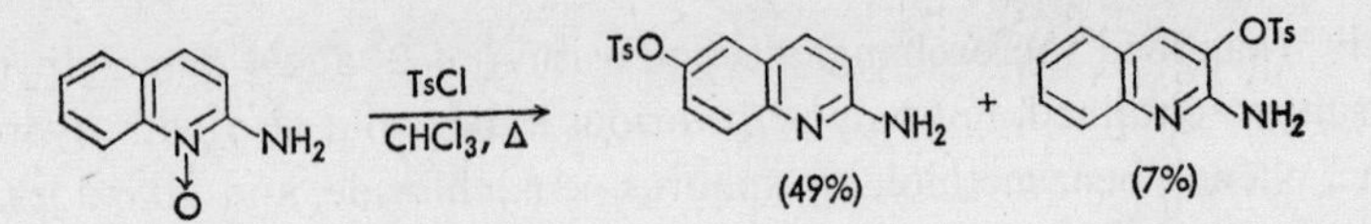

This reaction should be noted as indicating tosyloxylation of the position β to the aromatic *N*-oxide group or a position corresponding to it, but the same phenomenon has been found in later years to occur in many of the quinoline 1-oxide compounds.

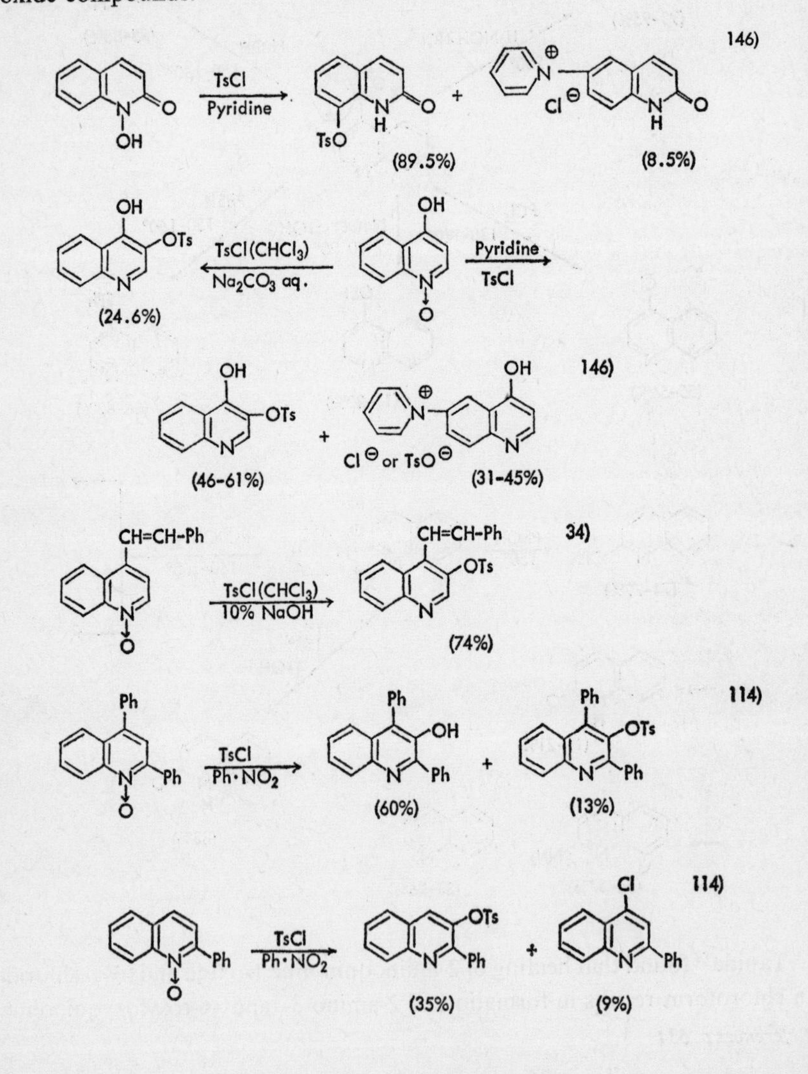

These compounds have in common that the α or γ position of the *N*-oxide group is substituted with an electron-releasing group. Their reaction mechanism will be discussed later (*cf. Section 7.4.6*).

Example 1. 3-Methyl-4-tosyloxyisoquinoline from 3-Methylisoquinoline 2-Oxide[147]

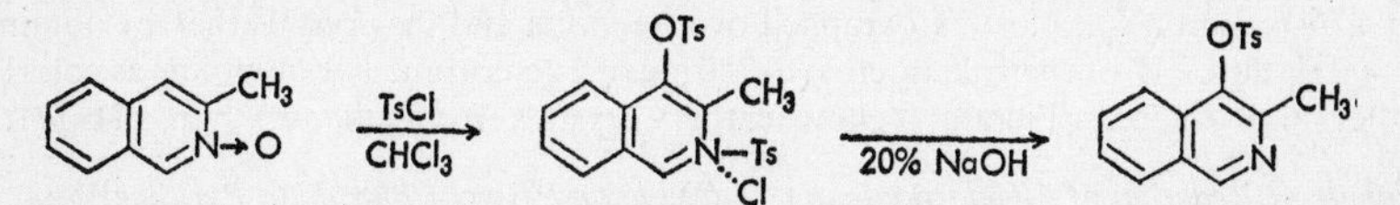

(i) A solution of 2 g of 3-methylisoquinoline 2-oxide hydrate (m.p. 138° from acetone) dissolved in 30 ml of chloroform was refluxed for 2 h with 4.0 g (2 mole equiv.) of tosyl chloride. During the first hour, hydrogen chloride gas escaped. After completion of the reaction, the mixture was evaporated *in vacuo* and the residue crystallized on addition of a small amount of ether. The crystals were washed with a large amount of ether; the tosyl chloride adduct (A) of 3-methyl-4-tosyloxyisoquinoline was obtained as colorless needles, m.p. 179°. Yield, 4.1 g (65%).
(ii) A mixture of 1.7 g of this adduct (A) in 40 ml of 20% sodium hydroxide solution and 40 ml of ether was shaken until the crystals dissolved completely. The ether layer was separated and the aqueous layer was extracted with two 20-ml portions of ether. The combined ether solution was dried over magnesium sulfate, the solvent was evaporated, and the crystalline residue (1.1 g) was recrystallized from hexane–benzene mixture to 1.03 g (quantitative) of 3-methyl-4-tosyloxyisoquinoline as colorless prisms, m.p. 99°.

Example 2. Carbostyril from Quinoline 1-Oxide[133]

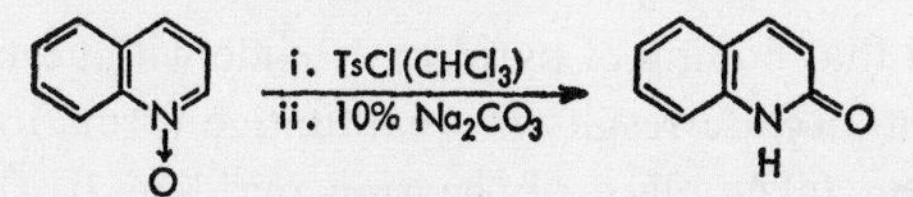

To a solution of 3 g of quinoline 1-oxide hydrate dissolved in 18 ml of chloroform, 3.6 g of tosyl chloride was added with shaking, by which the reaction occurred with evolution of heat. The mixture was refluxed on a water bath for 5 min, allowed to cool, and 10% sodium carbonate solution was added dropwise with shaking, by which the whole became turbid with evolution of carbon dioxide. When turbidity no longer occurred, the chloroform layer was separated, dried over anhydrous sodium sulfate, and the solvent was evaporated. The residue was extracted with 10% sodium hydroxide solution, the alkali solution was filtered, and the filtrate was neutralized with acetic acid. The crystals that precipitated were collected by filtration, washed with water, and recrystallized from methanol to 2 g of carbostyril, m.p. 194°.

Example 3. 2-(Chloromethyl)quinoline from Quinaldine 1-Oxide[128b]

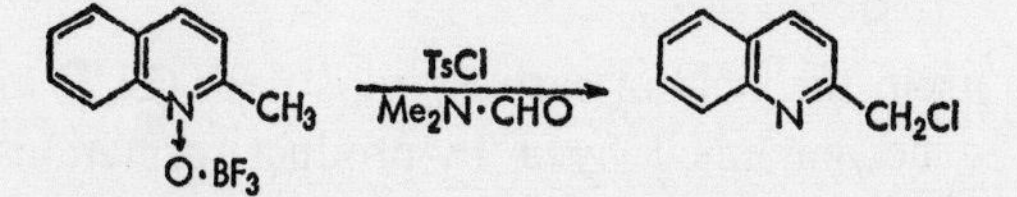

To a solution of 1 g of quinaldine 1-oxide (anhydrate) dissolved in 20 g of dimethylformamide, 1.8 g of boron trifluoride-etherate was added, followed by 1.45 g of tosyl chloride, and the whole mixture was warmed gently. Finally, the mixture was heated for 2.5 h on a boiling water bath, the solvent was evaporated under a reduced pressure, and the residue was dissolved in chloroform. The chloroform solution was shaken with 8% ammonia water to remove boron trifluoride and then with 8% hydrochloric acid solution to extract the basic matter. The acid solution was basified with sodium carbonate and extracted with ether. The ether extract was dried over potassium carbonate, the solvent was evaporated, and the residue (880 mg) was dissolved in petroleum ether–benzene mixture. This solution was chromatographed over alumina and the crystals thereby obtained were recrystallized from petroleum ether to 750 mg of 2-(chloromethyl)quinoline as colorless needles, m.p. 54–55.5°. Picrate: Yellow feathery crystals (from ethanol), m.p. 171–171.5°.

Example 4. Reaction of 4-Chloroquinoline 1-Oxide and Tosyl Chloride in Pyridine[146]

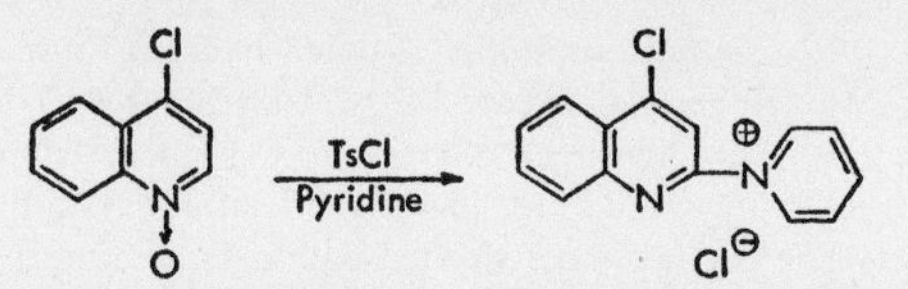

To a solution of 1.0 g of 4-chloroquinoline 1-oxide dissolved in 10 ml of pyridine, 1.2 g of tosyl chloride was added with shaking under ice-cooling. After allowing the mixture to stand overnight, the pyridine was evaporated under reduced pressure; the residue solidified on the addition of acetone. This solid was recrystallized from absolute ethanol–benzene mixture to 1.52 g (92.7%) of 1-(4-chloro-2-quinolyl)pyridinium chloride hydrate as scaly crystals, m.p. 248–251° (decomp.). Perchlorate: Leaflet crystals, m.p. 191–192°.

7.4.4 Reaction in the Presence of Acid Anhydride

Katada[5] found that boiling of pyridine 1-oxide with acetic anhydride and anhydrous sodium acetate results in quantitative formation of 2-pyridone. The same reaction takes place on heating pyridine 1-oxide with benzoic anhydride at 140–150°,[5] and it was also found to occur with *N*-oxides of the quinoline series.

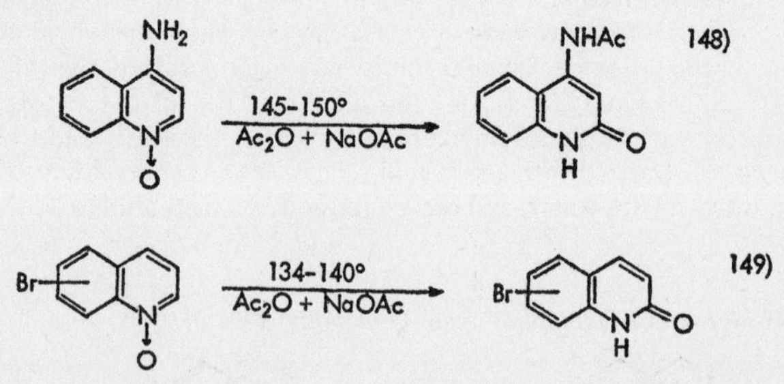

In quinoline itself, the yield of carbostyril is poor (33%) and the reaction is known to be accompanied by a by-product formation[149]. Ochiai and others[150] utilized this reaction on cinchona bases and converted its quinoline portion to a carbostyril type.

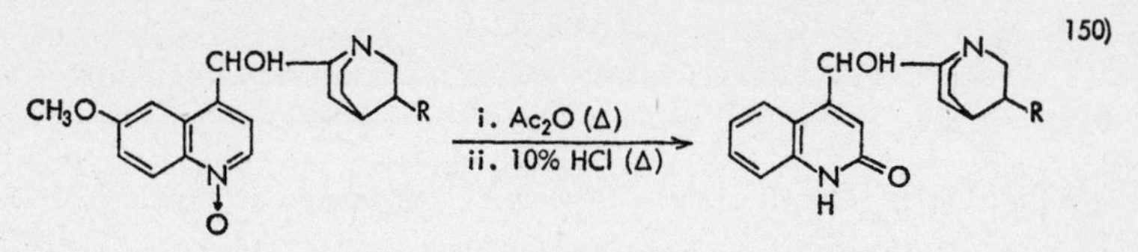

The same phenomena have been observed to date in many of the *N*-oxide compounds of the pyridine and quinoline series, as well as those of the diazine series. This is summarized in Table 7-4.

It should be noted from this Table 7-4 that, when there is a carboxyl group in 2-position of the pyridine system, decarboxylation occurs first. In such a case, preliminary esterification will prevent dealkoxycarbonylation and 6-alkoxycarbonyl-2-pyridone is formed. When there is an anionizable group such as chlorine, ethoxyl, and phenoxyl in the 2-position of pyridines, these groups are substituted by a hydroxyl and 1-hydroxy-2-pyridone is formed. When the 2-position is occupied, especially in the quinoline system, substitution with a hydroxyl occurs in the 4-position.

The fact that nicotinic acid 1-oxide forms a 2-ketone compound with retention of the *N*-oxide group is a specific phenomenon, and Bain and Saxton[156] have put forward the following explanation for this reaction.

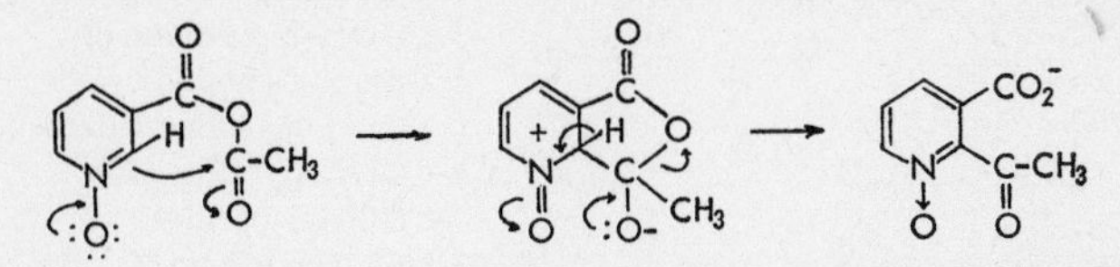

The cleavage of the pyrimidine ring by the reaction of adenine 1-oxide (7-25) and acetic anhydride at room temperature has been explained in the following manner[167b].

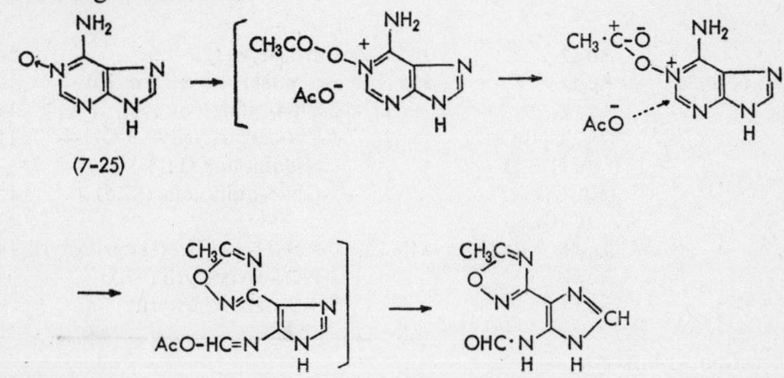

References p. 334

TABLE 7-4

REACTION OF AROMATIC AMINE OXIDES WITH ACETIC ANHYDRIDE

Substituent	Reagent	React. temp. (°C)	Reaction product (yield, %)	Ref.
Pyridine Series				
Nil	Ac_2O	Δ	2-Pyridone (quantitative)	5
2-X	Ac_2O	Δ	1-OH-2-pyridone X = Cl (71), OEt (61), OPh (84)	33
3-X	Ac_2O	Δ	3-X-2-acetoxypyridine X = Cl (61), Br (50), F (65)	152
3-CH_3	Ac_2O	Δ	3-CH_3-2-pyridone (35-40), 5-CH_3-2-pyridone (35–40), 1-(5-CH_3-2-pyridyl)-3-CH_3-2-pyridone (4)	14, 156
2-$COOCH_3$	Ac_2O	Δ	6-$COOCH_3$-2-pyridone (34)	153
2-COOH	Ac_2O	50	Pyridine 1-oxide, 2-pyridone + 1-(2-pyridyl)-2-pyridone	33, 153, 154
4-$COOCH_3$	Ac_2O	Δ	4-$COOCH_3$-2-pyridone (56)	153
3-$COOCH_3$	Ac_2O	Δ	3-$COOCH_3$-2-pyridone (28) 5-$COOCH_3$-2-pyridone (16)	153 155
3-COOH	Ac_2O	Δ	2-acetylnicotinic acid 1-oxide (30), 6-OH-3-COOH (3), 3-COOH-2-pyridone (18)	156
3-COOH	$(EtCO)_2O$	Δ	3-COOH-2-pyridone (18), 5-COOH-2-pyridone (2), 5,7-dioxo-6-methyl-cyclopenteno[*b*]pyridine 1-oxide (8)	156
4-COOH	Ac_2O	Δ	4-COOH-pyridine (22) 4-COOH-2-pyridone (7)	156
3-NO_2	Ac_2O	Δ	3-NO_2-2-pyridone	55
4-OR	Ac_2O	Δ	4-OR-2-pyridone R = Et (48), Me (50)	157 157
2-COOH-5-$COOCH_3$	Ac_2O	45	3-$COOCH_3$-pyridine 1-oxide	158
Quinoline Series				
Nil	Ac_2O	135–140	Carbostyril (20–35)	149
Br (3,4,5,6, or 7)	Ac_2O	135–140	Bromocarbostyril (*ca.* 70)	149
2-C_6H_5	Ac_2O	Δ	2-Ph-4-quinolone (50)	114
4-CN	$(F_3CCO)_2O$[a] (CH_2Cl_2)	Δ	4-CN-carbostyril + 3-OH-4-CN-quinoline (1:1)	159
2-CN	$(F_3CCO)_2O$ (CH_2Cl_2)	Δ	2-CN-4-quinolone (82.5)	159
4-NH_2	Ac_2O or Bz_2O	145–150	4-AcNH- or -BzNH-carbostyril	148
3-NO_2	Ac_2O		3-NO_2-carbostyril (94.5)	47
7,8-Benzo	Ac_2O		7,8-Benzocarbostyril	82

[a] No reaction with Ac_2O.

TABLE 7-4 (*continued*)

Substituent	Reagent	React. temp. (°C)	Reaction product (yield, %)	Ref.
Quinine ar. N→O	(*i*) Ac_2O (*ii*) 10% HCl	120–130	2′-OH-quinine (Over 50%)	150
Dihydroquinine ar. N→O	(*i*) Ac_2O (*ii*) 10% HCl	105	2′-OH-dihydroquinine (80)	150
4-Et_2N–CH_2CH_2O	Ac_2O	130–140	4-OH-quinoline 1-oxide	151
Isoquinoline Series				
Nil	(*i*) Ac_2O (*ii*) NaOH aq.	Δ	Isocarbostyril (50–64) + 4-OH-isoquinoline (8.9)	12, 160
3-CH_3	(*i*) Ac_2O (*ii*) 5% HCl	Δ	3-CH_3-isocarbostyril (40) + 3-CH_3-4-OH-isoquinoline (8.5)	160
3-Cl	Ac_2O	Δ	3-Cl-4-OH-isoquinoline (61) + 3-Cl-isocarbostyril (1)	48
Pyridazine Series				
3-OH 1-oxide	Ac_2O	Δ	3-OH-6-pyridazinone	161
3-OCH_3-6-CH_3 1-oxide		Δ	3-OCH_3-6-CH_2OAc-pyridazine (42–77)	94, 163
3-Ph-6-CH_3 1-oxide	Ac_2O	Δ	3-Ph-6-CH_2OAc-pyridazine(50)	162
3-OEt-6-CH_3 1-oxide	Ac_2O	100	3-OEt-6-CH_2OAc-pyridazine	162
3,6-$(OMe)_2$ 1-oxide	(*i*) Ac_2O (*ii*) Na_2CO_3 soln.	150–160	1-OAc-3-OMe-6-pyridazinone (68), 1,3-$(OMe)_2$-6-pyridazinone (23)	164 164
3,6-$(OMe)_2$-4-Me 1-oxide	(*i*) Ac_2O (*ii*) Na_2CO_3 soln.	95–100 room	3,6-$(OCH_3)_2$-4-R-pyridazine R = CH_2OAc (32), CH_2OH (23), CH_3 (2–3), 1-OH-3-OMe-4-CH_3-6-pyridazinone (4)	164
3-OCH_3-4-CH_3-6-Cl 1-oxide	(*i*) Ac_2O (*ii*) Na_2CO_3 soln.	95–100 room	3-OMe-4-R-6-Cl-pyridazine R = CH_2OH (8), CH_3 (1), CH_2Cl (11), N-R-3-OMe-4-CH_3-6-pyridazinone R = H (9.4), OH (8)	164
Pyrimidine and Quinazoline Series				
Pyrimidine 1-oxide	Ac_2O	132	6-AcO-pyrimidine	165
4-OR-quinazoline 1-oxide	Ac_2O	Δ	4-OR-2-quinazolone R = CH_3 (64), C_2H_5 (60)	102*a*
4-i-C_3H_7-quinazoline 1-oxide	Ac_2O	100	4-Isopropylquinazoline + 4-isopropyl-2-quinazolone	22

References p. 334

Continued overleaf

Table 7-4 (*continued*)

Substituent	Reagent	React. temp. (°C)	Reaction product (yield, %)	Ref.
Pyrazine and Quinoxaline Series				
2-$AcOCH_2$-5-CH_3-pyrazine 1,4-dioxide	Ac_2O	Δ	2,5-$(AcOCH_2)_2$-pyrazine 1-oxide + 2,5-$(AcOCH_2)_2$-pyrazine	166
3-Ph-quinoxaline 1-oxide	Ac_2O	Δ	3-Ph-2-quinoxalone	8
Benzimidazole Series				
1-CH_3-benzimidazole 3-oxide	Ac_2O	90–100	1-CH_3-3-Ac-2-benzimidazolone	111
Miscellaneous				
8-OH-purine 1-oxide	Ac_2O	Δ	2,8-$(OH)_2$-purine (*ca.* 15), 6,8-$(OH)_2$-purine (*ca.* 15)	167*a*
CH_3(H), N-CO, CH_2, Cl, C=N→O, Ph	Ac_2O	Δ	CH_3(H), N-CO, CH-OAc, Cl, C=N, Ph	168*a,b*
6-NH_2-purine 1-oxide	Ac_2O	room	CH_3, C=N, O, N=C, N, CH, (Ac)OHCNH, NH	167*b*

In 1953, Kobayashi and Furukawa[13] reported that refluxing of 2-picoline 1-oxide with acetic anhydride at 100–140°, followed by hydrolysis with 10% hydrochloric acid, afforded 2-(hydroxymethyl)pyridine in 45–50% yield, with a small amount of 6-methyl-2-pyridone as by-product. This latter product, however, was later found by Okuda[16] to be a mixture a of 3- and 5-hydroxy-2-picolines in approximately equal amount.

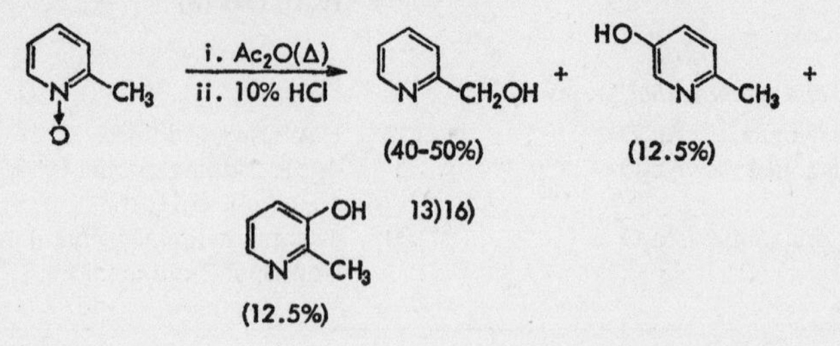

Subsequently, Boekelheide and Linn[14], and Bullit and Maynard[15], independently but at the same time, reported the same fact, which is represented below.

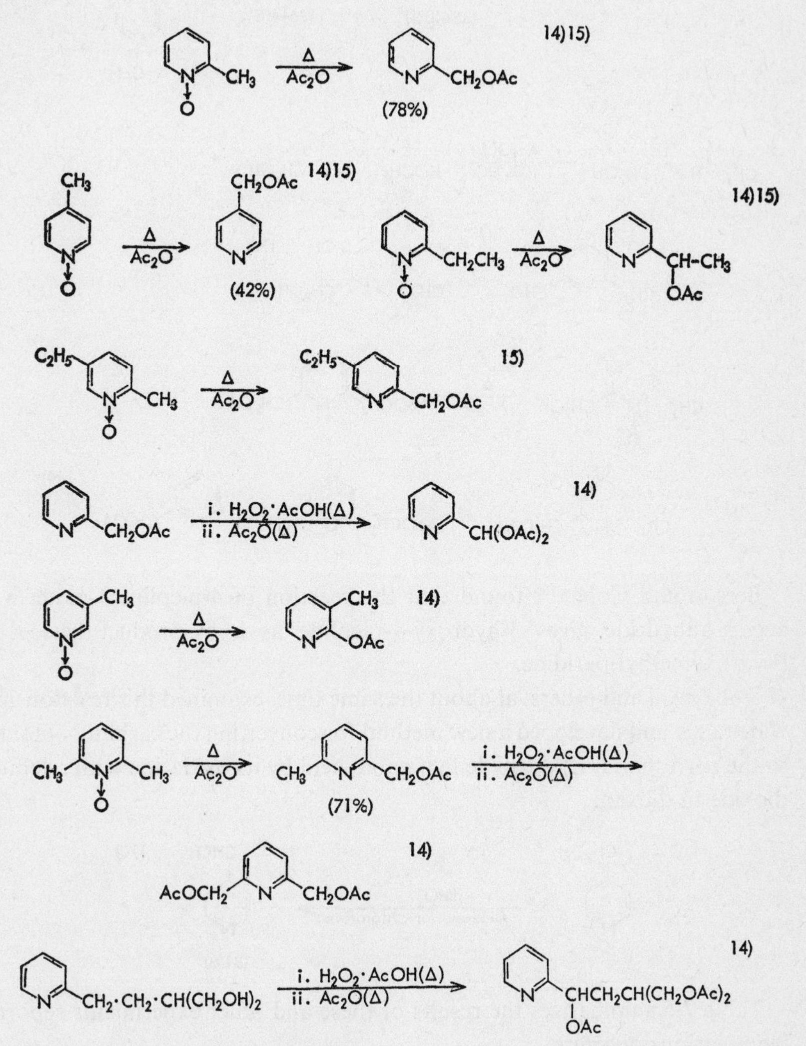

Among these reactions, that of 3-picoline l-oxide was further investigated by Bain and Saxton[156], and that of 6-(hydroxymethyl)-2-methylpyridine 1-oxide by Kato and others[176], and by Kaneko and others[184], respectively as follows.

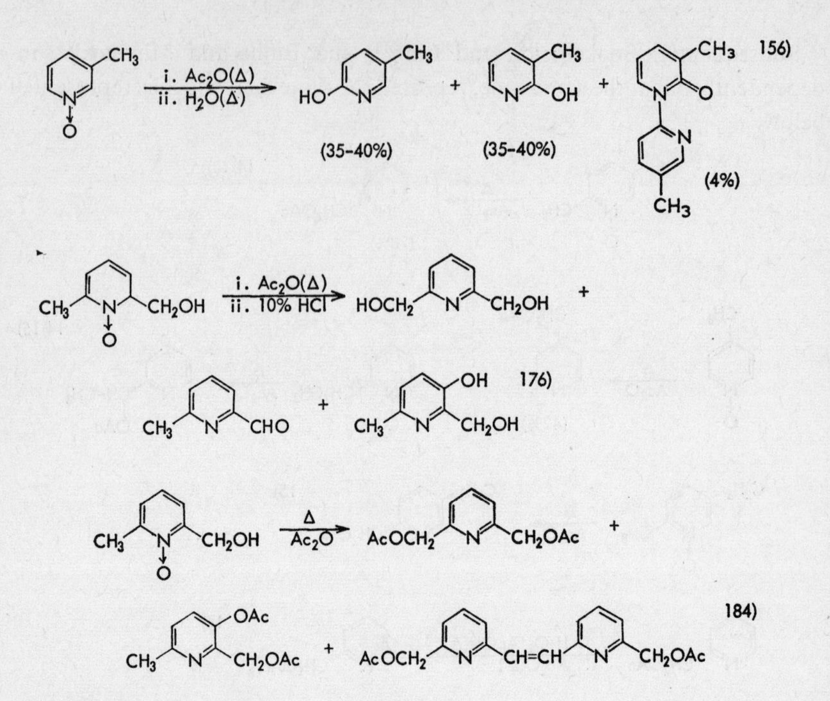

Berson and Cohen[169] found that the reaction of 4-picoline 1-oxide with acetic anhydride gives 3-hydroxy-4-picoline as a by-product besides 4-(hydroxymethyl)pyridine.

Kobayashi and others, at about the same time, examined this reaction in a wide range and developed a new method for converting the carbinol obtained to the corresponding aldehyde in a good yield by its oxidation with selenium dioxide in dioxan.

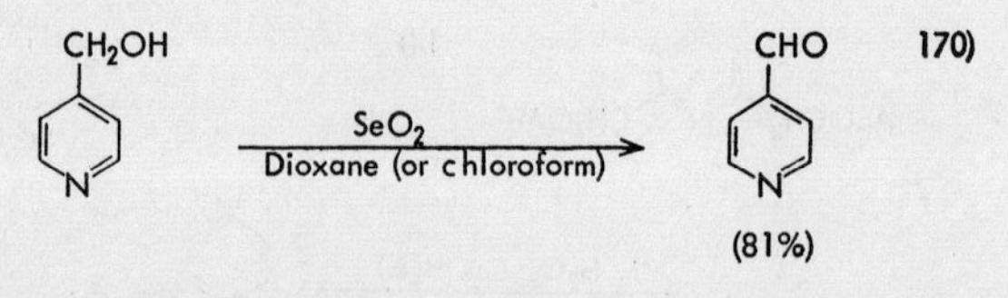

Table 7-5 summarizes the results of these and other experiments reported from various quarters.

In the afore-mentioned reaction, Kato used ketene in place of acetic anhydride, and carried out the reaction in acetone, in the presence of a catalytic quantity of conc. sulfuric acid. The results he obtained are listed in Table 7-6.

TABLE 7-5

REACTION OF *N*-OXIDES

Substituent	Reaction conditions	Reaction product (yield, %)	Ref.
Pyridine Series			
2,6-$(CH_3)_2$	*(i)* Ac_2O (Δ) *(ii)* 10% HCl (Δ)	2-CH_2OH-6-CH_3-pyridine (45) + 2,6-$(CH_3)_2$-3-OH-pyridine (5)	171
2,6-$(CH_3)_2$	*(i)* Ac_2O (Δ) *(ii)* 10% HCl (Δ)	2-CH_2OH-6-CH_3-pyridine + 2,6-$(CH_3)_2$-3-OH-pyridine + (6-CH_3-2-pyridyl)$_2$$(CH_2)_2$	176
2-CH_2OH-6-CH_3	*(i)* Ac_2O (Δ) *(ii)* 10% HCl (Δ)	2,6-$(CH_2OH)_2$-pyridine + 2-CHO-6-CH_3-pyridine + 2-CH_2OH-3-OH-6-CH_3-pyridine	176
2,6-$(CH_3)_2$-3-OH	*(i)* Ac_2O (Δ) *(ii)* 10% HCl (Δ)	2-CH_2OH-3-OH-6-CH_3-pyridine + 2-CH_3-3-OH-6-CH_2OH-pyridine	176
2-CH_2OH-6-CH_3	Ac_2O (Δ)	2,6-$(CH_2OAc)_2$-pyridine (79.5), 2-CH_2OAc-3-OAc-6-CH_3-pyridine (18.5), 2-CH_2OAc-5-OAc-6-CH_3-pyridine (2.0), 6-CH_2OAc-pyridine (2)-CH=CH-pyridine (2)-6-CH_2OAc (*ca.* 9)	184
2,4-$(CH_3)_2$	*(i)* Ac_2O (Δ) *(ii)* 10% HCl (Δ)	2-CH_2OH-4-CH_3-pyridine (30) + 2-CH_3-4-CH_2OH-pyridine (6) + 2,4-$(CH_3)_2$-3-OH-pyridine (2)	172*a, b*
2,3-$(CH_3)_2$	Ac_2O (Δ)	2-CH_2OAc-3-CH_3-pyridine (70)	173
4-C_2H_5	*(i)* Ac_2O (Δ) *(ii)* 10% HCl	4-CH_3CHOH-pyridine + 3-OH-4-C_2H_5-pyridine	155
2-CH=CH-Ph	*(i)* Ac_2O (Δ) *(ii)* 10% HCl	OH OH CH-CH (39) + 2-CH=CHPh-3-OH-pyridine (8) + 2-PhCH=CH-5-OH-pyridine (5) + 2-phCH=CH-pyridine (6.5)	174
2-CH_3-4-OCH_3	*(i)* Ac_2O (water bath) *(ii)* 10% HCl (Δ)	2-CH_2OH-4-OCH_3-pyridine (55.4) + 2-CH_3-4-OCH_3-5-OH-pyridine (4.2)	175
2-CH_3-4-Cl	*(i)* Ac_2O(water bath) *(ii)* 10% HCl (Δ)	2-CH_2OH-4-Cl-pyridine (50) + 2-CH_3-4-Cl-5-OH-pyridine(trace)	175
2-CH_3-6-Cl	*(i)* Ac_2O (Δ) *(ii)* 10% HCl	2-CH_2OH-6-Cl-pyridine (48.7) + 2-CH_3-3-OH-6-Cl-pyridine(trace) + 6-CH_3-2-pyridone (9)	33
2-CH_3-4-NH_2	Ac_2O (120–130°)	2-$AcOCH_2$-4-AcNH-pyridine (58.7)	177
2-CH_3-6-COOH	Ac_2O (Δ)	2-$AcOCH_2$-pyridine	178
2-CH_3-3-CH_2CO_2Et	Ac_2O (Δ)	2-$AcOCH_2$-3-CH_2CO_2Et	179*a*

References p. 334

Continued overlea

Table 7-5 (*continued*)

Substituent	Reaction conditions	Reaction product (yield, %)	Ref.
2-CH_3-3-$CONHCH_2Ph$	Ac_2O (Δ)	2-$AcOCH_2$-3-$CONHCH_2Ph$	179*a*
2-CH_3-3-COOEt	Ac_2O (Δ)	2-$AcOCH_2$-3-CO_2Et-pyridine + 2-CH_3-3-CO_2Et-5-OAc-pyridine	179*a, b*
2,6-$(CH_3)_2$-4-CH_2Ph	(*i*) Ac_2O (Δ) (*ii*) 10% HCl	2-CH_2OH-4-CH_2Ph-6-CH_3-pyridine	180
2,6-$(CH_3)_2$-4-CHOHPh	(*i*) Ac_2O (Δ) (*ii*) 10% HCl	2-CH_2OH-4-CHOHPh-6-CH_3-pyridine (20) + 2,6-$(CH_3)_2$-4-COPh-pyridine (65)	180, 154
2,6-$(CH_3)_2$-4-COPh	(*i*) Ac_2O (Δ) (*ii*) 10% HCl	2-CH_2OH-4-COPh-6-CH_3-pyridine + 2,6-$(CH_3)_2$-3-OH-4-COPh-pyridine	180
2,6-$(CH_3)_2$-4-CN	(*i*) Ac_2O (Δ) (*ii*) 10% HCl	2-CH_2OH-6-CH_3-4-CN-pyridine (74.5)	177
2,6-$(CH_3)_2$-4-NH_2	(*i*) Ac_2O(120–130°) (*ii*) EtOH–KOH (Δ)	2,6-$(CH_3)_2$-4-AcNH-pyridine(50)	177
2-CH_3-4-NH_2	(*i*) Ac_2O(120–130°) (*ii*) EtOH–KOH (Δ)	2-CH_2OH-4-AcNH-pyridine (58.7)	177
2,6-$(CH_3)_2$-4-Cl	(*i*) Ac_2O (Δ) (*ii*) H_2O (Δ)	2-CH_2OH-4-Cl-6-CH_3-pyridine (48) + 2,6-$(CH_3)_2$-4-Cl-3-OH-pyridine (6)	175
2,6-$(CH_3)_2$-4-NO_2	(*i*) Ac_2O (Δ) (*ii*) H_2O (Δ)	2-CH_2OH-4-NO_2-6-CH_3-pyridine (1.6) + 2-CH_2OH-4-OH-6-CH_3-pyridine (12)	175
2,6-$(CH_3)_2$-4-OCH_3	(*i*) Ac_2O (Δ) (*ii*) H_2O (Δ)	2-CH_2OH-4-OCH_3-6-CH_3-pyridine (main) + 2,6-$(CH_3)_2$-3-OH-4-OCH_3-pyridine	175
2-CH_3-3-OAc-4,5-$(CH_2OAc)_2$	Ac_2O (Δ)	2,4,5-$(CH_2OAc)_3$-3-OAc-pyridine	183
4-CH_3	Ac_2O (Δ)	4-CH_2OAc-pyridine (64.2) + 3-OAc-4-CH_3-pyridine (35.8) + 1,2-bis(4′-pyridyl)ethylene (*ca.* 13)	184
Quinoline Series			
2-CH_3	(*i*) Ac_2O(100°) (*ii*) 10% HCl (Δ)	2-$HOCH_2$-quinoline (40)	182
4-CH_3	(*i*) Ac_2O(100°) (*ii*) 10% HCl (Δ)	4-CH_3-carbostyril (6) + 4-$HOCH_2$-quinoline (8) + 3-OH-4-CH_3-quinoline (22)	182
4-CH=CH–Ph	Ac_2O($CHCl_3$) (Δ)	4-$(AcOCH)_2$Ph-quinoline (42.3) + 3-OH-4-CH=CHPh-quinoline (15.5)	34
2,4-$(CH_3)_2$	(*i*) Ac_2O(water bath) (*ii*) 10% HCl	2-CH_2OH-4-CH_3-quinoline (70) + 2,4-$(CH_3)_2$-3-OH-quinoline (15)	172

Table 7-5 (continued)

Substituent	Reaction conditions	Reaction product (yield, %)	Ref.
5,6,7,8-4H	(*i*) Ac_2O (Δ) (*ii*) OH^-	8-OH-5,6,7,8-4H-quinoline	181
Phenanthridine 5→O	Ac_2O (Δ)	phenanthridone	186
Pyrazine Series			
2-CH_3-1→O	Ac_2O (Δ)	2-$AcOCH_2$-pyrazine	185, 166
2-CH_3-1,4-di→O	Ac_2O (Δ)	1-$AcOCH_2$-pyrazine 4→O	187
3-CH_3-1→O	Ac_2O (Δ)	3-CH_3-6-OH-pyrazine	187, 166
2,5-$(CH_3)_2$-1→O	Ac_2O (Δ)	2-$AcOCH_2$-5-CH_3-pyrazine	187, 166
2,5-$(CH_3)_2$-1,4-di→O	Ac_2O (Δ)	2-CH_3-5-CH_2OAc-pyrazine 1→O + 2,5-$(CH_2OAc)_2$-pyrazine + 2,5-$(CH_3)_2$-pyrazine 1→O	187,166
2-CH_3-5-CH_2OAc-1,4-di→O	Ac_2O (Δ)	2,5-$(CH_2OAc)_2$-pyrazine 4→O + 2,5-$(CH_2OAc)_2$-pyrazine	166
2,6-$(CH_3)_2$-1→O	Ac_2O (Δ)	2-CH_2OAc-6-CH_3-pyrazine	187
2,6-$(CH_3)_2$-1,4-di→O	Ac_2O (Δ)	2-CH_2OAc-6-CH_3-pyrazine + 2,6-$(CH_3)_2$-pyrazine 1→O	187
Pyrimidine Series			
4,6-$(CH_3)_2$	Ac_2O (Δ)	6-CH_2OAc-4-CH_3-pyrimidine (probable)	126
Pyridazine Series			
2,4-$(CH_3)_2$	Ac_2O (Δ)	2-CH_2OAc-4-CH_3-pyridazine (probable)	94
Miscelleneous			
6-CH_3-purine 1→O	Ac_2O (Δ)	6-CH_2OAc-purine	167*a*

All the foregoing reactions may be considered to progress through acylation of the aromatic *N*-oxide group, and the acetate ion reacts with the position on the ring or side chain activated to S_N reaction. Hamana and others developed a series of substitution reactions with anionic groups by their simultaneous presence in the same reaction as shown below.

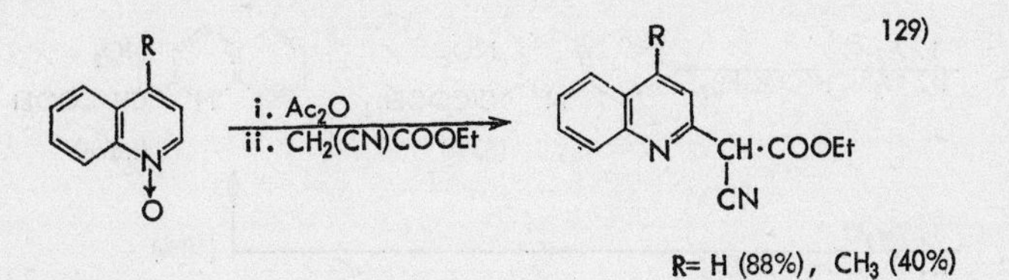

References p. 334

TABLE 7-6

REACTION OF METHYL DERIVATIVES OF PYRIDINE AND QUINOLINE SERIES *N*-OXIDES WITH KETENE[189,190]

N-Oxide of	Product (yield, %)			
	A	B	C	Recovery
4-Picoline	5	0.3	4	
3-Picoline			3	10
2-Picoline	50–60	trace		
2,6-Lutidine	33	17	4	
4-Nitro-2,6-lutidine	8	2[a]	3	41
4-Nitro-2-picoline				91.3
Quinaldine	16			
Lepidine		4	3	
4-Nitropyridine				91

[a] 4-Acetoxy-6-acetoxymethyl-2-picoline.

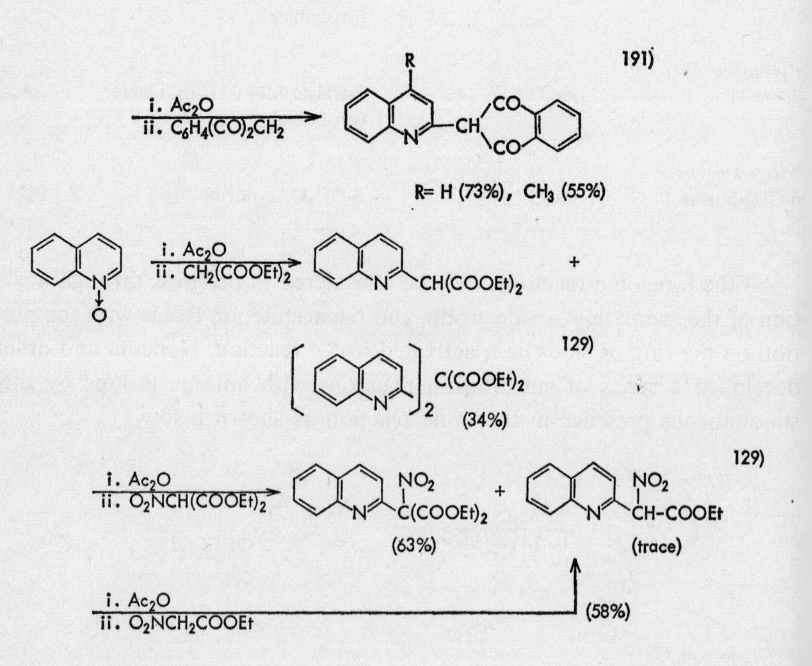

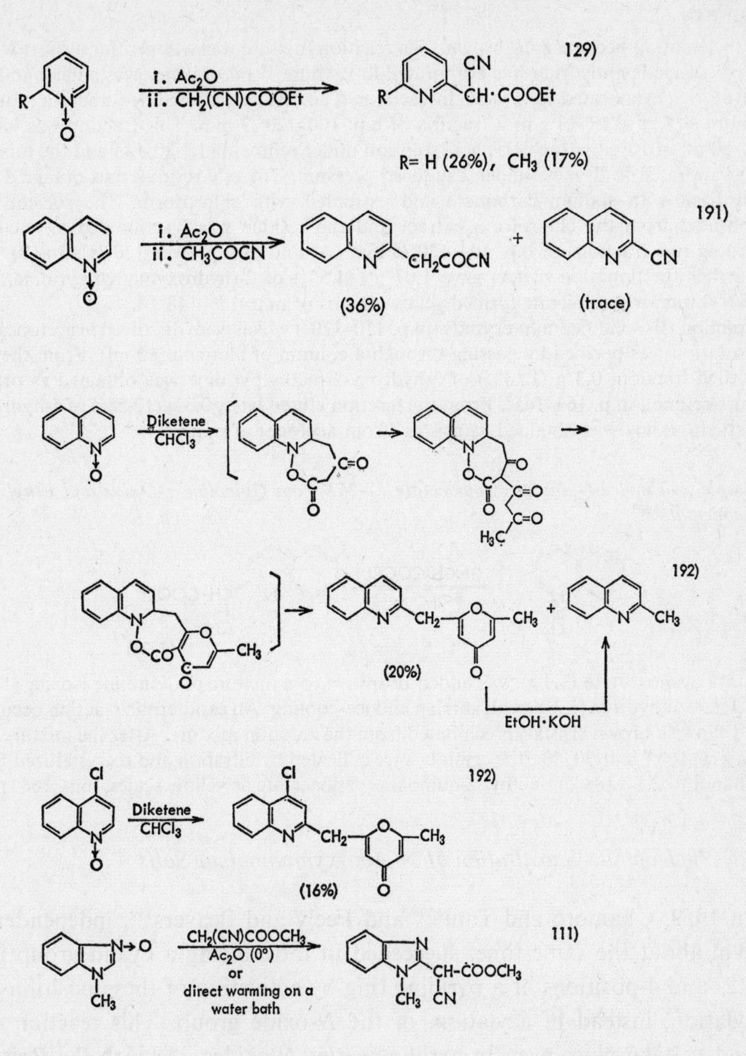

Example 1. Reaction of 2-Picoline 1-Oxide with Acetic Anhydride[16]

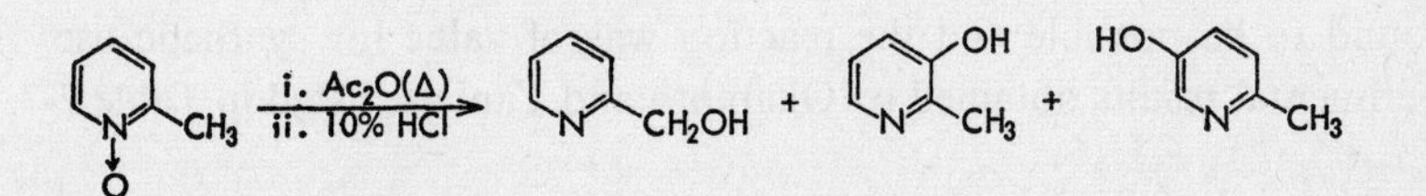

A solution of 2.4 g of 2-picoline 1-oxide dissolved in 3.6 ml of acetic anhydride was warmed on a boiling water bath, by which a violent reaction started within a few minutes

References p. 334

and the solution became dark brown. The reaction mixture was warmed for a short while, the excess acetic anhydride was evaporated in vacuum, 5 ml of water was added, and the mixture was evaporated to dryness in vacuum. The residue was distilled under a reduced pressure and gave 12–13 g of a fraction of b.p. 100–120°/3 mm. This fraction was boiled with 30 ml of 10% hydrochloric acid solution under reflux for 15–20 min and the mixture was evaporated to dryness under a reduced pressure. The oily residue thus obtained was neutralized with sodium carbonate and extracted with chloroform. The solvent was evaporated from the chloroform extract and the residue was fractionated in vacuum, affording two fractions of b.p. 100–120°/5 mm (A) and of b.p. over 120°/5 mm (B).

Further fractionation of (A) gave 1.07 g (44.5%) of 2-(hydroxymethyl)pyridine, b.p. 84–87°/3 mm, whose picrate formed yellow prisms of m.p. 156–148°.

Fraction (B) solidified into crystals (m.p. 110–120°) which were dissolved in acetone, and the solution was purified by passing through a column of alumina (50 ml). From the less adsorbed fraction, 0.3 g (12.5%) of 3-hydroxy-2-methylpyridine was obtained as prisms (from acetone), m.p. 163–165°. From the fraction eluted later, 0.3 g (12.5%) of 5-hydroxy-2-methylpyridine was obtained as prisms (from acetone) of m.p. 167°.

Example 2. Ethyl 2-Quinolinecyanoacetate (7-26) from Quinoline 1-Oxide and Ethyl Cyanoacetate[129]

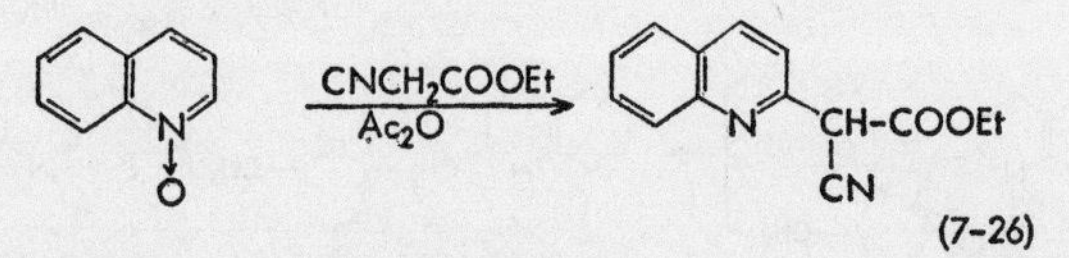

Ethyl cyanoacetate (1.3 g) was added dropwise to a mixture of quinoline 1-oxide (1.5 g) and acetic anhydride (1.3 g) with stirring and ice-cooling. An exothermic reaction occurred and yellowish brown crystals precipitated from the reaction mixture. After the mixture had been kept for 7 h at 30–40°, the crystals were collected by filtration and recrystallized from methanol to 2.1 g (88%) of ethyl 2-quinolinecyanoacetate as yellow scales, m.p. 166–167°.

7.4.5 *Nucleophilic Substitution of N-Alkoxyammonium Salts*

In 1959, Okamoto and Tani[193], and Feely and Beavers[194], independently and at about the same time, succeeded in introducing a cyano group into the 2- and 4-positions of a pyridine ring by activation of these positions by alkylation, instead of acylation, of the *N*-oxide group. This reaction was found to take place even in pyridine series *N*-oxides in which the Reissert reaction with benzoyl chloride and cyanide anion was almost impossible. Moreover, cyanation of not only the 2-position, but also the 4-position was found to be possible and the reaction was of value for synthetic use. Experimental results obtained by Okamoto and Tani are listed in Table 7-7.

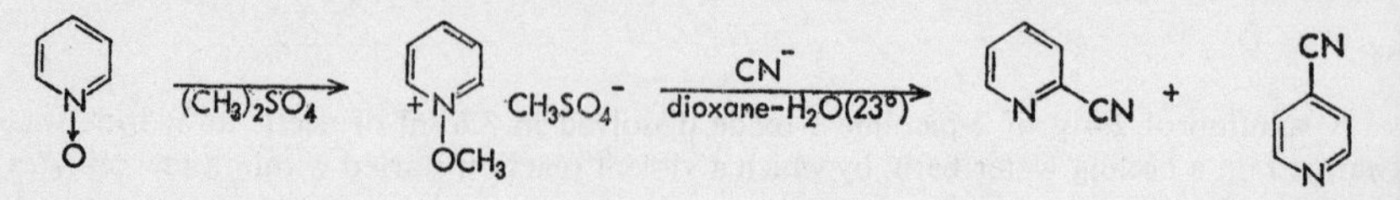

TABLE 7-7

REACTION OF *N*-ALKOXYAMMONIUM SALTS OF PYRIDINE AND QUINOLINE SERIES WITH CYANIDE ANION[193]

N-Oxide of	Yield (%) of product	
	2-Cyano deriv.	4-Cyano deriv.
Pyridine	48	24
2-Picoline	45 (6-CN)	18
3-Picoline	30	15
2,6-Lutidine	33 (6-CH_2CN)	13
Quinoline	71	trace
Isoquinoline	49 (1-CN)	

Feely and Beavers further prepared the following nitriles by this process.

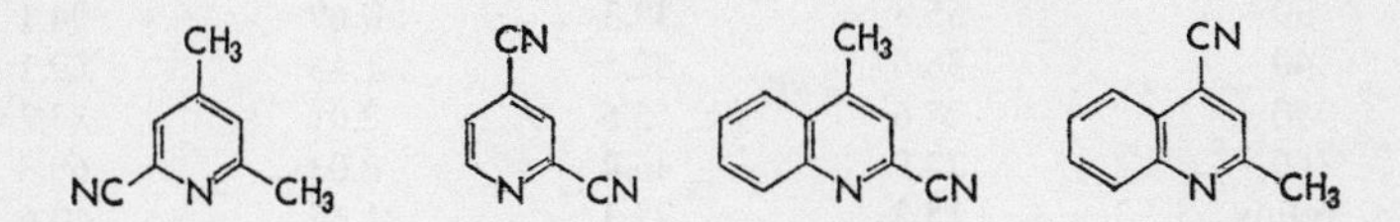

Tani[195] made a detailed examination of the reaction between *N*-alkoxypyridinium salts and potassium cyanide and obtained the results listed in Table 7-8. He found that this reaction occurs at a pH above 8, that its optimum pH is about 11, and that the ratio of 4-cyano to 2-cyano in the product is greatly dependent on the reaction temperature, polarity of the solvent used, and size of the alkyl in the *N*-alkoxyl group. He showed that, in general, the higher the reaction temperature, the greater the polarity of the solvent, and the bulkier the alkyl group, the greater was the value of the 4-cyano to 2-cyano ratio.

By selecting the reaction conditions, taking these factors into consideration, 2- and 4-cyanopyridines can be obtained in more than 80 and 70% yield, respectively.

Progress of the same reaction was followed by measuring the ultraviolet spectrum of the methosulfate of ethyl nicotinate 1-oxide, and it was found that the absorption band at 382 mμ, which appeared immediately after the addition of potassium cyanide, began to disappear with passage of time (*cf.* Fig. 7-1). Tani[195] assumed the formation of a 1,4- or 1,6-dihydropyridine-type compound in the intermediate step and proposed the following reaction mechanism.

TABLE 7-8

FACTORS INFLUENCING THE REACTION OF *N*-ALKOXYPYRIDINIUM SALT WITH CYANIDE ANION[195]

		Product (yield, %)			
		2-CN	4-CN	Ratio of 4-/2-	Total
1-Methoxypyridinium iodide[a]					
pH		Effect of pH			
11.91		43.1	21.3	0.49	64.4
10.99		21.8	34.6	1.59	56.4
10.07		13.7	22.9	1.67	36.6
9.03		3.6	4.3	1.19	7.9
8.06					trace
Temp.(°C)		Effect of temperature[b]			
2		75.7	4.5	0.06	80.2
10		82.1	9.0	0.11	91.1
20		72.4	16.7	0.23	89.1
30		55.8	38.3	0.69	94.1
40		36.6	52.5	1.43	89.1
50		27.6	55.6	2.01	83.2
60		22.9	46.4	2.03	69.3
reflux		15.3	25.3	1.65	40.6
Solvent		Effect of solvents[c]			
H_2O		23.9	10.8	0.45	34.7
H_2O:EtOH (1:1)		55.1	14.2	0.26	69.3
H_2O:EtOH (3:7)		73.4	12.7	0.17	86.1
H_2O:EtOH (2:8)		82.1	9.0	0.11	91.1
EtOH		77.0	6.2	0.08	83.2
Dioxan		79.4	8.7	0.11	88.1
N-Alkoxypyridinium salts[d]					
R	X^-	Steric effect of N–OR group			
CH_3	I	72.4	16.7	0.23	89.1
C_2H_5	I	20.5	58.3	2.84	78.8
n-C_4H_9	TsO	23.0	49.8	2.17	72.8

[a] To the solution of the salt (2.3 g in 10 ml of the buffer), 1.3 g of KCN (in 12 ml of the buffer) was added at 40° and stirred at this temperature for 30 min. Britton–Robinson buffer (H. T. S. Britton and R. A. Robinson, *J. Chem. Soc.*, (1931) 1456) was used and the pH value was determined immediately after the addition of KCN.

[b] To the solution of 1-methoxypyridinium iodide (2.30 g) in dil. EtOH (EtOH:H_2O = 8:2), KCN solution (1.3 g in 3 ml of H_2O) was added dropwise and the mixture was stirred for 30 min.

[c] To the solution of 2.3 g of the quaternary salt in the solvent, 1.3 g of KCN solution (in 3 ml of H_2O) was added at 10° and the mixture was stirred for 30 min at this temperature.

[d] To the solution of the quaternary salt in EtOH (8 ml)–H_2O (2 ml) mixture, 1.3 g of KCN (in 2 ml of H_2O) was added at 20° and the mixture was stirred for 30 min at this temperature.

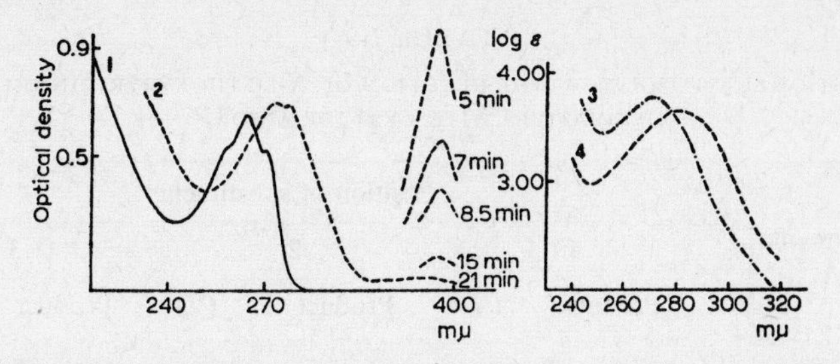

Fig. 7-1. Ultraviolet spectra of 3-ethoxycarbonyl-1-methoxypyridinium methosulfate and its reaction products with potassium cyanide.
(1) 1.66×10^{-4} mole 3-ethoxycarbonyl-1-methoxpyridinium methosulfate (in H_2O);
(2) reaction of 3-ethoxycarbonyl-1-methoxypyridinium methosulfate (1.66×10^{-4} mole) with KCN (1 mole) in H_2O at 15°;
(3) 6-cyano-3-ethoxycarbonylpyridine in ethanol;
(4) 4-cyano-3-ethoxycarbonylpyridine in ethanol.

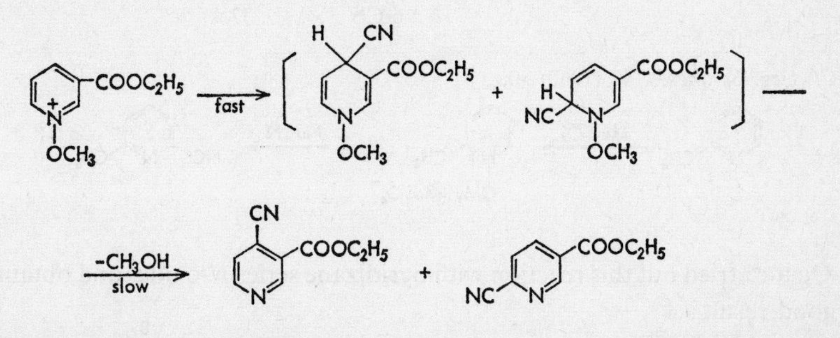

Tani[196] further examined the effect of substituent in the pyridine ring in this reaction and his results are given in Table 7-9. It may be seen from this table that an electron-drawing group in the 2- or 4-position accelerates this reaction and an electron-releasing group has the reverse effect. This is consistent with the assumed mechanism given above.

Tani carried out this reaction with the *N*-oxide of ethyl 6-methylnicotinate and obtained ethyl 4-cyano-6-methylnicotinate (7-29) in a good yield and developed a new synthetic method for vitamin B_6, starting with 5-ethyl-2-methylpyridine (aldehyde-collidine)[37].

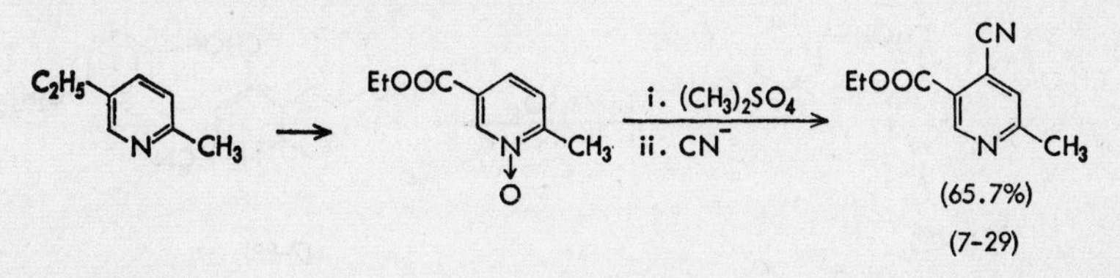

References p. 334

TABLE 7-9

EFFECT OF SUBSTITUENT ON THE REACTION OF *N*-METHOXYPYRIDINIUM SALT DERIVATIVES WITH CYANIDE ANION[196a]

Substituent	Position of substituent					
	4		2		3	
	Product	(%)	Product	(%)	Product	(%)
NO_2	2-CN	53.7				
CN	2,4-$(CN)_2$	70	6-CN	83.7	2-CN	27.8
					6-CN	17.6
CO_2CH_3	2-CN	69.2	6-CN	50.3	4-CN	31.6
					2-CN	19.0
Cl	2-CN	55.6	6-CN	46.7		
CH_3	2-CN	30.0	4-CN	18	2-CN	30
			6-CN[a]	45	4-CN	15
OCH_3	2,4-$(CN)_2$	40	2,6-$(CN)_2$	14.6	2-CN	67.8
			6-CN	37.4		

[a] *Cf. Org. Syntheses*, 42 (1962) 30.

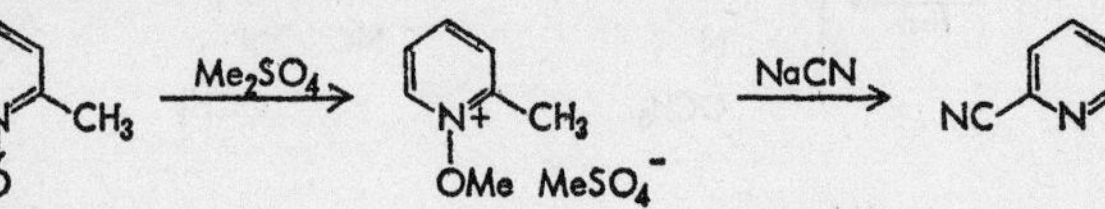

Ogata[89] tried out this reaction with pyridizane series *N*-oxides and obtained a good result.

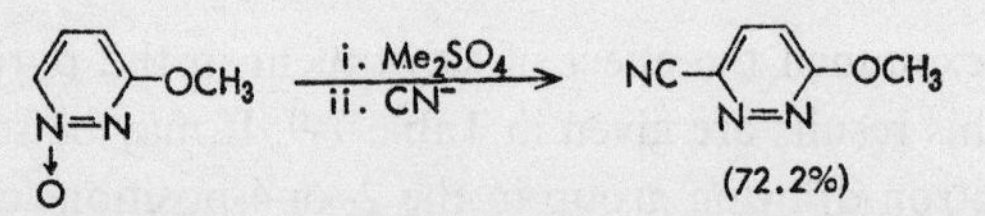

Okamoto and Takayama[197] confirmed that this reaction took place readily (yield, 67%) in 6-methoxyquinoline l-oxide and showed that application of this reaction to dihydroquinine *N,N'*-dioxide (7-27) resulted in cyanation of the 2'-position and deoxygenation of the two nitrogens at the same time, producing 2'-cyanodihydroquinine (7-28).

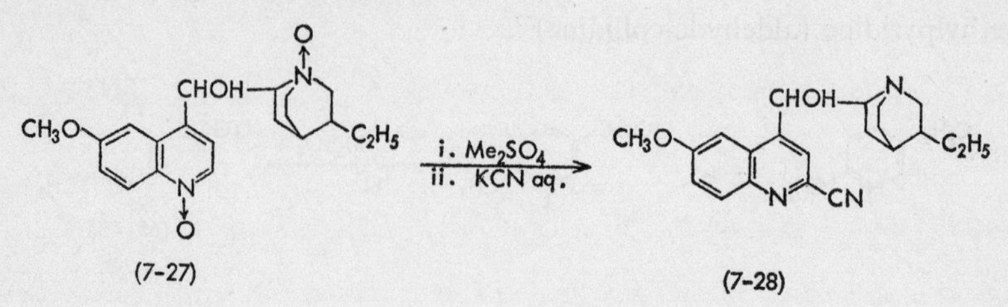

Okamoto and Takayama[198] further found that the application of methyl ketones to the methosulfate of quinoline 1-oxide, in the presence of alkali, results in the formation of quinaldyl ketones with regeneration of quinoline and assumed its mechanism to be as follows.

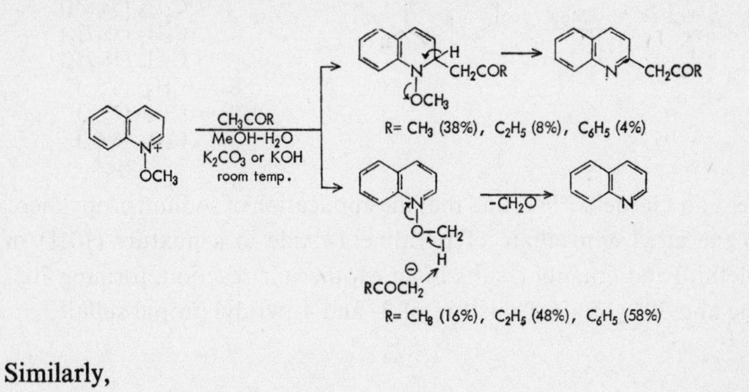

Similarly,

$$\text{R-CH}_2\text{NO}_2 \,/\, \text{NaOMe} \; ; \; \text{H}^+ \qquad \text{R} = \text{H, CH}_3, \text{C}_2\text{H}_5 \qquad [199]$$

This reaction is characterized by the fact that the pyridine ring is cleaved to form a polyene derivative in a good yield, instead of deoxygenation of the *N*-oxide group.

Cervinka[200a] reported that the application of the Grignard reagent to the quaternary ammonium salt of pyridine and methylpyridine *N*-oxides chiefly gives the corresponding 2-alkylpyridine, with formation of the original pyridine base as a by-product.

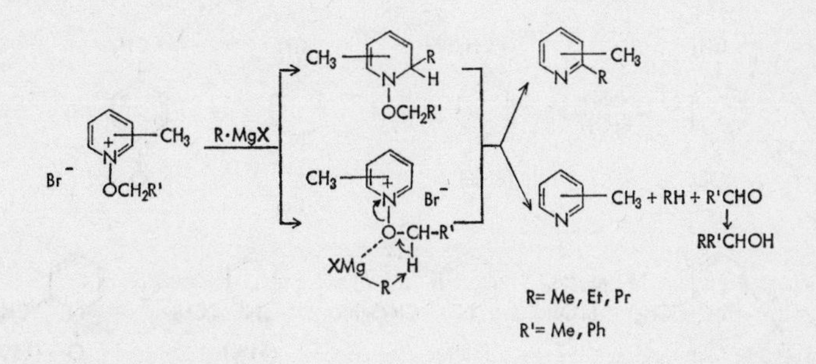

References p. 334

She further obtained the same result with quinoline series *N*-oxides[200b].

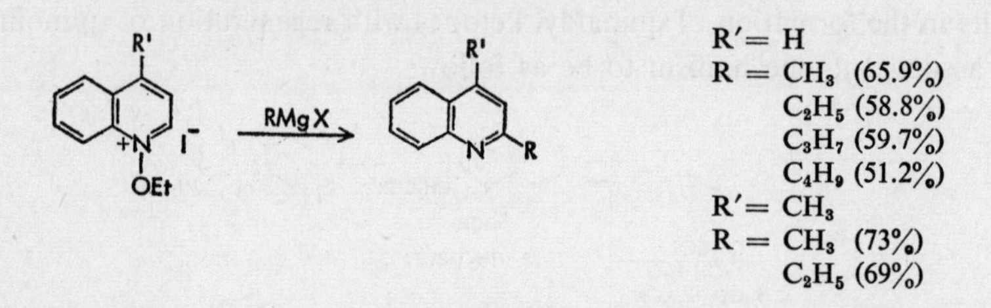

Bauer and Gardella[201a] found that the application of sodium propylmercaptide to the ethyl ethosulfate of pyridine 1-oxide in a mixture (10:1) of 1-propanethiol and ethanol results in an exothermic reaction, forming 70% of pyridine and 30% of a 6:1 mixture of 3- and 4-pyridyl propyl sulfide.

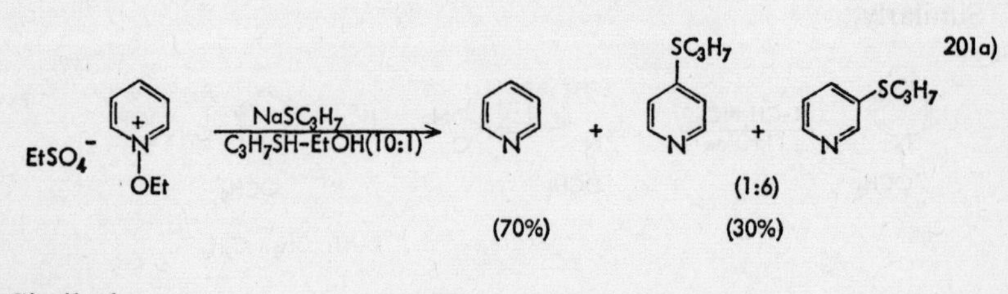

Similarly

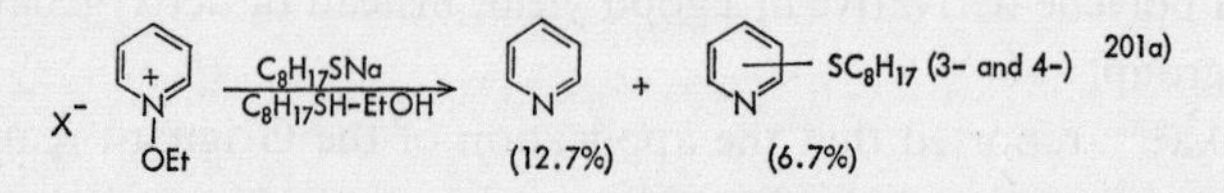

This reaction did not take place with sodium thiophenoxide but the same reaction with methylpyridine series *N*-oxides gave the following results[201b].

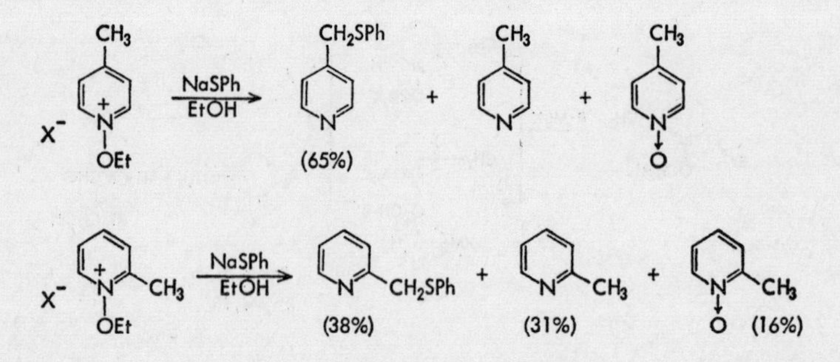

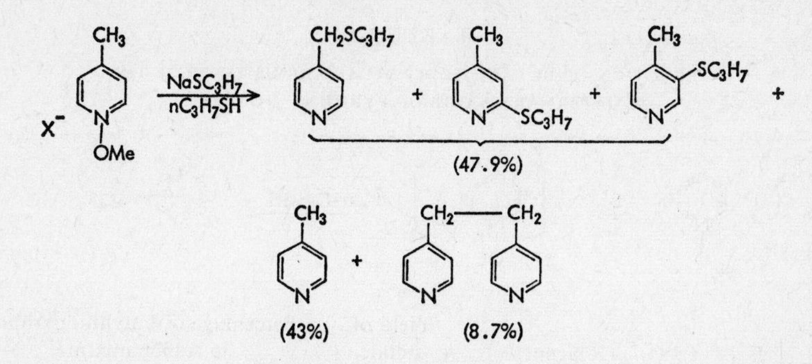

Further, Bauer and Dickerhofe[201c] examined the reaction of butanethiol with various types of the quaternary salts of pyridine 1-oxide and obtained some interesting results, as listed in Table 7-10.

Example 1. Ethyl 4-Cyano-6-methylnicotinate (7-29) from Ethyl 6-Methylnicotinate 1-Oxide[37]

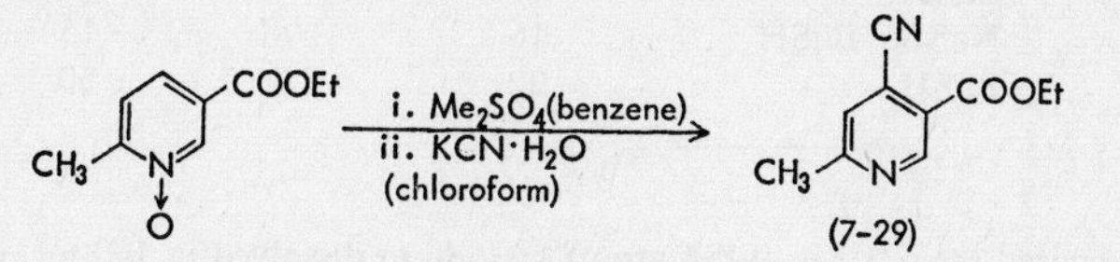

A solution of 28 g of dimethyl sulfate dissolved in 20 ml of dehydrated benzene was added to a solution of 29 g of ethyl 6-methylnicotinate 1-oxide dissolved in 50 ml of dehydrated benzene at room temperature and the mixture was allowed to stand at room temperature for 20 h. The lower oily layer was separated, the benzene layer was extracted with 60 ml of water, and 150 ml of ethanol was added to the combined oily layer and aqueous extract. While stirring this mixture at 23–25°, a solution of 21 g of potassium cyanide in 40 ml of water was added dropwise and the mixture was stirred for a further 3 h at the same temperature. The mixture was then extracted with chloroform, the extract was dried over anhydrous sodium sulfate, and the chloroform was evaporated. The residue was submitted to vacuum distillation and the fraction of b.p. 125–140°/5.5 mm was recrystallized from diisopropyl ether to 20 g (65.7%) of ethyl 4-cyano-6-methylnicotinate of m.p. 90.5–92.5°. From the recrystallization mother liquor, an oily ethyl 6-methylnicotinate (picrate, yellow needles, m.p. 168–170°) was obtained in a small amount.

Example 2. 1-Methoxyimino-6-aci-nitrohexa-2,4-diene (7-30) from 1-Methoxypyridinium Iodide and Nitromethane[199]

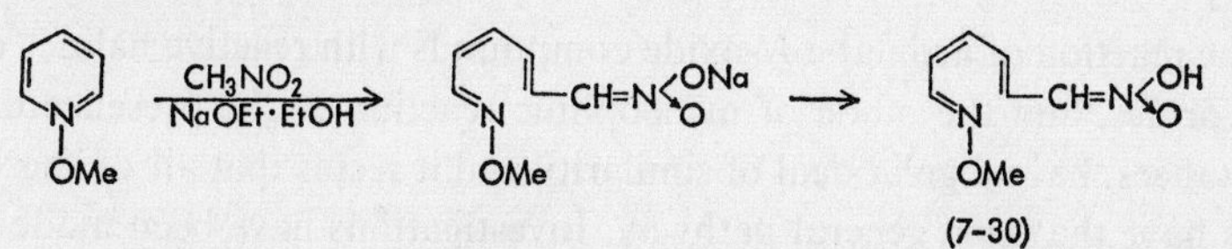

To a suspension of 32.2 g (0.136 mole) of 1-methoxypyridinium iodide in 300 ml of dehydrated ethanol, 8.3 g (0.136 mole) of nitromethane was added and sodium ethoxide

References p. 334

TABLE 7-10

REACTION OF BUTANETHIOL WITH VARIOUS TYPES OF THE QUATERNARY SALTS OF PYRIDINE 1-OXIDE[201c]

E	Reagent	Yield of sulfide (%)	Percentage of butylthiopyridine in sulfide mixture		
			2-	3-	4-
Et_2SO_4	NaSBu–BuSH	15	16	60	24
TsOEt	NaSBu–BuSH	11	11	74	15
Ac_2O	BuSH	67	61	39	
AcCl	NaSBu–BuSH	10	89	9	2
PhCOCl	BuSH	19	81	18	1
PhCOCl	NaSBu–BuSH	16	81	15	4
$PhSO_2Cl$	BuSH	32	50	50	

solution, prepared from 3.2 g (0.136 atom) of sodium dissolved in 100 ml of dehydrated ethanol, was added dropwise during about 1 h. The flask was cooled from outside to about 5°. The suspended compound dissolved gradually and gave way to a pale yellow precipitate. The mixture was stirred for 3 h at room temperature, cooled, and the precipitate was collected by filtration. The dried yellow filter cake weighed 15 g. The filtrate was evaporated to dryness under a reduced pressure and 7.3 g of brownish yellow solid was obtained. Both these solids are sodium salts. The solid was dissolved in a small quantity of water and cautiously neutralized with 2N hydrochloric acid after addition of cracked ice. The precipitate thereby formed was extracted with ether, the extract was dried over anhydrous sodium sulfate, and the ether was evaporated, leaving 14.08 g (60%) of crude crystals of 1-methoxyimino-6-*aci*-nitrohexa-2,4-diene. Recrystallization from methanol gave pale yellow needles, m.p. 91–92° (decomp.).

7.4.6 Theories regarding the Reaction Mechanism in Nucleophilic Substitutions

The reaction of aromatic *N*-oxide compounds with reactive halides or acid anhydrides, and the mode of nucleophilic reaction in the presence of these substances, have a great deal of similarity and it seems that all of these reactions have the same general pathway. Investigations have been made chiefly of the reaction with acetic anhydride, but none of the mechanisms fits all of the facts.

Katada[5] proposed the following mechanism, based on the fact that while the catalytic reduction of the reaction mixture of pyridine 1-oxide and acetic anhydride gives 2-piperidone in a quantitative yield, the solution in which 2-pyridone and acetic anhydride had been boiled resists catalytic reduction under the same reaction condition.

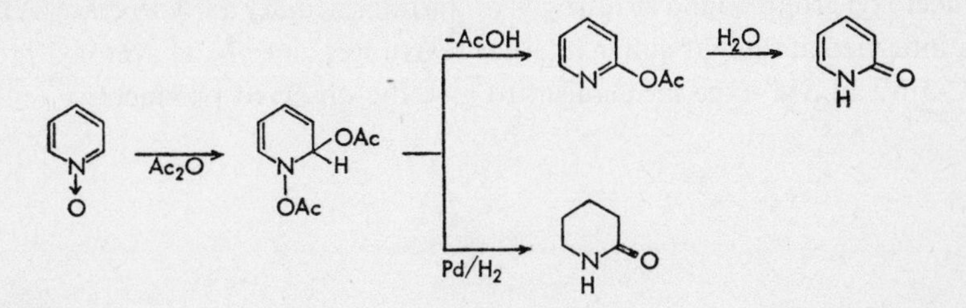

Kobayashi and others[13] considered anionotropy as the mechanism for the reaction of 2-picoline 1-oxide and acetic anhydride, and Okuda[16] proposed the following mechanism by considering the formation of β-acetoxyl compounds.

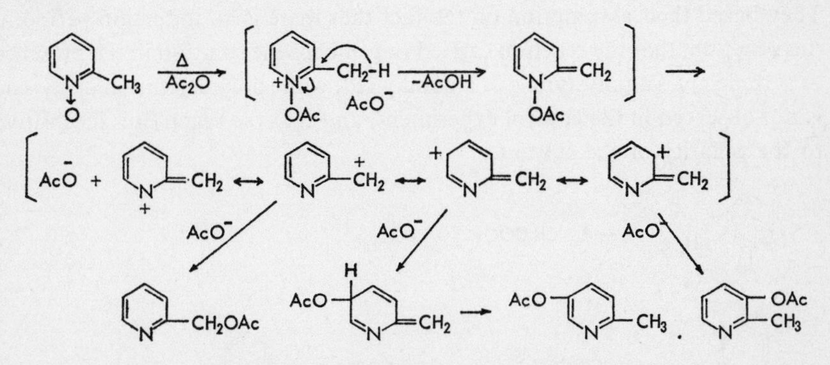

On the other hand, Bullit and others[15] stated that Pachter's interpretation[118] of the reaction of quinaldine 1-oxide and benzoyl chloride (*cf. Section 7.4.2*) was in agreement with the following cyclic and ionic mechanisms.

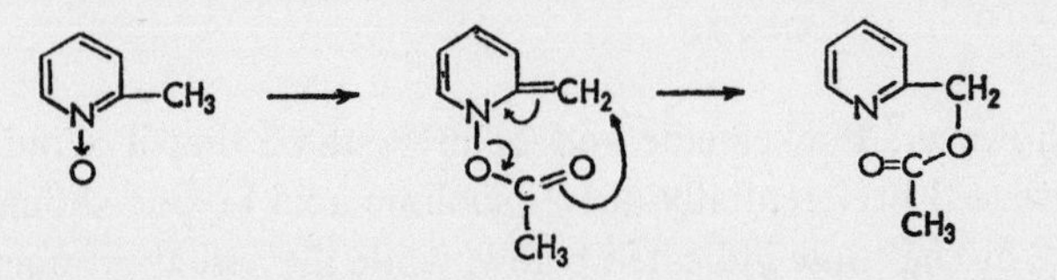

This explanation seems ingenious but it cannot *per se* explain the by-product formation of 3- and 5-acetoxy-2-picoes or thelin reaction of 4-picoline

References p. 334

1-oxide. The radical process suggested by Boekelheide and others[202] for the reaction of 4-picoline 1-oxide and acetic anhydride cannot be eliminated, but Berson and Cohen[169] stated that the reaction could be explained by the two heterolytic mechanisms in which the attack of the external acetate ion takes place on the C-3 position or a methylene group, accompanied by extrusion of the acetoxyl group bound to nitrogen of the intermediary anhydro base (A), or this intermediate might suffer internal rearrangement of the acetoxyl group to C-3 by an *S*Ni′-type mechanism to give the observed products.

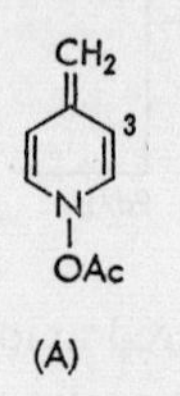

On the other hand, Boekelheide and Harrington[202] at first assumed the following radical chain mechanism for the reaction of 2-picoline 1-oxide. They based their assumption on the fact that there is an induction period in this reaction, that the reaction carried out in benzene solution in the presence of styrene gives a polystyrene in a good yield, while polymerization of styrene is not observed in the control experiment, and that the reaction is insensitive to the polarity of the solvent.

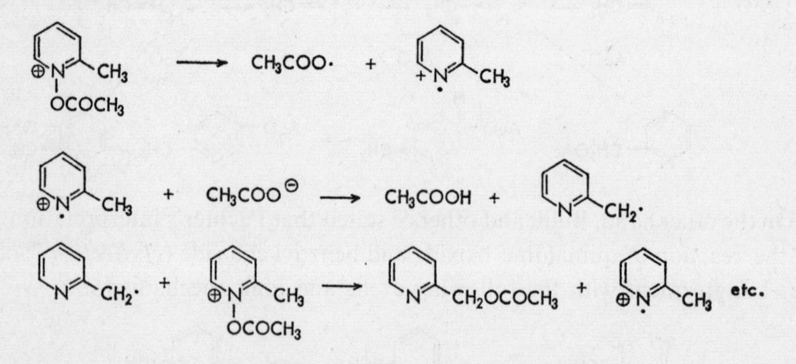

Later, however, Boekelheide and Lehn[153] stated that if a radical process were to proceed preferentially, (*a*) 2-picolinic acid l-oxide should change to 2-pyridone by the route indicated below, while the actual product is pyridine 1-oxide, and (*b*) heating of 2-picolinic acid 1-oxide with acetophenone results in the formation of a pyridine 1-oxide adduct, as was observed by Hammick and others[204] with picolinic acid, so that the reaction with acetic anhydride is predominantly an ionic reaction; they failed, however, to give any conclusion.

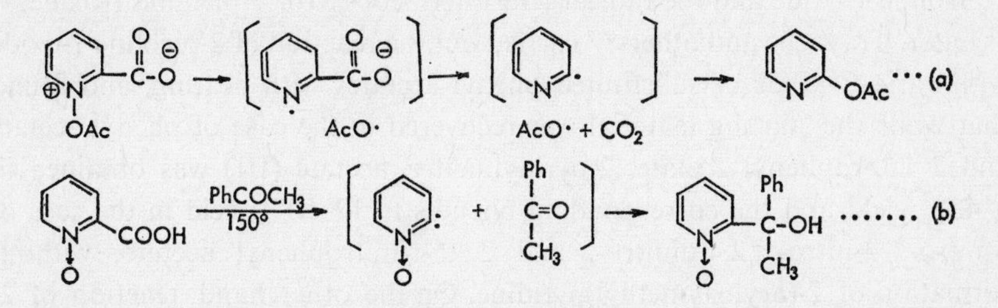

Traynelis and others proved the formation of carbon dioxide, methane, and methyl acetate, all originating from the acetate radical, in the reaction of 2-picoline 1-oxide[203a] and 4-picoline 1-oxide[203b] with acetic anhydride. Further, in the reaction of 4-picoline 1-oxide with acetic anhydride, they obtained a total of 65% of 4-pyridylmethyl acetate and 3-acetoxy-4-methylpyridine in 88–89:12–11 ratio, in addition to a basic fraction containing 4-picoline (2.9%), 2,4-dimethylpyridine (0.2%), and 4-ethylpyridine (0.6%), and proved the formation of a 4-picolyl radical in the intermediate step. However, since there are several pieces of evidence which cannot be explained by the radical chain mechanism, such as the fact that there is little difference in the formation of picolyl acetate even on the addition of a radical scavenger like *m*-dinitrobenzene, and that treatment of 4-picoline 1-oxide with butyric anhydride in the presence of sodium acetate gives a butyrate instead of an acetate, Traynelis and others[203b] excluded the free radical chain mechanism, the nucleophilic attack of acid anions on the intermediate anhydro base, and proposed an intramolecular rearrangement in the intermediarily formed anhydro base, pointing out the possibility of the following allylic rearrangement.

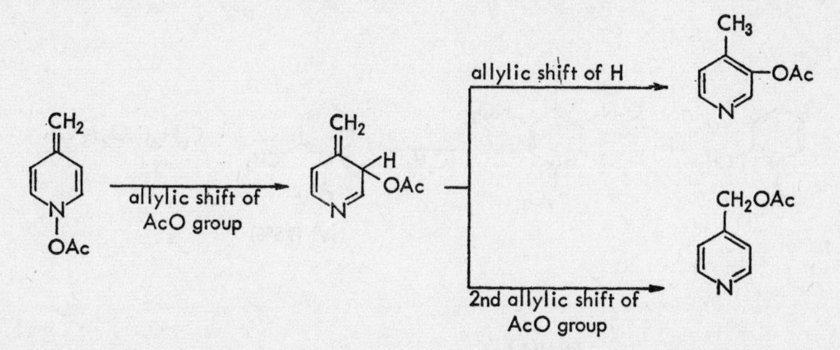

They stated that this rearrangement might proceed via an ion pair or radical pair, which resulted from a heterolytic or homolytic cleavage, respec-

References p. 334

tively, of the nitrogen—oxygen bond in the anhydro base or the internal cyclic rearrangement, and they assumed that the ionic path would be favorable for 2-picoline 1-oxide and the radical-pair intermediate for 4-picoline 1-oxide.

Later, Traynelis and others[207] carried out the reaction of 2-picoline 1-oxide with various kinds of substituted phenyl acetates with heating, and found that while the starting material was recovered in the case of phenyl acetate and 2-chlorophenyl acetate, 2-pyridylmethyl acetate (III) was obtained in 5–43% yield and the corresponding phenols in 12–50% yield in the case of 8-nitro-, 4-nitro-, 2,4-dinitro-, and 2,4,6-trinitrophenyl acetate, without formation of 2-(aryloxymethyl)pyridine. On the other hand, reaction of 2-picoline 1-oxide and picryl acetate in benzene easily gives 1-acetoxy-2-methylpyridinium picrate (IV) in a good yield, but this (IV) does not form (III) even when heated with sodium acetate in acetic acid and forms (III) in 20% yield when heated with triethylamine in dehydrated dioxan. Traynelis and others considered these results from the basicity of the phenoxide ion (ArO^-) and assumed that, as shown below, 2-picoline 1-oxide first forms (I), which changes into (II) by the action of a base (phenoxide ion, triethylamine), and (III) is formed by the intramolecular rearrangement of (II).

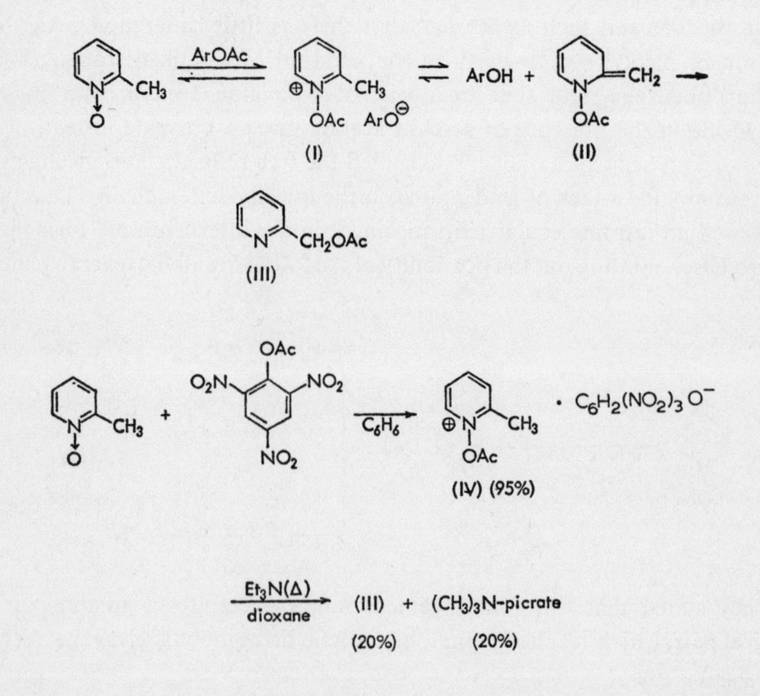

Traynelis and others further carried out analyses of the products from the reaction of 4-picoline 1-oxide with 2,4,6-trichlorophenyl acetate, found the products to be as shown below, and claimed that the result is in agreement with the foregoing mechanism[203a,b].

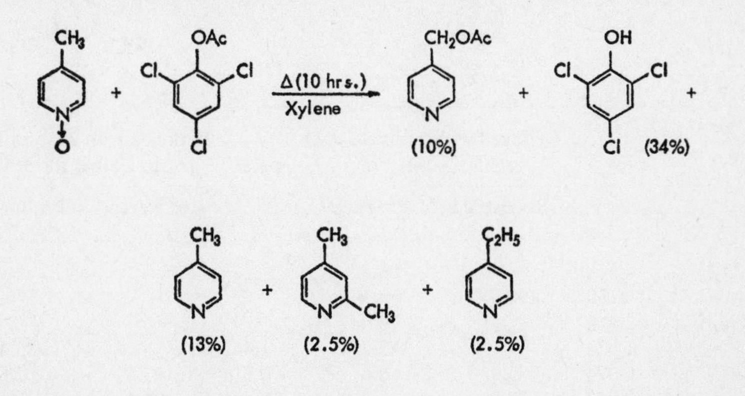

Furukawa[205] carried out this reaction of 2-picoline 1-oxide derivatives with acetic anhydride in chloroform solution and found that the reaction progresses more smoothly when there is an electron-releasing group in the 4-position than in the presence of an electron-drawing group; he calculated the relationship between the pKa′ values of these derivatives and the yield of the reaction products, as summarized in Table 7-11.

These results show that, with the exception of the 4-amino derivative, the reaction progresses smoothly with the compounds having a positive pKa′ value and the reaction hardly occurs in compounds having a negative pKa′ value. This fact indicates that the reaction with acetic anhydride begins with acetylation of the oxygen in the *N*-oxide group.

Figures 7-2 and 7-3 are ultraviolet absorption spectra of 2-picoline 1-oxide derivatives, having a methoxyl or a nitro group in the *para*-position, respectively, during addition of acetic anhydride in dioxan solution, measured by Furukawa. Changes of the absorption in the spectra of the nitro derivative [λ_{max} 351 mμ (log ε 4.193)], which is resistant to the addition of acetyl ion, are much smaller than that of the methoxyl derivative [λ_{max} 279 mμ (log ε 4.152)]. Examination of this change in the ultraviolet absorption curve according to the concentration of acetic anhydride shows that this change is a reversible shift, indicating that the addition of CH_3CO^+ to the *N*-oxide group is in equilibrium and is dependent on the concentration of acetic anhydride.

References p. 334

TABLE 7-11

REACTION OF 2-PICOLINE 1-OXIDE DERIVATIVES WITH ACETIC ANHYDRIDE IN CHLOROFORM SOLUTION

X	pKa′	Refluxed with 2 moles Ac_2O in $CHCl_3$ for 3 h[b] (4-X-2-picoline 1-oxide)		pKa′	Refluxed with 2 moles Ac_2O in $CHCl_3$ for 3 h[b] (4-X-2,6-lutidine 1-oxide)	
		Recovered(%)	Product[a](%)		Recovered(%)	Product[a](%)
$N(CH_3)_2$	4.370		86.8			
NH_2	4.100	(90.0)[c]				
OCH_3	2.414		76.0			
H	1.022		85.5	1.442		83.4
CN	−0.674	83.9	8.0	−0.614	93.6	2.6
NO_2	−0.968	92.3		−0.861	91.5	
$CO_2C_2H_5$				−0.126	82.8	

[a] Including ω-hydroxy and β-hydroxy compounds.
[b] The best conditions for 2-picoline 1-oxide.
[c] 4-Acetamido-2-picoline 1-oxide. When the reaction temperature is raised to 120–130°, 4-acetamido-2-(acetoxymethyl)pyridine is formed. It was assumed by Furukawa[205] that resistance to rearrangement in chloroform solution is due to stabilization by formation of AcN=⟨ ⟩=N–OAc.

Furukawa measured the periodical change in the ultraviolet absorption of quinaldine 1-oxide* with addition of acetic anhydride in various concentrations, as shown in Fig. 7-4.

Reaction rate constants calculated from these data were found to agree with the pseudo-first-order reaction. It may therefore be concluded that the first step in this reaction is the addition of the acetyl group to the oxygen in the *N*-oxide group, but this is in equilibrium and is dependent on the concentration of acetic anhydride, the rate-determining step being the cleavage of the nitrogen–oxygen bond. It is not certain whether this cleavage is an ionic or

* The ultraviolet absorption maximum of the 2-picoline 1-oxide series overlaps the absorption of acetic anhydride itself and measurement of absorption in the wave-length region below 280 mμ is impossible. For this reason, the absorption of quinaldine 1-oxide was measured.

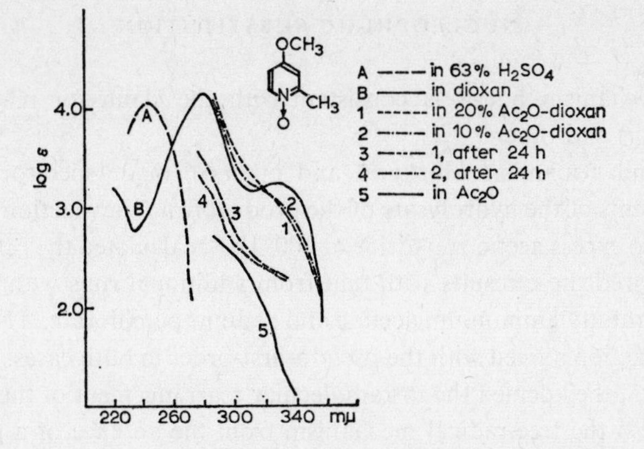

Fig. 7-2. Ultraviolet spectra of 4-methoxy-2-picoline 1-oxide.

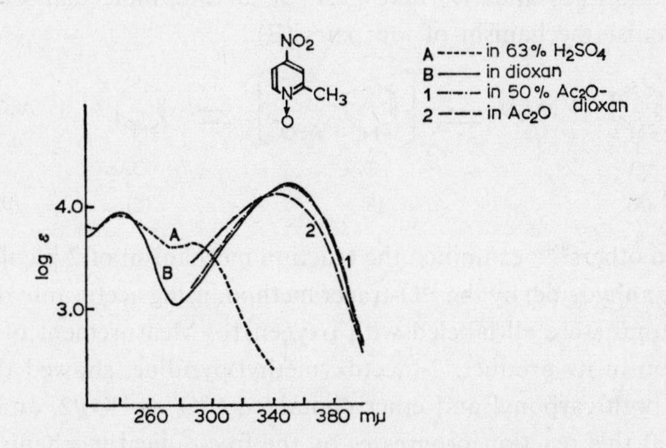

Fig. 7-3. Ultraviolet spectra of 4-nitro-2-picoline 1-oxide.

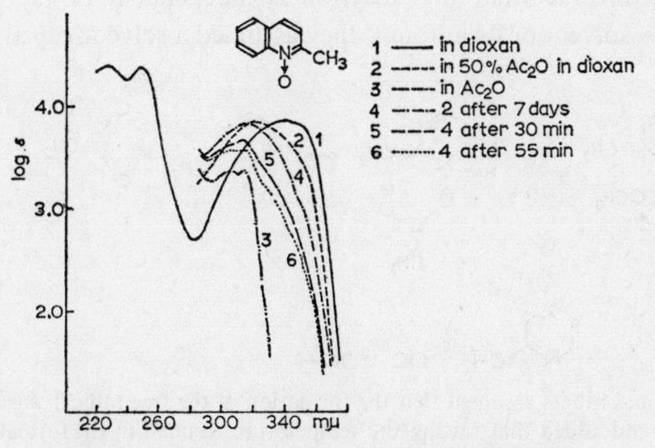

Fig. 7-4. Ultraviolet spectra of quinaldine 1-oxide.

References p. 334

radical mechanism but it is consistent with the claim for rearrangement by Traynelis and others.

In a similar manner, Markgraf and others[206] made spectrophotometric measurements of the hydrolysate of the product from the reaction of pyridine 1-oxide and excess acetic anhydride at 100–130°, calculated the rate constants and compared these results with that from additional runs with added salts such as tetrabutylammonium acetate and sodium perchlorate. They revealed that the reaction agreed with the pseudo-first-order in both cases. From these kinetic data, they denied the intramolecular rearrangement of the free cation (C) and also the free-radical mechanism from the absence of a gaseous decomposition product*, and suggested an intermolecular mechanism in which the free cation (C) and (D) take part, or an intermolecular or concerted intramolecular mechanism of ion pairs (B).

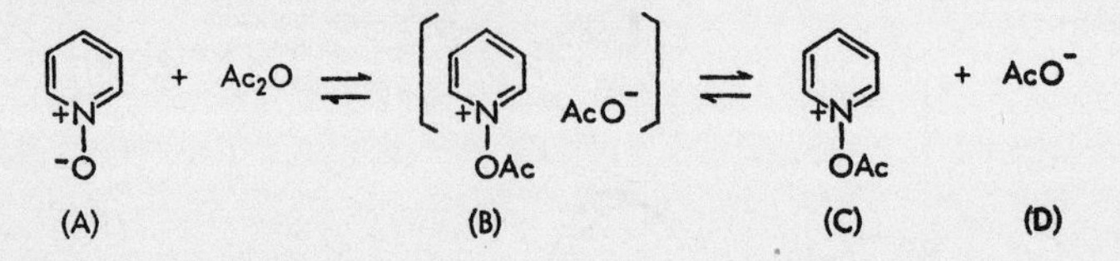

Oae and others[208a] examined the reaction mechanism of 2-picoline 1-oxide and acetic anhydride, by the ^{18}O-tracer method, using acetic anhydride whose oxygen atoms were all labeled with oxygen-18. Measurement of oxygen-18 distribution in its product, 2-(acetoxymethyl)pyridine, showed that oxygen atoms in both carbonyl and ether contained $(^{18}O + {}^{16}O)/2$, and they concluded that this reaction progresses by the free-radical mechanism and not by intramolecular rearrangement or intermolecular mechanism. Further, considering the fact that this reaction is independent of the presence or absence of a solvent or its amount, they assumed a solvent-caged radical pair

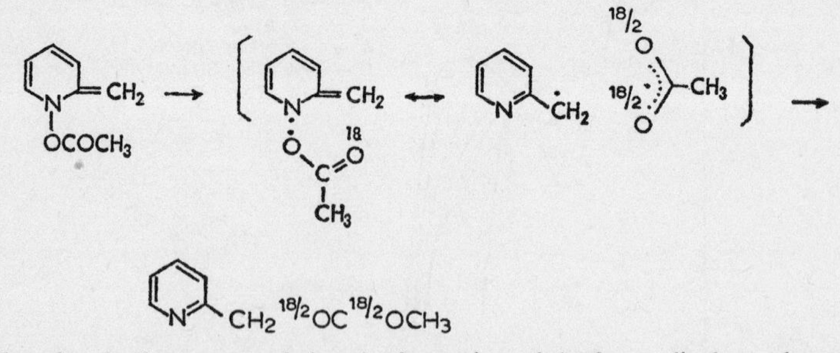

* Markgraf and others assumed that the formation of the free radical was dependent on temperature and added that raising the temperature resulted in the formation of carbon dioxide, and 2- and 4-picolines.

process, as shown below, and not the radical chain process suggested by Boekelheide and others[202].

Oae and others[208b] carried out the same examination with 4-picoline 1-oxide, separated 4-(acetoxymethyl)pyridine and 3-acetoxy-4-picoline in the product, and measured excess oxygen-18 in these products and in their hydrolyzates. Distribution of oxygen-18 from these measurements agreed well with intermolecular mechanism due to nucleophilic attack by the acetate ion, and proposed the following mechanism.

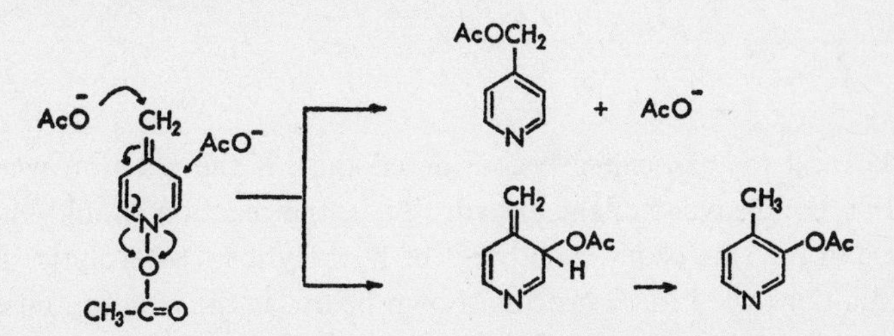

In other words, a reaction mechanism different from that of 2-picoline 1-oxide has been proposed for 4-picoline 1-oxide by these workers, who presumed that this difference is due to the difference in the strength of the nitrogen–oxygen bond in the anhydro base and to some steric factors.

Kaneko and his associates[184] carried out the reaction of 2- and 4-picoline 1-oxide, and their homologs, with acetic anhydride, separated their reaction products by Megachrome (gas chromatograph for separation), and isolated and purified each fraction by low-pressure distillation. Results are shown in Table 7-12.

TABLE 7-12

REARRANGEMENT PRODUCTS FROM THE REACTION OF METHYLPYRIDINE 1-OXIDES AND ACETIC ANHYDRIDE[184]

Pyridine 1-oxide	Rearrangement product (yield, %)				Hydroxyl ratio[a]
	Alcohol acetate	Phenol acetate		Dipyridyl-ethylene	
		3	5		
4-Picoline	64.2	35.8		*ca.* 1.3	1.8
2-Picoline	68.9	12.6	18.5		2.2
2,6-Lutidine	66.6	33.3		?	2.0
6-(Hydroxymethyl)-2-picoline	79.5	2.0	18.5	*ca.* 9	3.9

[a] Hydroxyl ratio = alcohol acetate/phenol acetate

Kato and others[176] had already observed the by-product formation of a small amount of 1,2-bis(6-methyl-2-pyridyl)ethane (A) from the reaction of 2,6-lutidine 1-oxide and acetic anhydride, but Kaneko and others did not obtain this compound. The latter workers obtained, from the same reaction of 4-picoline 1-oxide and 6-(hydroxymethyl)-2-picoline 1-oxide, a small amount of the corresponding dipyridylethylene derivative (such as B).

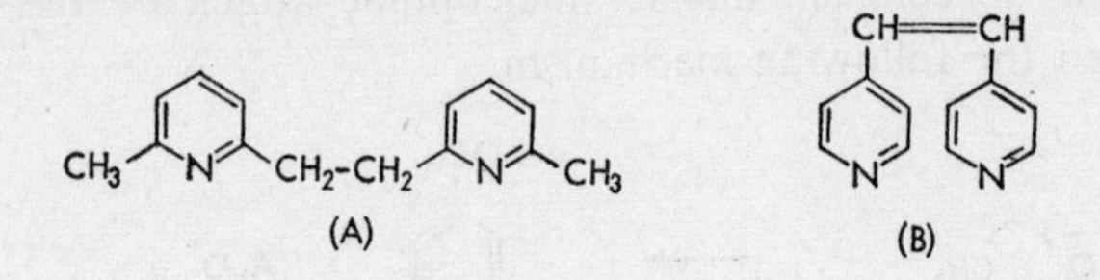

Kaneko and his associates[184] considered that, if the reaction were to be initiated by the cleavage of the anhydro base, the reaction should invariably take place in the side chain, whether it be homolytic or heterolytic, and they proposed a concerted ionic process shown below as the reaction mechanism which would account for the formation of the B-type compound and the value of the hydroxyl ratio (ratio of alcohol acetate to phenol acetate).

Formation of alcohol acetate and phenol acetate

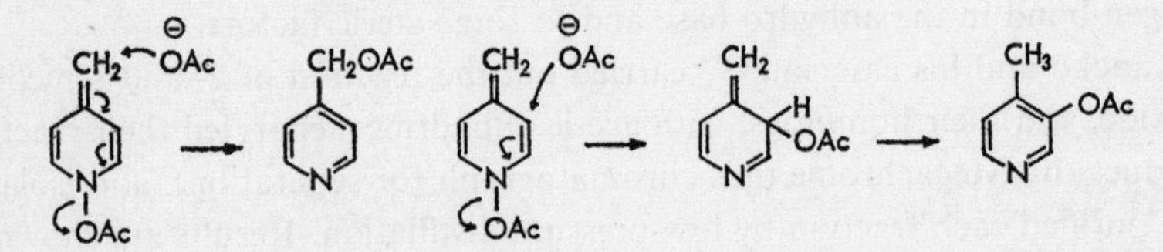

Formation of dipyridylethylene

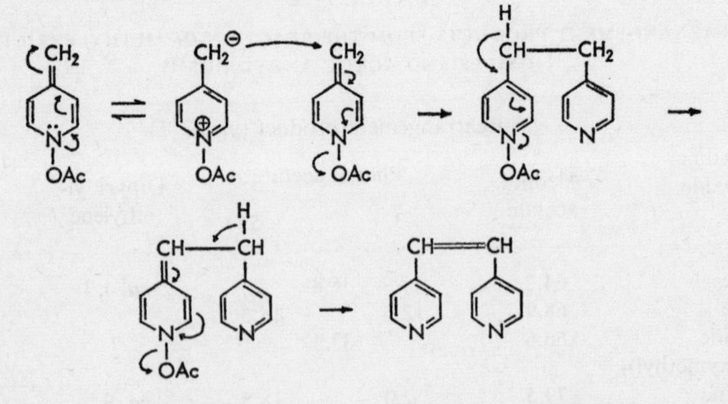

However, this explanation alone would not be sufficient to explain the formation of dipyridylethane derivatives from 2,6-lutidine 1-oxide, and

Kaneko and others also assumed the partial contribution of a concerted radical process.

Ochiai and Ikehara[12] suggested the following ionic process as the mechanism for the formation of 4-tosyloxyisoquinoline and a small amount of isocarbostyril from the reaction of isoquinoline 2-oxide and tosyl chloride.

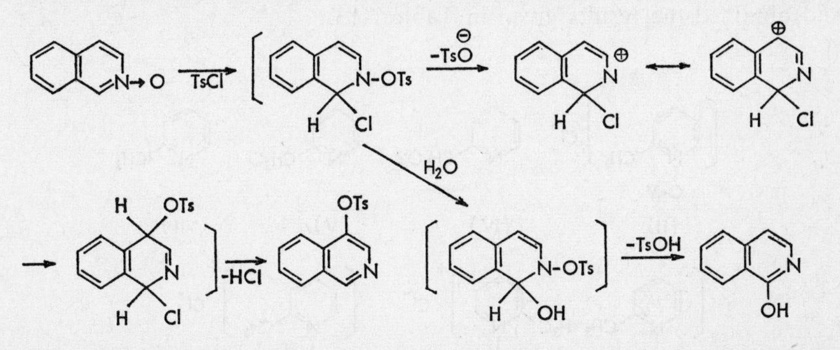

Oae and others[209] examined this reaction by the technique described above, using tosyl chloride uniformly labeled with oxygen-18 in the two oxygen atoms. Measurement of the distribution of oxygen-18 in the 4-tosyloxyisoquinoline obtained as the product showed that oxygen-18 was hardly found in the ether-oxygen. They concluded that this result is contrary to the aforementioned solvent-separated ion pair process (A) and to the intimate ion pair (B) rearrangement for this reaction, and that the probable mechanism is the α,γ-intimate ion pair (C) rearrangement.

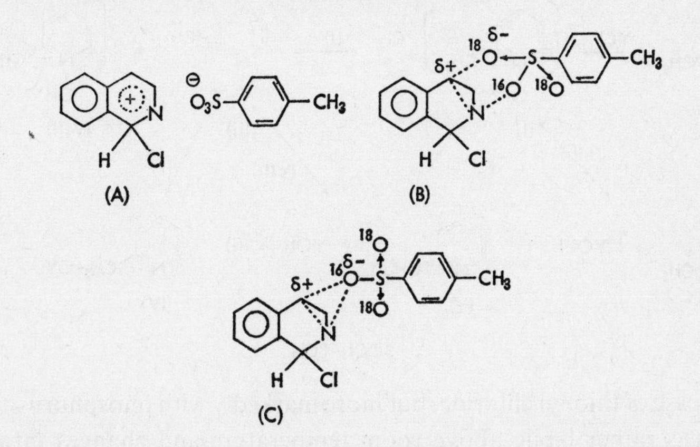

Oae and others also examined the formation of 3-tosyloxypyridine from the reaction of pyridine l-oxide and tosyl chloride, and put forward the same

References p. 334

reaction mechanism, because there was no oxygen-18 in the ether-oxygen bonded to the 3-position.

Vozza[210] examined the reaction of 2-picoline 1-oxide with various reactive halides, such as benzoyl, acetyl, thionyl, and benzenesulfonyl chlorides, phosphorus trichloride, and hydrogen chloride, under various conditions and obtained the results given in Table 7-13.

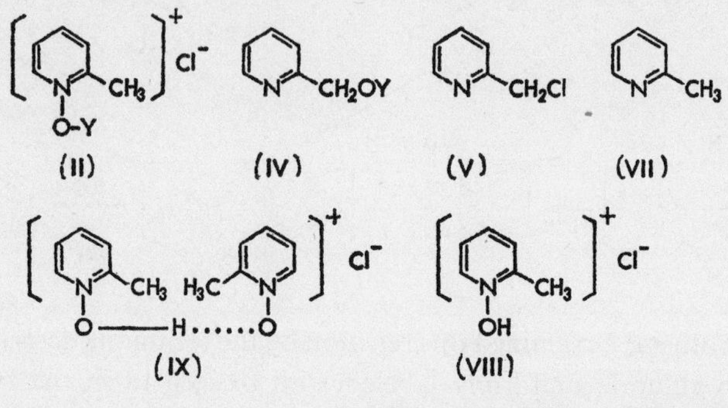

Vozza proposed the following mechanism for these reactions with reactive halides and assumed that the difference in the direction of this reaction was due to the difference in the relative stability of (II) (affected by temperature and kinds of Y used).

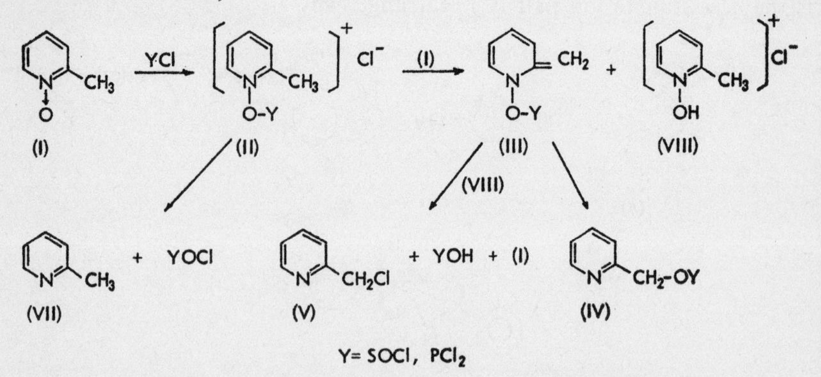

In the case of thionyl chloride, but more markedly with phosphorus trichloride, (II) is rather labile above room temperature and changes into (VII). In the case of benzoyl and acetyl chlorides, and also with thionyl chloride, though much weaker, (II) is sufficiently stable at temperatures below 20° to

TABLE 7-13

REACTION OF 2-PICOLINE 1-OXIDE WITH REACTIVE HALIDES

Halide	Reaction conditions		Product (yield, %)
	Solvent	Temp. (°C)	
PhCOCl	dioxan	reflux	IV (56.2), V (15.6)
PhCOCl	benzene	15–20	VIII (39.6), IV (17.8)
AcCl	AcOH	reflux	IV (53.7), V (7.0)
AcCl	benzene	15–20	VIII (37.0)
$SOCl_2$	benzene	15–20	VIII (34.2)
$SOCl_2$	benzene	50–reflux	VII (92.0)
$PhSO_2Cl$	benzene	25	V (47.6)
$PhSO_2Cl$	benzene	reflux	V (71.8)
PCl_3	benzene	15–reflux	VII (94.0)
HCl (gas)	benzene	15–20	IX (55.01)
conc. HCl	H_2O	25	VIII (84.5)

form (III) by withdrawal of a proton by another molecule of (I), and at the same time forms the main product (VIII). At a refluxing temperature, (II) and (III) are both labile, and (III) undergoes rearrangement to (IV). Reaction with benzenesulfonyl chloride results in the formation of (V), and not (IV) or (VIII). Vozza assumed that this was due to the fact that the nucleophilic attack of a chlorine ion on an exocyclic carbon atom of (III) was more rapid than that of benzoyl or acetyl chloride, and liberates the benzenesulfonate ion.

The common theory regarding the mechanism of the reaction with reactive halides or acid anhydride among these various assumptions is that there is a formation of an intermediary adduct and that an anhydro base is formed by the base-catalyzed withdrawal of a proton if there is a methyl group in 2- or 4-position to the *N*-oxide group. Various theories have been proposed for the transformation of these adducts or anhydro bases, such as the free-radical process, solvent-caged free radical process, intramolecular ionic process, concerted intermolecular ionic process, and α,γ-intimate ion-pair process. It still seems difficult to conclude which of these is the true mechanism.

On the other hand, it should be noted that there have been many instances where, if an anion is present, a nucleophilic attack by that anion takes place and, if the α- or γ-position of the *N*-oxide group is occupied by an electron-

References p. 334

releasing group, such a nucleophilic attack occurs in the position β to the *N*-oxide group or positions corresponding to it.

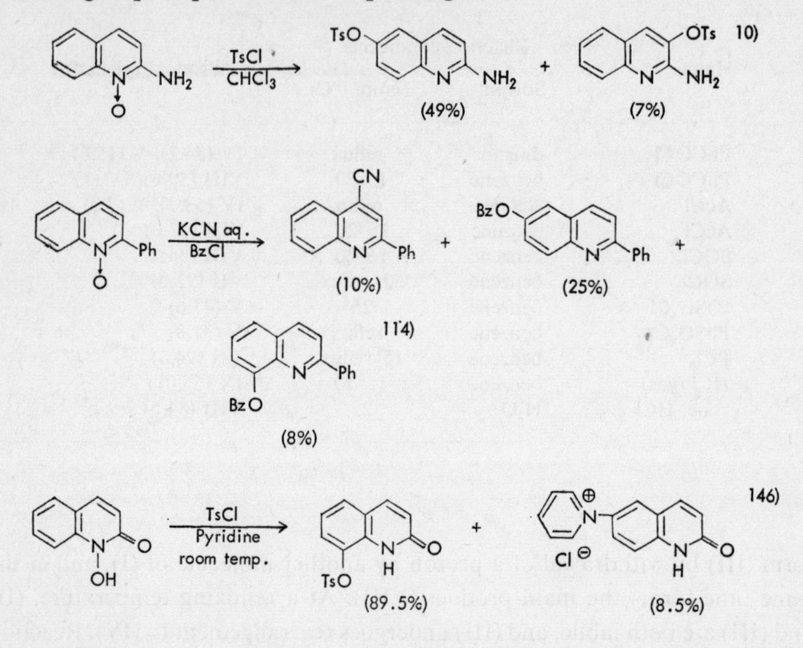

These facts suggest that, irrespective of whether we are dealing with a radical or ionic process, resonance systems having a certain life-time take part in the intermediate stage. Examples are given below. Synthetic reactions may be planned on the basis of this idea.

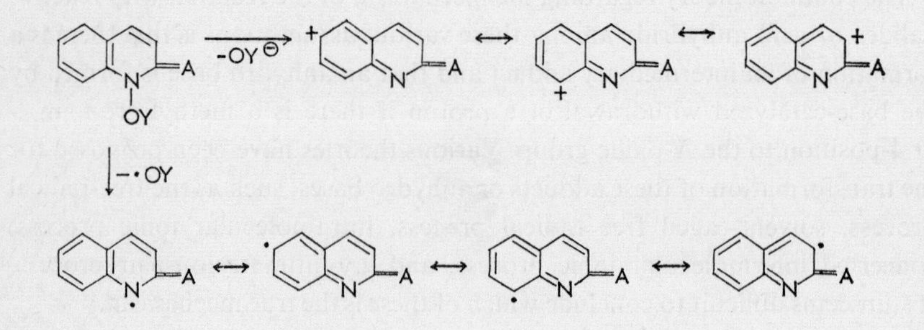

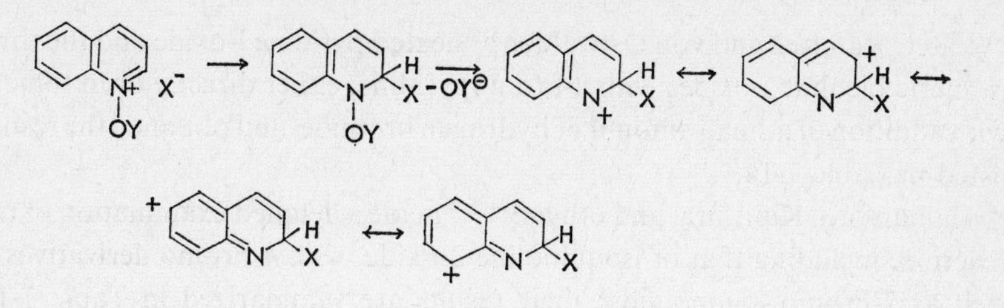

7.4.7 Reaction with Reactive Aryl Halides

Takeda and Hamamoto[211] found that heating of an equivalent amount of pyridine 1-oxide with 2-bromopyridine or 2-bromoquinoline directly on a water bath resulted in the formation of *N*-(2-pyridyl)-2-pyridone (7-31) and its monobromo derivative, or *N*-(2-pyridyl)carbostyril (7-32), respectively. The same reaction of quinoline 1-oxide with 2-bromopyridine or 2-bromoquinoline was found to give *N*-(2-quinolyl)-2-pyridone (7-33) and its monobromide, or *N*-(2-quinolyl)carbostyril (7-34)[212].

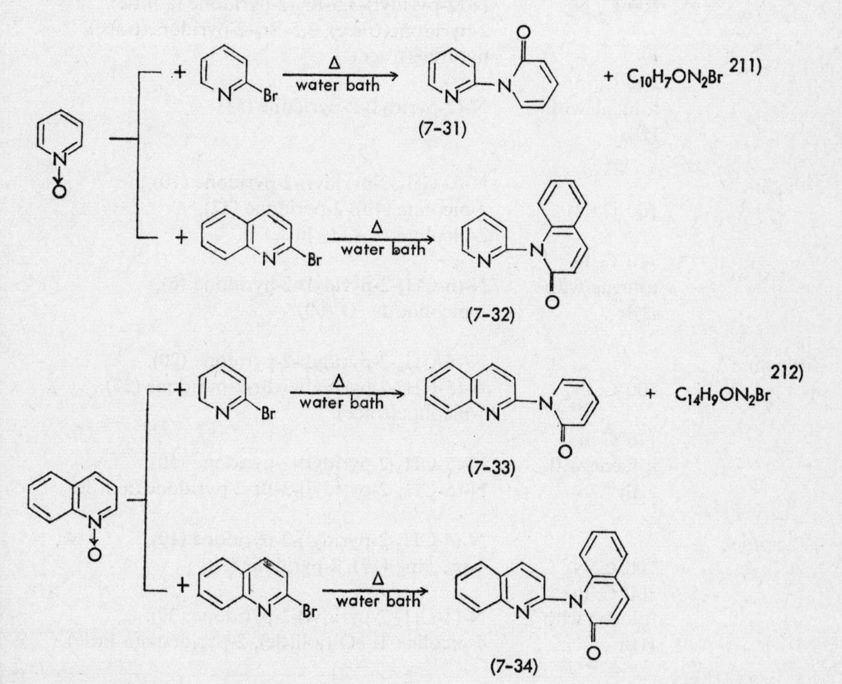

References p. 334

Later, Ramirez and von Ostwalden[213] heated pyridine l-oxide and the three isomeric picoline l-oxides with 2-bromopyridine, either directly or in toluene with addition of a small amount of hydrogen bromide, and obtained the results listed in Table 7-14.

Hamamoto, Kajiwara, and others[214,215] made a detailed examination of this reaction, including that of isoquinoline 2-oxide, with 4-bromo derivatives as well as 2-bromo compounds; their results are summarized in Table 7-15. The formation of 1-(4-pyridyl)-4-pyridone from the reaction of pyridine 1-oxide and 4-bromopyridine shows that the *N*-oxide group does not take

TABLE 7-14

REACTION OF PYRIDINE AND METHYLPYRIDINE 1-OXIDES WITH 2-BROMOPYRIDINE

N-Oxide of	Reaction conditions	Product (yield, %)
Pyridine	100°C, N_2	N-(2-pyridyl)-2-pyridone (20–30), N-(2-pyridyl)-3-Br-2-pyridone (20–30), N-(2-pyridyl)-3,5-Br_2-2-pyridone (a little), 2-pyridone(trace), 3,5-Br_2-2-pyridone(trace), pyridine(trace)
	100°C, in toluene with HBr	N-(2-pyridyl)-2-pyridone (53)
2-Picoline	100°C, N_2	N-(6-CH_3-2-pyridyl)-2-pyridone (10), 2-picoline (40), 2-pyridone (22), 2-picoline 1→O (a little)
	110°C, in toluene with HBr	N-(6-CH_3-2-pyridyl)-2-pyridone (6), 2-picoline 1→O (60)
3-Picoline	100°C, N_2	N-(5-CH_3-2-pyridyl)-2-pyridone (20), N-(5-CH_3-2-pyridyl)-3-Br-2-pyridone (20), 3-picoline(trace)
	110°C in toluene with HBr	N-(5-CH_3-2-pyridyl)-2-pyridone (50), N-(5-CH_3-2-pyridyl)-3-Br-2-pyridone (a little)
4-Picoline	100°C, N_2	N-(4-CH_3-2-pyridyl)-2-pyridone (10), 4-picoline (17), 2-pyridone (18)
	100°C, in toluene with HBr	N-(4-CH_3-2-pyridyl)-2-pyridone (30), 4-picoline 1→O (a little), 2-pyridone (a little)

part in this reaction and suggests that this product is a hydrolyzate of the dimer of 4-bromopyridine, since this pyridylpyridone is formed by the reaction of quinoline 1-oxide and 4-bromopyridine. Kajiwara and others proved its formation by obtaining this compound as the main product from the reaction of 3-picoline 1-oxide and 4-bromopyridine.

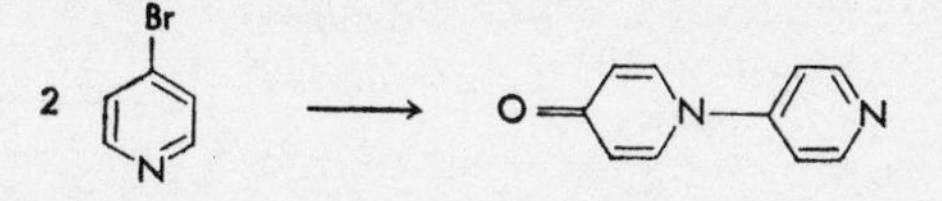

These reactions of aromatic *N*-oxide compounds with reactive aryl bromides give results very similar to those from the reaction of pyridine 1-oxide with tosyl chloride[130a] or with 2-pyridyl *p*-toluenesulfonate, etc.[130b] reported by den Hertog and others.

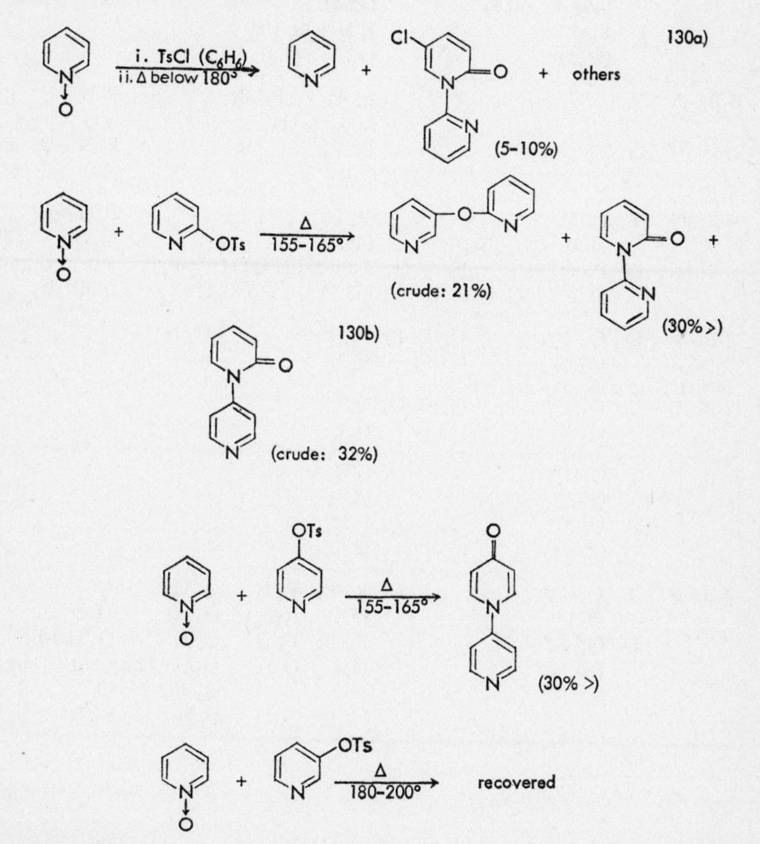

References p. 334

TABLE 7-15

REACTION OF *N*-OXIDES OF PYRIDINE, QUINOLINE, AND ISOQUINOLINE WITH REACTIVE ARYL BROMIDES

N-Oxide compd.	React. bromo compds.	Reaction condition	Product (yield, %)					
			Dimer			From bromo compound	From *N*-oxide compound	Others
			Normal	Bromide	Ether			
Pyridine 1-oxide	2-Br–P	D	$P_2–P_2'$ (60.2)	(0.7)	$P_3–O–P_2$ (2.6)	$P_2'H$ (3.1)		
	2-Br–Q	D	$P_2–Q_2'$ (61.7)		$P_3–O–Q_2$ (0.9)	$Q_2'H$ (18.9)		
	1-Br–I	T	$P_2–I_1'$ (45)	(0.9)		$I_1'H$ (15.4)		
		D	$P_2–I_1'$ (76.5)			$I_1'H$ 4-Br–$I'H$ (12.3) (7.3)		
	4-Br–P	T				$P_4–P_4'$ $P_4–P_4'.Br$ (61) (1)		P–N→O P (41) (1)
		D				$P_4–P_4'$ (64)		P–N→O P (40) (2)
	4-Br–Q	T	$P_2–Q_4'$ (0.6)			$Q_4'H$ (46.7)		P–N→O P (4.8) (2)
		D	$P_2–Q_4'$ (0.6)	(1.4)	$P_3–O–Q_4$ (0.3)	3-Br–$Q_4'H$ $Q_4'H$ (33.1) (15.8)	$P_2–P_2'$ (16.1)	P (a little)
Quinoline 1-oxide	2-Br–P	D	$Q_2–P_2'$ (57.1)		$Q_3–O–P_2$ (11.9)	$P_2'H$ (0.2)	$Q_2'H$ (0.8)	
	2-Br–Q	D	$Q_2–Q_2'$ (70.8)		$Q_3–O–Q_2$ (13.7)	$Q_2'H$ (1.1)		
	1-Br–I	T	$Q_2–I_1'$ (70)	(1.7)		$I_1'H$ (4.4)		
		D	$Q_2–I_1'$ (77.9)		$Q_3–O–I_1$ (5.4)	$I_1'H$ (2.6)		
	4-Br–P	T	$Q_2–P_4'$ (32)			$P_4–P_4'$ $P_4'H$ (5.4) (trace)	$Q_2'H$ (3.28)	Q (1.4)
		D	$Q_2–P_4'$ (40.5)			$P_4–P_4'$ $P_4'H$ (11) (5.5)	$Q_2'H$ Q_2Br (10.7) (2.8) $Q_2–Q_2'$ (1.7)	$Q_4'H$ 3-Br–$Q_4'H$ (3.6) (5.6) Q (4.7)

Table 7-15 (*continued*)

N-Oxide compd.	React. bromo compds.	Reaction condition	Dimer: Normal	Dimer: Bromide	Dimer: Ether	Product (yield, %): From bromo compound	From *N*-oxide compound	Others
	4-Br–Q	T	Q_2–Q_4' (13)	(3.8)		$Q_4'H$ (8)	$Q_2'H$ (5.4)	
		D	Q_2–Q_4' (23.5)	(1.3)	Q_3–O–Q_4 (1)	$Q_4'H$ (12)	$Q_2'H$ (10.6)	Q (10.6)
Isoquinoline N-oxide	2-Br–P	T	I_1–P_2' (52)	(0.5)		$P_2'H$ (2.6)		4-Br–I (8.4)
		D	I_1–I_1' (57.4)			$P_2'H$ (3.1)		4-Br–I (4.6); I (0.6)
	2-Br–Q	T	I_1–Q_2' (30)	(1)		$Q_2'H$ (13.2)		4-Br–I (3.7); (1.6)
		D	I_1–Q_2' (28.6)			$Q_2'H$ (47.8)	4-Br–$I_1'H$ (7.2)	4-Br–I (7.2); I (1.4)
	1-Br–I	T	I_1–I_1' (62.2)			$I_1'H$ (8.5)		4-Br–I (9.5)
		D	I_1–I_1' (55.7)			$I_1'H$ (11.3)		4-Br–I (5.5)
	4-Br–P	T	I_1–P_4' (17.3)			P_4–P_4' (23.6)	$I_1'H$ (7.7); I_1–I_1' (5.0)	4-Br–I (9.5)
		D	I_1–P_4' (31.4)			P_4–P_4' (26.4); $P_4'H$ (11.2)	$I_1'H$ (2.5)	4-Br–I (3.7)
	4-Br–Q	T	I_1–Q_4' (18)				$I_1'H$ (8.2); I_1–I_1' (1.5)	4-Br–I (4.4)
		D	I_1–Q_4' (22.3)			$Q_1'H$ (23)	$I_1'H$ (9.3)	4-Br–I (10.5); I (1)

T : Refluxed in toluene with catalytic amount of hydrogen bromide-acetic acid.

D : Refluxed in dioxan

P : Pyridine, P_2: 2-Pyridyl, P_3: 3-Pyridyl, P_4: 4-Pyridyl,

P_2': P_4':

Q: Quinoline, Q_2: 2-Quinolyl, Q_3: 3-Quinolyl, Q_4: 4-Quinolyl,

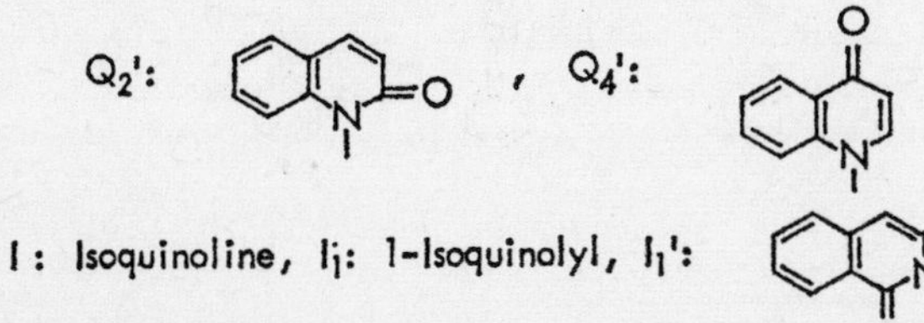

I : Isoquinoline, I_1: 1-Isoquinolyl, I_1':

The chemical reactivity of 2- and 4-tosyloxypyridine may be considered as similar to that of 2- and 4-bromopyridine, so that the reactions illustrated above may be taken to be of the same type as that of bromopyridines.

For the mechanism of these reactions, Takeda and others[211] at first assumed that 2-bromopyridine acted in the same way as alkyl halides and that the co-ordination compound formed by this reaction underwent decomposition due to the lability of the nitrogen–oxygen bond, and the resonance system of the decomposed fragments reunited to form the dimeric product.

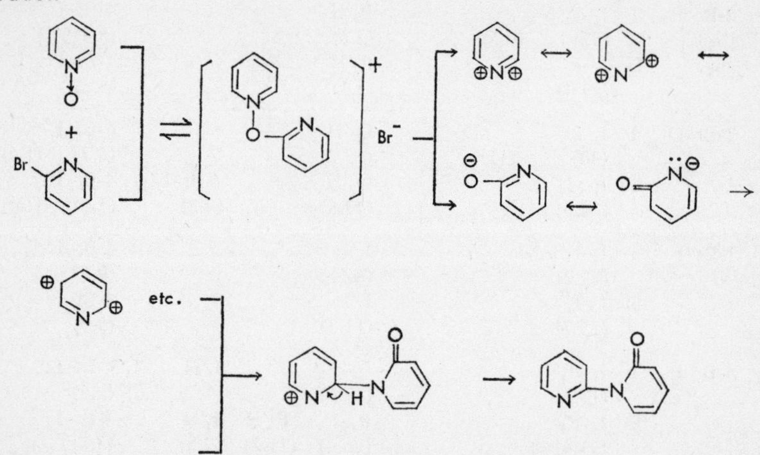

Ramirez and von Ostwalden[213] assumed the following cyclic process and considered that the by-product formation of the bromo derivatives was due to bromination of intermediate products by hypobromous acid formed by the reaction of hydrogen bromide generated in the medium and pyridine 1-oxide.

$$\geqslant N^{+}\text{–}OH \cdot Br^{-} \rightarrow \geqslant N \cdot HOBr$$

The exclusive formation of the 5′-methyl derivative in the reaction of 3-picoline l-oxide, without formation of the 3′-methyl compound, was assumed by these workers to be due to the steric hindrance of the cyclization by the methyl group to its *ortho* position.

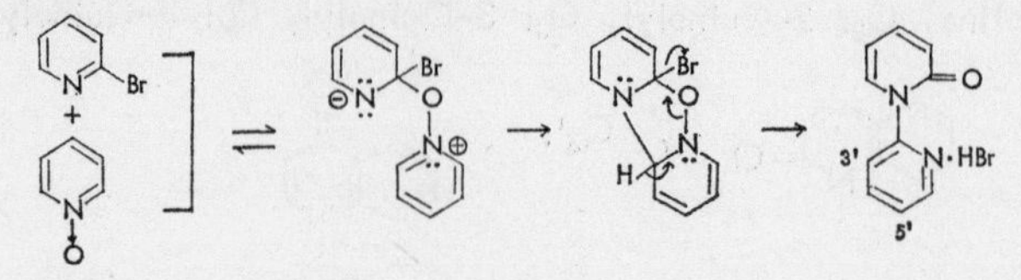

The hypothesis put forward by Takeda and others has a weak point in the assumption of an intermediary di-cation, and there are several details in these reactions which cannot be explained entirely by this hypothesis, such as the formation of an ether-type dimer only in 3-position and not in 2- or 4-position, no formation of an ether dimer by the reaction of isoquinoline 2-oxide but bonding of bromine ion in 4-position, and the fact that, considering the resonance system of the pyridyl di-cation, bonding of pyridone nitrogen is possible also in 3- and 4-positions, so that a larger number of products should have been formed.

The mechanism assumed by Ramirez and Ostwalden seems reasonable in that it can explain the formation (below) of an ether dimer and of the bromo derivative and pyridone. However, their hypothesis, *per se*, makes it difficult to explain the formation of a normal dimer by the reaction of pyridine 1-oxide with 4-bromoquinoline, and of quinoline 1-oxide with 4-bromopyridine or 4-bromo-quinoline, and the formation of a considerable amount of *N*-(4-pyridyl)-2-pyridone from the reaction of pyridine 1-oxide with 2-tosyloxypyridine.

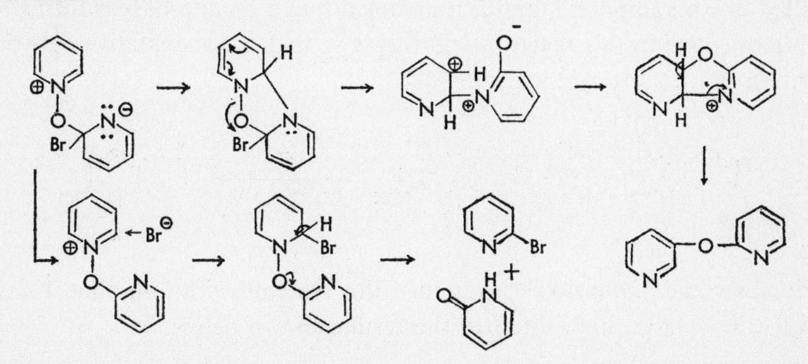

A mechanism which would explain all the data of this reaction has not been put forward as yet and this is one of the points which will have to await further examination.

7.5 Oxidative Hydroxylation

As has already been stated (*cf. Section 3.1.3.b*), direct *N*-oxidation of π-electron-deficient aromatic *N*-heterocycles is accompanied by hydroxylation, and a hydroxyl derivative or hydroxylated *N*-oxide compound is sometimes formed. In the case of 6-nitroquinoline, hydroxylation appears to precede *N*-oxidation. In the majority of cases, however, the reaction seems to result from increased activation of the 2-position to hydroxylation accompanying *N*-oxidation, as shown below.

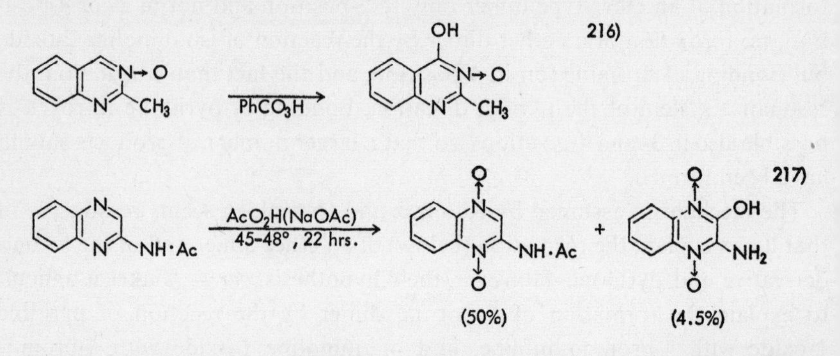

It is well known that quaternary ammonium salts of π-electron-deficient aromatic *N*-heterocycles like the alkylpyridinium and alkylquinolinium salts are oxidized by ferricyanides in alkaline medium to the corresponding *N*-alkyl-2-oxo compound, and it is assumed that a pseudo-base is formed as an intermediate in this reaction and this base undergoes oxidative dehydrogenation by the ferricyanide ion[218,219].

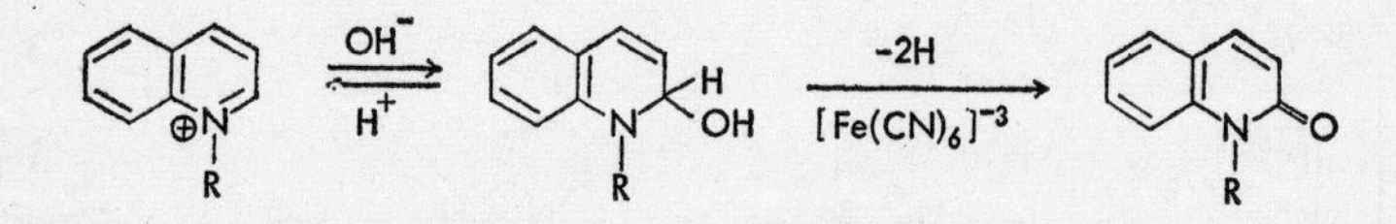

Hamana and Yamazaki[220] examined this reaction with quinoline 1-oxide and its derivatives, and obtained the results shown below.

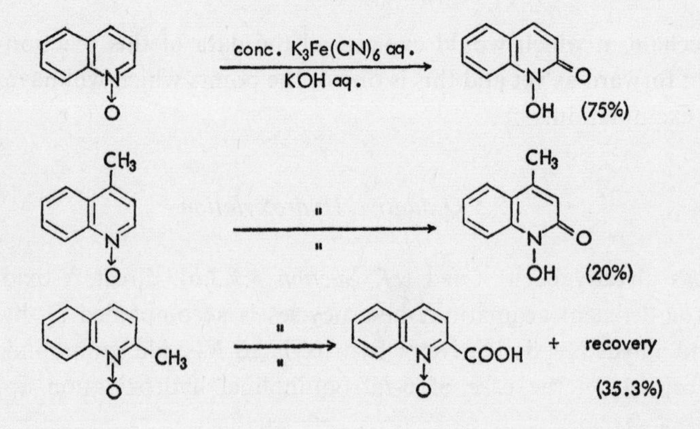

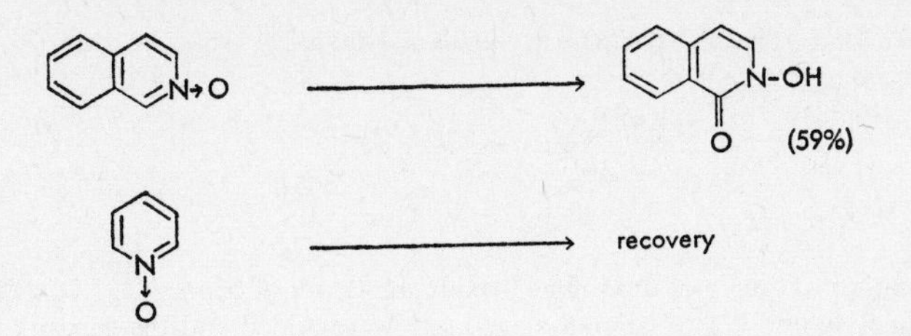

In lepidine 1-oxide, in which the electron-withdrawing effect of the *N*-oxide group has been decreased by inductive effect of the methyl group, the yield of the corresponding 1-hydroxycarbostyril is markedly lowered, while only the oxidation of the methyl group occurs in quinaldine 1-oxide, in which the 2-position is occupied. Moreover, the fact that this reaction does not take place in pyridine l-oxide, contrary to the case of the alkylpyridinium salt, and the starting material is recovered, indicates that activation of the 2-position to nucleophilic substitution by the *N*-oxide group is less than that by the nitrogen of quaternary ammonium salts.

Ochiai and Ohta[221] found that the heating of quinoline 1-oxide with lead tetraacetate in benzene or chloroform solution results in the formation of an amorphous compound, assumed to be *N*-acetyloxycarbostyril, and by allowing this product to stand in air or by its hydrolysis with dilute hydrochloric acid, 1-hydroxycarbostyril (7-35) is formed in a satisfactory yield.

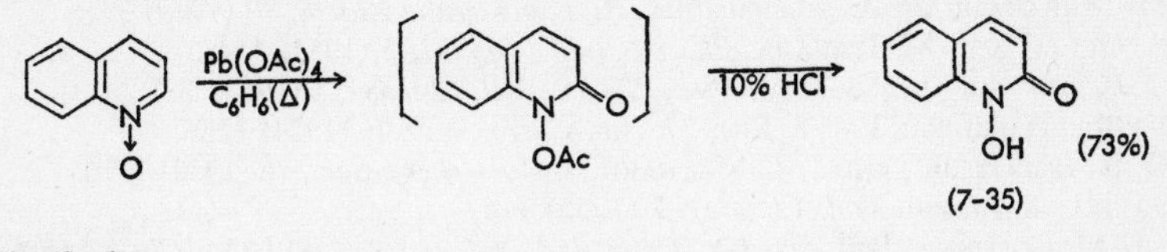

Similarly

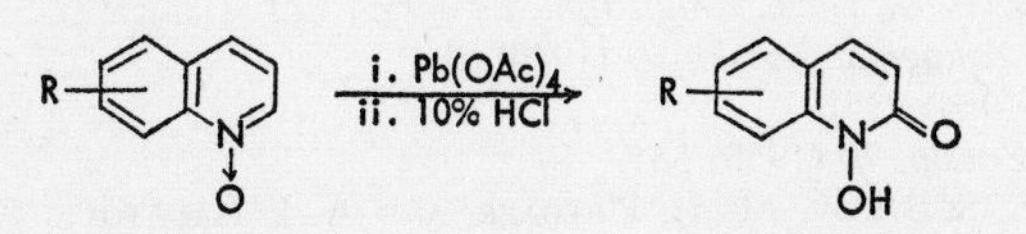

R= 4-CH_3 (81%), 6-CH_3 (78.8%), 3-Br (62.6%), 3-Cl (66.4%)

This reaction does not take place with pyridine 1-oxide, and is not effected in 4-nitroquinoline 1-oxide, which facts suggest that acylation of the oxygen in the *N*-oxide group and the naphthoid reactivity of quinoline are important factors affecting the progress of the reaction. Ohta[222] proved the foregoing assumption by carrying out this reaction with lead tetrabenzoate, obtaining *N*-benzoyloxycarbostyril as prisms of m.p. 128–131°.

References p. 334

Example: *N-Hydroxycarbostyril from Quinoline 1-Oxide*[221]

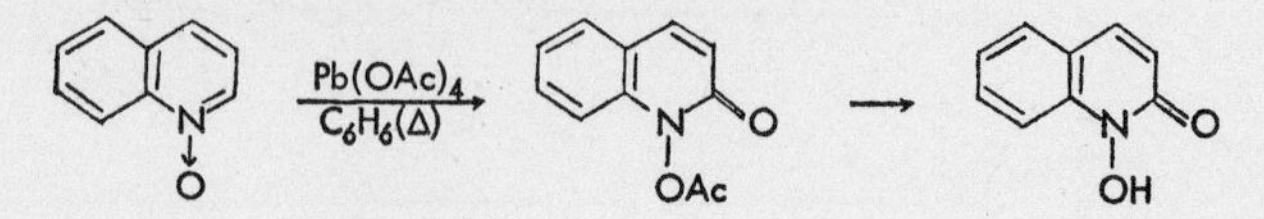

To a solution of 4.65 g of quinoline 1-oxide in 300 ml of benzene, 21.28 g (1.5 moles) of lead tetraacetate and 1 g of calcium carbonate were added and the mixture was refluxed for 1.5 h. The separated precipitate was collected by filtration and washed with chloroform. The combined filtrate and washings were concentrated to a small volume and further evaporated under a reduced pressure. The syrupy residue was mixed with 20 ml of 10% hydrochloric acid solution and warmed on a water bath for 30 min, by which the residue became crystalline. The crystals were collected by filtration and recrystallized from ethanol to pale brown prisms, m.p. 188–189°. Yield, 3.88 g (73.4%).

REFERENCES

1 J. Meisenheimer, *Ber.*, 59 (1926) 1848.
2 B. Bobranski, *Ber.*, 71 (1938) 578.
3 B. Bobranski, L. Kochansko, and A. Kowalewska, *Ber.*, 71 (1938) 2385.
4 M. Henze, *Ber.*, 69 (1936) 1566.
5 M. Katada, *Yakugaku Zasshi*, 67 (1947) 51.
6 T. Higashino, *Chem. Pharm. Bull. (Tokyo)*, 9 (1961) 635.
7 K. Lehmstedt and H. Klee, *Ber.*, 69 (1936) 1157.
8 E. Hayashi and Ch. Iijima, *Yakugaku Zasshi*, 82 (1962) 1093.
9 M. Colonna and A. Risaliti, *Gazz. Chim. Ital.*, 86 (1956) 688, 705.
10 H. Tanida, *Chem. Pharm. Bull. (Tokyo)*, 7 (1959) 887.
11 M. Murakami and E. Matsumura, *Nippon Kagaku Zasshi*, 70 (1949) 393.
12 E. Ochiai and M. Ikehara, *Pharm. Bull. (Japan)*, 3 (1955) 454.
13 G. Kobayashi and S. Furukawa, *Pharm. Bull. (Japan)*, 1 (1953) 347.
14 V. Boekelheide and W. J. Linn, *J. Am. Chem. Soc.*, 76 (1954) 1286.
15 O. H. Bullit, Jr., and J. T. Maynard, *J. Am. Chem. Soc.*, 76 (1954) 1370.
16 S. Okuda, *Pharm. Bull. (Japan)*, 3 (1955) 316.
17 *(a)* M. Colonna, *Boll. Sci. Fac. Chim. Ind. Bologna*, 4 (1940) 134 [*Chem. Abstr.*, 34 (1940) 7290]. *(b)* M. Colonna and A. Risaliti, *Gazz. Chim. Ital.*, 83 (1953) 58. *(c)* M. Colonna and S. Fatutta, *Gazz. Chim. Ital.*, 83 (1953) 622.
18 E. Ochiai and K. Arima, *Yakugaku Zasshi*, 69 (1949) 51.
19 A. Risaliti, *Ric. Sci.*, 24 (1954) 2352.
20 G. Kobayashi, *Yakugaku Zasshi*, 70 (1950) 381.
21 O. S. Otroshchenko, A. S. Sodykov, M. U. Urebaer, and A. I. Isametova, *Zh. Obshch. Khim.*, 33 (1963) 1038.
22 E. Hayashi and T. Higashino, *Chem. Pharm. Bull. (Tokyo)*, 12 (1964) 43.
23 T. Kato and H. Yamanaka, *J. Org. Chem.*, 30 (1965) 910.
24 J. H. Blumenthal and N. J. Plainfield, *U.S. Pat.*, 2,874,162 (February 17, 1959) [*Chem. Abstr.*, 53 (1959) 12311].
25 Ch. Kaneko and T. Miyasaka, *Shika Zairyo Kenkyusho Hokoku*, 2 (1963) 269.
26 T. Kato, H. Yamanaka, and M. Hikichi, *Yakugaku Zasshi*, 85 (1965) 331.
27 V. Boekelheide and R. Scharrer, *J. Org. Chem.*, 26 (1961) 3802.
28 B. Umezawa, *Chem. Pharm. Bull. (Tokyo)*, 8 (1960) 698.

29 *(a)* R. HUISGEN, *Proc. Chem. Soc.*, (1961) 357. *(b)* R. HUISGEN, *Angew. Chem.*, 75 (1963) 742. *(c)* R. HUISGEN, *Angew. Chem.*, 75 (1963) 628.
30 E. HAYASHI, *Yakugaku Zasshi*, 81 (1961) 1030.
31 H. SEIDEL AND R. HUISGEN, *Tetrahedron Letters*, 29 (1963) 2023.
32 *(a)* S. TAKAHASHI AND H. KANÔ, *Tetrahedron Letters*, 25 (1963) 1687. *(b)* S. TAKAHASHI AND H. KANÔ, *Chem. Pharm. Bull. (Tokyo)*, 12 (1964) 1290.
33 M. HAMANA AND M. YAMAZAKI, *Yakugaku Zasshi*, 81 (1961) 574.
34 M. HAMANA AND H. NODA, *Yakugaku Zasshi*, 83 (1963) 342.
35 H. GILMAN AND J. T. EDWARD, *Can. J. Chem.*, 31 (1953) 457.
36 *(a)* T. KATO, *Yakugaku Zasshi*, 75 (1955) 1239. *(b)* T. KATO, *Yakugaku Zasshi*, 75 (1955) 1236.
37 H. TANI, *Yakugaku Zasshi*, 81 (1961) 182.
38 R. W. GOULEY, G. W. MOERSCH, AND H. S. MOSHER, *J. Am. Chem. Soc.*, 69 (1947) 303.
39 G. B. BACHMANN AND D. E. COOPER, *J. Org. Chem.*, 9 (1944) 302.
40 L. BERNARDI, G. PALAMIDESSI, A. LEONE, AND G. LARINI, *Gazz. Chim. Ital.*, 91 (1961) 1431.
41 M. COLONNA, A. RISALITI, AND R. SERRA, *Gazz. Chim. Ital.*, 85 (1955) 1508.
42 G. BUCHMANN. *Chem. Tech. (Berlin)*, 9 (1957) 388 [*Chem. Abstr.*, 53 (1959) 2229].
43 *(a)* T. ITAI, *Yakugaku Zasshi*, 65B (1945) 4. *(b)* H. J. DEN HERTOG AND J. OVERHOFF, *Rec. Trav. Chim.*, 69 (1950) 468. *(c)* T. ITAI, *Eisei Shikensho Hokoku*, 67 (1950) 125.
44 E. OCHIAI, *J. Org. Chem.*, 18 (1953) 549.
45 M. HAMANA, Y. HOSHIDE, AND K. KANEDA, *Yakugaku Zasshi*, 76 (1956) 1337.
46 H. J. DEN HERTOG, J. MAAS, C. R. KOLDER, AND W. P. COMBÉ, *Rec. Trav. Chim.*, 74 (1955) 59.
47 CH. KANEKO, *Chem. Pharm. Bull. (Tokyo)*, 7 (1959) 273.
48 M. M. ROBISON AND B. L. ROBISON, *J. Am. Chem. Soc.*, 80 (1958) 3443.
49 I. SUZUKI, *Yakugaku Zasshi*, 71 (1951) 789.
50 *(a)* E. OCHIAI, G. KOBAYASHI, M. HAMANA, AND T. SUGASAWA, *Yakugaku Zasshi*, 71 (1951) 260. *(b)* M. HAMANA AND K. UZU, *Yakugaku Zasshi*, 74 (1954) 1315. *(c)* E. OCHIAI AND M. Y. ISHIKAWA, *Chem. Pharm. Bull. (Tokyo)*, 6 (1958) 212. *(d)* E. OCHIAI, M. Y. ISHIKAWA, AND Y. OKA, *Ann. Rept. Itsuu Lab.*, 12 (1962) 29. *(e)* E. OCHIAI, H. KATAOKA, T. DODO, AND M. TAKAHASHI, *Ann. Rept. Itsuu Lab.*, 12 (1962) 11.
51 G. R. REDFORD, A. R. KATRITZKY, AND H. M. WUEST, *J. Chem. Soc.*, (1963) 4600.
52 H. J. DEN HERTOG AND C. HOOGZAND, *Rec. Trav. Chim.*, 76 (1957) 261.
53 Y. SUZUKI, *Pharm. Bull. (Japan)*, 5 (1957) 78.
54 L. REPPEL, H. D. EILHAUER, AND P. KRETSCHMER, *Arch. Pharm.*, 296 (1963) 169.
55 E. C. TAYLOR AND J. S. DRISCOLL, *J. Org. Chem.*, 25 (1960) 1716.
56 E. OCHIAI AND CH. KANEKO, *Chem. Pharm. Bull. (Tokyo)*, 8 (1960) 28.
57 H. J. DEN HERTOG AND N. A. I. M. BOELRIJK, *Rec. Trav. Chim.*, 70 (1951) 578.
58 E. C. TAYLOR AND A. J. CROVETTI, *J. Org. Chem.*, 19 (1954) 1633.
59 K. PALAT, M. CELADNIK, L. NOVACEK, AND M. POLSTEV, *Cesk. Farm.*, 6 (1957) 369 [*Chem. Abstr.*, 52 (1958) 10071].
60 E. C. TAYLOR, JR., AND A. J. CROVETTI, *J. Am. Chem. Soc.*, 78 (1956) 214.
61 R. JÛJO, *Yakugaku Zasshi*, 66B (1946) 49.
62 M. VAN AMMERS AND H. J. DEN HERTOG, *Rec. Trav. Chim.*, 74 (1955) 1160.
63 M. O. ISHIKAWA, *Yakugaku Zasshi*, 65 (1945) 105.
64 M. VAN AMMERS AND H. J. DEN HERTOG, *Rec. Trav. Chim.*, 75 (1956) 1259.
65 G. BÜCHI, R. E. MANNING, AND F. A. HOCHSTEIN, *J. Am. Chem. Soc.*, 84 (1962) 3394.
66 A. RISALITI, *Ric. Sci.*, 24 (1954) 2351 [*Chem. Abstr.*, 49 (1954) 15902].
67 E. OCHIAI, M. O. ISHIKAWA, AND Z. SAI, *Yakugaku Zasshi*, 63 (1943) 280.

68 O. J. Magidson and M. W. Rubzow, *Zh. Obshch. Khim.*, 7 (1937) 1896 [*Chem. Zentr.*, 1 (1938) 3774].
69 E. Ochiai, A. Ohta, and H. Nomura, *Pharm. Bull. (Japan)*, 5 (1957) 310.
70 M. Katada, *Yakugaku Zasshi*, 68 (1948) 123.
71 M. Colonna and A. Risaliti, *Boll. Sci. Fac. Chim. Ind. Bologna*, 9 (1951) 82 [*Chem. Abstr.*, 46 (1952) 7102].
72 M. Colonna, *Boll. Sci. Fac. Chim. Ind. Bologna*, 1 (1941) 86 [*Chem. Zentr.*, I (1942) 2266].
73 E. Ochiai and Ch. Kaneko, *Chem. Pharm. Bull. (Tokyo)*, 7 (1959) 267.
74 E. Ochiai, Ch. Kaneko, and J. Inomata, *Yakugaku Zasshi*, 78 (1958) 584.
75 S. Yoshida, *Yakugaku Zasshi*, 66 (1946) 158.
76 M. O. Ishikawa, *Proc. Imp. Acad., Tokyo*, 20 (1944) 599.
77 *(a)* G. Buchmann and R. Ernst, *Ger. (East) Pat.* 20,182 (Nov. 1, 1960) [*Chem. Abstr.*, 56 (1962) 462]. *(b)* G. Buchmann, *Z. Chem.*, 1 (1961) 84.
78 *(a)* M. O. Ishikawa, *Yakugaku Zasshi*, 65B (1945) 98. *(b)* M. O. Ishikawa and I. Kikkawa, *Yakugaku Zasshi*, 75 (1955) 33.
79 T. Naito, *Yakugaku Zasshi*, 67 (1947) 144.
80 Y. Kobayashi, *Chem. Pharm. Bull. (Tokyo)*, 6 (1958) 273.
81 I. Iwai, *Yakugaku Zasshi*, 71 (1951) 1291.
82 I. Iwai, *Yakugaku Zasshi*, 71 (1951) 1288.
83 N. H. Cromwell and R. A. Mitsch, *J. Org. Chem.*, 26 (1961) 3812.
84 M. Ikehara, *Pharm. Bull. (Japan)*, 2 (1954) 111.
85 E. Ochiai and M. Ikehara, *Pharm. Bull. (Japan)*, 2 (1954) 109.
86 E. Ochiai and M. Ikehara, *Pharm. Bull. (Japan)*, 2 (1954) 72.
87 G. Buchmann and J. Schumann, *Wiss. Z. Tech. Hochsch. Chem. Leuna-Merseburg*, 4(No. 1) (1961-62) 1 [*Chem. Abstr.*, 58 (1962) 4520].
88 E. P. Hart, *J. Chem. Soc.*, (1954) 1879.
89 M. Ogata, *Chem. Pharm. Bull. (Tokyo)*, 11 (1963) 1522.
90 H. Igeta, *Chem. Pharm. Bull. (Tokyo)*, 7 (1959) 938.
91 H. Igeta, *Chem. Pharm. Bull. (Tokyo)*, 8 (1960) 368.
92 S. Sako, *Yakugaku Zasshi*, 82 (1962) 1208.
93 T. Itai and S. Kamiya, *Chem. Pharm. Bull. (Tokyo)*, 11 (1963) 348.
94 M. Ogata and H. Kanô, *Chem. Pharm. Bull. (Tokyo)*, 11 (1963) 29.
95 I. Suzuki, T. Nakajima, and T. Itai, *Chem. Pharm. Bull. (Tokyo)*, 11 (1963) 268.
96 M. Ogata, H. Kanô, and K. Tôri, *Chem. Pharm. Bull. (Tokyo)*, 11 (1963) 1527.
97 G. T. Newbold and F. S. Spring, *J. Chem. Soc.*, (1947) 1183.
98 G. Palamidessi and L. Bernardi, *Gazz. Chim. Ital.*, 93 (1963) 339.
99 G. Karmas and P. E. Spoer, *J. Am. Chem. Soc.*, 78 (1956) 4071.
100 J. K. Landquist, *J. Chem. Soc.*, (1953) 2816.
101 E. Hayashi, Ch. Iijima, and Y. Nagasawa, *Yakugaku Zasshi*, 84 (1964) 171.
102 *(a)* T. Higashino, *Yakugaku Zasshi*, 79 (1959) 699. *(b)* T. Higashino, *Chem. Pharm. Bull. (Tokyo)*, 9 (1961) 636.
103 E. Hayashi and T. Higashino, *Chem. Pharm. Bull. (Tokyo)*, 12 (1964) 59.
104 A. Reissert, *Ber.*, 38 (1905) 1603.
105 *(a)* E. Ochiai and I. Nakayama, *Yakugaku Zasshi*, 65B (1945) 582. *(b)* I. Nakayama, *Yakugaku Zasshi*, 70 (1950) 355.
106 T. Takahashi, J. Okada, and Y. Hamada, *Yakugaku Zasshi*, 77 (1957) 1243.
107 M. Hamana and I. Kumadaki, *Yakugaku Zasshi*, 86 (1966) 1090.
108 E. Ochiai and Z. Sai, *Yakugaku Zasshi*, 65 (1945) 418.
109 V. Georgian, R. J. Harrison, and L. L. Skaletzky, *J. Org. Chem.*, 27 (1962) 4577.
110 *(a)* E. Ochiai and H. Yamanaka, *Pharm. Bull. (Japan)*, 3 (1955) 175. *cf.* M. Ogata

H. WATANABE, K. TÔRI, AND H. KANÔ, *Tetrahedron Letters*, (1964) 19. *(b)* H. YAMANAKA, *Chem. Pharm. Bull. (Tokyo)*, 6 (1958) 633. *(c)* H. YAMANAKA, *Chem. Pharm. Bull. (Tokyo)*, 7 (1959) 297.
111 S. TAKAHASHI AND H. KANÔ, *Chem. Pharm. Bull. (Tokyo)*, 12 (1964) 783.
112 H. IGETA, *Chem. Pharm. Bull. (Tokyo)*, 11 (1963) 1473.
113 E. OCHIAI, M. HAMANA, Y. KOBAYASHI, AND CH. KANEKO, *Chem. Pharm. Bull. (Tokyo)*, 8 (1960) 487.
114 M. HAMANA AND K. SHIMIZU, *Yakugaku Zasshi*, 86 (1966) 59.
115 CH. KANEKO, *Chem. Pharm. Bull. (Tokyo)*, 8 (1960) 286.
116 M. HAMANA AND K. FUNAKOSHI, *Yakugaku Zasshi*, 80 (1960) 1031.
117 M. HENZE, *Ber.*, 69 (1936) 534.
118 I. J. PACHTER, *J. Am. Chem. Soc.*, 75 (1953) 3026.
119 T. ITAI AND S. NATSUME (née SUZUKI), *Chem. Pharm. Bull. (Tokyo)*, 12 (1964) 228.
120 M. HAMANA AND H. NODA, *Chem. Pharm. Bull. (Tokyo)*, 11 (1963) 1331; 13 (1965) 912.
121 I. NAKAYAMA, *Yakugaku Zasshi*, 70 (1950) 423.
122 E. MATSUMURA, *Nippon Kagaku Zasshi*, 74 (1953) 446, 547; *Mem. Osaka Univ. Liberal Arts Educ.*, 1 (1952) 1.
123 *(a)* E. MATSUMURA, *Nippon Kagaku Zasshi*, 74 (1953) 363. *(b)* E. MATSUMURA, T. HIROOKA, AND K. IMAGAWA, *Nippon Kagaku Zasshi*, 82 (1961) 104.
124 E. MATSUMURA, *Mem. Osaka Univ. Liberal Arts Educ.*, 10 (1962) 191.
125 E. MATSUMURA AND H. TAKEDA, *Nippon Kagaku Zasshi*, 81 (1960) 515.
126 R. R. HUNT, J. F. W. MCOMIE, AND E. R. SAYER, *J. Chem. Soc.*, (1959) 525.
127 M. MURAKAMI AND E. MATSUMURA, *Nippon Kagaku Zasshi*, 72 (1951) 509.
128 *(a)* H. TANIDA, *Yakugaku Zasshi*, 78 (1958) 1083. *(b)* H. TANIDA, *Yakugaku Zasshi*, 78 (1958) 611.
129 M. HAMANA AND M. YAMAZAKI, *Chem. Pharm. Bull. (Tokyo)*, 11 (1963) 415.
130 *(a)* P. A. DE VILLIERS AND H. J. DEN HERTOG, *Rec. Trav. Chim.*, 75 (1956) 1303. *(b)* P. A. DE VILLIERS AND H. J. DEN HERTOG, *Rec. Trav. Chim.*, 76 (1957) 647.
131 E. OCHIAI, T. WATANABE, AND S. SUZUKI, *Yakugaku Zasshi*, 76 (1956) 1421.
132 E. OCHIAI AND T. NAKAGOME, *Chem. Pharm. Bull. (Tokyo)*, 6 (1958) 495.
133 E. OCHIAI AND T. YOKOKAWA, *Yakugaku Zasshi*, 75 (1955) 213.
134 T. YOSHIKAWA, *Yakugaku Zasshi*, 81 (1961) 1601.
135 H. YAMANAKA, *Chem. Pharm. Bull. (Tokyo)*, 7 (1959) 152.
136 H. TANIDA, *Yakugaku Zasshi*, 78 (1956) 608.
137 H. TANIDA, *Chem. Pharm. Bull. (Tokyo)*, 7 (1959) 944.
138 M. HAMANA AND O. HOSHINO, *Yakugaku Zasshi*, 84 (1964) 35.
139 M. HAMANA AND K. FUNAKOSHI, *Yakugaku Zasshi*, 82 (1962) 512.
140 T. ZINCKE, G. HEUSER, AND W. MÖLLER, *Ann.*, 333 (1904) 296.
141 *(a)* D. JERCHEL, H. FISCHER, AND K. THOMAS, *Ber.*, 89 (1956) 2921. *(b)* D. JERCHEL AND L. JACOB, *Ber.*, 91 (1958) 1226.
142 M. HAMANA AND K. FUNAKOSHI, *Yakugaku Zasshi*, 82 (1962) 518.
143 M. HAMANA AND K. FUNAKOSHI, *Yakugaku Zasshi*, 82 (1962) 523.
144 M. HAMANA AND K. FUNAKOSHI, *Yakugaku Zasshi*, 84 (1964) 42.
145 M. HAMANA AND K. FUNAKOSHI, *Yakugaku Zasshi*, 84 (1964) 23.
146 M. HAMANS AND K. FUNAKOSHI, *Yakugaku Zasshi*, 84 (1964) 28.
147 CH. KANEKO, T. MIYASAKA, AND I. YOKOE, *Shika Zairyo Kenkyusho Hokoku*, 2 (1963) 473.
148 K. ODA, *Yakugaku Zasshi*, 64B (1944) 76.
149 E. OCHIAI AND T. OKAMOTO, *Yakugaku Zasshi*, 68 (1948) 88.
150 E. OCHIAI, T. OKAMOTO, AND Y. KOBAYASHI, *Yakugaku Zasshi*, 68 (1948) 109.
151 E. OCHIAI AND M. IKEHARA, *Yakugaku Zasshi*, 68 (1948) 139.

152 M. C. Cava and B. Weinstein, *J. Org. Chem.*, 23 (1958) 1616.
153 V. Boekelheide and W. L. Lehn, *J. Org. Chem.*, 26 (1961) 428.
154 W. Sauermilch, *Arch. Pharm.*, 293 (1960) 452.
155 T. Takahashi and K. Kariyone, *Chem. Pharm. Bull. (Tokyo)*, 8 (1960) 1106.
156 B. M. Bain and J. E. Saxton, *Chem. Ind. (London)*, (1960), 402; *J. Chem. Soc.*, (1961) 5216.
157 K. Takeda and K. Igarashi, *Shionogi Kenkyusho Nempo*, 1 (1951) 1.
158 K. R. Huffman, F. C. Schaefer, and G. A. Peters, *J. Org. Chem.*, 27 (1962) 554; *cf.* M. L. Peterson, *J. Org. Chem.*, 25 (1960) 565.
159 H. U. Doeniker and J. Druey, *Helv. Chim. Acta*, 41 (1958) 2148.
160 M. M. Robison and B. L. Robison, *J. Org. Chem.*, 21 (1957) 1337.
161 H. Igeta, *Chem. Pharm. Bull. (Tokyo)*, 7 (1959) 938.
162 K. Kumagai, *Nippon Kagaku Zasshi*, 81 (1960) 350, 1148.
163 T. Nakagome, *Yakugaku Zasshi*, 82 (1962) 249.
164 T. Nakagome, *Yakugaku Zasshi*, 83 (1963) 934.
165 H. Bredereck, R. Grompper, and H. Herlinger, *Ber.*, 91 (1958) 2832.
166 C. F. Koelsch and W. H. Gumprecht, *J. Org. Chem.*, 23 (1958) 1603.
167 *(a)* M. A. Stevens, A. Giner-Sorolla, H. W. Smith, and G. B. Brown, *J. Org. Chem.*, 27 (1962) 567. *(b)* M. A. Stevens, H. W. Smith, and G. B. Brown, *J. Am. Chem. Soc.*, 82 (1960) 1148.
168 *(a)* S. C. Bell and S. J. Childress, *J. Org. Chem.*, 27 (1962) 1691. *(b)* S. C. Bell, S. J. Childress, T. S. Sulkowski, and C. Gochman, *J. Org. Chem.*, 27 (1962) 562.
169 J. A. Berson and T. Cohen, *J. Am. Chem. Soc.*, 77 (1955) 1281.
170 S. Furukawa and Y. Kuroiwa, *Pharm. Bull. (Japan)*, 3 (1955) 232.
171 G. Kobayashi, S. Furukawa, and Y. Kawada, *Yakugaku Zasshi*, 74 (1954) 790.
172 *(a)* S. Furukawa, *Pharm. Bull. (Japan)*, 3 (1955) 413. *(b)* S. Furukawa, *Yakugaku Zasshi*, 76 (1956) 900.
173 S. Ginsberg and I. B. Wilson, *J. Am. Chem. Soc.*, 79 (1957) 481.
174 S. Furukawa, *Yakugaku Zasshi*, 79 (1959) 487.
175 S. Furukawa, *Yakugaku Zasshi*, 77 (1957) 11.
176 T. Kato, T. Kitagawa, T. Shibata, and K. Nakai, *Yakugaku Zasshi*, 82 (1962) 1647.
177 S. Furukawa, *Yakugaku Zasshi*, 79 (1959) 77.
178 W. Baker, K. M. Buggle, J. F. W. McOmie, and D. A. M. Watkins, *J. Chem. Soc.*, (1958) 3594.
179 *(a)* Y. S. Sato, T. Iwashige, and T. Miyadera, *Chem. Pharm. Bull. (Tokyo)*, 8 (1960) 427. *(b)* H. J. Rimek, *Ann.*, 670 (1963) 69.
180 T. Kato, *Yakugaku Zasshi*, 75 (1955) 1233.
181 F. Zymalkowski and H. Rimek, *Arch. Pharm.*, 294 (1961) 759.
182 G. Kobayashi, S. Furukawa, Y. Akimoto, and T. Hoshi, *Yakugaku Zasshi*, 74 (1954) 791.
183 G. R. Redford, A. R. Katritzky, and H. M. Wuest, *J. Chem. Soc.*, (1963) 4601.
184 Ch. Kaneko, S. Yamada, and I. Yokoe, *Shika Zairyo Kenkyusho Hokoku*, 2 (1963) 475.
185 M. Asai, *Yakugaku Zasshi*, 79 (1959) 1273.
186 E. Hayashi and Y. Hotta, *Yakugaku Zasshi*, 80 (1960) 834.
187 B. Klein, J. Berkowitz, and N. E. Hetman, *J. Org. Chem.*, 26 (1961) 126.
188 T. Nakagome, *Yakugaku Zasshi*, 81 (1961) 1048.
189 T. Kato, F. Hamaguchi, and T. Oiwa, *Yakugaku Zasshi*, 78 (1958) 422.
190 T. Kato, Y. Goto, and Y. Yamamoto, *Yakugaku Zasshi*, 82 (1962) 1649.
191 M. Hamana and M. Yamazaki, *Chem. Pharm. Bull. (Tokyo)*, 11 (1963) 411.
192 T. Kato and H. Yamanaka, *Chem. Pharm. Bull. (Tokyo)*, 12 (1964) 18.
193 T. Okamoto and H. Tani, *Chem. Pharm. Bull. (Tokyo)*, 7 (1959) 130, 925.

194 F. Feely and E. M. Beavers, *J. Am. Chem. Soc.*, 81 (1959) 4004.
195 H. Tani, *Chem. Pharm. Bull. (Tokyo)*, 7 (1959) 930.
196 *(a)* H. Tani, *Yakugaku Zasshi*, 80 (1960) 1418. *(b)* H. Tani, *Yakugaku Zasshi*, 81 (1961) 141.
197 T. Okamoto and H. Takayama, *Yakugaku Zasshi*, 82 (1962) 1076.
198 T. Okamoto and H. Takayama, *Chem. Pharm. Bull. (Tokyo)*, 11 (1963) 514.
199 T. Okamoto, private communication.
200 *(a)* O. Cervinka, *Collection Czech. Chem. Commun.*, 27 (1962) 567. *(b)* O. Cervinka, A. Fábryová, and L. Matouchová, *Collection Czech. Chem. Commun.*, 28 (1963) 535.
201 *(a)* L. Bauer and L. A. Gardella, *J. Org. Chem.*, 28 (1963) 1320. *(b)* L. Bauer and L. A. Gardella, *J. Org. Chem.*, 28 (1963) 1323. *(c)* L. Bauer and T. E. Dickerhofe, *J. Org. Chem.*, 29 (1964) 2183.
202 V. Boekelheide and D. L. Harrington, *Chem. Ind. (London)*, (1955) 1423.
203 *(a)* V. J. Traynelis and R. F. Martello, *J. Am. Chem. Soc.*, 80 (1958) 6590. *(b)* V. J. Traynelis and R. F. Martello, *J. Am. Chem. Soc.*, 82 (1960) 2744.
204 M. R. F. Ashworth, R. P. Daffern, and D. L. L. Hammick, *J. Chem. Soc.*, (1939) 809.
205 S. Furukawa, *Yakugaku Zasshi*, 79 (1959) 492.
206 J. H. Markgraf, H. B. Brown, Jr., S. C. Mohr, and R. G. Peterson, *J. Am. Chem. Soc.*, 85 (1963) 958.
207 V. J. Traynelis, A. I. Gallagher, Sr., and R. F. Martello, *J. Org. Chem.*, 26 (1961) 4365.
208 *(a)* S. Oae, T. Kitao, and Y. Kitaoka, *J. Am. Chem. Soc.*, 84 (1962) 3359. *(b)* S. Oae, T. Kitao, and Y. Kitaoka, *J. Am. Chem. Soc.*, 84 (1962) 3362.
209 S. Oae, T. Kitao, and Y. Kitaoka, *Tetrahedron*, 19 (1963) 827.
210 J. F. Vozza, *J. Org. Chem.*, 27 (1962) 3856.
211 K. Takeda, K. Hamamoto, and H. Tone, *Yakugaku Zasshi*, 72 (1952) 1427.
212 K. Takeda and K. Hamamoto, *Yakugaku Zasshi*, 73 (1953) 1158.
213 F. Ramirez and P. W. von Ostwalden, *J. Am. Chem. Soc.*, 81 (1959) 156.
214 *(a)* S. Kajiwara, *Nippon Kagaku Zasshi*, 85 (1964) 672. *(b)* S. Kajiwara, *Nippon Kagaku Zasshi*, 86 (1965) 93. *(c)* S. Kajiwara, *Nippon Kagaku Zasshi*, 86 (1965) 1060.
215 K. Hamamoto and S. Kajiwara, private communication.
216 J. Meisenheimer and A. Dietrich, *Ber.*, 57 (1924) 1715.
The original article gave this reaction as:

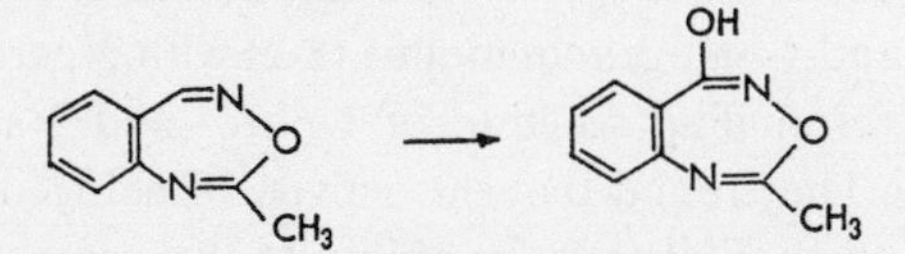

but this was corrected as shown by E. Hayashi (private communication).
217 A. S. Elina and L. G. Tsirulnikova, *Zh. Obshch. Khim.*, 33 (1963) 1544 [*Chem. Abstr.*, 59 (1963) 12807].
218 F. Kröhnke and K. Ellenast, *Ann.*, 600 (1957) 176.
219 B. S. Thyagarajan, *Chem. Revs.*, 58 (1958) 439.
220 M. Hamana and M. Yamazaki, *Chem. Pharm. Bull. (Tokyo)*, 10 (1962) 51.
221 E. Ochiai and A. Ohta, *Sci. Papers Inst. Phys. Chem. Res. (Tokyo)*, 56 (1962) 290; *Chem. Pharm. Bull. (Tokyo)*, 10 (1962) 1260.
222 A. Ohta, *Chem. Pharm. Bull. (Tokyo)*, 11 (1963) 1586.

Chapter 8

Effect of the N-oxide Group on Substituents

8.1 General

In π-electron-deficient aromatic *N*-heterocycles, the strong $-T$ effect of the ring nitrogen causes a decrease of electron density in the positions *ortho* and *para* to the nitrogen atom, and substituents in these positions show reactivity different from that of ordinary aromatic substituents (so-called "active substituents"). This activity is markedly increased by changing the ring nitrogen to a quaternary ammonium salt. In the corresponding *N*-oxides, the strong $-T$ and $-I$ effects of their N^+ tend to increase such activity over that of the parent tertiary amines. On the other hand, the $+T$ effect of the lone-pair electrons of O^- in the *N*-oxide group makes inverse polarization possible; the reactivity of such a compound sometimes differs from that of the corresponding tertiary amines. This is closely related to the fact that the positions *ortho* and *para* to the *N*-oxide group in aromatic *N*-oxide compounds can undergo both S_N and S_E reactions (*cf. Section 2.2.1*, p. 14), and consequently are the cause of multiplicity of the reaction of such compounds. The most marked example of the S_N reaction is the activity of electro-negative substituents (halogen, nitro, etc.) towards nucleophilic reagents. Okamoto and his collaborators[1] made kinetic analyses of the reaction of 4-halogenoquinoline 1-oxide (8-1) and 4-halogenoquinoline (8-2) with piperidine, in conjunction with the corresponding reaction of 1-nitro- and 1-acetyl-4-halogenonaphthalene (8-3). They found that the activity of halogen decreased in the following order and showed that the ability to S_N reaction increases with *N*-oxidation.

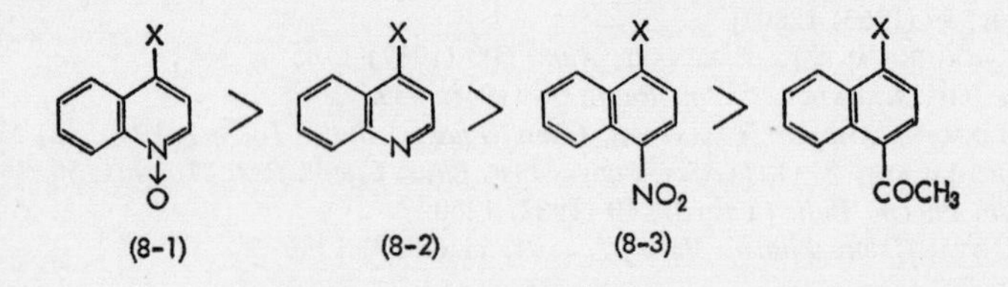

Coppens and others[2] examined the reaction of 2-, 3-, and 4-chloro-pyridine 1-oxides with piperidine and found that the reactivity of chlorine decreased in the order of $2>4\gg3$. Liveris and Miller[3] measured the reaction rate of substitution in the 2-, 3-, and 4-chloro derivatives of pyridine, pyridine 1-oxide, and pyridinium salt in methanol solution with sodium methoxide (in the case of *N*-methylpyridinium salt, the rate was computed from the substitution with sodium *p*-nitrophenoxide in methanol solution) and found the following facts regarding the activity of the chlorine atom.

(a) The rate or free-energy order of position reactivity is $4>2>3$ for the pyridines and their *N*-oxides, and $2>4>3$ for the pyridinium compounds.

(b) The order of activation energy is $4<2<3$ in all cases.

(c) The order of the reactivity of ring substituents (=X–) is $\geqslant N^+ \cdot CH_3 > \geqslant N^+–O^- > \geqslant N > \geqslant CH$ in all positions.

(d) The order of activation energy is in the reverse order except for the approximate equality of $N^+ \cdot CH_3$ and $N^+–O^-$ in the three positions.

These facts all point to the increased activity of the halogen atom in the positions *ortho* and *para* to ring nitrogen as a result of oxygenation of the nitrogen. Sako[4] examined the activity of the chlorine atom in chloropyridazine *N*-monoxides in the S_N reaction and found a large number of experimental examples which indicate that the chlorine atom in the position *para* to the *N*-oxide group is less reactive than that in the position *para* to the non-oxygenated tertiary ring-nitrogen. He explained this phenomenon as the partial neutralization of S_N-activation by the *N*-oxide group through the electromeric effect of the oxygen atom in the *N*-oxide group. This conjecture was supported by the result of studies on the NMR spectra by Kawazoe and Natsume[5], who found that the chemical shifts of ring protons in pyridazine *N*-oxides can be explained as the increased π-electron density in the positions *ortho* and *para* to the *N*-oxide group and decreased density in the *meta* position, *i.e.* in the positions *ortho* and *para* to the non-oxygenated ring nitrogen.

8.2 Methyl Group

A methyl group in the positions *ortho* and *para* to the ring-N in π-electron-deficient *N*-heteroaromatic compounds is far more active than that in the *meta* position, and it is known that the activity increases further when the ring-nitrogen is converted to a quaternary ammonium salt. It has been shown that oxygenation of the ring-nitrogen also increases this activity to a certain extent.

Adams and Miyano[6] found that whereas the ester condensation of 2- (8-4) and 4-picoline 1-oxide (8-5) with ethyl oxalate progressed by the use of potassium ethoxide catalyst, the corresponding non-oxygenated picolines failed to undergo this reaction, indicating the increased activity of the methyl group by *N*-oxidation.

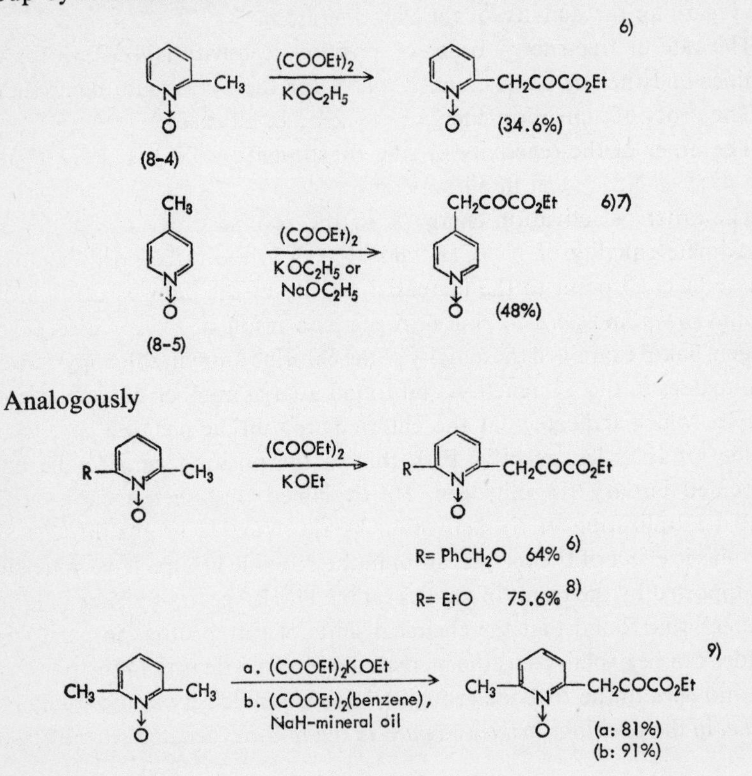

Jerchel and Heck[10] made a comparative qualitative examination of the activity of methyl group in the condensation reaction of 2- and 4-picolines, their methiodides, and their *N*-oxides, with benzaldehyde in acid (acetic anhydride–acetic acid) and basic (piperidine–methanol) conditions. According to their results, the reactivity of tertiary amines and methiodides in these condensation reactions was in the decreasing order of methiodide (basic)> picoline (acid)>methiodide (acid), and the activity of the methyl group, both in acid and basic media, decreased in the order of 4-, 2-, *N*-methyl, and 3-

position. On the other hand, *N*-oxide compounds failed to undergo condensation in the acid medium and the reaction progressed only in the presence of a basic catalyst (potassium methoxide), the activity being much smaller than that in quaternary salts. These workers assumed that both the *ortho* and *para* positions in the *N*-oxide compounds are under the influence of both plus and minus electromeric effects of the *N*-oxide group and the proton withdrawal from the methyl groups in these positions by the base catalyst is partially blocked by the electron-releasing effect ($+E$) of the *N*-oxide group.

Ochiai and Kato[11] carried out the reaction of 4-nitro- and 4-methoxy-2,6-lutidine 1-oxides with benzaldehyde, in the presence of zinc chloride, and obtained only a minute amount of the condensation products. Parker and Furst[12] found that the condensation of 2- and 4-picoline 1-oxides with *p*-(dimethylamino)benzaldehyde did not progress in the presence of an acid catalyst like zinc chloride or conc. hydrochloric acid, but did if the reactants were heated in toluene in the presence of piperidinium acetate with azeotropic distillation of the water formed by the reaction; the corresponding condensation of quinaldine l-oxide progressed with a fair yield under the same conditions. Further, Pentimalli[13] reported that a good result was obtained by the use of pyridine–potassium hydroxide as a catalyst in a similar condensation of 2-picoline l-oxide (8-4).

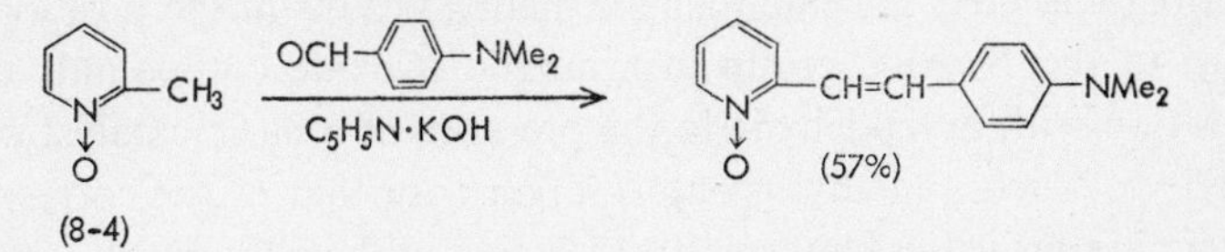

Similar reactions have been carried out by Katritzky and Monro[14], and by Herz and Murty[15] on picoline 1-oxides, by Hamana and Noda[16] on lepidine 1-oxide (8-6), and by Pentimalli[17] on quinaldine 1-oxide.

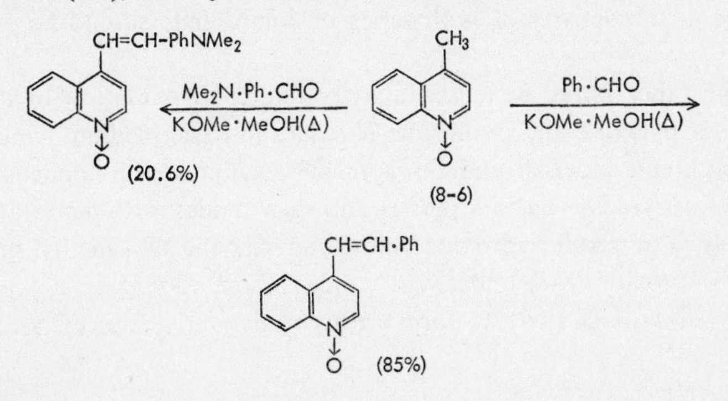

Colonna carried out the condensation reactions indicated below in order to determine the structure of two isomeric bisquinolylethylene 1,1′-dioxides (8-7 and 8-7′) to be described later.

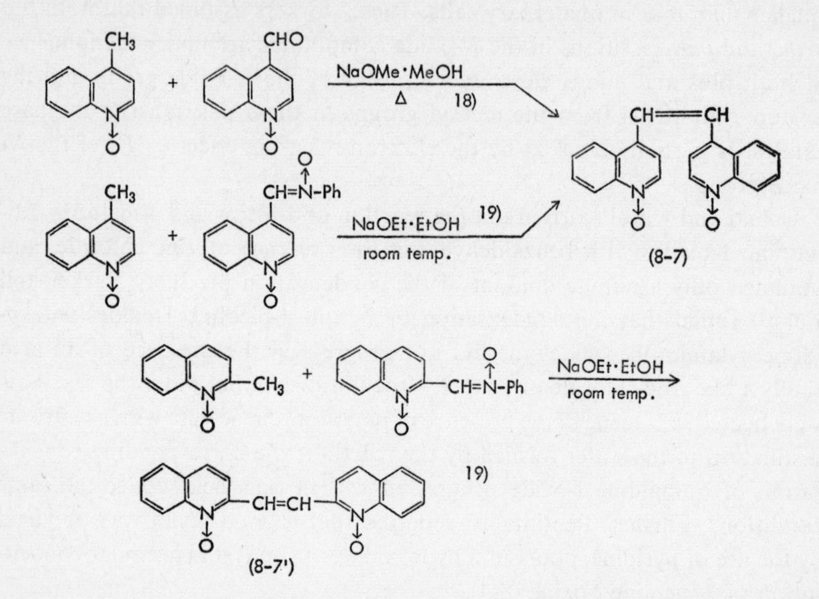

The foregoing facts indicate that the methyl groups in the positions *ortho* and *para* to the *N*-oxide group in aromatic *N*-oxide compounds undergo condensation with arylaldehyde in the presence of a basic catalyst and that the activity of such a methyl group is larger than that of the corresponding tertiary amine derivatives but smaller than that of the quaternary ammonium compounds. In the presence of an acid catalyst, however, the activity of the methyl group is greater in the parent tertiary amines than in the corresponding *N*-oxide group. Correlation between the difference in reaction conditions and that in the activity of both series of compounds should be examined further.

On the other hand, increased activity due to *N*-oxidation in the same reaction of pyrazine and pyridazine *N*-oxides in basic medium is not necessarily apparent. Koelsch and Gumprecht[20] examined the condensation of 2,5-dimethylpyrazine and its mono- and di-*N*-oxides with benzaldehyde in the presence of alkali hydroxide and found that the two methyl groups in the mono-*N*-oxide (8-8) underwent condensation, though in a poor yield, just as in di-*N*-oxide (8-9), to form a distyryl derivative. Since 2,5-dimethylpyrazine failed to undergo the reaction under the same conditions, they concluded that although activation by *N*-oxidation is apparent, there is no evidence for superior reactivity of one of the methyl groups in the mono-*N*-oxide.

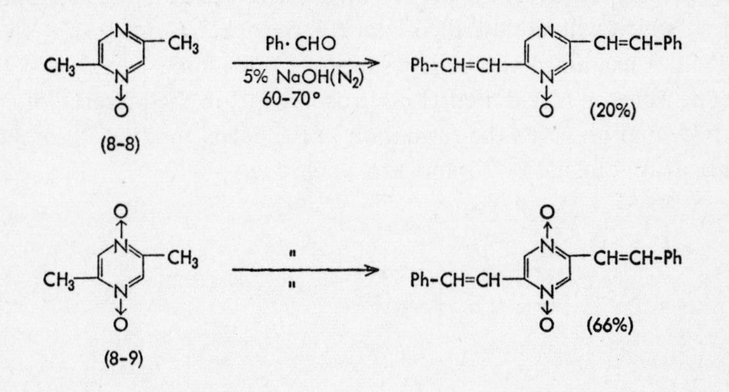

Itai and Sako[21] found that condensation of 3,6-dimethylpyridazine 2-oxide (8-10) and excess of aldehydes, with sodium methoxide as a catalyst, resulted in the formation of a distyryl derivative (8-11) but not a monostyryl derivative, indicating that the two methyl groups possessed the same reactivity.

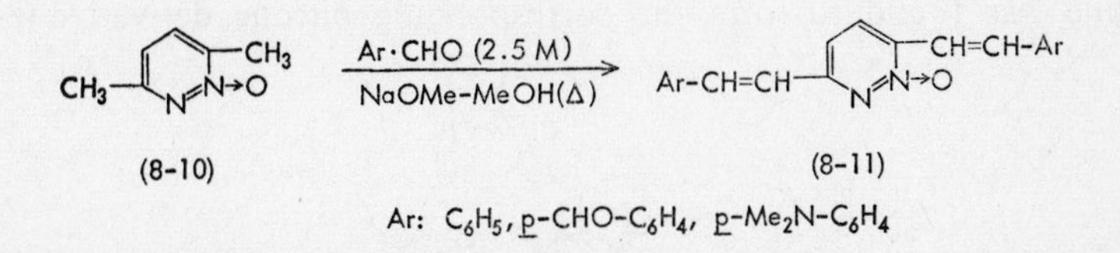

The reaction of 3-, 4-, 5-, and 6-methylpyridazine l-oxides with benzaldehyde was also examined under similar conditions at various temperatures, and it was found that the reaction progressed to the styryl compound through an intermediate adduct. It was assumed, from the yield of their product and the amount of the starting compound recovered, that the activity of the methyl group decreased in the order of 5>4>6>3 positions.

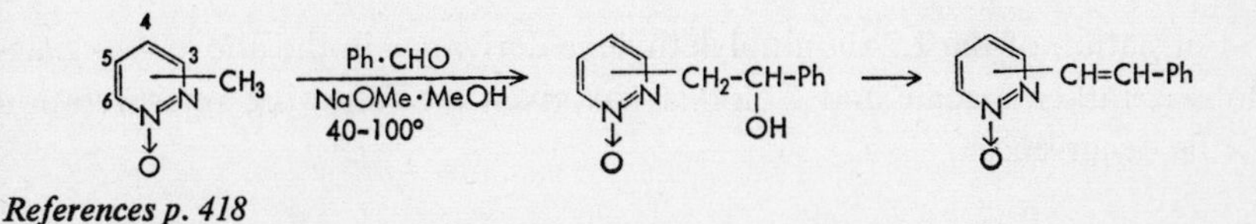

It seems that the activity of methyl groups in the positions *ortho* and *para* to the *N*-oxide group and to the free nitrogen atom towards the aldehyde is about the same, and that of the methyl group *para* to the free nitrogen atom is greater than that *para* to the *N*-oxide group. Such notable difference in activity is more markedly apparent with chlorine atoms in the corresponding position, which will be described later (*cf. Section 8.4*, p. 391).

Colonna[22] examined the condensation of quinaldine 1-oxide (8-12) with nitrosobenzene or *N,N*-dimethyl-*p*-nitrosoaniline, in the presence of sodium methoxide and observed the formation of the corresponding 1,2-diquinolyl-ethylene *N,N'*-dioxide (8-7') and azoxybenzene.

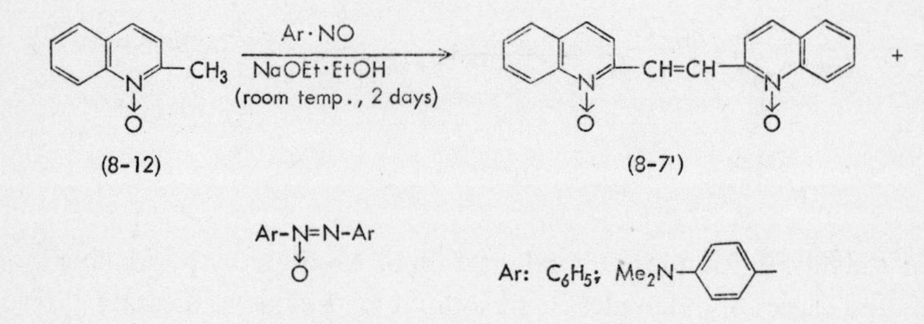

The same reaction of lepidine 1-oxide (8-6) with nitrosobenzene gives 1,2-bis(1-oxido-4-quinolyl)ethylene (8-7), while that with *N,N*-dimethyl-*p*-nitroso-aniline was found to form the corresponding nitrone derivative (8-13)[16].

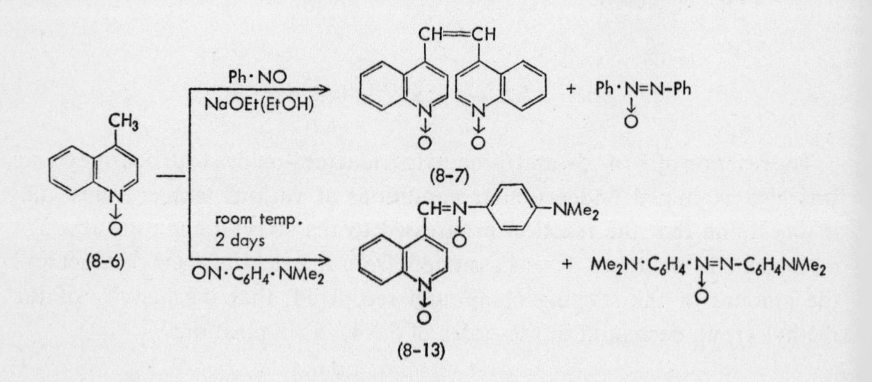

Formation of the 1,2-diquinolylethylene derivative in the above two reactions is rather specific and Colonna has given the following interpretation for its occurrence.

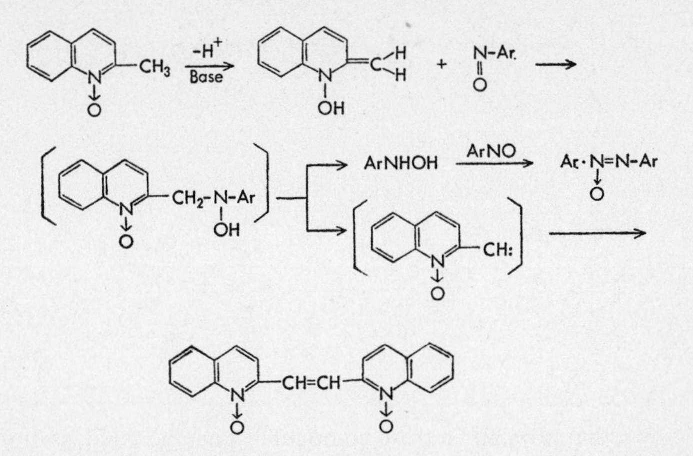

Colonna and Zamparella[19] found the same reaction to take place between methylquinoline 1-oxides and aromatic nitro compounds, and gave the following explanation for its reaction mechanism.

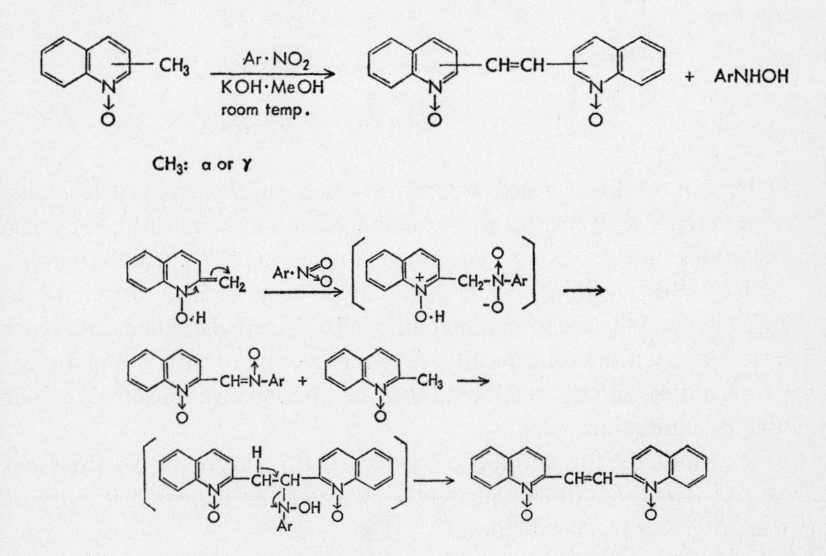

Later, Mishina and Efros[23] reported the formation of the corresponding anil from 9-methylacridine 10-oxide (8-14) and *N*,*N*-dimethyl-*p*-nitroso-aniline, and Bedford and others[24] that from tri-*O*-acetyladermine l-oxide (8-15) and the same aniline derivative.

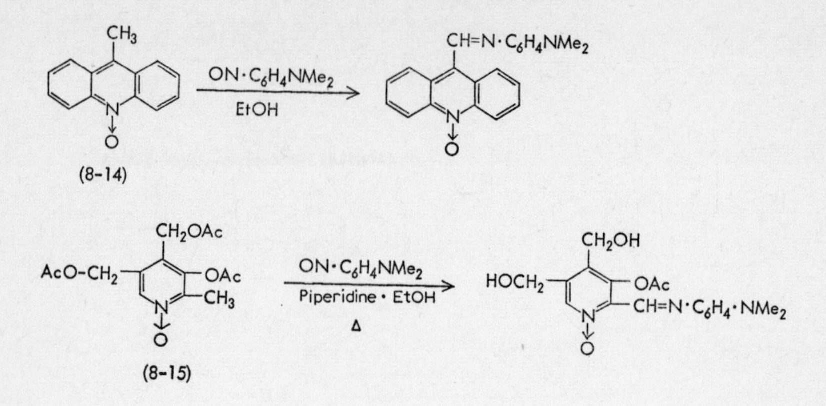

It is known that aromatic nitroso compounds undergo condensation with active methyl compounds, in the presence of a base, and form nitrone and/or anil compounds.

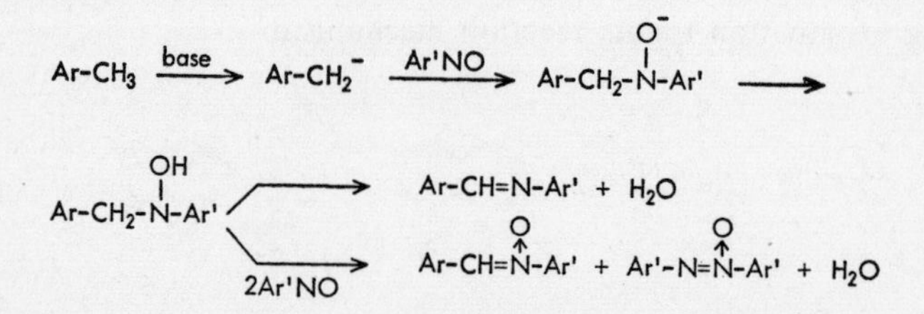

The kind of product formed depends not only on the types of reactants and bases used but also varies with reaction conditions, even with the same reactants and bases[25]. As an active methyl compound, 2-nitrotoluene does not undergo this reaction[27], while 2,4-dinitrotoluene forms a nitrone or an anil[26,27]. Of the *N*-heteroaromatic compounds, 9-methylacridine undergoes this reaction[28], but the condensation does not take place with 2- and 4-picoline[29], quinaldine, and lepidine[28] until these compounds are converted to their quaternary ammonium salts.

Consequently, the formation of a condensate from aromatic *N*-oxides and nitrosoaryl described above apparently indicates the increased activity of the methyl group by *N*-oxidation.

Taylor and his collaborators[30] found that 4-nitro-3-picoline 1-oxide (8-16) and butyl nitrite undergo oxidative dimerization in the presence of sodium methoxide and form 1,2-bis(4-nitro-1-oxido-3-pyridyl)ethane (8-17). This reaction seems similar to the foregoing reaction of methylquinoline 1-oxides and nitrosoaryl, but further examination revealed that the reaction did not require butyl nitrite and the reaction of (8-16) with sodium ethoxide, even in a nitrogen stream, gave (8-17). This fact suggests that the nitro group or the *N*-oxide group acts as an oxidizing agent, like butyl nitrite.

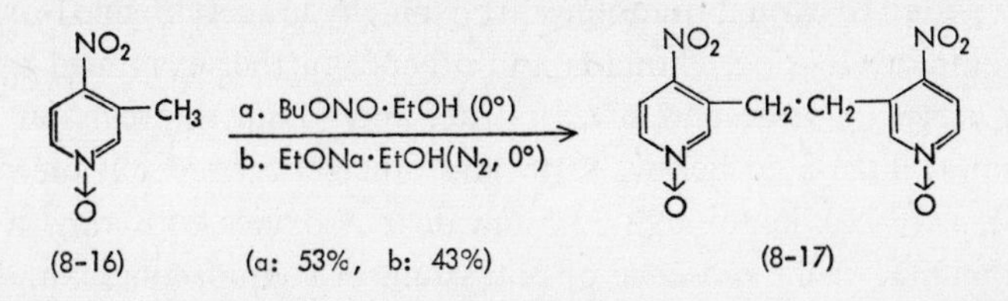

If this reaction is carried out in the presence of oxygen, in 30% methanolic potassium hydroxide solution, dehydrogenation accompanied by substitution of the nitro group with methoxyl takes place and 1,2-bis(4-methoxy-1-oxido-3-pyridyl)ethylene (8-18) is formed.

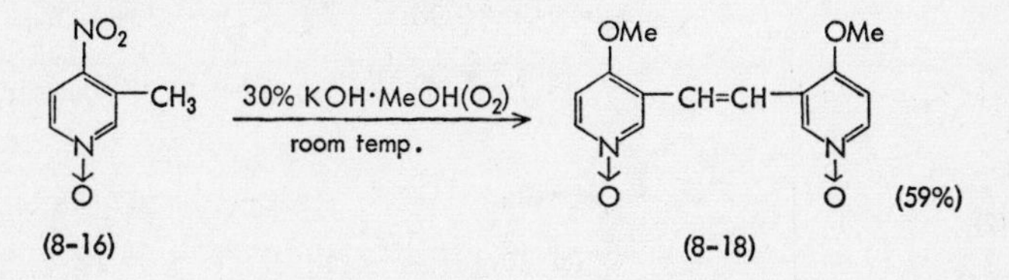

The same reaction of 4-nitro-3-picoline (8-19) under the same condition (b) as above does not progress and the starting compound is completely recovered. This indicates the rôle of the *N*-oxide group in the activation of the methyl group. Reaction of 4-nitro-3-picoline with 30% ethanolic potassium hydroxide at 0°, in the presence of oxygen, results in oxidative dimerization and 1,2-bis(4-nitro-3-pyridyl)ethane (8-20) is formed in 79% yield, while the same reaction carried out at room temperature gives 1,2-bis(4-ethoxy-3-pyridyl)ethylene (8-21). (8-21) is also formed by the same treatment of (8-20).

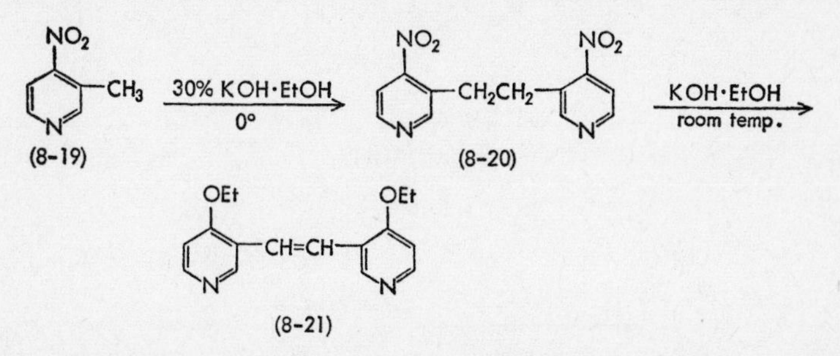

It has been proved that 4-ethoxy-3-picoline 1-oxide does not undergo dimerisation under the same conditions and it has become clear that activation of the methyl group by a nitro group is also responsible for this reaction.

During the nitration of quinaldine l-oxide (8-12) with acetyl chloride and silver nitrate (*cf. Section 6.2.4*), Tanida[31] found that the methyl group is partly nitrosated to form 2-quinolinecarboxaldehyde acetyloxime l-oxide and 2-cyanoquinoline 1-oxide. Hamana and others[32] further examined this reaction with acetyl chloride and silver nitrite, and found the reaction to take the course (a) illustrated below. Kato and others[33] carried out the reaction of quinaldine (8-22), lepidine (8-23), and their *N*-oxides with amyl nitrite in liquid ammonia, in the presence of potassium or sodium amide, and found that the methyl group underwent the same reaction, irrespective of the oxygenation of ring nitrogen, and that the reaction progressed more smoothly in the absence of *N*-oxidation.

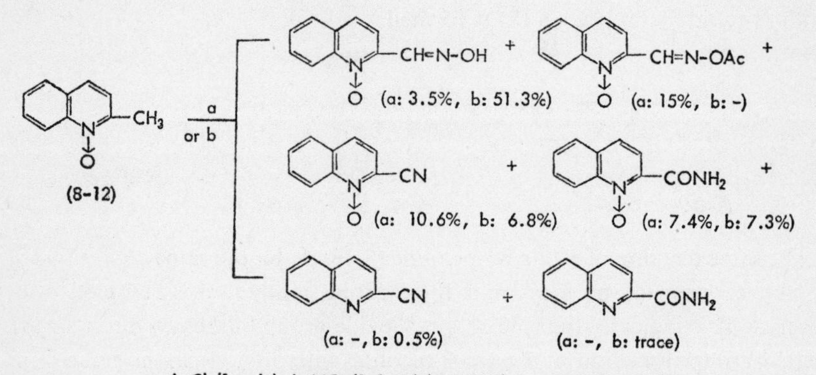

a: AcCl (1 mole)·$AgNO_2$(1.1 mole) in $CHCl_3$, room temp., 4 hrs.

b: AmONO·KNH_2 (liq. NH_3).

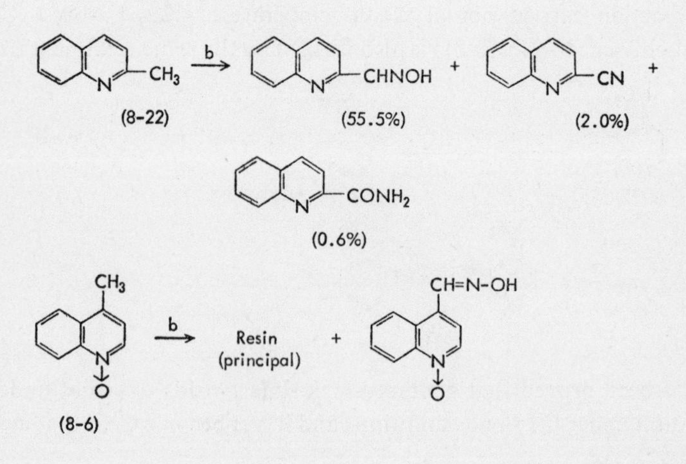

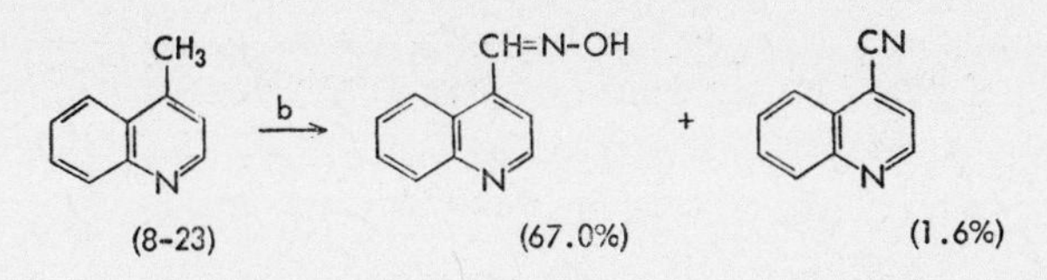

Cislack[34] carried out this reaction with 2-picoline 1-oxide (8-4) and obtained a patent on the manufacture of 2-cyanopyridine by the application of acyl nitrite or acetyl chloride and silver nitrite at room temperature. Kato and others[35] found that the application of amyl nitrite to 2-picoline 1-oxide in liquid ammonia, in the presence of sodium or potassium amide, or sodium hydroxide, gave 2-pyridinecarboxaldehyde oxime 1-oxide (8-24) and picolinamide 1-oxide (8-25). This reaction does not take place with 3-picoline 1-oxide, while the reaction progresses smoothly with 4-picoline 1-oxide (8-5), with formation of the nitrile as a by-product.

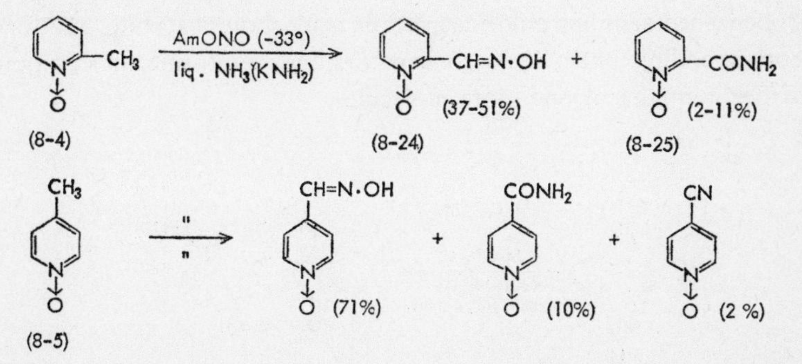

Ogata[36] carried out the same reaction with methylpyridazine *N*-oxides and showed that not only the methyl group in the positions *ortho* and *para* to the *N*-oxide group, but those *ortho* and *para* to the free nitrogen, *i.e. meta* to the *N*-oxide group, also undergo reaction to form the corresponding oxime.

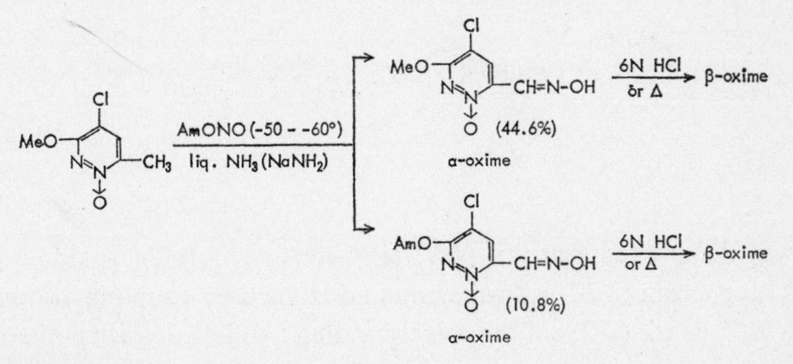

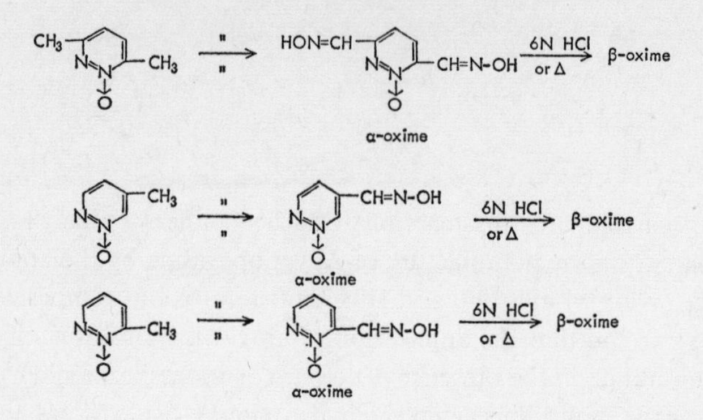

Mishina and Efros[23] proved that 9-methylacridine 10-oxide (8-14) and 6-methylphenanthridine 5-oxide (8-26) reacted, though quite slowly, with *p*-nitrobenzenediazonium chloride in acetic acid–sodium acetate, and formed the corresponding phenylhydrazone compounds or formazan compounds by further coupling of one more molecule.

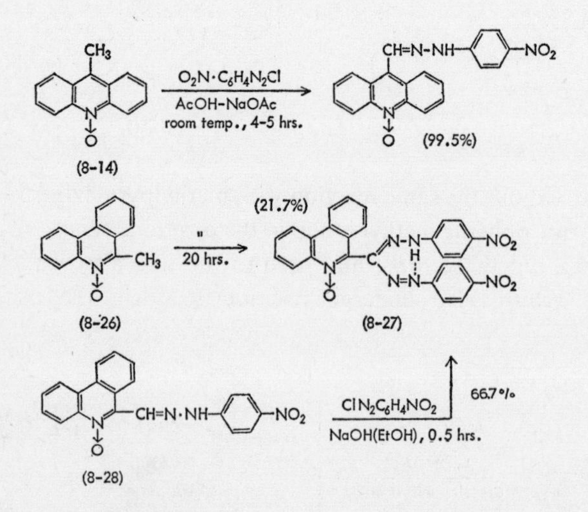

It was also proved that this *N*,*N'*-bis(4-nitrophenyl)-*C*-(5-oxido-6-phenanthridyl)formazan (8-27) is formed in a good yield by coupling of the *p*-nitrophenylhydrazone (8-28) of the corresponding 5-aldehyde with *p*-nitrobenzene diazonium salt. However, this reaction does not progress with *N*-oxides of quinaldine and lepidine until they are converted to their quaternary salts from which the formazan compounds are formed with attendant deoxygenation. Quaternary salts of 2- and 4-picoline 1-oxides do not undergo this reaction, indicating that a highly reactive methyl group is required for this purpose.

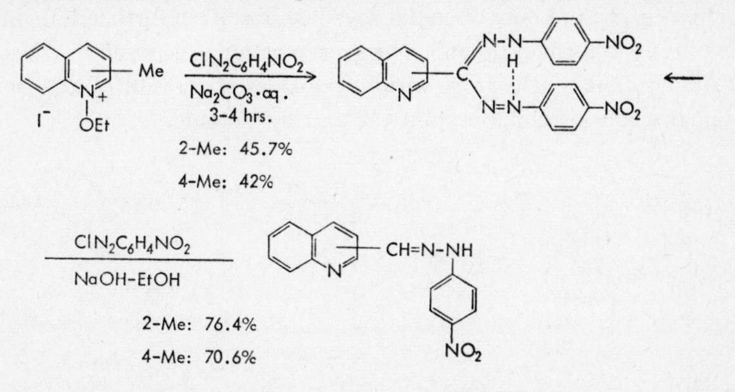

Hamana and Umezawa[37] found that refluxing of 2-picoline 1-oxide (8-4) with iodine and excess pyridine resulted in the formation of 1-(1-oxido-2-pyridylmethyl)pyridinium iodide (8-29). Colonna[38] carried out this reaction with quinaldine and lepidine 1-oxides, and obtained the corresponding pyridinium salts.

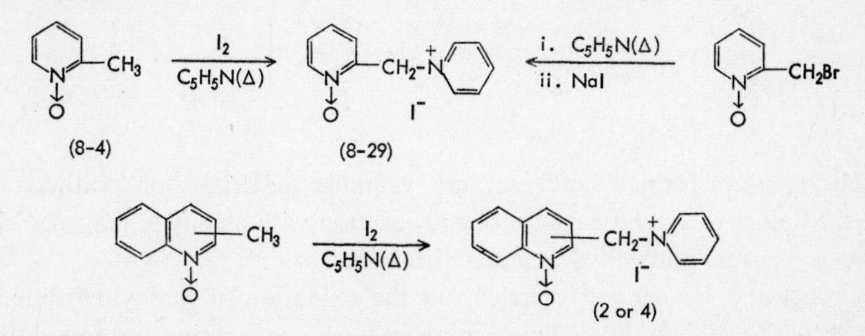

Originally, this reaction was found by Ortoleva[39] and developed by King[40] for the reaction of an active methylene or methyl group in malonic acid, acetophenone, etc., and should hence be called the Ortoleva-King reaction[38,41a]. Kröhnke[41b] reported that this reaction progressed with quinaldine, lepidine, and 9-methylacridine but not with 2- and 4-picoline unless they are converted to their methiodides. The fact that the reaction takes place in the case of *N*-oxide compounds of the above-mentioned picolines indicates that *N*-

References p. 418

oxidation is responsible for the activation of the methyl group in 2- or 4-position.

Kröhnke[41b] found that the pyridinium salt formed by this reaction undergoes condensation with *N,N*-dimethyl-*p*-nitrosoaniline in the presence of an alkaline catalyst to form the corresponding nitrone and that this reaction, carried out in the presence of sodium cyanide, results in further 1,3-addition of the nitrone to form cyanoanil. The same reaction was found to take place with the pyridinium salts from various *N*-oxides as a result of experiments by Hamana and his collaborators[37,42], and by Colonna[38].

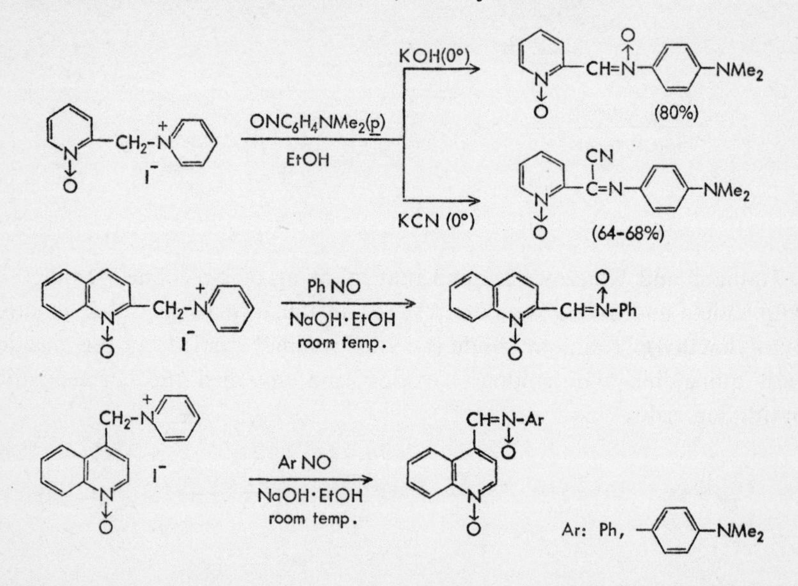

The nitrones formed here are very valuable materials for synthesis as masked aldehydes, and easily undergo reaction with aldehyde reagents like hydroxylamine, phenylhydrazine, and semicarbazide[43].

Jerchel and his school[44] carried out the oxidation of methylpyridine 1-oxides with selenium dioxide by heating in dioxan or pyridine and found that whereas the methyl group in 3- and 5-positions is resistant to this oxidation, that in the 2-position is oxidized to an aldehyde, and the methyl group in 4-position is oxidized to a carboxylic acid. In this case, the reaction is more facile in pyridine than in dioxan, and the methyl group in 4-position is more reactive than that in the 2-position. Further, this reaction progresses even in the corresponding tertiary amine derivatives to form a carboxylic acid, and the reactivity is said to be better than that of the corresponding *N*-oxide compounds. The oxidation of the methyl group in 2-position to an aldehyde is valuable in synthetic chemistry.

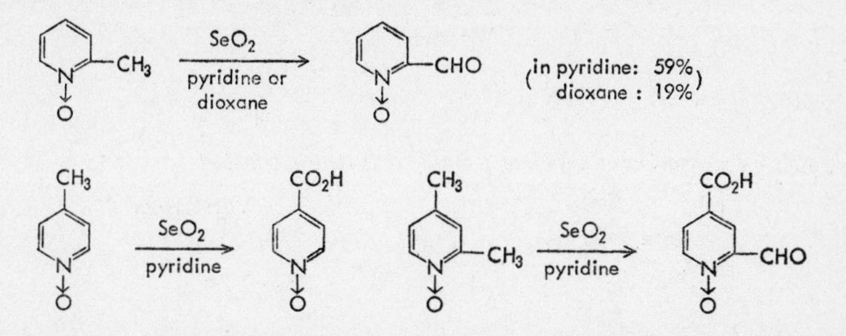

In a similar manner, quinaldine 1-oxide[32] (8-12) and 6-methylphenanthridine 5-oxide[23] (8-26) are also oxidized to the corresponding aldehyde compounds. If the reaction of quinaldine 1-oxide is carried out in ethanolic solution, 1,2-di(2-quinolyl)hydroxyethane *N,N'*-dioxide (8-30) and 1,2-di-(2-quinolyl)ethylene *N,N'*-dioxide (8-7') are formed as by-products[45]. Oxidation of lepidine 1-oxide (8-6) with selenium dioxide has also been reported[46].

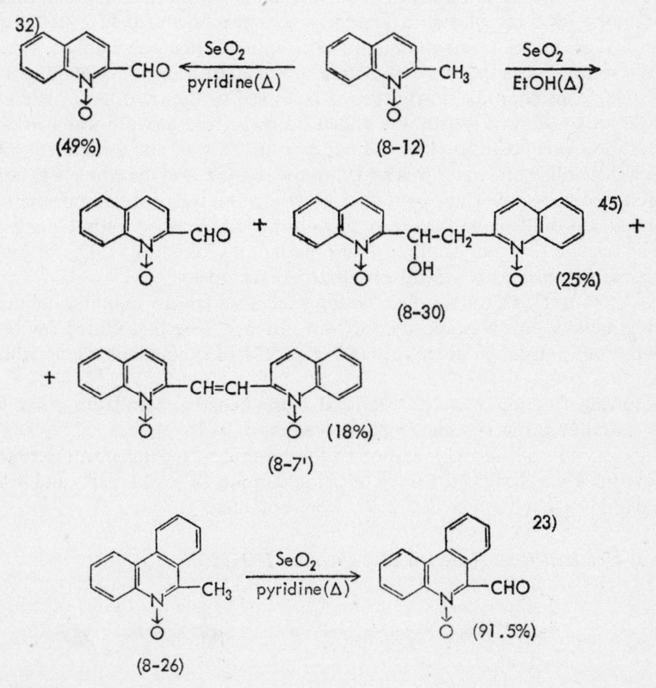

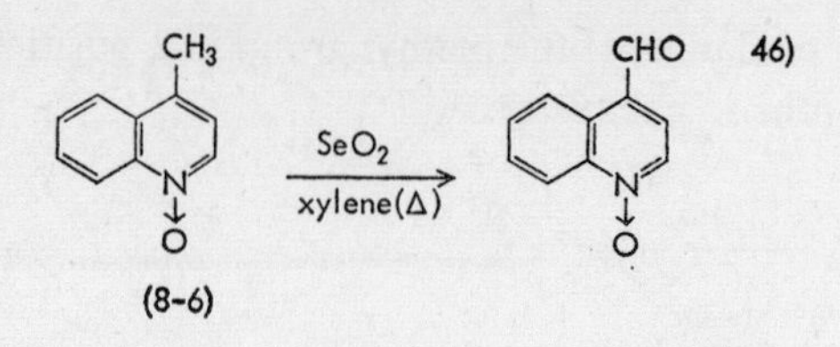

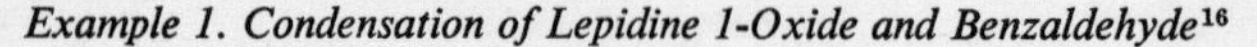

Example 1. Condensation of Lepidine 1-Oxide and Benzaldehyde[16]

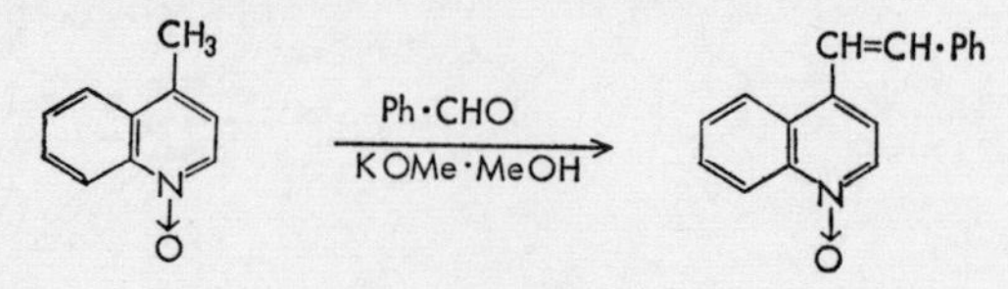

A solution of 2 g of lepidine 1-oxide and 2.7 g of benzaldehyde dissolved in a solution of 5 ml of methanol containing 0.33 g of potassium was refluxed on a water bath for 10 h and unreacted aldehyde was distilled off by steam distillation. When cooled, the solidified crystalline mass was collected by filtration, washed several times with water, dried in air, and recrystallized from ethyl acetate–chloroform to 2.5 g of 4-styrylquinoline 1-oxide as yellow prisms, m.p. 178–179°.

Example 2. Reaction of Quinaldine 1-Oxide and Amyl Nitrite[33]

In a 500-ml three-necked flask provided with a stirrer and a condenser, and chilled in dry ice–acetone, 300 ml of liquid ammonia was placed and 0.32 g (0.011 g-atom) of potassium was added with a small amount of iron trichloride as a catalyst. When the blue color of the solution disappeared completely, 1.50 g (0.01 mole) of quinaldine 1-oxide was added by which the solution slowly turned from red to deep red. After stirring for 1 h, 2.46 g (0.02 mole) of amyl nitrite was added dropwise, the mixture was stirred for 2.5 h, and the solution was neutralized by the addition of 0.8 g of ammonium chloride. Liquid ammonia was distilled off, water was added to the residue, and the whole was concentrated under a reduced pressure. The by-product amyl alcohol was distilled off, the residue was extracted with chloroform, and the insoluble residue was washed with a small amount of water. This residue was recrystallized from methanol to 0.98 g (51.4%) of quinaldehyde 1-oxide oxime as white scaly crystals, m.p. 215°(decomp.).

The extracted chloroform solution was passed through an alumina column and the column was eluted with benzene–chloroform mixture. The first eluted fraction was recrystallized from petroleum ether to 0.007 g (0.5%) of quinaldonitrile as white needles, m.p. 94°.

The following fraction was recrystallized from benzene–petroleum ether to 0.013 g (0.8%) of quinaldonitrile 1-oxide as white needles, m.p. 167–168°.

The crude crystals obtained by elution of the column with chloroform were fractionally recrystallized from acetone and a trace of quinaldamide, m.p. 126–128°, and 0.14 g (7.3%) of quinaldamide 1-oxide, m.p. 216–217°, were obtained.

Example 3. Ortoleva-King Reaction of 2-Picoline 1-Oxide[37a]

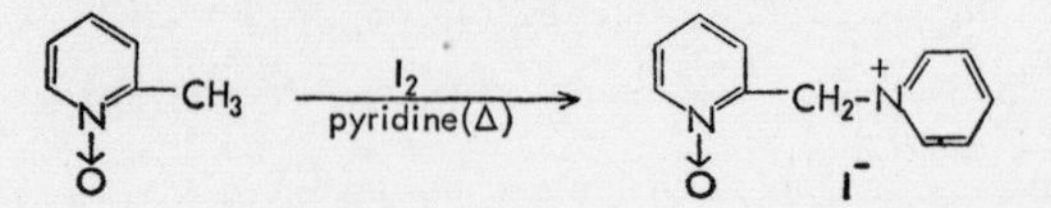

A solution of 109 g (1 mole) of 2-picoline 1-oxide and 254 g (1 mole) of iodine in 500 ml of pure pyridine was refluxed for 5 h in an oil bath (bath temp., 140°). The reaction mixture was allowed to stand overnight in a refrigerator and the crystals that precipitated out were collected. Treatment of the residue with activated charcoal and recrystallization from ethanol afforded 126 g (40.1%) of 1-(1-oxido-2-pyridylmethyl)pyridinium iodide as light yellowish white needles, m.p. 180–181°(decomp.).

Example 4. N-(p-Dimethylaminophenyl)-2-(1-oxido-2-pyridyl)nitrone Hemihydrate[37a]

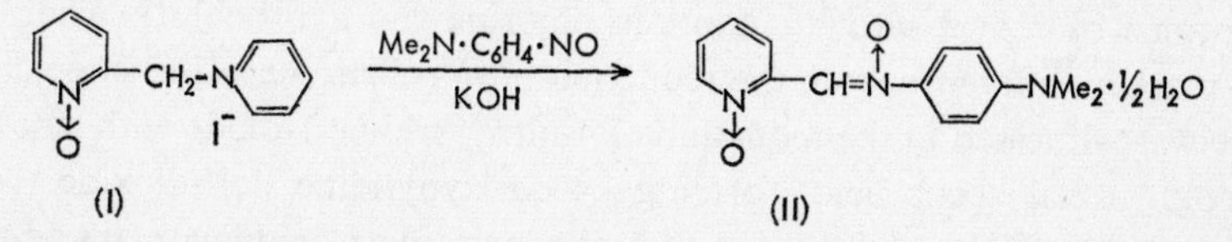

To a solution of 27 g (1 mole) of (I) in 25 ml water, an ethanolic solution of 13 g (1 mole) of *N,N*-dimethyl-*p*-nitrosoaniline was added and a solution of 6.5 g (*ca.* 1.5 moles) of potassium hydroxide in 130 ml of water was added dropwise with stirring and ice-cooling. After a short time, red crystals began to precipitate out which were collected and recrystallized from ethyl acetate to 18 g (80%) of (II), m.p. 160–161°

8.3 Nitro Group

The nitro groups in the positions *ortho* and *para* to the *N*-oxide group in aromatic *N*-oxide compounds are markedly more active to S_N reaction than that in the *meta* position, and the nitro group in the *para* position shows characteristic properties to reduction. Reduction of the nitro group in the *ortho* position of the *N*-oxide group is virtually undeveloped.

8.3.1 Activity to Reducing Agents

The nitro group in the 4-position of pyridine l-oxide shows a specific behavior to reducing agents. At the time that reduction of 4-nitropyridine 1-oxide was first carried out by Ochiai and Katada[47] in 1943, there was comparatively little literature on the reduction of nitropyridines, chiefly dealing with the reduction of 3-nitropyridine reported by Kirpal and Reiter[48], and that of 2-nitropyridine by Kirpal and Böhm[49] and by Bysstitzkaja and Kirssanow[50]. Such brief and scattered reports seemed to suggest that the reduction of nitropyridines took the usual course of aromatic nitro compounds in general, forming an amino group in acid reaction, a hydroxyamino group in neutral reaction, and reduction to the hydrazo group via the azoxy group in alkaline reaction.

However, treatment of 4-nitropyridine 1-oxide (8-31) with various reducing agents showed that the compound is reduced by the route shown below, *i.e.*

References p. 418

a bimolecular reduction. Reduction with hydrogen sulfide in cold ethanol solution containing ammonia or with zinc dust in cold dilute acetic acid solution resulted in the formation of 4,4′-azopyridine 1,1′-dioxide (8-32). The most noteworthy fact was that 4-nitropyridine 1-oxide is reduced to 4,4′-azopyridine 1,1′-dioxide by treatment with sodium nitrite in alkali hydroxide solution at 40–50°, while 4,4′-hydrazopyridine 1,1′-dioxide (8-33) is formed by reduction with stannous chloride and hydrochloric acid. This is a marked difference from the reduction of nitrobenzenes. The same difference was witnessed in the reduction of 4-nitropyridine 1-oxide with zinc dust in cold glacial acetic acid, forming 4,4′-azoxypyridine, 1,1′-dioxide (8-34), which is also formed by heating of 4-nitropyridine 1-oxide with zinc dust in water at 100°. Passage of hydrogen sulfide through a boiling solution of 4-nitropyridine 1-oxide in ethanol containing ammonia results in the formation of 4-aminopyridine 1-oxide (8-35), which is also formed by the reduction of 4-nitropyridine 1-oxide with stannous chloride in hydrochloric acid, accompanying the formation of 4,4′-hydrazopyridine 1,1′-dioxide (8-33). Catalytic reduction of 4-nitropyridine 1-oxide over palladium–carbon in hydrochloric acid gives 4-aminopyridine (8-36), but the same reduction in ethanol or alkaline solution gives 4-aminopyridine 1-oxide.

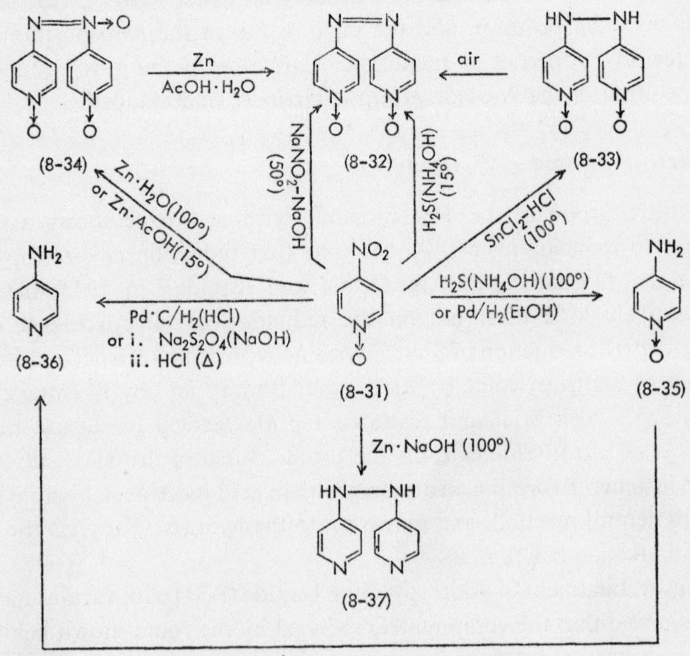

Ochiai, Teshigawara, and Naito[51] found that the electrolytic reduction of 4-nitropyridine 1-oxide in dilute hydrochloric acid, with a copper plate as a cathode, results in the formation of 4,4′-azopyridine 1,1′-dioxide (8-32) in 78% yield. Den Hertog and Overhoff[52] found that the reduction of 4-nitropyridine l-oxide with iron powder in acetic acid effects smooth deoxygenation as well and 4-aminopyridine (8-36) is formed in a good yield, while boiling of 4-nitropyridine 1-oxide in 10% sodium hydroxide solution with arsenic trioxide gives 4,4′-azopyridine 1,1′-dioxide (8-32) and 4,4′-azopyridine 1-oxide (8-38). The latter was found to be formed, together with 4,4′-azopyridine (8-39), on heating 4-nitropyridine 1-oxide with zinc powder or sodium stannite in alkaline solution[53]. At the same time, den Hertog and others found that 4,4′-azoxypyridine (8-41) is formed by warming 4-nitropyridine (8-40) with arsenic trioxide in sodium hydroxide solution at 70°, and they determined the structure of the azoxypyridine by *N*-oxygenation of the azoxypyridine (8-41) with perbenzoic acid, obtaining 4,4′-azoxypyridine 1-oxide (8-42) and 1,1′-dioxide (8-34)[53].

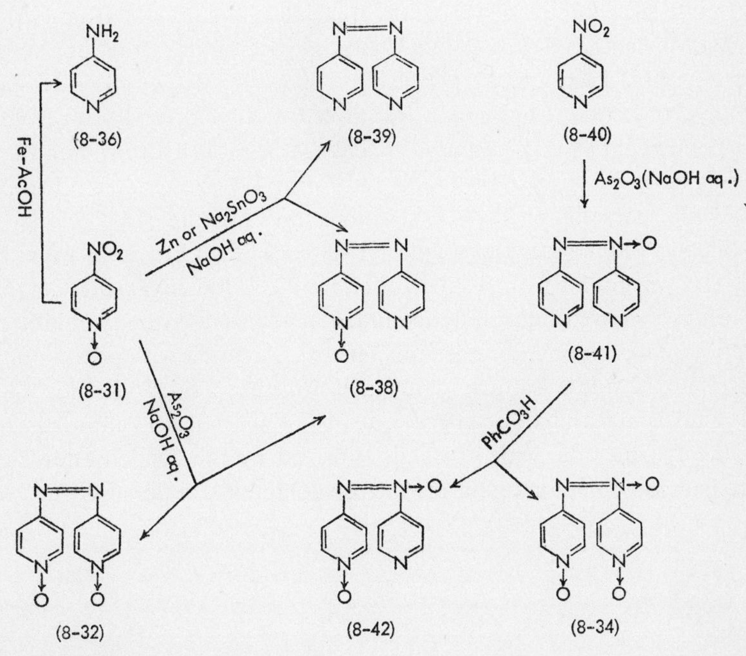

References p. 418

The behavior of 4-nitro-2-picoline l-oxide[54], 4-nitro-3-picoline l-oxide[55], and 4-nitro-2,6-lutidine l-oxide[55] to various reducing agents was examined and the same bimolecular reaction was found to take place.

In this connection, Brown[56] reported that the reduction of 2-nitropyridine 1-oxide (8-43) in ethanol over palladium–carbon catalyst by passing hydrogen gas resulted in the formation of 2-aminopyridine (8-44) and that the reduction could not be stopped at the stage of 2-aminopyridine 1-oxide. There is no other report on the reduction of 2-nitropyridine 1-oxide and this reaction has as yet not been sufficiently investigated.

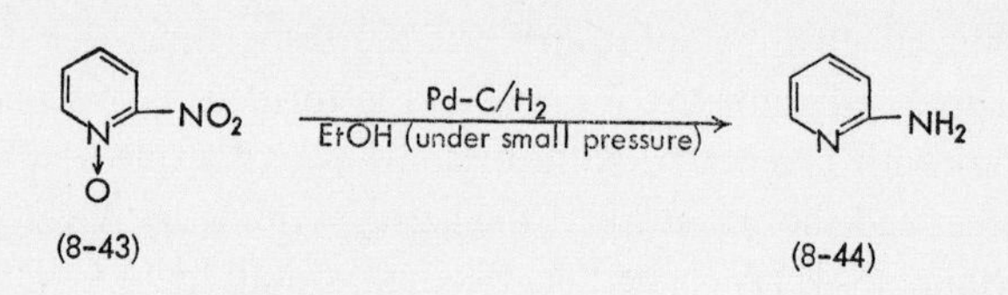

Recently, Ochiai and Mitarashi[57] found that reduction of 4-nitropyridine 1-oxide (8-31) with phenylhydrazine in ethanol solution afforded 4-(hydroxyamino)pyridine 1-oxide (8-45) in a very good yield. This substance agrees with the compound synthesised by Quilico and others[58] from γ-pyrone and hydroxylamine. The same reduction is effected in methyl homologs of 4-nitropyridine 1-oxide and in 4-nitroquinoline 1-oxide and the corresponding 4-hydroxyamino compounds are formed in a good yield of over 90%[59].

As pointed out by Quilico and his collaborators[58], 4-(hydroxyamino)pyridine 1-oxide is comparatively labile and forms 4,4′-azopyridine 1,1′-dioxide in cold 10% potassium hydroxide solution, and a mixture of 4,4′-azopyridine 1,1′-dioxide and 4,4′-azoxypyridine 1,1′-dioxide when heated in water. The same fact was found with 4-(hydroxyamino)-2,6-lutidine 1-oxide[60]. Quilico and others[58] showed that catalytic reduction of this 4-(hydroxyamino) compound gave 4-aminopyridine 1-oxide, and its oxidation with potassium permanganate in dilute sulfuric acid gave 4-nitrosopyridine 1-oxide (8-46), while oxidation with hydrogen peroxide in glacial acetic acid afforded azoxypyridine 1,1′-dioxide, which was also formed by the condensation of the hydroxyamino compound and the nitroso compound in dilute sulfuric acid.

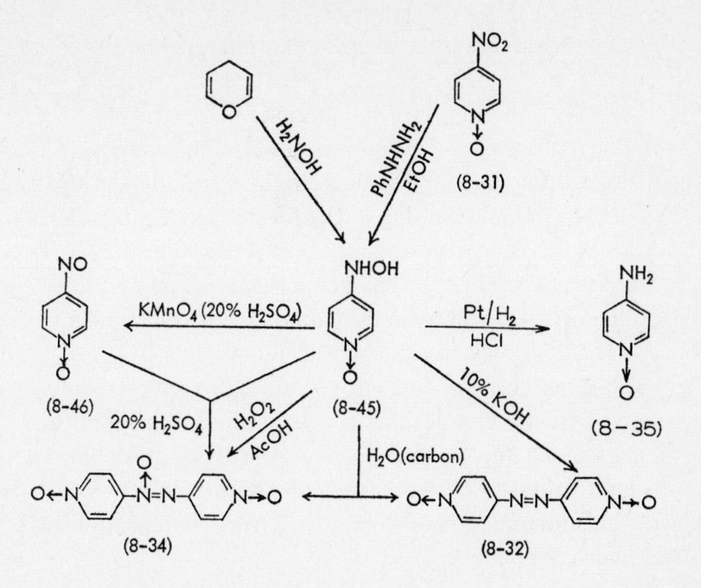

Yates and others[61] obtained 4-(hydroxyamino)-2,6-lutidine 1-oxide by heating 2,6-dimethyl-4-pyrone with excess of hydroxylamine hydrochloride in the presence of pyridine and reported that the azo compound is formed when the hydroxyamino compound is brought into a solution of sodium hydroxide of pH 12–13, and the azoxy compound is formed when the solution is brought to pH 7–9 in the presence of oxygen (air) or is irradiated in sunlight at pH 5–6.

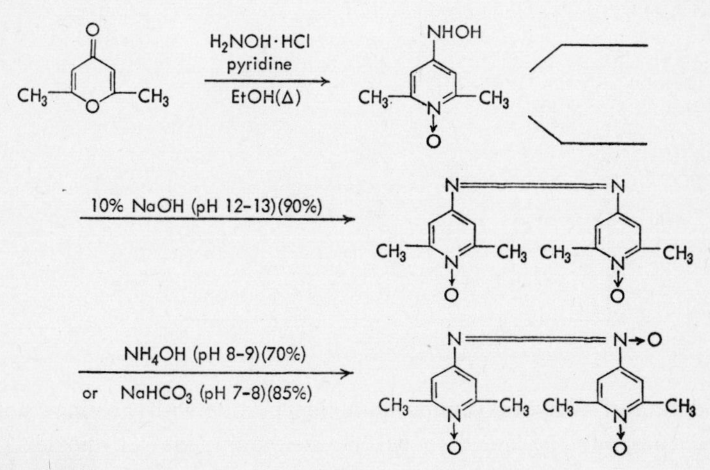

References p. 418

These facts indicate that the reduction of 4-nitropyridine 1-oxide proceeds to the formation of the amino compound via the nitroso and hydroxyamino compounds, as in aromatic nitro compounds in general, but the hydroxyamino compound formed as an intermediate is very active and reacts with the nitroso or amino compound undergoing condensation in the form of an azoxy or azo compound, resulting in the formation of a bimolecular reduction product. This assumption is endorsed by the fact that the 4-(hydroxyamino) compound is easily isolated when it is comparatively stable. The most marked example is 4-(hydroxyamino)quinoline 1-oxide, which is comparatively stable and sparingly soluble, so that the reduction of 4-nitroquinoline 1-oxide (8-47) tends to form 4-(hydroxyamino)quinoline 1-oxide (8-48) and does not give rise to bimolecular condensation[62–64]. Ochiai and Naito[62] had at first assumed the product to be 4,4′-hydrazoquinoline 1,1′-dioxide but it was later found to be 4-(hydroxyamino)quinoline l-oxide by identification with an authentic product prepared from 4-chloroquinoline l-oxide (8-49) and hydroxylamine[63].

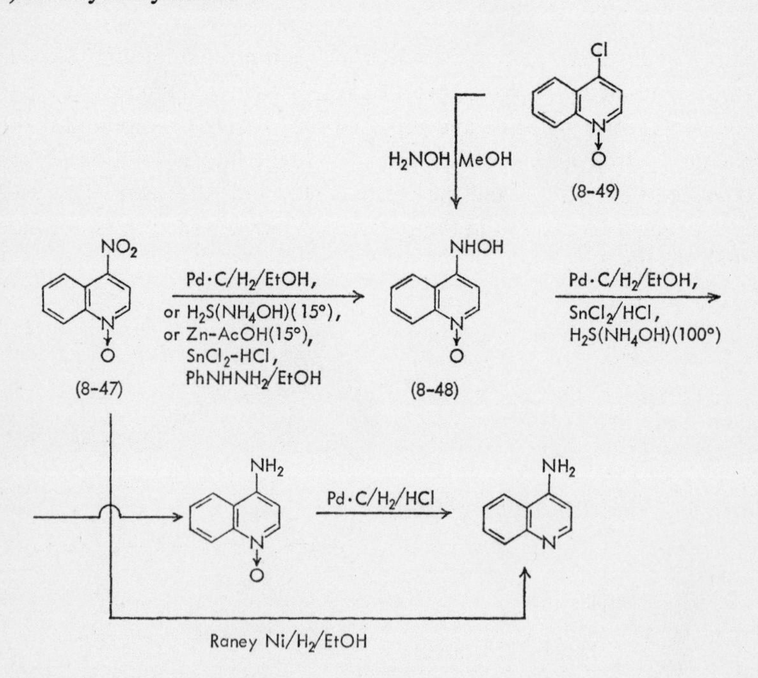

Oxidation of 4-(hydroxyamino)quinoline 1-oxide with Fehling's solution gives a compound assumed to be 4,4′-azoxyquinoline 1,1′-dioxide (8-50), while heating with phosphoryl chloride gives 2-chloro-4-aminoquinoline (8-51)[63].

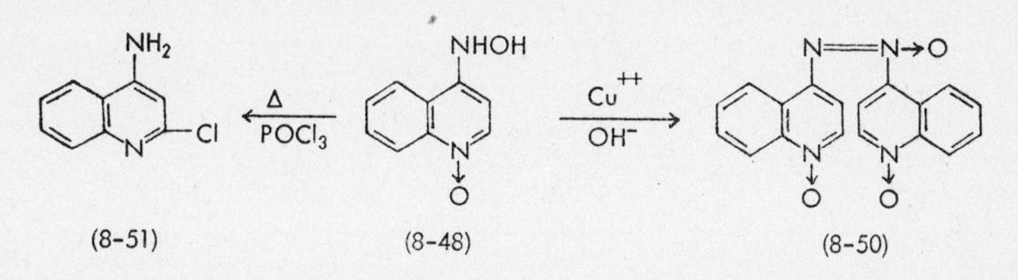

Itai and Kamiya[65] obtained 4-(hydroxyamino)quinoline 1-oxide by treatment of 4-nitroquinoline 1-oxide with hydrazine hydrate, but the reaction was too drastic and the yield was 49%. This is not a general reaction and is not any better than reduction with phenylhydrazine[59].

3-Nitroquinoline 1-oxide is reduced to 3-aminoquinoline 1-oxide (8-53) even by catalytic reduction in alkaline medium. Catalytic reduction of 3-nitroquinoline 1-oxide (8-52) over palladium–carbon in ether, in the presence of sodium hydrogen carbonate, the reaction being stopped after absorption of two moles of hydrogen, affords 3-(hydroxyamino)quinoline 1-oxide (8-54) in a satisfactory yield. If the reaction is not stopped at this stage, the reduction proceeds to the formation of 3-aminoquinoline[66].

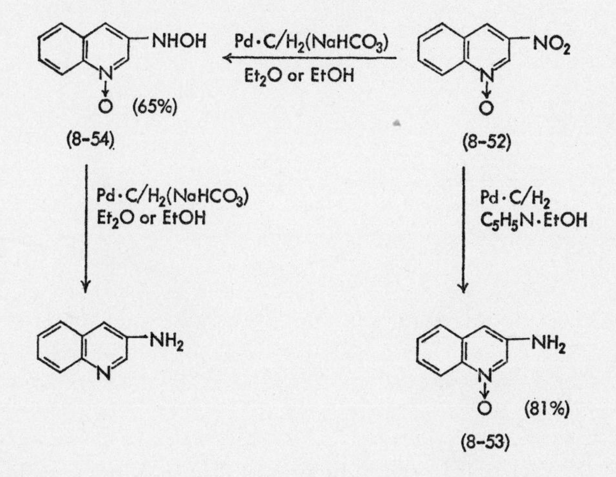

The nitro group in 1,2-bis(1-oxido-4-nitro-3-pyridyl)ethane (8-55) is greatly resistant to an S_N reaction (*cf. Section 8.2*, p. 349), but Taylor and others[67] obtained the corresponding 4-(hydroxyamino) compound (8-56) by catalytic reduction of (8-55) over palladium–carbon in ethanolic hydrochloric acid solution. This is probably because the hydroxyamino group, like the nitro group, has been inactivated by the same cause, presumably by steric hindrance.

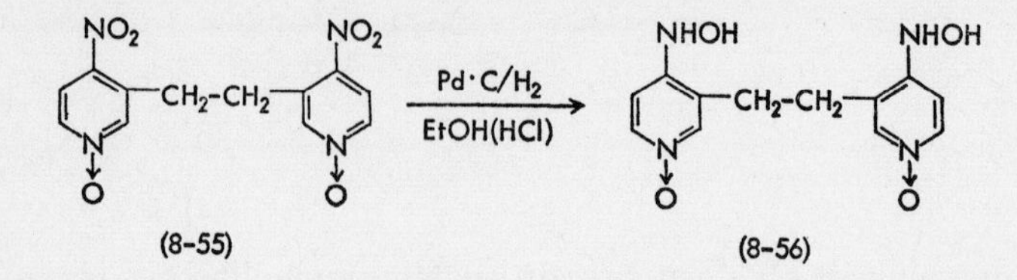

In a similar manner, Taylor and Ehrhart[68] obtained the corresponding 4-(hydroxyamino) compound (8-59) from 4-nitronicotinic acid 1-oxide (8-57), having a nitro group *ortho* to the carboxyl, by heating its solution in ammonia-alkaline medium and passing hydrogen sulfide through the solution. However, catalytic reduction of 4-nitronicotinic acid 1-oxide over Raney nickel gives 4-aminonicotinic acid (8-58) and reduction with hydrazine hydrate in the presence of Raney nickel results mainly in substitution of the nitro group with a hydrazino group, and the formation of 4-aminonicotinic acid l-oxide as a by-product.

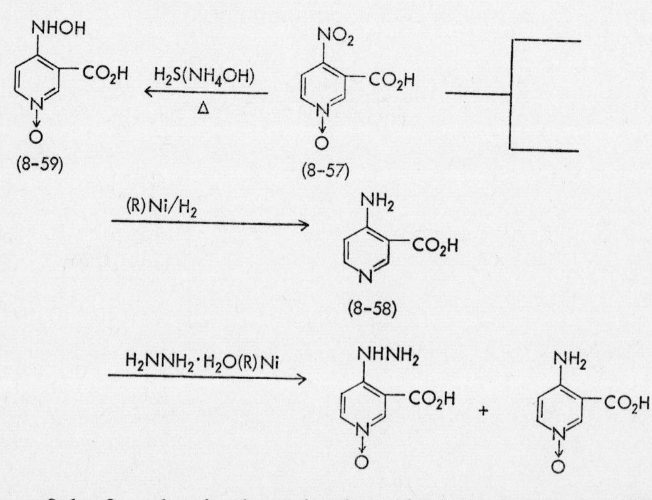

In view of the fact that in the reduction of 2-iodo-4-nitropyridine l-oxide (8-60) with hydrazine hydrate, Talik[69] found that iodine did not react with hydrazine and the product turned out to be 2-iodo-4-aminopyridine l-oxide (8-61), the foregoing example indicates that the nitro group in 4-nitronicotinic acid l-oxide causes some hindrance against reduction.

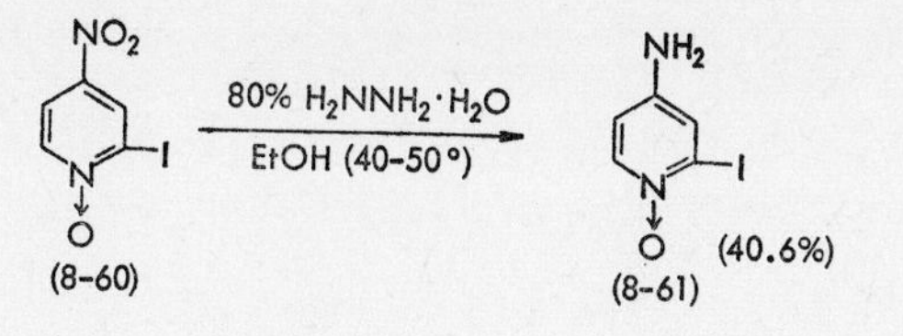

In the foregoing examples of reduction, there are indications that the *N*-oxide group has in some reactions activated the nitro group in its *para*-position against reduction. Brown and Evans[70] found that 4-nitro-2,6-lutidine 1-oxide (8-62) is reduced to 4,4′-azo-2,6-lutidine 1,1′-dioxide (8-63) by treatment with sodium borohydride in cold diglyme solution, and concluded that the nitro group had been activated by the *N*-oxide group, since sodium borohydride does not usually react with a nitro group.

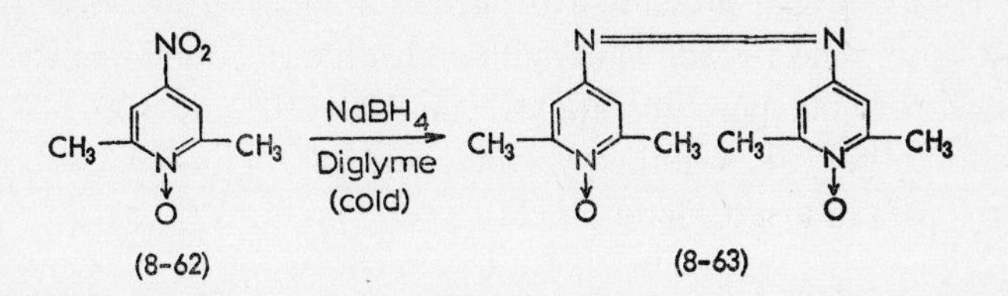

Ochiai and Mitarashi[59] found that the reduction of 3-nitroquinoline 1-oxide, with the nitro group *meta* to the *N*-oxide group, with phenylhydrazine in ethanol solution, did not progress sufficiently at room temperature, as in the case of the corresponding 4-nitro compound, but required warming, and that the yield was poor. Itai and Natsume[71] carried out the reduction of 4-(8-64) and 3-nitropyridazine 1-oxide (8-65) and found, as shown in the following scheme, that with the 3-nitro compound, interruption of neutral catalytic reduction over palladium–carbon resulted in the formation of 3-(hydroxyamino)pyridazine 1-oxide (8-66), while the hydroxyamino compound was not obtained with the 4-nitro group and that the reduction progressed to form the 4-amino compound.

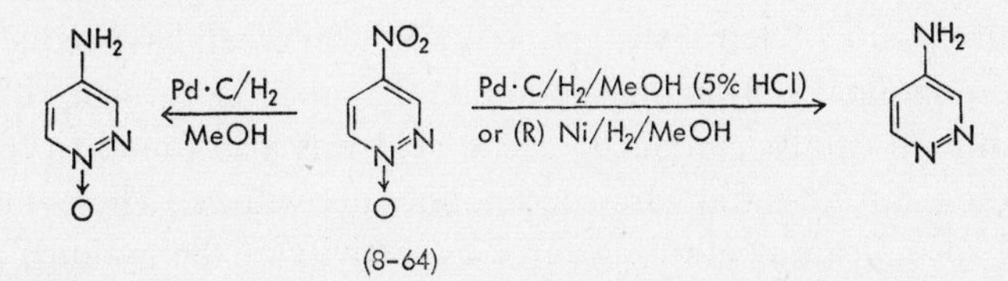

References p. 418

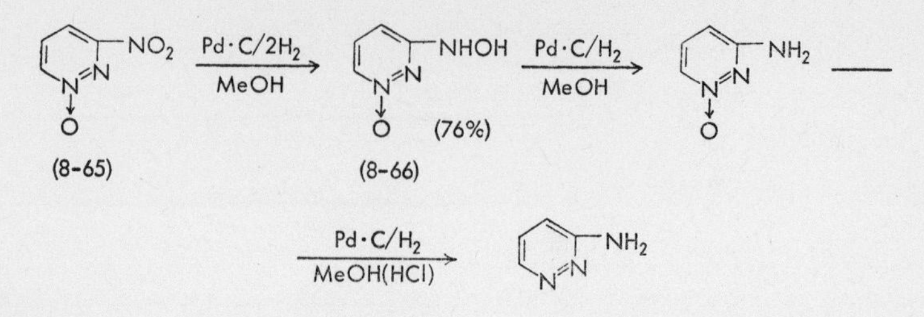

The marked fact found in these reduction reactions is the resistance of the *N*-oxide group to any form of reduction, as was stressed in a separate section (*cf. Section 5.2.1*); this still left much to be clarified in the problem of deoxygenation of the *N*-oxide group. However, the matter was solved by the discovery of Hayashi and others[64] that deoxygenation can be effected smoothly, in a quantitative yield, by catalytic reduction over Raney nickel (*cf. Section 5.2.2*).

Ishii carried out high-pressure hydrogenation of 4-nitropyridine 1-oxide[72a] and 4-nitroquinoline 1-oxide[72b] in methanol solution containing acetic acid, using nickel formate–paraffin catalyst[73], at 100–110° and 120–140°, respectively, and obtained 4-aminopyridine and 4-amino-5,6,7,8-tetrahydroquinoline (8-67) in a satisfactory yield.

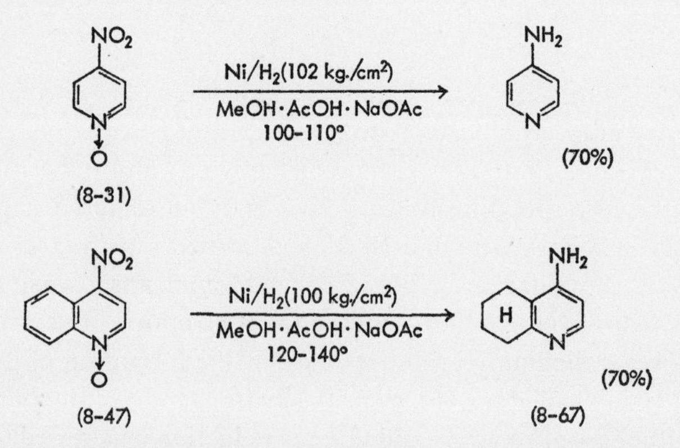

The formation of 4-amino-5,6,7,8-tetrahydroquinoline is an interesting fact. The same high-pressure hydrogenation of 4-hydroxyquinoline 1-oxide gave 4-hydroxy-5,6,7,8-tetrahydroquinoline[72b], while catalytic reduction of 4-aminoisoquinoline[74], 1-aminoisoquinoline[75], and isocarbostyril[75] over Adams platinum catalyst in glacial acetic acid was also found to form the corresponding 5,6,7,8-tetrahydroisoquinoline derivatives. These facts are considered to indicate neutralization of the activity of the pyridine portion of the quinoline or isoquinoline ring against hydrogenation by the polar effect (+*T* effect) of the amino or hydroxyl group[76] (*cf. Section 1*, p. 2).

Example. Formation of 4-(Hydroxyamino)pyridine 1-oxide by Reduction of 4-Nitropyridine 1-Oxide with Phenylhydrazine[57b]

One gram of 4-nitropyridine 1-oxide was added in small portions to a solution of 4.4 g of phenylhydrazine dissolved in 10 ml of ethanol, by which the former dissolved with red coloration, which became darker as the solution was warmed on a water bath. When the bath temperature reached 40°, there was evolution of nitrogen which became more active at a bath temperature of 50–52° but ceased after about 1 h. The reaction mixture became orange-yellow and deposited yellow scaly crystals. After the mixture was cooled in an ice bath, the mixture was filtered and the filtrate was evaporated under reduced pressure, in nitrogen stream, leaving some scaly crystals and orange-red viscous fluid. The two lots of crystals were combined (0.9 g) and recrystallized from water in nitrogen atmosphere. The colorless crystals became orange at 180° and black at 200°, and decomposed at 237°.

8.3.2 Activity to Nucleophilic Reagents

A nitro group in the position *para* to the *N*-oxide in aromatic *N*-oxide compounds is active to nucleophilic substitution and is known to undergo easy substitution by halogens, alkoxyl, aryloxyl, arylthio, hydroxyl, and amino groups. Such activity of the nitro group can also be seen in 2-[77b] and 4-nitropyridines[77a] without *N*-oxidation, but the activity is much less in these cases and is especially so in acid reaction. Katada[77a] compared this reaction with that of 4-nitropyridine 1-oxide[77c] and found that the corresponding *N*-oxide compound showed a more facile reaction under milder conditions, revealing increased activity of the nitro group due to *N*-oxidation.

(a) Halogen Substitution

In 1943, at the time of structural determination of 4-nitroquinoline l-oxide, Ochiai, Ishikawa, and Sai[78] recognized the formation of 2,4-dichloroquinoline on heating 4-nitroquinoline l-oxide with phosphoryl chloride on a water bath. Later, Ochiai, Itai, and Yoshino[79] found that the same reaction of 4-nitropyridine l-oxide mainly gave 4-chloropyridine l-oxide. Further, Ishikawa carried out the same reaction on 4-nitroquinoline 1-oxides[80], with the nitro group in the benzene portion, 4-nitro-2-picoline l-oxide, and 4-nitro- (8-68) and 3-nitro-2,6-lutidine 1-oxide[81] (8-69) and found that only the nitro group in the 4-position underwent substitution with a chlorine atom.

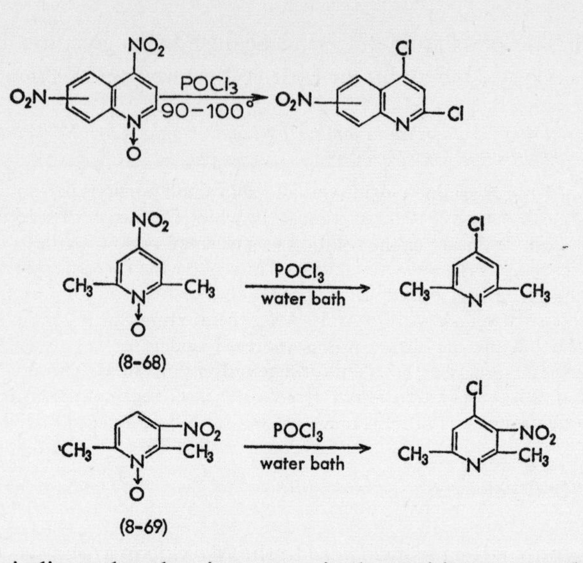

These facts indicate that the nitro group in the position *para* to the *N*-oxide group is especially active to chlorine substitution. Itai[82] examined this reaction with respect to reaction temperature and time, and found that the same reaction progressed with sulfuryl chloride and acetyl chloride; that the formation of 4-chloroquinoline 1-oxide increased under milder reaction conditions; that the formation of 2,4-dichloro compound increased as reaction conditions became more drastic and that this tendency was especially marked in quinoline compounds; and that the best conditions for obtaining 4-chloroquinoline l-oxide were to carry out the reaction with acetyl chloride at 0° for the quinoline system and at around 50° for the pyridine system.

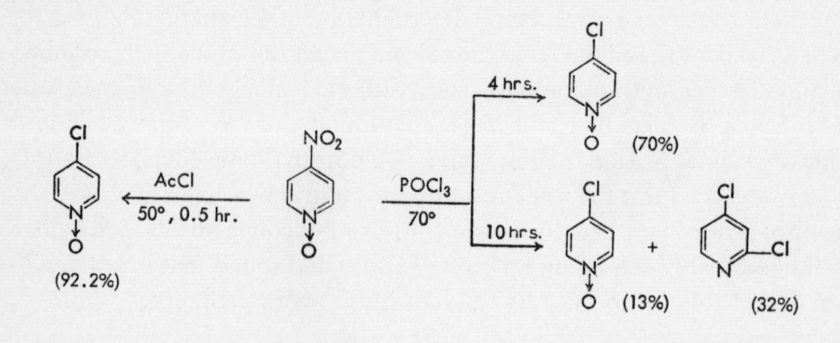

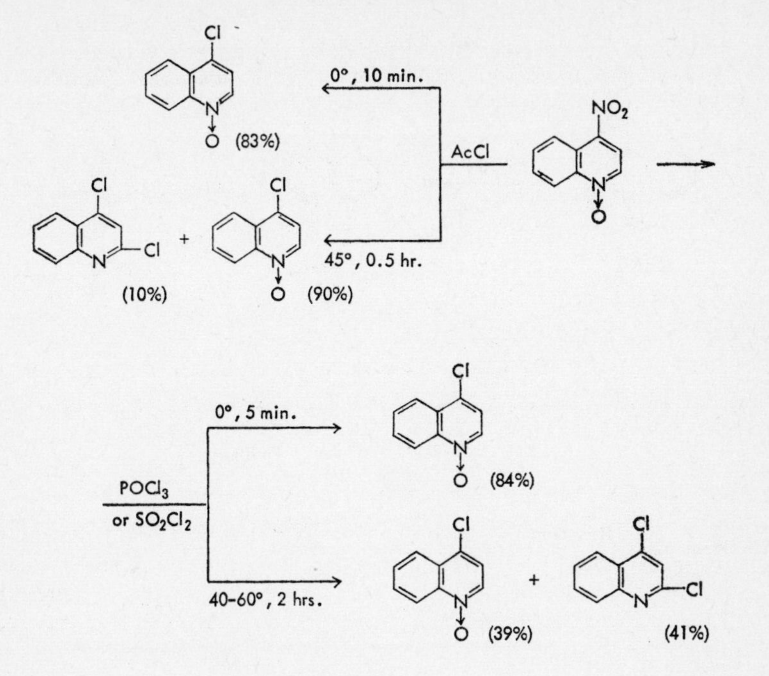

The use of acetyl bromide in place of acetyl chloride in this reaction was found to give 4-bromopyridine 1-oxide.

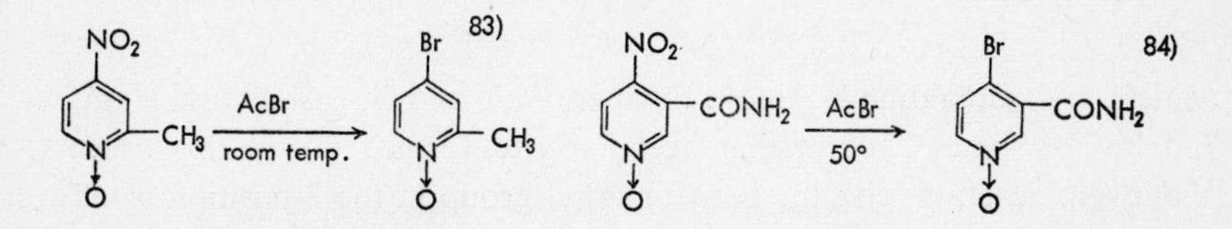

Hamana and his associates[85] carried out the reaction of 4-nitroquinoline 1-oxide and phosphoryl bromide in chloroform or ethyl acetate and obtained the results shown in Table 8-1.

It is seen from these results that phosphoryl bromide first attacks the *N*-oxide group at a low temperature and then reacts with the nitro group when warmed.

Of these reactions, the most widely used is that with acetyl chloride, which has been reported for 4-nitro derivatives of 2-picoline 1-oxide,[83] 3-bromopyridine 1-oxide[86], 3-picoline 1-oxide[87,88], nicotinic acid *N*-oxide[88], nicotinamide *N*-oxide[84], 1,2-bis(1-oxido-3-pyridyl)ethane[30a], 7-chloroquinoline 1-

References p. 418

TABLE 8-1

REACTION OF 4-NITROQUINOLINE 1-OXIDE WITH PHOSPHORYL BROMIDE

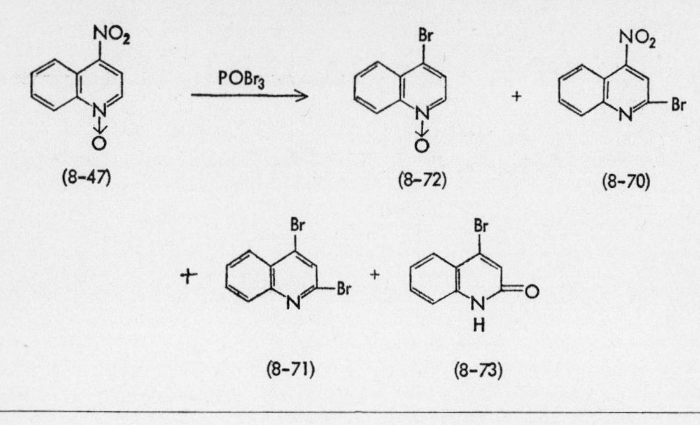

Solvent	Reaction temp.(°C)	Yield of product (%)				
		(8-70)	(8-71)	(8-72)	(8-73)	(8-47) (recovery)
$CHCl_3$	5–10	73	2.0	0.8	0.4	4.0
	61	25.6	51	0.8	0.8	2.5
AcOEt	5	10.5			0.8	78.0
	64	3.8	64	5.0	0.8	

oxide[89], and quinaldine 1-oxide[32]. Further, Kato and Hayashi[90], and Hamana and his associates[32] found that acetyl nitrite, formed during this process, underwent reaction with the active methyl group in the 2-position to effect its nitrosation, or that further dehydration occurred to form an isonitroso compound or a nitrile as a by-product (*cf. Section 8.2*).

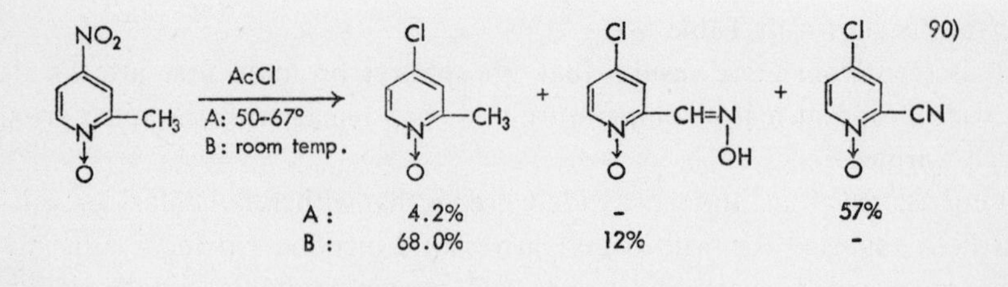

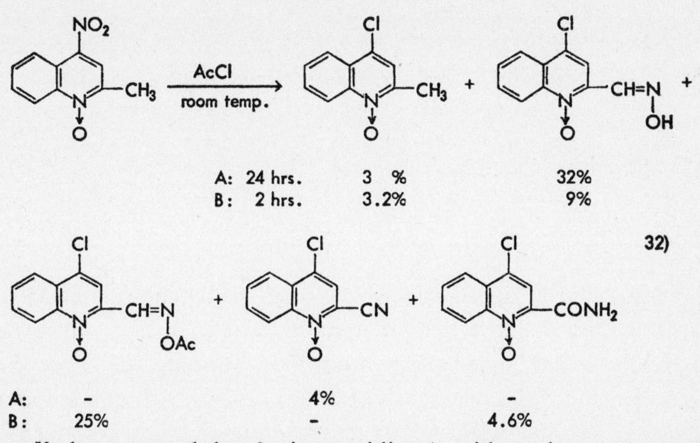

Brown[56] also reported that 2-nitropyridine 1-oxide underwent smooth substitution in the cold by treatment with acetyl chloride to form the 2-chloro derivative and that the activity of this reaction was greater than that of 4-nitropyridine 1-oxide.

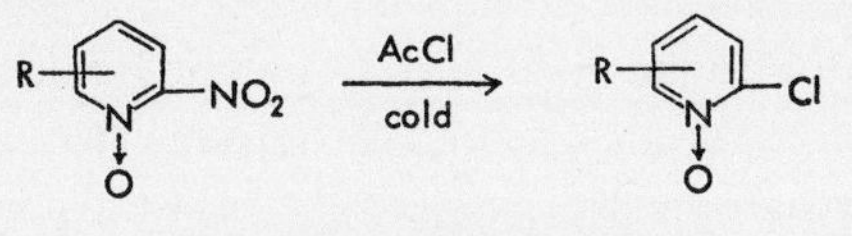

Itai and Natsume[71a] found that this chlorine substitution reaction progressed with 4-nitropyridazine l-oxide (8-64), and Suzuki and Nakajima[91] showed that it took place with 4-nitrocinnoline l-oxide (8-74). Natsume and Itai observed that the nitro group in the position *meta* to the *N*-oxide group in pyridazine *N*-oxide was also active, though slightly less than that in the *para* position.

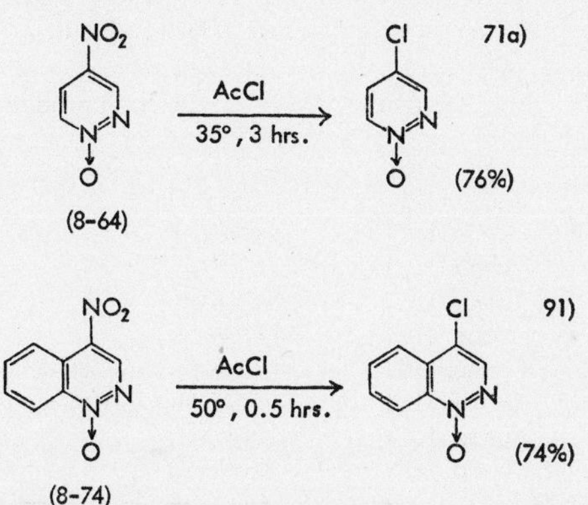

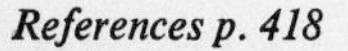
References p. 418

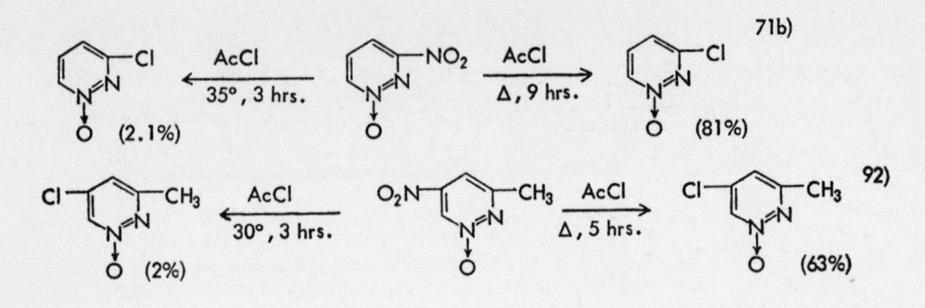

While 4-nitropyridazine 1-oxide reacts with acetyl chloride at 35° to give the corresponding 4-chloro derivative in 76% yield, 3- or 5-nitropyridazine 1-oxide hardly reacts under these conditions, forming the 3- or 5-chloro compound in about 2% yield, whilst such a 3- or 5-chloro compound is formed in *ca.* 60–80% yield when the reaction mixture is boiled. These facts indicate that the *N*-oxide group in pyridazine series *N*-oxides activates the nitro group in its *para* position markedly, while the nitro group in the position *ortho* or *para* to the free nitrogen is also activated to a certain extent through the adjacent *N*-oxide group.

In 1951, Okamoto[93] found that 4-nitroquinoline 1-oxide and its derivatives, when boiled in conc. hydrochloric or hydrobromic acid, underwent substitution of their 4-nitro group with halogen to form the corresponding 4-haloquinoline 1-oxide (Table 8-2). At about the same time, den Hertog and Combé[94] found that the nitro group in 4-nitropyridine 1-oxide underwent substitution with a halogen when heated with conc. hydrochloric or hydrobromic acid under a stronger condition than that in the case of quinoline

TABLE 8-2

REACTION OF 4-NITROQUINOLINE 1-OXIDE AND ITS DERIVATIVES IN BOILING HYDROCHLORIC OR HYDROBROMIC ACID[93]

Quinoline 1-oxide	Reagent	Reaction product	Yield (%)
4-NO_2–	conc. HCl	4-Cl–	95
	30% HBr	4-Br–	60
4-NO_2-3-Br–	conc. HCl	4-Cl-3-Br–	79
4-NO_2-6-Cl–	conc. HCl	4,6-Cl_2–	78
4-NO_2-6-Br–	conc. HCl	4-Cl-6-Br–	78
4,6-$(NO_2)_2$–	conc. HCl	4-Cl-6-NO_2–	52
4,5-$(NO_2)_2$-6-CH_3–	conc. HCl	4-Cl-5-NO_2-6-CH_3–	78
4,5-$(NO_2)_2$-6-Br–	conc. HCl	4-Cl-5-NO_2-6-Br–	54
5-NO_2–	conc. HCl	recovered	90
6-NO_2–	conc. HCl	recovered	90

compound. However, this reaction with hydrobromic acid was found to be accompanied by side reactions such as hydrolysis, bromination of the 3-position, or deoxygenation[94,95].

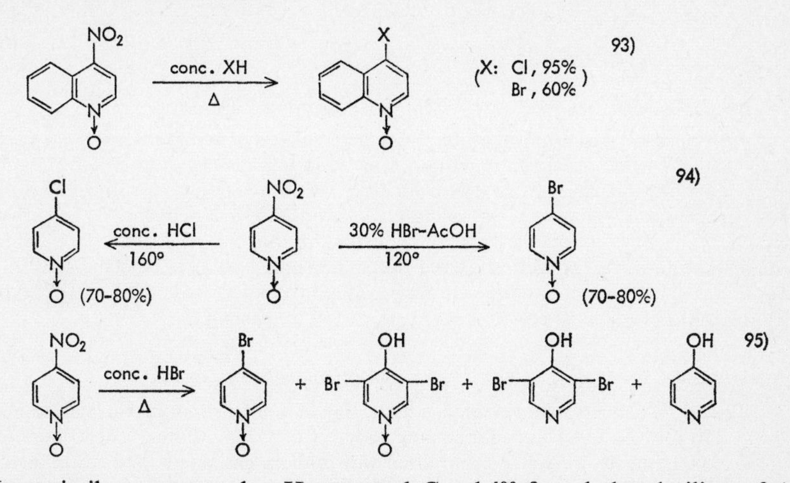

In a similar manner, den Hertog and Combé[96] found that boiling of 4-nitropyridine 1-oxide with 5% hydrochloric acid at 180–190° resulted in the formation of 3-chloro-4-hydroxypyridine 1-oxide, while Acheson and others[97] obtained 10-hydroxyacridone by heating 9-nitroacridine 10-oxide with 2N hydrochloric acid at 100°. Krause and Langenbeck[98], and Profft and Schulz[99] both obtained good results by carrying out this reaction in boiling methanol or ethanol and passing dry hydrogen chloride gas through the solution.

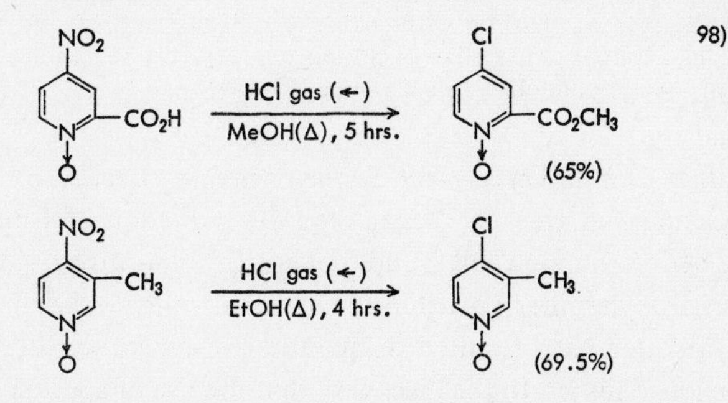

Kanô and his associates[100] found that heating of 3-methoxy-4-nitrocinnoline 1-oxide with conc. hydrochloric acid resulted in preferential substitution of the nitro group over hydrolysis of the methoxyl group to form 4-chloro-3-methoxycinnoline 1-oxide.

References p. 418

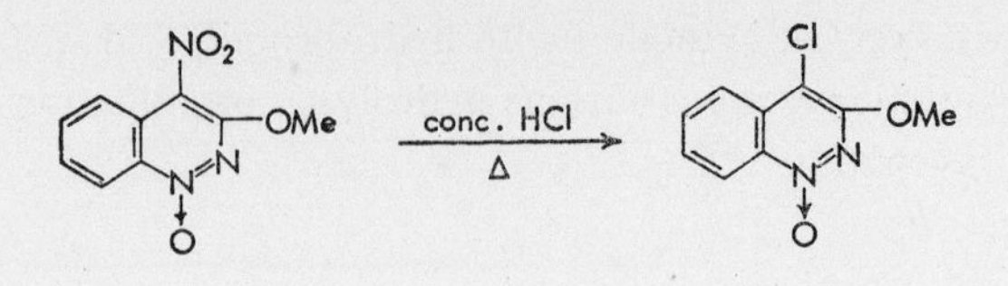

Example 1. 4-Chloropyridine 1-Oxide from 4-Nitropyridine 1-Oxide[82b]

A mixture of 1.0 g of 4-nitropyridine 1-oxide and 5.0 ml of acetyl chloride was warmed at 50° and the whole solidified in 30 min. This solid was dissolved in ice water, basified with sodium carbonate, and extracted with ether. The ether extract was dried over potassium carbonate and the solvent was evaporated, leaving 0.05 g of pungent oil (2,4-dichloropyridine). The aqueous layer left after ether extraction was further extracted with chloroform, the chloroform extract was dried over potassium carbonate, and the solvent was evaporated. The residual crystals were recrystallized from acetone to 0.85 g (92.2%) of 4-chloropyridine 1-oxide as colorless scales, m.p. 169.5° (decomp.).

Example 2. 4-Chloroquinoline 1-Oxide from 4-Nitroquinoline 1-Oxide

(i)[82b] To 50 ml of acetyl chloride chilled to 0°, 10 g of 4-nitroquinoline 1-oxide was added slowly. The mixture was warmed gradually and kept at 45° for 40 min. The reaction mixture was poured into ice water, neutralized with sodium carbonate, and steam-distilled. The distillate was extracted with ether and the ether solution was dried over anhyd. sodium sulfate. Evaporation of ether left 0.11 g of 3,4-dichloroquinoline as colorless needles, m.p. 62–65.5°.

The aqueous layer left after ether extraction was basified with sodium carbonate and extracted with chloroform. The extract solution was dried over potassium carbonate and the solvent was evaporated. The residual crystals were recrystallized from acetone to 9.35 g (98.9%) of 4-chloroquinoline 1-oxide as pale yellow needles, m.p. 133–133.5°.

(ii)[93] A mixture of 1 g of 4-nitroquinoline 1-oxide and 10 ml of conc. hydrochloric acid was warmed on a water bath by which evolution of nitrogen dioxide gas became evident. After 30 minutes' warming, the reaction mixture was evaporated to dryness on a water bath, the residue was basified with sodium carbonate, and extracted with chloroform. The residue left after evaporation of chloroform was recrystallized from benzene to 0.95 g (95%) of 4-chloroquinoline 1-oxide, m.p. 133–134°.

(b) Substitution with an Oxygen- or Sulfur-containing Group

In 1943, Ochiai and Katada[77c] examined the reduction of 4-nitropyridine 1-oxide with hydroxylamine and found that heating of 4-nitropyridine 1-oxide (8-31) with hydroxylamine hydrochloride in methanolic potassium hydroxide solution on a water bath resulted in quantitative formation of 4-methoxypyridine 1-oxide. This led to the discovery that the use of sodium ethoxide in ethanol or sodium phenoxide in phenol resulted in the formation of 4-ethoxy- or 4-phenoxy-pyridine 1-oxide in a good yield. Later, Ochiai, Ishikawa, and Sai[78] found that the reaction of 4-nitroquinoline 1-oxide with sodium ethoxide in ethanol gave 4-ethoxyquinoline 1-oxide, and Ochiai, Itai, and Yoshino[79] proved the formation of 4-(*p*-tolylthio)pyridine 1-oxide from the reaction of 4-nitropyridine 1-oxide and sodium *p*-thiocresol. Further, Ishikawa[81] examined the same reaction of 3- and 4-nitro derivatives of 2,6-lutidine 1-oxide and found that whereas the nitro group in the 4-position almost quantitatively underwent substitution with an ethoxyl group, the 3-nitro derivative became resinous under the same reaction conditions and did not form the ethoxyl derivative. Okamoto[93] carried out the reaction of 5- and 6-nitroquinoline 1-oxides with sodium methoxide in methanol and found that the corresponding methoxyquinoline 1-oxides were formed in 22 and 27% yield, respectively. The same reaction of 5- and 6-nitroquinoline without *N*-oxide group results in almost total recovery of the 5-nitro compound and resinification of the 6-nitro compound, neither forming the methoxyl derivative. Okamoto concluded from this reaction that the activation of the nitro group by *N*-oxidation was due to the inductive effect of the *N*-oxide group. This is noteworthy in view of the fact that the nitro group in the benzene portion of quinoline is resistant to halogen substitution (*cf. Section 8.3.2* p. 367). These methoxyl substitutions are attended with great resinification and the yield is very poor, as opposed to the smooth substitution of the nitro group in the 4-position.

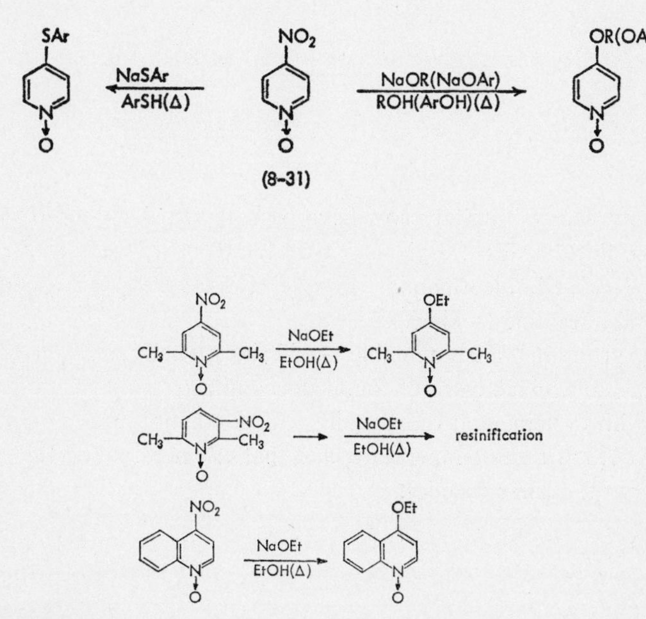

References p. 418

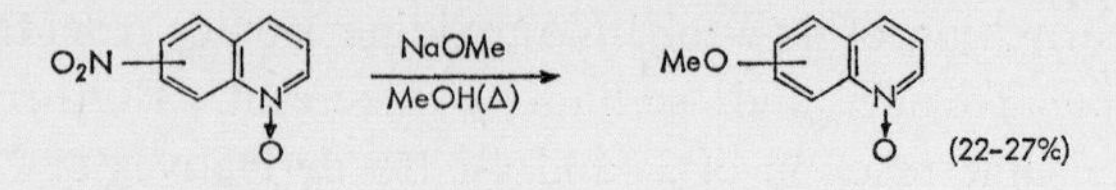

Katada[101] examined the reaction of 4-nitropyridine 1-oxide with alkoxides or phenoxides and found that alkoxides are more active the smaller the number of carbon atoms; that the yield decreased rapidly with increasing number of carbon atoms, though benzyloxide and phenoxide are very active; that the corresponding alcohol or phenol is advantageous as the solvent; that nitrobenzene was effective when the reaction is difficult; and that the reaction does not progress in acetone or benzene. Ikehara[102] found the same tendency with 4-nitroquinoline l-oxide. The following examples show that the nitro group in 4-position of pyridine 1-oxide is more active towards alkoxides than the chlorine atom in 2-position[69].

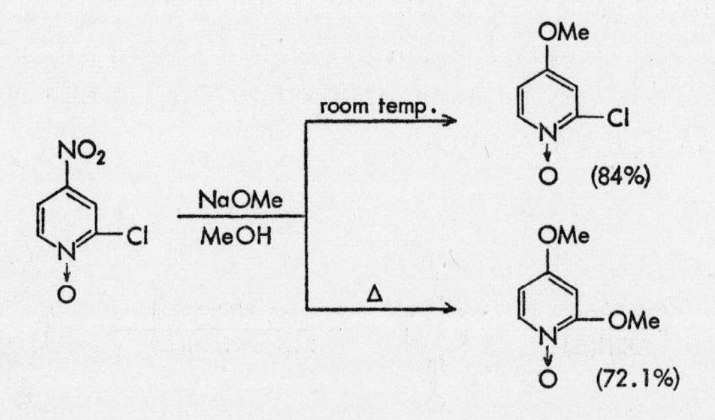

A large number of similar cases have been reported for 4-nitro derivatives of pyridine 1-oxide[94], 2-picoline 1-oxide[83,99,103], 2,6-lutidine 1-oxide[11], 3-picoline 1-oxide[87], nicotinamide 1-oxide[104], 5-ethyl-2-methylpyridine 1-oxide[103], and quinoline 1-oxide[105].

The nitro group in 4-nitroquinoline 1-oxide is especially readily substituted with alkanethiols. Okabayashi[106] found that 4-nitroquinoline 1-oxide reacts smoothly with cysteine and thioglycolic acid in ethanol, under mild conditions of pH 7.2 at room temperature, and that ethanethiol and thioglycerol both undergo the same reaction.

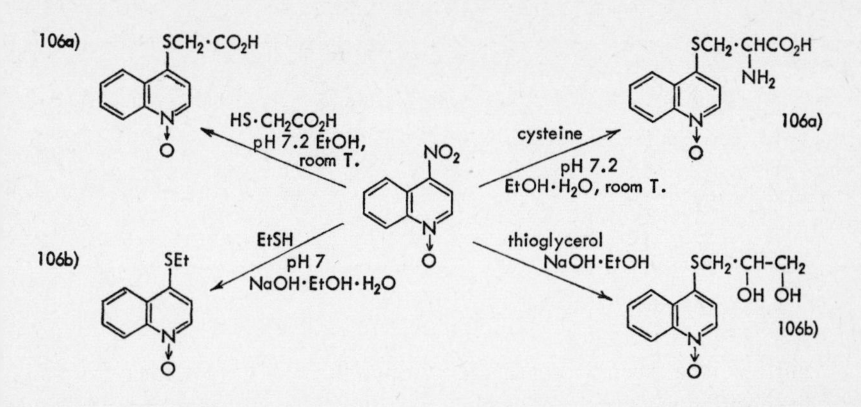

Okamoto and Itoh[107] made a kinetic examination of the nucleophilic exchange between the 4-nitro group in 4-nitroquinoline l-oxide, its 2-methyl, 7-chloro, 6-nitro, and 8-nitro derivatives, 4-nitropyridine 1-oxide, and 4-nitroquinoline, with the SH group in thioglycolic acid, by amperometric titration in a dioxan–water system of pH 7 and revealed the following points:
(i) 4-Nitroquinoline 1-oxides reacted 5–30 times faster than 4-nitroquinoline, while the reaction of 4-nitropyridine 1-oxide was about 1/10 that of 4-nitroquinoline 1-oxide.
(ii) Derivatives of 4-nitroquinoline 1-oxide having an electron-attracting group in their benzene portion reacted somewhat faster than 4-nitroquinoline 1-oxide itself.

(iii) In general, the greater the polarity of the solvent, the larger becomes the K and the smaller $\Delta E^{\neq}$ and $\Delta S^{\neq}$. In 4-nitroquinoline, the reaction rate becomes a little faster with increasing polarity of the solvent, but its effect is unexpectedly small. This situation is somewhat different in 4,8-dinitroquinoline 1-oxide.

In addition, Okamoto and Itoh[107] gave some considerations on the characteristics of the *N*-oxide group as the accelerating factor for this reaction and its relation to solvation.

It has been confirmed that this reaction of the nitro group also takes place in the nitrated derivatives of diazine series *N*-oxides.

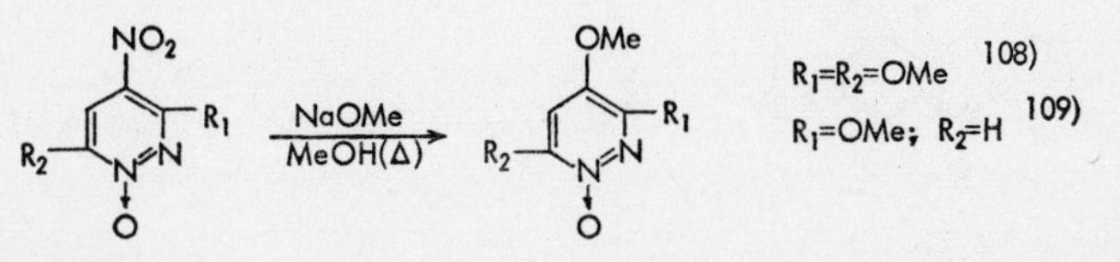

References p. 418

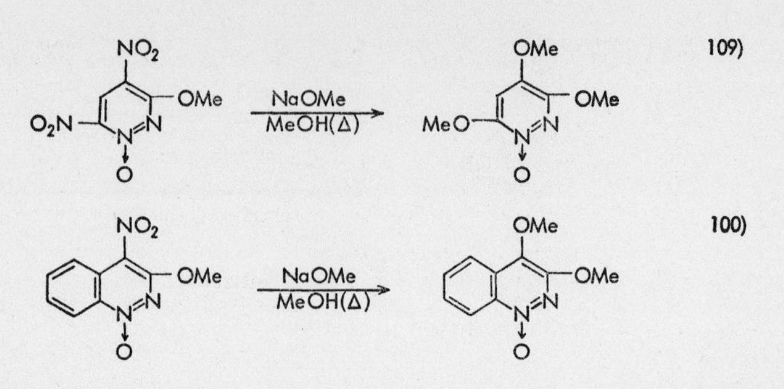

Further, it has been shown that in nitropyridazine *N*-oxides, the nitro group in the position *meta* to the *N*-oxide group is about as active towards alkoxides as the nitro group in the *para*-position[71a]. This fact is noteworthy in view of the behavior of these nitro groups towards acetyl chloride. As already mentioned (*cf. Section 8.3.2*, p. 371) both of them are active towards acetyl chloride, but that in the *para*-position has a greater activity than that in the *meta*-position.

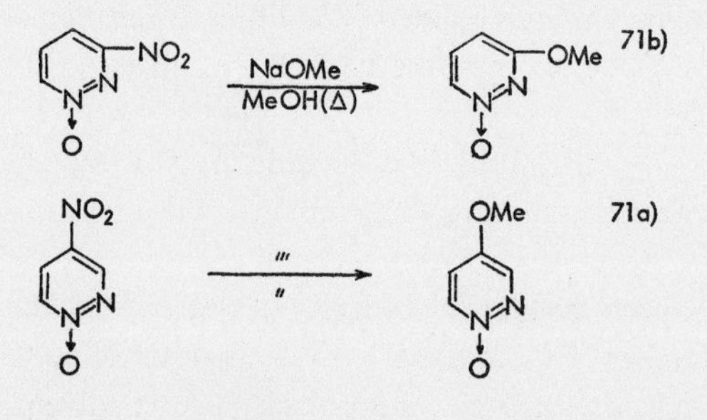

Otomasu[110] found that the nitro group in 3-nitrophenazine 5-oxide (8-75) underwent facile substitution by alkoxyl groups. It should be noted that the nitro group that undergoes this reaction is the one in the position *para* to the non-oxygenated ring-nitrogen. This nitro group is in conjugation both with the *N*-oxide group and the non-oxygenated ring-nitrogen so that it might be considered to be under the indirect polar effect of the *N*-oxide group.

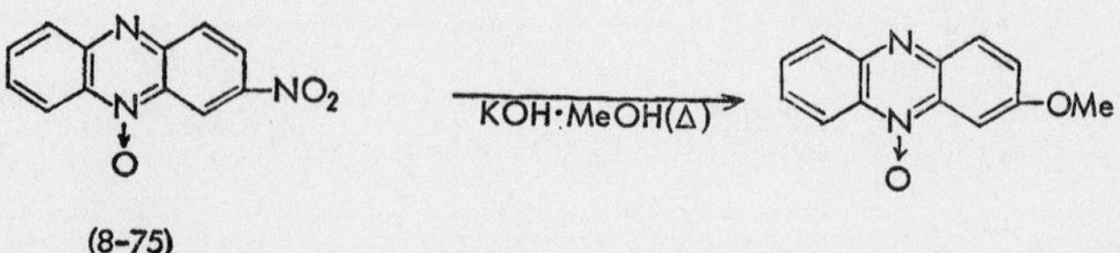

(8-75)

References p. 418

When 4-nitropyridine 1-oxide is warmed in aqueous alkali hydroxide solution, the nitro group undergoes substitution with a hydroxyl group, but the sodium nitrite formed by the reaction acts as a reducing agent to produce 4,4′-azopyridine 1,1′-dioxide and it is difficult to obtain 4-hydroxypyridine 1-oxide in a pure state[47]. Ochiai and his collaborators synthesized 4-hydroxypyridine 1-oxide[111] and 4-hydroxyquinoline 1-oxide[112] by the treatment of 4-nitropyridine 1-oxide and 4-nitroquinoline 1-oxide with sodium benzyloxide and catalytic reduction of the 4-benzyloxyl derivatives thereby formed, using palladium–carbon catalyst in ethanol solution. Later, den Hertog and Combé[96] succeeded in obtaining 4-hydroxypyridine 1-oxide in over 60% yield by warming 4-nitropyridine 1-oxide with sodium hydroxide solution at 40–50° while adding hydrogen peroxide to prevent reduction by sodium nitrite.

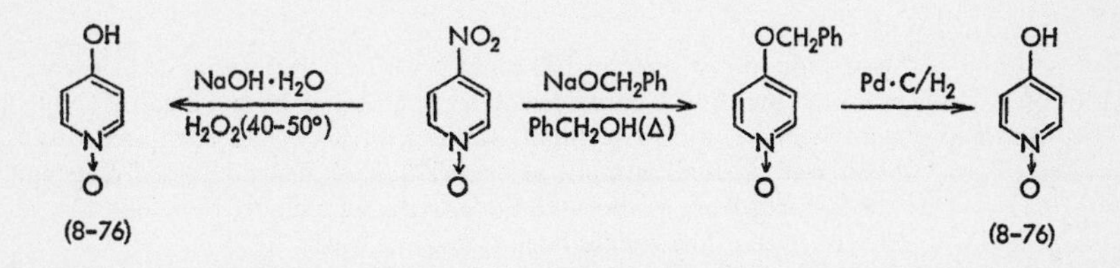

Okamoto[93] obtained 4-hydroxyquinoline 1-oxide in about 71% yield by refluxing 4-nitroquinoline 1-oxide in 40% sulfuric acid. He also proved that only the nitro group in 4-position underwent substitution to form 4-hydroxy derivatives by the same treatment of 4-nitroquinaldine 1-oxide, 4,8-dinitroquinoline 1-oxide, and 6-bromo-4,5-dinitroquinoline 1-oxide.

Hayashi[113] found that heating of 4-nitropyridine 1-oxide with acetic anhydride on a water bath resulted in the formation of *N*-acetoxy-4-hydroxy-3-nitropyridine betaine, and treatment of the mother liquor left after its isolation afforded a small amount of 4-hydroxypyridine l-oxide. This fact indicates that the nitration progresses *via* the acetyl nitrite formed in this reaction, and its removal outside of the reaction system by the concurrent addition of dimethylaniline was found to inhibit nitration, 4-hydroxypyridine 1-oxide being obtained in 63% yield. The same reaction was found to take place with 4-nitroquinoline 1-oxide[114] and 4-nitro-3-picoline 1-oxide[87a].

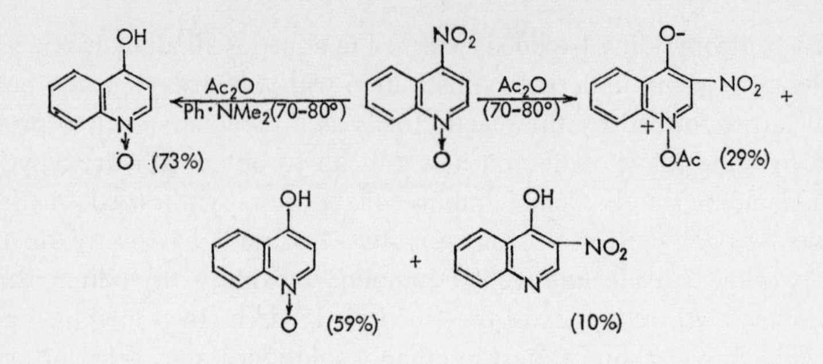

Example 1. Preparation of 4-Benzyloxypyridine 1-Oxide[115]

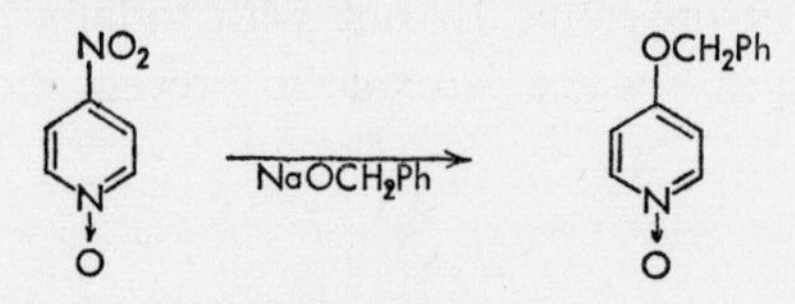

A solution of 2 g of sodium dissolved in 100 ml of benzyl alcohol was added slowly to a solution of 12 g of 4-nitropyridine 1-oxide dissolved in 80 ml of benzyl alcohol (Note 1). This mixture was allowed to stand overnight, sodium nitrite that precipitated out was filtered off and washed with a small amount of benzyl alcohol. The combined pale yellow filtrate (Note 2) and washings were evaporated under reduced pressure by which the solution turned from orange to red. The residual solution was allowed to cool, ether was added to it, and it was allowed to stand overnight. The white needle crystals that precipitated were collected by filtration, washed first with ether and then with acetone, and dried. Yield, 13.3 g.

The filtrate was concentrated under a reduced pressure, ether was added to the residual solution, and when allowed to stand further 1.5 g of crystals precipitated. Total yield of crude crystals, m.p. 173–176°, 14.8 g (*ca.* 85%). This was recrystallized from methanol–acetone mixture to colorless prismatic crystals of m.p. 175.5–177°.

Note 1: Addition of sodium benzyloxide solution slowly to the solution of the nitro compound gives better crystals than mixing *vice versa.*

Note 2: If freshly distilled benzyl alcohol is used, this filtrate is pale yellow, but if hydrous alcohol is used, the filtrate becomes orange, the final product becomes reddish, and the yield lower.

Example 2. Preparation of 4-Hydroxypyridine 1-Oxide

(i)[115] A solution of 3 g of sodium dissolved in 40 ml of methanol was added to the solution of 23 g of 4-benzyloxypyridine 1-oxide dissolved in 60 ml of methanol, 0.5 g of palladium–carbon (40% palladium) was added to this mixture, and this was submitted to catalytic reduction. Hydrogen was absorbed rapidly to an amount of 1 mole. After filtration of the catalyst, the solvent was evaporated and the sodium salt of 4-hydroxypyridine 1-oxide precipitated out as white needle crystals. The crystals were collected with addition of a small amount of acetone and dried. Yield, 15.9 g (quantitative). The sodium salt was dissolved in as small an amount of hot water as possible, the solution was acidified with acetic acid, and 4-hydroxypyridine 1-oxide precipitated out as colorless columnar crystals which were recrystallized from ethanol, m.p. 239–241° (decomp.).

(ii)[113] A mixture of 12 g of dimethylaniline and 70 ml of acetic anhydride was heated on a boiling water bath, and 14 g of 4-nitropyridine 1-oxide was added to it in small portions, by which the solution became red and green fumes began to drift slowly. This addition was completed in 40 min and the mixture was heated for a further 20 min. The reaction mixture was evaporated as much as possible under a reduced pressure and 20 ml of methanol was added. The crystals thereby precipitating out were collected by filtration, washed with acetone, and dried. Yield, 7.3 g. This was purified as the sodium salt, as in *(i)*, and liberation of the free base gave 63% yield of 4-hydroxypyridine 1-oxide as needles, m.p. 239–241° (decomp.).

Example 3. Preparation of 4-Hydroxyquinoline 1-Oxide by Decomposition of 4-Nitroquinoline 1-Oxide with Sulfuric Acid[93]

A solution of 1 g of 4-nitroquinoline 1-oxide in 10 ml of 40% sulfuric acid was heated under reflux for 5 h (the solution boiling at 130–140°). The bright brown reaction mixture was cooled, and saturated sodium carbonate solution was added until the separated crystals dissolved again. The solution, mixed with a small amount of carbon, was boiled, filtered from carbon, and acetic acid was added to the filtrate until slightly acid. The precipitated crystals were recrystallized from methanol to needles, m.p. 238° (decomp.). Yield, 0.6 g (71%).

(c) Substitution with an Amino Group

Nucleophilic substitution of the nitro group with amino group is accompanied by reduction with the amine and the reaction generally does not progress smoothly. If this nitro group is first substituted with a halogen and then reacted with an amine, as will be described in the following section (*cf. Section 8.4*, p. 383), the reaction progresses smoothly. Ochiai and Katada[77c] found that heating of 4-nitropyridine 1-oxide with ethanolic ammonia in a sealed tube at 100° resulted in 75% recovery of the starting material with main formation of 4,4′-azopyridine 1,1′-dioxide, and 4-aminopyridine 1-oxide was isolated only as its picrate. They found that heating of 4-nitropyridine 1-oxide with benzylamine in a sealed tube at 100° gave 4,4′-azopyridine 1,1′-dioxide and benzaldehyde, and there was no evidence for the formation of 4-benzylaminopyridine 1-oxide. Ochiai and others[79] found 4-morpholinopyridine l-oxide to be formed by the reaction of 4-nitropyridine 1-oxide and morpholine, but its yield was very small. As has been stated earlier (*cf. Section 8.3.1*, p. 360), reaction of the 4-nitro derivative of pyridine and quinoline 1-oxide with phenylhydrazine or hydrazine does not result in substitution but only in reduction.

Talik[69] found that heating of 2-bromo-4-nitropyridine 1-oxide and diethylamine in ethanol solution resulted in the formation of 2-diethylamino-4-nitropyridine 1-oxide. This contrasts with the fact, as stated earlier, that in the reaction of 2-chloro-4-nitropyridine 1-oxide with sodium methoxide, the nitro group is more active than the chlorine atom and 2-chloro-4-methoxypyridine l-oxide is formed. Similarly, reaction of an amine with methyl 4-nitropicolinate results in reaction of the ester group only, the nitro group remaining intact[98].

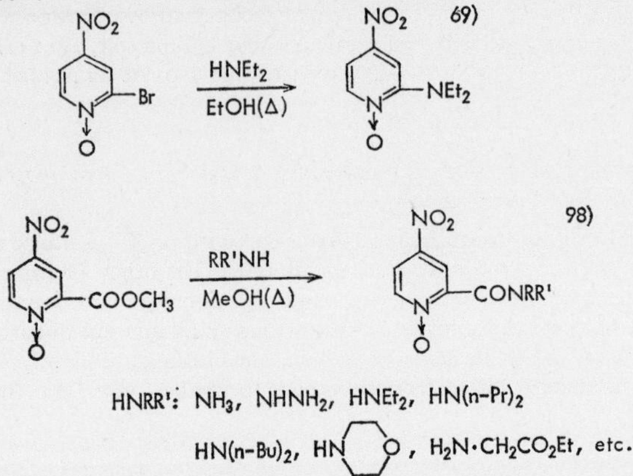

The nitro group in 4-nitroquinoline 1-oxide is more active than that in pyridine 1-oxide. Itai[116] has found that heating of 4-nitroquinoline 1-oxide with morpholine or piperidine at 120–130° results in concurrent substitution of the nitro group with an amino group and deoxygenation of the *N*-oxide group, 4-morpholino- and 4-piperidino-quinolines being formed in a considerable yield.

8.4 Halogens

The halogen atoms in the positions *ortho* and *para* to the ring nitrogen in π-deficient *N*-heteroaromatic compounds show activity to nucleophilic substitution by amines and alkoxides, and this activity increases further by *N*-oxidation.

In 1944, Ochiai and his school[79] found that heating of 4-chloropyridine 1-oxide (8-77) with morpholine or diethylamine in a sealed tube at 130–135° resulted in the respective formation of morpholino- (8-78) and 4-diethylaminopyridine 1-oxide (8-79) in considerably good yields and that the addition of copper powder in this reaction caused deoxygenation. Heating of 4-chloropyridine 1-oxide with sodium hydrogen sulfide in methanol, on a water bath, gave 4,4′-dipyridyl sulfide 1,1′-dioxide (8-80), which, when allowed to stand at room temperature in acetic acid with 30% hydrogen peroxide, underwent oxidation to form the corresponding sulfone (8-81).

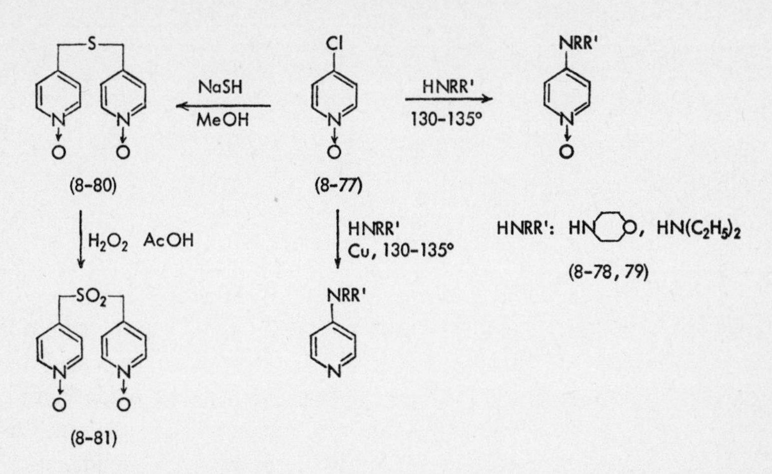

Itai[116] carried out the same reaction with 4-chloroquinoline 1-oxide and obtained the following result.

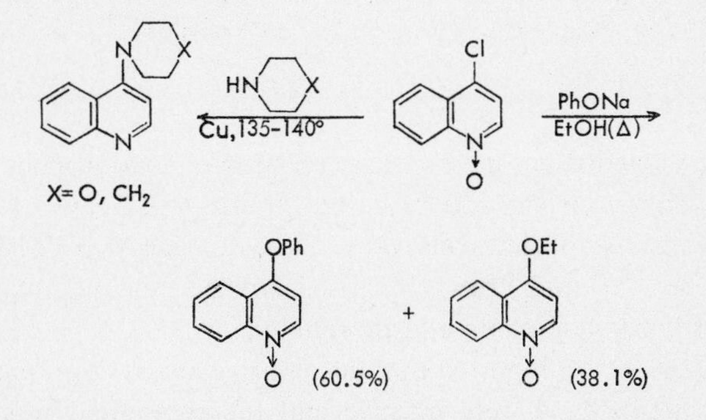

As shown above, reaction of 4-chloroquinoline 1-oxide with amines, without addition of copper dust, resulted in concurrent substitution and deoxygenation of the *N*-oxide group. This substitution reaction with amines was used extensively as the most suitable reaction for introducing an amino group into 4-position, because substitution of a nitro group in the 4-position with amines was not progressing smoothly at that time. Utilizing this reaction, Yoshida[89] observed that heating of 4,7-dichloroquinoline 1-oxide (8-82), obtained by the treatment of 7-chloro-4-nitroquinoline 1-oxide with acetyl chloride, with 4-diethylamino-1-methylbutylamine, in the presence of copper dust, resulted in the reaction of the chlorine atom in 4-position only, 7-chloro-4-(4-diethylamino-1-methylbutylamino)quinoline (8-83) being formed in a satisfactory yield.

References p. 418

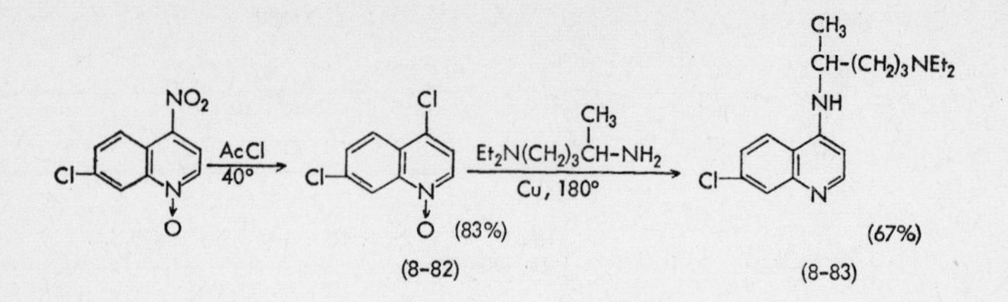

The same reaction with amines has been carried out with 4-chloropyridine 1-oxide[87b], 2-chloropyridine 1-oxide[117–119], 2-chloroisonicotinic acid 1-oxide[117], 4-chloro-3-picoline 1-oxide[87b], 4-chloronicotinic acid 1-oxide[120], and 4-chloroquinaldine l-oxide[121]. The exchange reaction with hydroxylamine[63] and hydrazine[65] has also been carried out on 4-chloroquinoline 1-oxide.

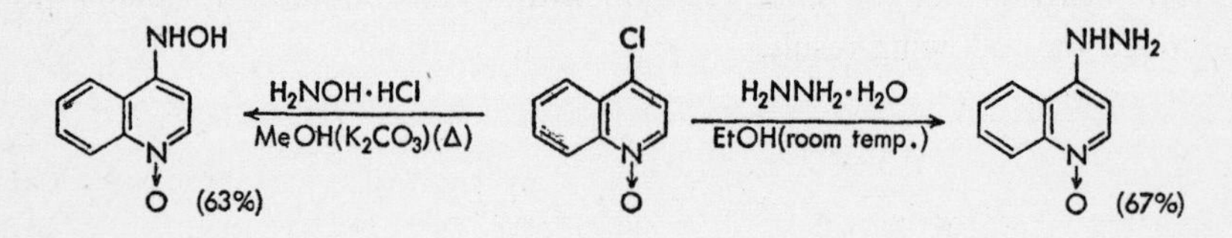

Landquist[122] carried out the exchange reaction of 2-chlorophenazine 5,10-dioxide and piperidine. As will be described later (*cf. Section 8.4*, p. 389), the exchange reaction of chloro derivatives of pyridazine *N*-oxide has also been examined. Cresswell and Brown[123] carried out the exchange reaction of 2-chloro-6-aminopurine 1-oxide and morpholine.

Taylor and others[30a] reported that the chlorine atoms in 1,2-bis(4-chloro-1-oxido-3-pyridyl)ethane were inactive and the compound was recovered when heated with glacial acetic acid, conc. ammonia, or liquid ammonia, and that even when heated with liquid ammonia at 150°, only deoxygenation of the *N*-oxide took place and substitution of the chlorine did not occur. The same tendency had been observed in the reaction of the corresponding 4-nitro compounds and alkoxides (*cf. Section 8.2*, p. 349).

Itai[124a] examined the reaction of 4-chloropyridine 1-oxide and 4-chloroquinoline 1-oxide with thiourea in ethanol solution and found that both compounds formed the corresponding thiouronium chlorides (8-84, 8-85) in a good yield, but that their decomposition by alkali differed slightly, as shown below, and the mercapto compound formed by this reaction tended to change to the sulfide (8-86, 8-87) more easily in the pyridine series than in the quinoline series.

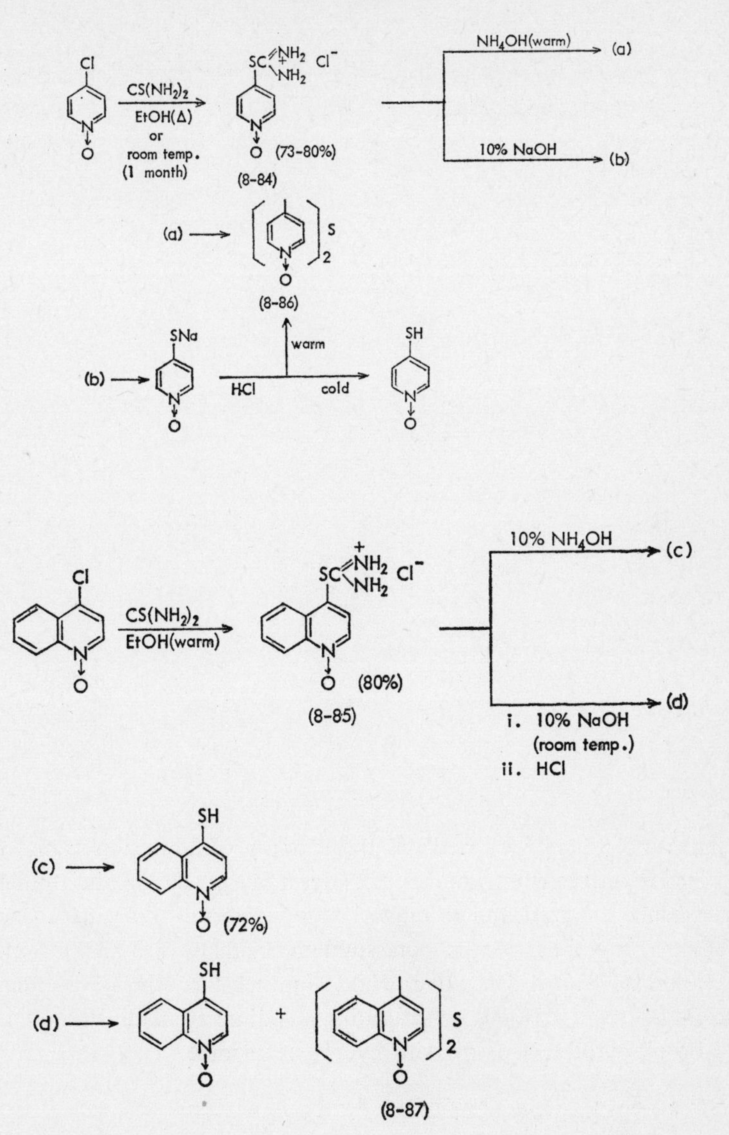

Itai[124b] further examined the reaction of these chloro compounds with thiols and found that the reaction proceeded smoothly in ethanol solution, either directly or in the presence of alkali, to form the corresponding sulfides. Oxidation of these sulfides by hydrogen peroxide in glacial acetic acid gave the corresponding sulfones.

References p. 418

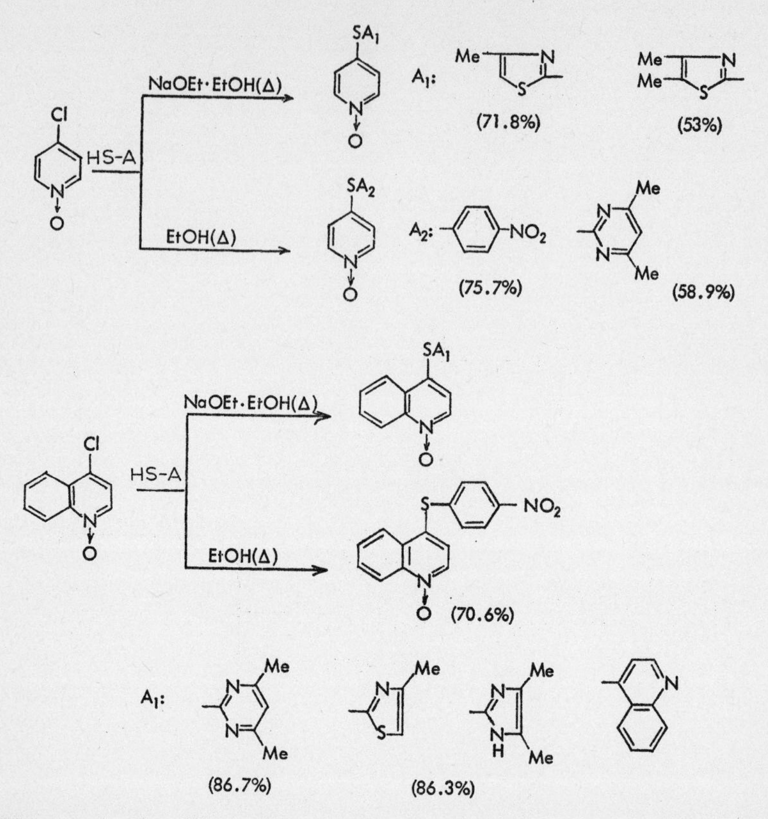

Shaw and others[125] carried out the reaction of 2-bromopyridine 1-oxide and its derivatives with thiourea or sodium benzyl sulfide and obtained the same result. Leonard and Wajngurt[126] obtained a 2-thiocyanato derivative (8-88) by the application of potassium thiocyanate to 2-bromopyridine 1-oxide and its 4-nitro derivative, while Mautner and others[127] synthesized a 2-selenol derivative (8-89) by saturating a sodium ethoxide-ethanol solution of 2-bromopyridine 1-oxide with hydrogen selenide.

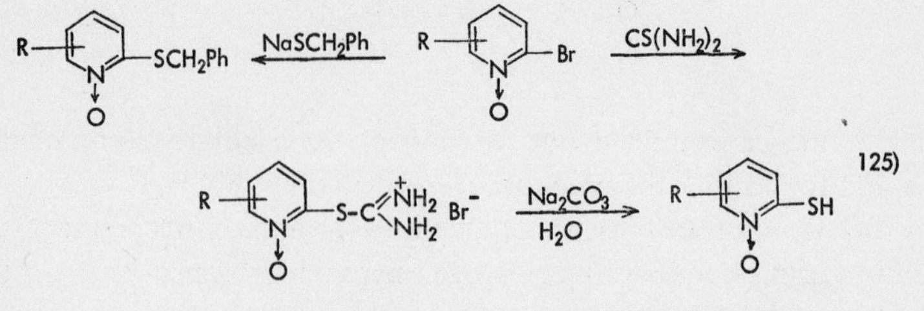

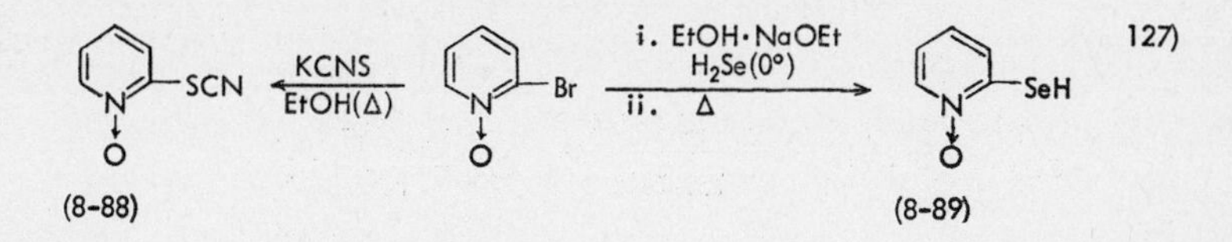

Many reports have been made on the reaction of halogen atoms in the 2-[119,128] or 4-position[120] of amine oxides of the pyridine series with thiol compounds. The most interesting among these is the formation of thioglucosides by the reaction of 4-chloropyridine 1-oxide, 4-chloroquinoline 1-oxide, and 2-bromopyridine 1-oxide with thio-sugars[129].

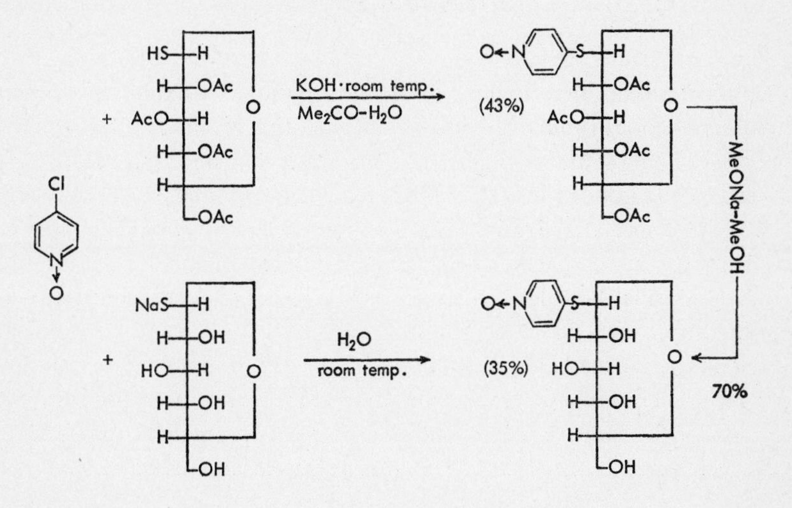

Landquist[122] observed that the reaction of 2-chlorophenazine 5,10-dioxide and sodium thioglycolate resulted in deoxygenation to form 2-(carboxymethylthio)phenazine.

In parallel with the foregoing reaction of 2- and 4-halogeno *N*-oxide compounds with thiol compounds, their reaction with sodium alkoxides, phenoxides, and alkali hydroxides has also been examined and formation of alkoxy, phenoxy, and hydroxy substitution compounds has been proved to take place with pyridine[99,119], quinoline[116,121], pyrazine[130], and phenazine[122] *N*-oxides. Fujimoto[131] found that the reaction of 4-chloro-3,5-diiodo-2,6-lutidine 1-oxide and *p*-nitrophenol resulted in the exchange of chlorine at the 4-position alone, with the iodine atoms in the 3- and 5-positions remaining intact.

References p. 418

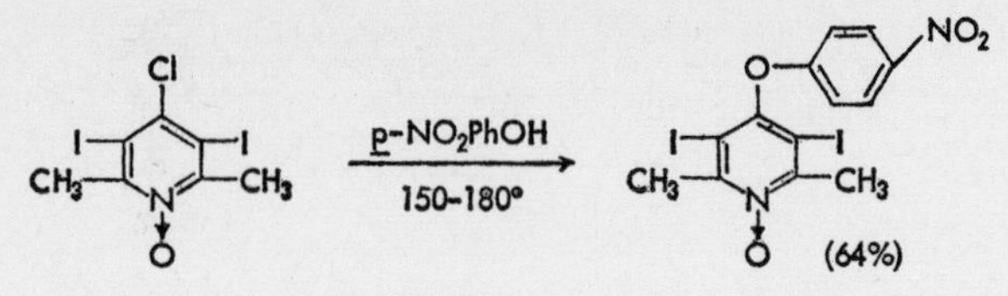

Profft and Schulz[99] found that the amino group in 2-aminoethanol did not take part in the reaction of 4-chloro-3-picoline 1-oxide and 2-aminoethanol, and that exchange with the hydroxyl group mainly took place.

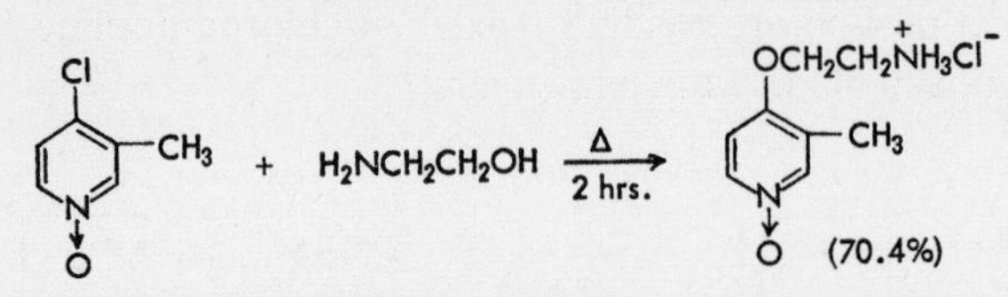

Suzuki[132] found that boiling of 4-chloropyridine 1-oxide or 4-chloroquinoline 1-oxide in aqueous sodium sulfite solution resulted in the formation of corresponding sodium 4-sulfonate compounds, but since this reaction also proceeds with a good yield with non-oxygenated compounds[133], there does not seem to be a special activation of the halogen atom by *N*-oxygenation. Itai and Kamiya found that the reaction of 4-chloropyridine 1-oxide[65] and 4-chloropyridazine 1-oxide[134] with sodium azide caused substitution of the chlorine atom with the azide group.

S_N reaction of the halogen in the position *meta* to the *N*-oxide group has been reported in some cases, but the same reaction has been found to take place under the same reaction conditions in non-oxygenated compounds, so that special activation by *N*-oxygenation seems to be uncertain.

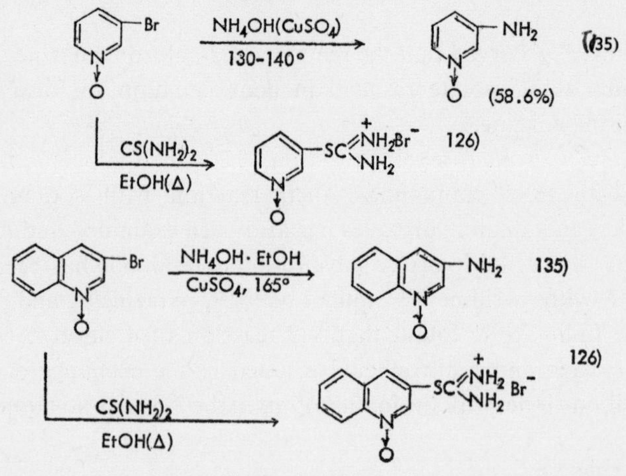

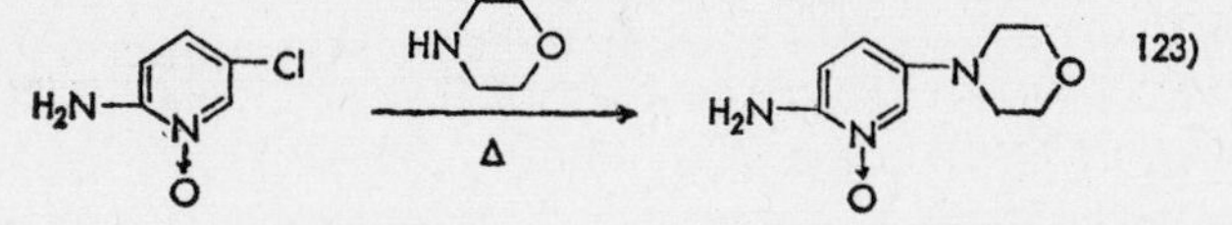

Kato[136] carried out the reaction of 2-, 3-, and 4-chloropyridine 1-oxides with potassium amide in liquid ammonia and found that 3- and 4-chloro compounds gave the corresponding 3- and 4-amino derivatives, and further confirmed the absence of other isomers by thin-layer and gas chromatography. In this case, 2-chloropyridine 1-oxide underwent resinification and gave a small amount of 2- and 3-aminopyridine 1-oxide. He concluded from these results that while the reaction of 3- and 4-chloropyridine 1-oxide with potassium amide follows the S_{N2} mechanism, that of the 2-chloro compound proceeds by the benzyne mechanism.

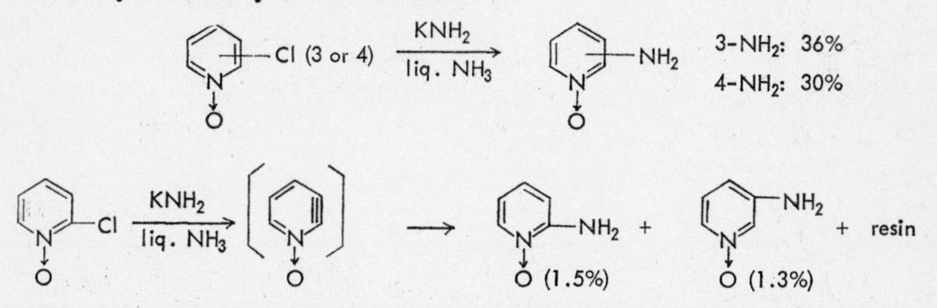

This is opposed to the fact that the amination of non-oxygenated 2-, 3-, and 4-chloropyridines with sodium amide in liquid ammonia[137] proceeds by the benzyne mechanism for the 3- and 4-chloro compounds and by the S_{N2} reaction for the 2-chloro compound.

As for the halogen atom in the halogen derivatives of pyridazine and cinnoline series *N*-oxides, the polar effect of the *N*-oxide group and non-oxygenated ring-nitrogen atom is overlapped in these compounds and it is difficult to judge the effect of the *N*-oxide group on the halogen atom only from their S_N reaction, as was also the case with the nitro group (*cf. Section 8.3.2.b*, p. 378) and methyl group (*cf. Section 8.2*, p. 345).

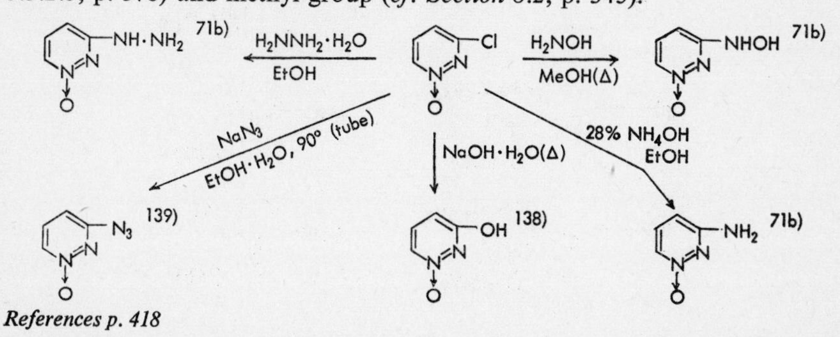

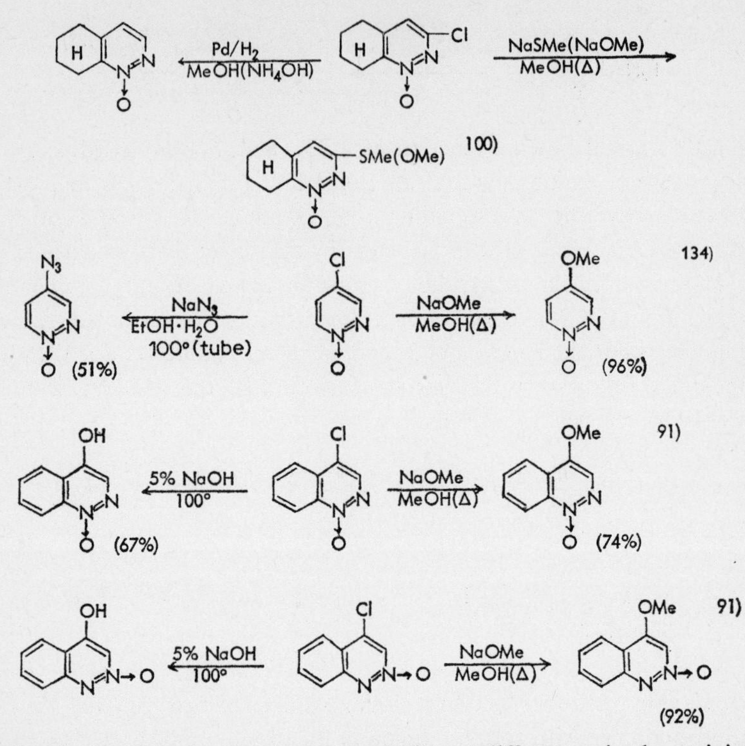

In the above examples, there are practically no differences in the activity of the positions *para* and *meta* to the *N*-oxide group. When a substituent is introduced into these compounds, its polar effect adds interference and the situation becomes still more complicated.

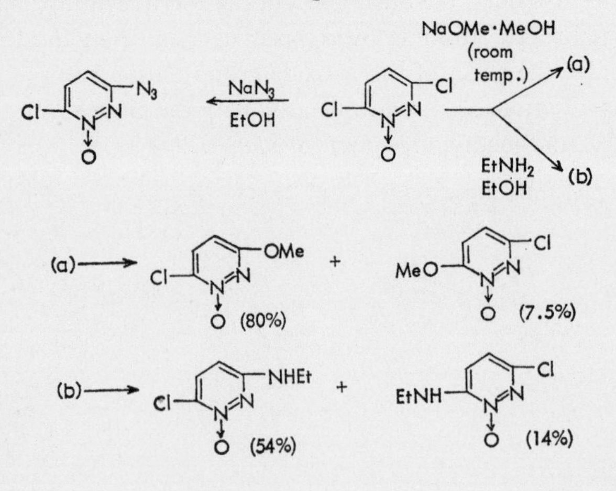

In the above example, the chlorine atom in the 3-position, *i.e. meta* to the *N*-oxide group, has greater S_N reactivity than that in the 6-position, *i.e. ortho* to the *N*-oxide group[140]. In the following example[100], oxygenation of the ring-nitrogen atom has clearly activated the chlorine atom in the 6- or *ortho*-position.

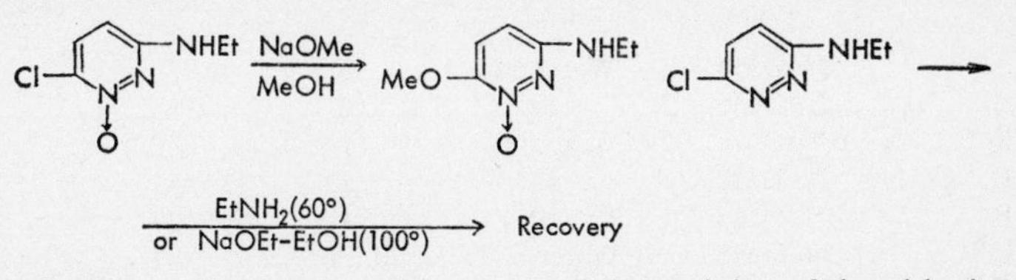

Sako[4] further carried out a comparison of the activity of the chlorine atom in the following three kinds of chloropyridazine derivatives (A, B, C) against nucleophilic reagents, and obtained the result shown in Table 8-3. As shown in this table, the reactivity decreased in the order of C, A, and B, *i.e.* the chlorine atom in the position *meta* to the *N*-oxide group and *para* to the free nitrogen had the greatest activity, while that in the position *para* to the *N*-oxide group and *meta* to the free nitrogen had the least activity.

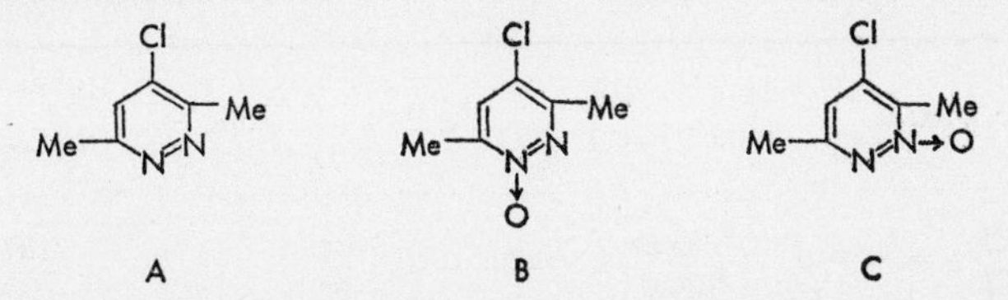

Itai and Sako[141] carried out spectrometric determinations of the reaction rates of 3-, 4-, 5-, and 6-chloropyridazine 1-oxides with piperidine, calculated their rate constants, and obtained the values listed in Table 8-4. These data indicate that the reactivity of chlorine atom falls in the order of the 5, 3, 6, and 4 positions.

These results suggest that, of the two nitrogen atoms in these pyridazine compounds, the electromeric effect of the tertiary nitrogen is further fortified by the inductomeric effect of N^+ in the adjacent *N*-oxide group and increases the activity of its *para*-position to nucleophilic reagents, while activation of the *para*-position by the *N*-oxide group is likely to be neutralized partially by the electromeric effect of the O^- in the *N*-oxide group.

Kawazoe and Natsume[5] assumed, from the shift of the ring proton in the NMR spectrum of pyridazine series *N*-oxides, that the π-electron density was increased in the α- and γ-positions, and decreased in the β-position to the *N*-oxide group. This fact confirms the foregoing possibility.

References p. 418

TABLE 8-3

REACTION OF CHLOROPYRIDAZINE DERIVATIVES WITH NUCLEOPHILIC REAGENTS[4]

Chloro-pyridazine	Reagent	Reaction conditions		Yield of product (%)	Recovery (%)
		Temp. (°C)	Time h()		
A	NaOMe	Water bath	1	80	
A	NaOMe	19	2	55	
B	NaOMe	Water bath	1	82	
B	NaOMe	40	2	59	33
B	NaOMe	19	4	20	16
C	NaOMe	Water bath	20 min	87	
C	NaOMe	19	2	80	15
A	$EtNH_2$	120–130	6	10.5	34
B	$EtNH_2$	150	6		82
C	$EtNH_2$	120–130	6	80	

TABLE 8-4

REACTION OF CHLOROPYRIDAZINE 1-OXIDES WITH PIPERIDINE[141]

Pyridazine 1-oxide	$K_2^{50°} \times 10^5$ mole^{-1}/sec^{-1}	Pyridazine 1-oxide	$K_2^{50°} \times 10^5$ mole^{-1}/sec^{-1}
5-Cl	288.0	3-Br	187.0
3-Cl	126.0	4-Br	7.34
6-Cl	39.4	5-Cl, 3,6-Me_2	3.15
4-Cl	7.08	4-Cl, 3,6-Me_2	0.0694

The above is the nucleophilic reaction of a halogen atom with active oxygen, sulfur, or nitrogen compounds. These reactions progress smoothly, in general, while the substitution reaction with carbanions meets with great resistance, except in special cases.

Nakayama[142] examined the reaction of 4-chloro- and 4-bromoquinoline 1-oxide with potassium cyanide, copper cyanide, ethyl sodiomalonate, and α-cyanobenzylsodium under various conditions. He found that all the starting materials were recovered in the case of *N*-oxide compounds, while non-oxygenated 4-bromoquinoline gave 4-cyanoquinoline when heated with copper cyanide over a direct flame and the reaction of this 4-bromoquinoline with benzyl cyanide in the presence of sodium amide at room temperature gave only α-(4-quinolyl)phenylacetonitrile. Later, Takahashi[143] succeeded in the preparation of 4-[2-pyridyl(cyanomethyl)]quinoline 1-oxide (8-90) in 73% yield by the condensation of 4-chloroquinoline 1-oxide and 2-pyridylacetonitrile with sodium amide.

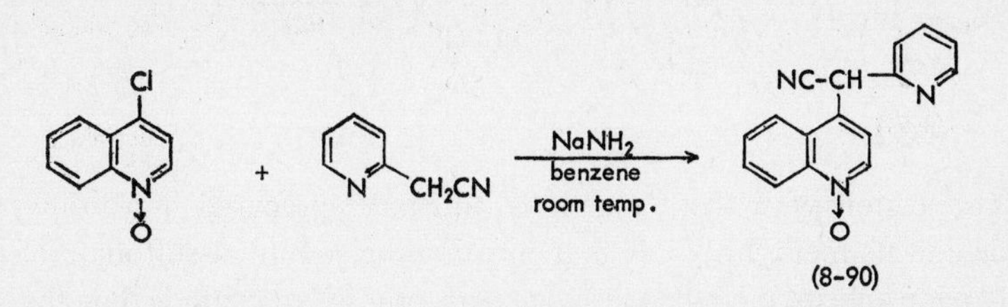

Adams and Reifschneider[144] carried out the condensation of 2-bromopyridine 1-oxide and 2-bromo-6-methylpyridine 1-oxide with sodium salts of diethyl malonate, β-ketocarboxylic acid esters, and ethyl cyanoacetate and successfully prepared pyridine l-oxide derivatives with the carbon-containing group in 2-position by the hydrolysis of the condensates.

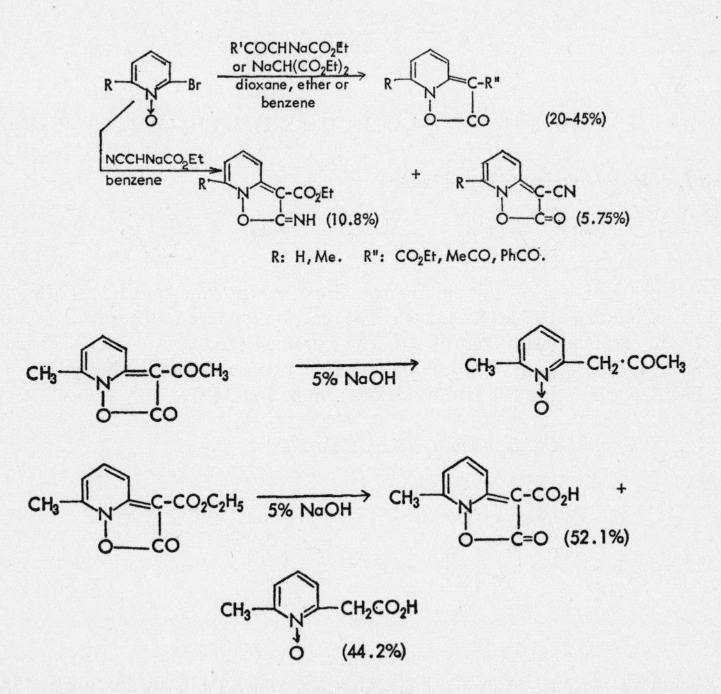

In a similar manner, Efros and Mishina[23] succeeded in the condensation of

References p. 418

9-chloroacridine 10-oxide (8-91) with diethyl sodiomalonate and conversion of this condensate to 9-methylacridine 10-oxide (8-92) by its hydrolysis.

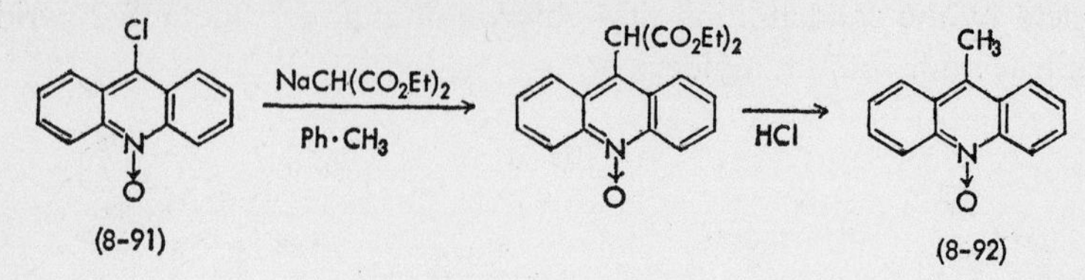

In connection with these reactions, Suzuki[132] prepared 4-cyanopyridine 1-oxide and its methyl homologs by heating the sodium 4-sulfonate of pyridine 1-oxide and its methyl homologs with potassium cyanide, but the yield from this reaction was only around 25–44%, in contrast to the yield of 56–86% in a similar reaction of non-oxygenated compounds[133]. This fact shows that this reaction is unsuitable as a method for synthesis of 4-cyanopyridine 1-oxide and a better method would be the *N*-oxidation of 4-cyanopyridine.

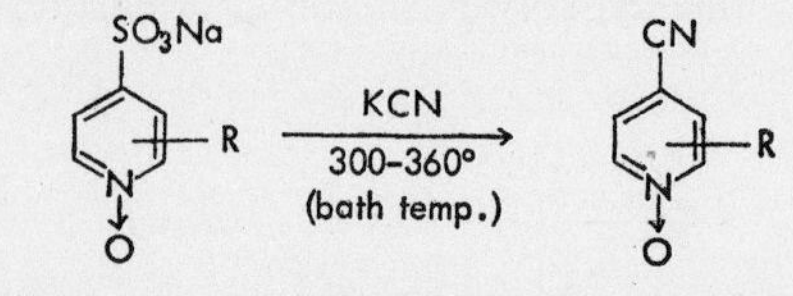

R (yield, %) = H (44.3), 2–CH_3 (25.2), 3–CH_3 (36.2), 2,6–$(CH_3)_2$ (27.1)

Example 1. 4-Morpholinopyridine 1-Oxide[145]

A mixture of 1 g of 4-chloropyridine 1-oxide, 1 g of morpholine, and 1.5 ml of water was heated in a sealed tube at 130–140° for 5 h. The reaction mixture was basified with sodium carbonate and extracted with chloroform. The extracted product was recrystallized from acetone to 0.74 g (53.2%) of 4-morpholinopyridine 1-oxide as prismatic crystals of m.p. 75–78°. Picrate, yellow needles (from acetone), m.p. 168–169°.

Example 2. Quinoline 1-Oxide 4-Thiouronium Chloride[146]

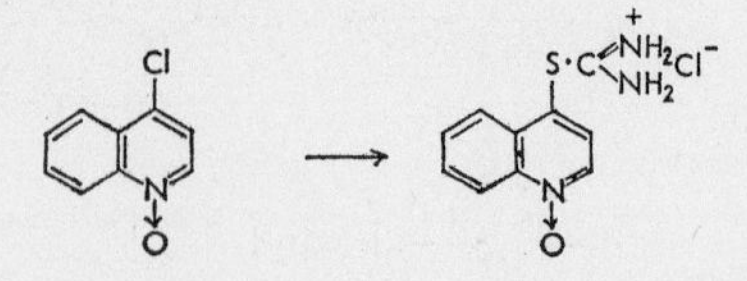

A mixture of 1 g of 4-chloroquinoline 1-oxide and 0.45 g of thiourea in 10 ml of dehydrated ethanol was warmed on a water bath. After refluxing for 1 h, the mixture was allowed to cool and the precipitated crystals were collected to 1.28 g (89%) of thiouronium salt as pale yellow needles, m.p. 165.5° (decomp.). Recrystallization from methanol failed to raise its decomposition point.

Example 3. 4-Mercaptoquinoline 1-Oxide[146]

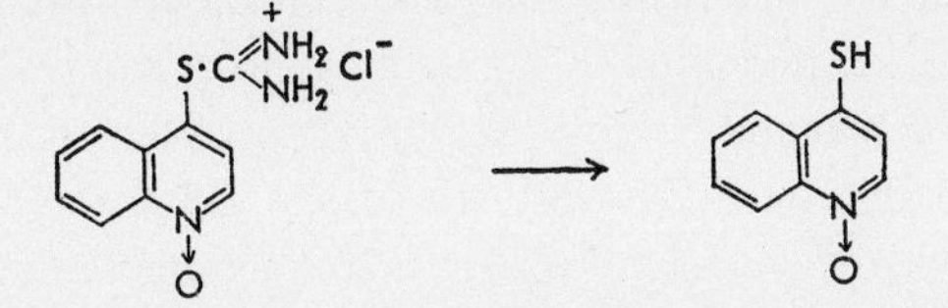

To a suspension of 1 g of quinoline 1-oxide 4-thiouronium chloride in 10 ml of water, 3.0 ml of 10% sodium hydroxide solution was added dropwise, by which the solution turned reddish orange and the salt dissolved gradually. A minute quantity of undissolved matter was filtered off and the filtrate was acidified with acetic acid. The reddish orange crystals that separated out were collected by filtration and washed with water to give 0.55 g (80%) of 4-mercaptoquinoline 1-oxide, m.p. 140–150.0° (efferv.).

Example 4. 4-[2-Pyridyl(cyanomethyl)quinoline 1-Oxide[143] *(8-90)*

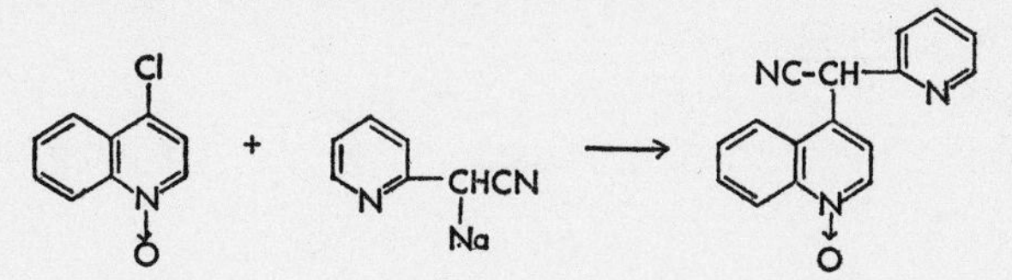

To a suspension of 2 g of finely powdered sodium amide in 100 ml of benzene, a solution of 5.95 g of 2-pyridylacetonitrile in 10 ml of benzene was added dropwise during 5 min and the mixture was stirred for 10 min. To this mixture, a suspension of 4.5 g of powdered 4-chloroquinoline 1-oxide in 50 ml of benzene was added dropwise with stirring and the whole mixture was stirred at room temperature for 1 h. The mixture was stirred further after addition of 50 ml of water and the upper benzene layer was decanted. The lower aqueous layer was mixed with 100 ml of water, made ammonia-alkaline with ammonium chloride, and extracted with chloroform. The separated benzene layer was shaken with 5% sodium hydroxide solution and this aqueous solution, after addition of ammonium chloride, was extracted with chloroform. The combined chloroform solution was washed with water and the solvent was evaporated. The syrupy residue thereby obtained was treated with a few drops of ethanol and the leaflet crystals formed were collected, and washed first with a small quantity of ethanol and then with ether. Orange plates, m.p. 154–156°. Yield, 4.75 g (73%). Recrystallization from ethanol gave colorless plates, m.p. 130–132°, as a semihydrate.

8.5 Amino Group

The diazonium salt of an amino group in the position *ortho* or *para* to the ring nitrogen in *N*-heteroaromatic compounds is very labile and undergoes decomposition by the application of nitrous acid in the cold, evolving nitrogen. When the diazotization is carried out in dilute sulfuric acid or strong

References p. 418

hydrohalogen acid, such an amino group is known to be easily substituted by a hydroxyl[147] or a halogen[148]. Ochiai and Teshigawara[111] showed that the diazonium salt of 4-aminopyridine 1-oxide is comparatively stable and that the compound can be converted to an azo dye, and also substituted by chlorine (53% yield), bromine (less than 30%), or a cyano group (65%) by the Sandmeyer reaction. Ochiai and Naito[112] similarly converted 4-aminoquinoline 1-oxide to an azo dye by diazotization and showed that the amino group can be substituted by chlorine (42% yield), bromine (6.6%), iodine (66%), cyano (13%), and thiocyanato (39%). In all these cases, substitution with a mercapto or arsenious acid group has not succeeded as yet. Substitution with a hydroxyl group has not materialized, because the hydroxyl derivative

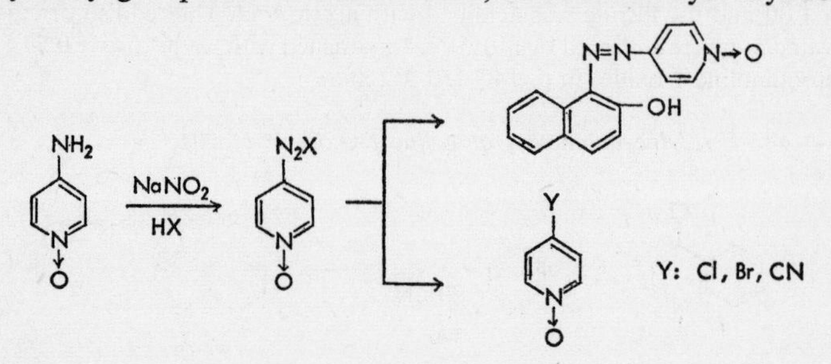

so formed immediately undergoes coupling with the diazonium salt to form an azo dye. Leonard and Wajngurt[126] have succeeded in converting 4-aminopyridine 1-oxide to 4-thiocyanatopyridine 1-oxide (8-93), though in a poor yield, by the addition of a cobalt salt besides potassium thiocyanate.

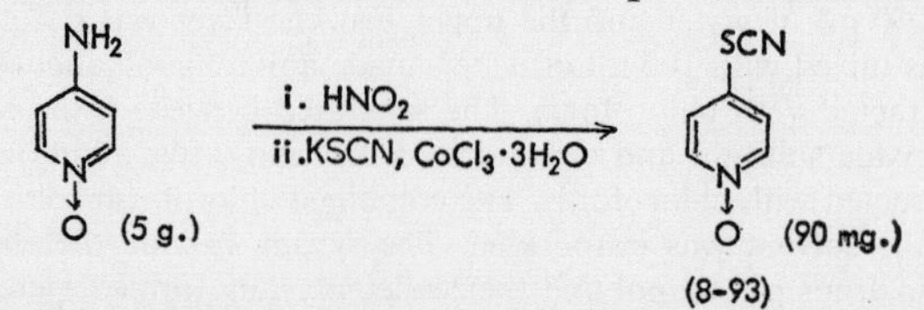

Naito[149] proved that diazotization of 4,6-diaminoquinoline 1-oxide (8-94) with 1 mole of nitrous acid results in preferential diazotization of the amino group in 6-position, as in the case of 4,6-diaminoquinoline, by converting it to 4-amino-6-chloroquinoline 1-oxide (8-95).

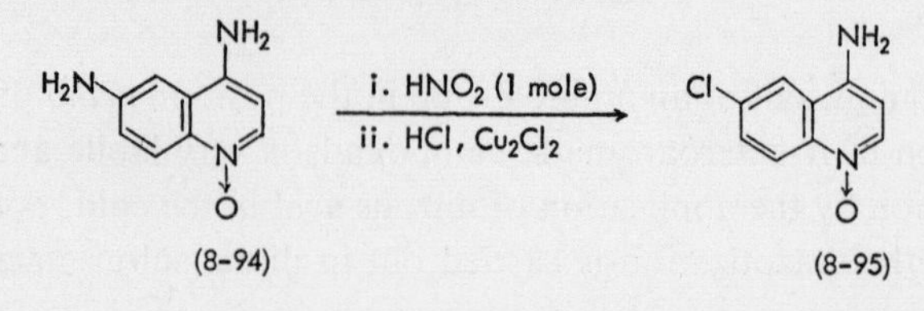

Later, Katritzky[118] carried out diazotization of 2-aminopyridine 1-oxide (8-96) and converted it to an azo dye by coupling with 2-naphthol. At the same time, he compared the ultraviolet spectrum of 2-aminopyridine 1-oxide with those of its alkylated compounds (8-97, -98, -99, -100) and concluded that the amino group in 2-position is present in an amino form and not as its tautomeric imino form.

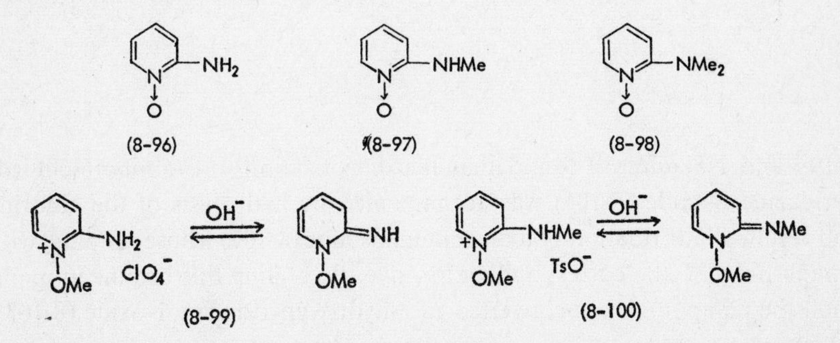

The fact that the diazonium salt of 2-aminopyridine 1-oxide is more stable than that of 2-aminopyridine, as in the case of 4-aminopyridine 1-oxide, was assumed to be due to the contribution of the following structures (A and B), and further to their valency-bond tautomeric form (C).

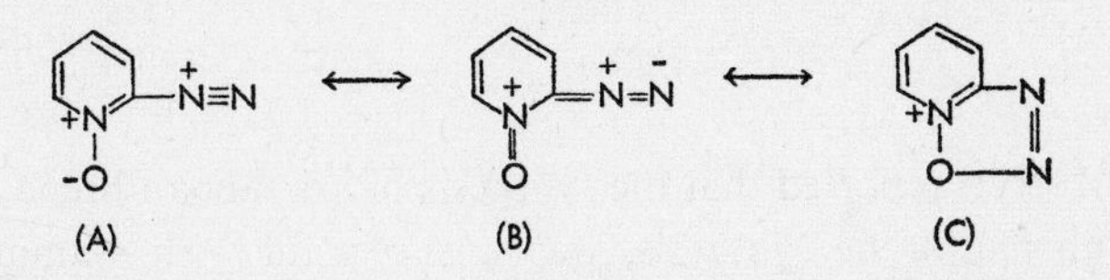

Tanida[150] found that while 2-aminoquinoline 1-oxide (8-101) and 1-aminoisoquinoline 2-oxide (8-102) are diazotized to form an azo dye, they both undergo deoxygenation during substitution with chlorine by the Sandmeyer reaction to form 2-chloroquinoline (8-103) and 1-chloroisoquinoline (8-104). He also obtained the diazonium sulfate of 4-aminoquinoline 1-oxide (8-105) in crystalline form and converted it to 4-nitroquinoline 1-oxide by the Sandmeyer reaction. The same conversion, however, was not successful with 2-aminoquinoline 1-oxide and 1-aminoisoquinoline 2-oxide.

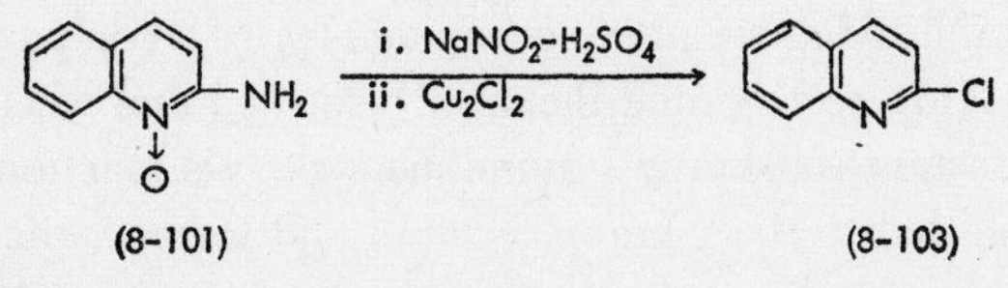

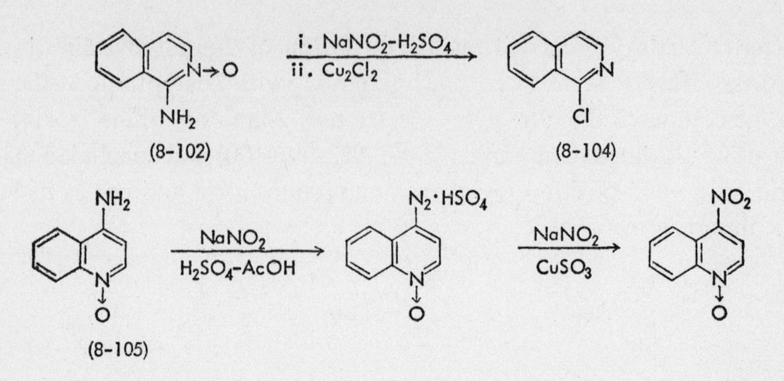

Itai and Nakajima[151] found that the diazotization of 3-amino-6-chloropyridazine 2-oxide (8-106) was accompanied by hydrolysis of the chlorine and resulted in formation of a diazonium betaine (8-107) whose coupling with 2-naphthol gave the corresponding azo dye. By boiling this betaine in methanol, the compound was converted to 3-hydroxypyridazine 1-oxide (8-108).

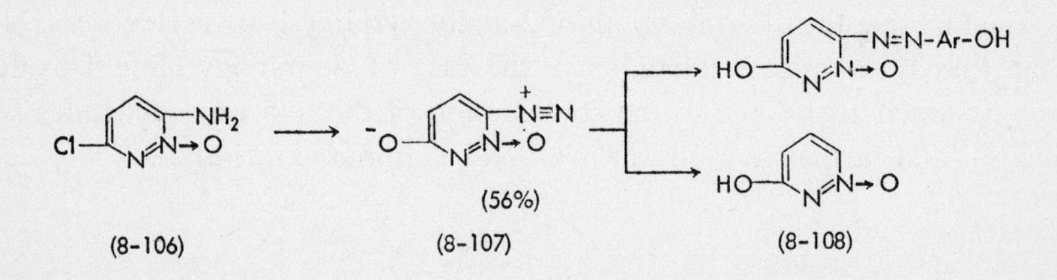

Herz and Murty[15] reported that the synthesis of azaphenanthrene, following Pschorr's phenanthrene synthesis, was unsuccessful with 4-aminopyridine derivatives, but was effected when the compounds were converted to their *N*-oxides.

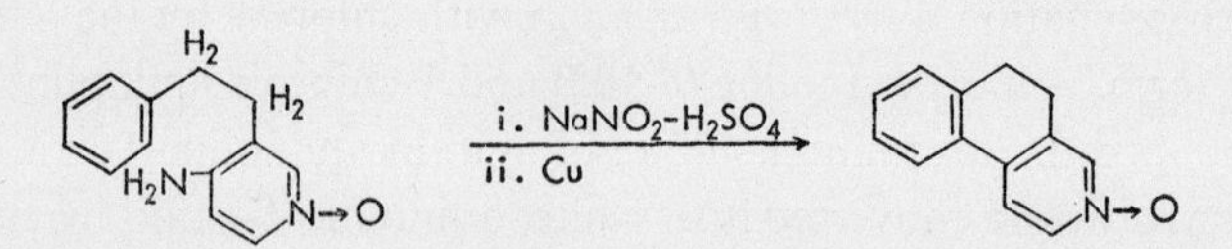

These facts indicate that the amino group in the *ortho* or *para* position of the *N*-oxide group in aromatic *N*-oxide compounds does not take an imino form[118] but is stabilized as an amino group and forms a comparatively stable diazonium salt. Kato *et al.*[152] and Ikekawa[153] showed that whereas 4-aminopyridine was resistant to Skraup's quinoline synthesis reaction, its *N*-oxide compound submitted to this Skraup reaction and formed, though in a low yield, the corresponding naphthyridine *N*-monoxide. Kato and others had assumed that the Skraup reaction of 4-amino-2-picoline 1-oxide (8-109) would result in cyclization at the position *para* to the methyl group, *i.e.* at the 5-position, but Ikekawa proved that the cyclization had taken place in 3-position (8-110).

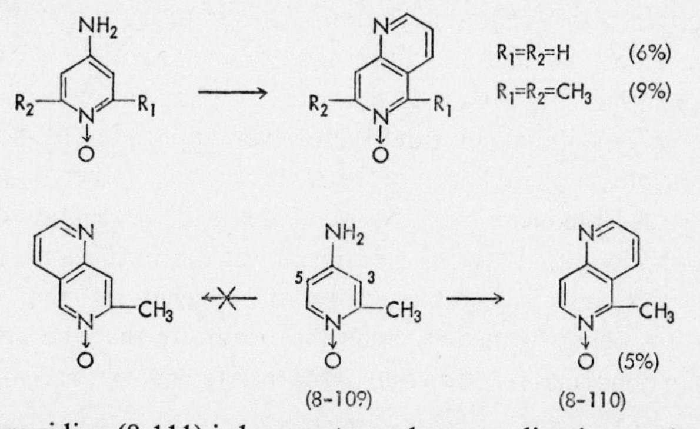

3-Aminopyridine (8-111) is known to undergo cyclization at the 2-position during condensation with ethoxymethylenemalonic acid ester to form a 1,5-naphthyridine derivative (8-112)[154]. Hauser and Murray[135] showed that in this reaction, if carried out with 3-aminopyridine 1-oxide, the cyclization occurs in the 4-position to form the 1,7-naphthyridine 7-oxide derivative (8-113) and also observed that the reaction occurs somewhat more readily. This fact indicates that the cyclization to the *ortho* position is inhibited by *N*-oxygenation and the *para* position is activated.

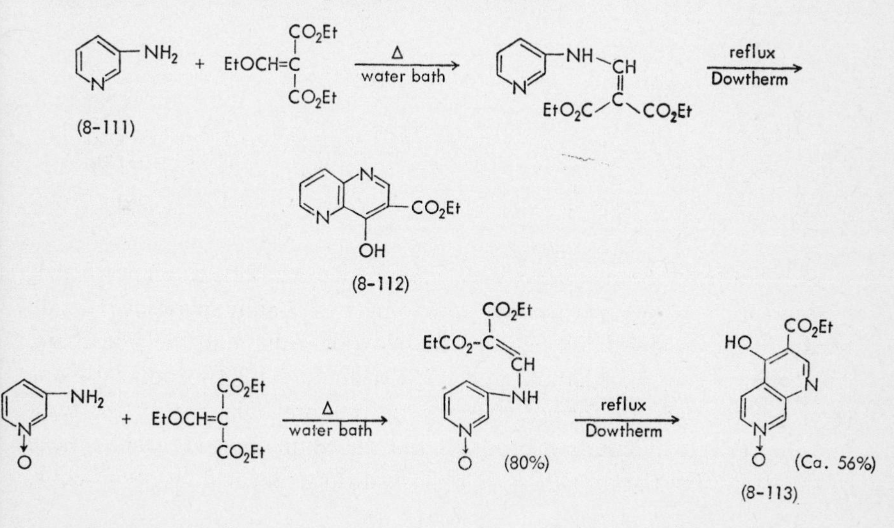

Other evidence for the ordinary reaction of the amino group in 2- or 4-position of aromatic *N*-oxide compounds is provided by the formation of 4-pyridylthiourea 1-oxide (8-114) from 4-aminopyridine 1-oxide by the addition of thiocyanic acid[155], oxidation of 2-aminopyridine 1-oxide and its methyl homologs by monopersulfuric acid to the corresponding 2-nitro derivatives[56], preferential acylation of the amino group in the acylation of 4-aminopyridine 1-oxide and 4-aminoquinoline 1-oxide[118,156], and transition of these amino compounds to the corresponding sulfonamides by treatment with *p*-acetylsulfanyl chloride[157]. However, application of methyl iodide in this case results in methylation of the *N*-oxide group rather than the amino group[158] and, as stated earlier (*cf. Section 7.4.3*, p. 287), reaction of 2-aminoquinoline 1-oxide and tosyl chloride results in the formation of 3- and 6-tosyloxyl compounds accompanied by deoxygenation, the tosylamino compound not being formed[159]. Such facts indicate that the amino group undergoes an abnormal reaction due to the interference of the *N*-oxide group.

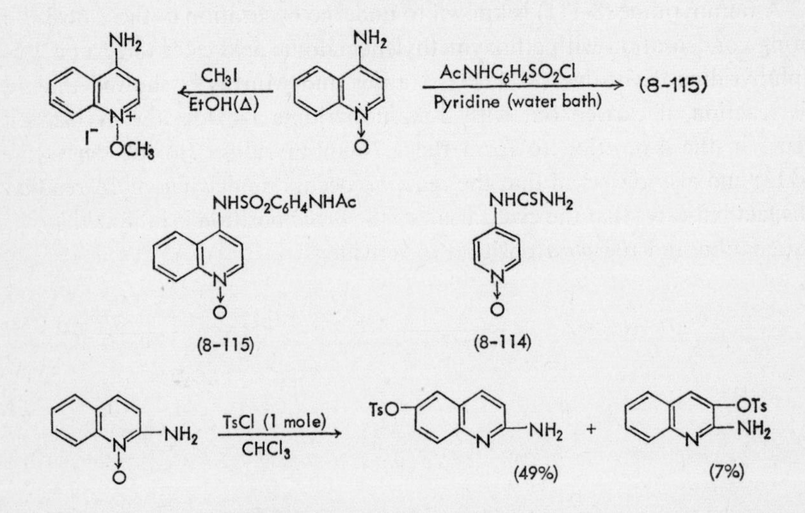

As has already been stated (*cf. Section 5.1.4.b*, p. 182), it is known that heating of the *N*-ethoxycarbonyl compounds of 2-aminopyridine 1-oxide results in the condensation between the *N*-oxide group and the side chain, with elimination of ethanol, and 2*H*-pyridino[*b*]-(1,2,4)-oxadiazol-2-one (8-116) derivatives are formed. Numerous structures take part in the resonance of these condensation products and the compounds are stable, some of them showing the properties of a masked carbamic acid, as seen below.

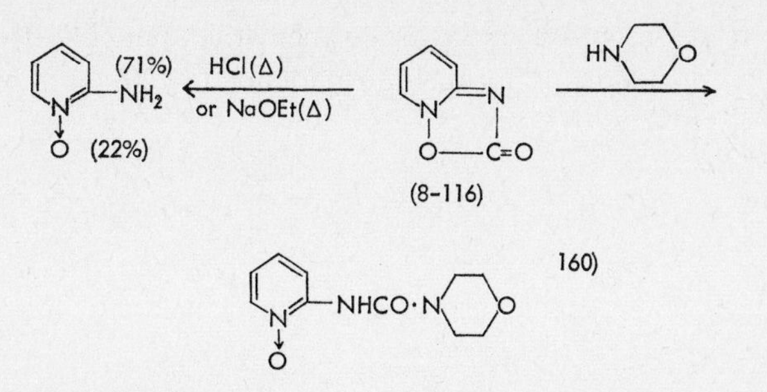

Example. 4-Iodoquinoline 1-Oxide[112]

A solution of 0.5 g of 4-aminoquinoline 1-oxide dissolved in a mixture of 2 ml of conc. sulfuric acid and 5 ml of water with warming (sulfate of the compound will precipitate out in the cold) was cooled to about 5° in an ice-water bath and a solution of 0.25 g of sodium nitrite dissolved in 2 ml of water was added dropwise with efficient stirring to effect diazotization, by which the precipitated sulfate dissolved gradually. The whole became a yellow solution after about 2 h. While stirring this solution vigorously, a concentrated aqueous solution of 1 g of potassium iodide was added by which evolution of nitrogen was observed and the solution turned dark purple. The solution was rendered alkaline with potassium carbonate and extracted with chloroform. The chloroform solution was dried over anhydrous sodium sulfate, the solvent was evaporated, and 0.55 g of pale reddish crystals thereby obtained were recrystallized from acetone with the aid of activated carbon and 4-iodoquinoline 1-oxide was obtained as colorless plates, m.p. 166°.

8.6 Oxygen- or Sulfur-containing Groups

8.6.1 Alkoxyl or Aryloxyl Groups, and Alkylthio or Arylthio Groups

The alkyloxyl, aryloxyl, and arylthio groups in the positions *ortho* and *para* to the *N*-oxide group in aromatic *N*-oxide compounds are active to substitution with amines. In 1944, Ochiai and his group[79] found that heating of 4-phenoxypyridine 1-oxide with morpholine in a sealed tube at 190–205° resulted in the formation of 4-morpholinopyridine 1-oxide (8-117) in about 77% yield, and the same treatment of 4-(*p*-tolylthio)pyridine 1-oxide (8-118) with morpholine was accompanied by deoxygenation to form 4-morpholinopyridine (8-119).

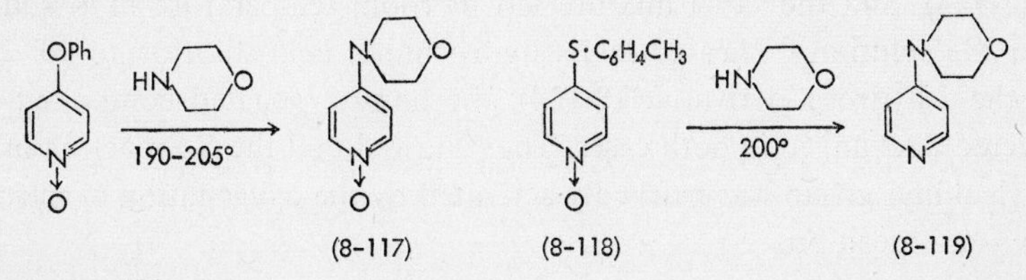

References p. 418

The same reaction of 4-phenoxy- or 4-(phenylthio)quinoline 1-oxide also results in deoxygenation to form 4-morpholino- (8-120) or 4-piperidino-quinoline (8-121)[116,161].

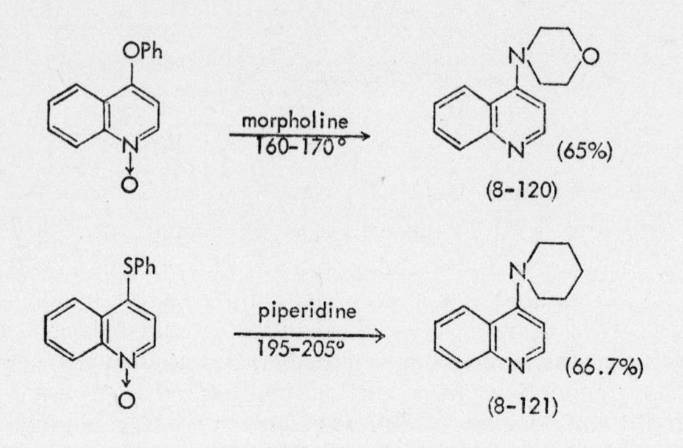

In 9-phenoxyacridine 10-oxide, this reaction takes place only by warming on a water bath and the substitution is effected without accompanying deoxygenation of the *N*-oxide group[162].

This reaction occurs more readily in diazine series *N*-oxides. Yamanaka[163] found that refluxing of 6-methyl-4-phenoxypyrimidine 1-oxide with sodium methoxide in methanol, sodium ethoxide in ethanol, sodium thiophenoxide in ethanol, or with piperidine resulted in smooth substitution of the phenoxyl group by methoxyl, ethoxyl, phenylthio, or piperidino groups, respectively, and that substitution with morpholine took place at below 100°, while raising of this temperature to above 100° was accompanied by deoxygenation of the *N*-oxide group. However, this reaction is known to take place with non-oxygenated 6-methyl-4-phenoxypyrimidine so that special activation by the *N*-oxide group is uncertain. Crosswell and Brown[123] showed that whereas heating of 2-methylsulfinyladenine 1-oxide (8-122) with ethanolamine in dimethylformamide solution easily gave the 2-(2-hydroxyethylamino) derivative (8-123), and the same mixture left at room temperature in N sodium hydroxide solution or heated in 2N hydrochloric acid solution to 90° both gave the 2-hydroxy derivative (8-124), the non-oxygenated compound was recovered unchanged in both cases. They showed that the S_N reaction of the methylsulfinyl group was markedly activated by the oxygenation of nitrogen in the *ortho* position.

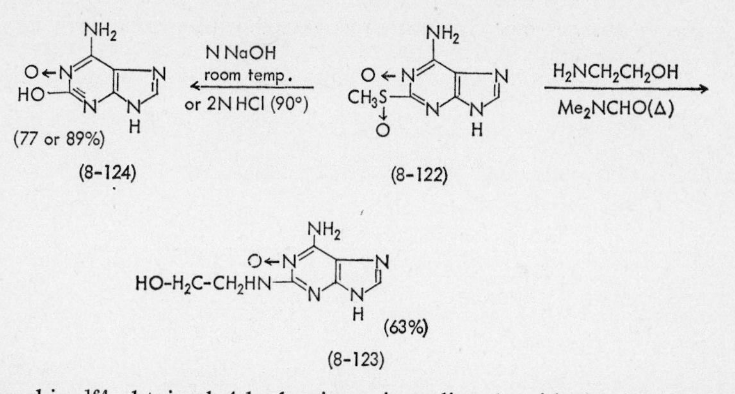

Higashino[164] obtained 4-hydrazinoquinazoline 1-oxide by heating 4-alkoxyquinazoline 1-oxide with hydrazine hydrate.

Sako[165] found that heating of 4-methoxy-3,6-dimethylpyridazine 1-oxide with 5% sodium hydroxide resulted in substitution of the methoxyl group in the position *para* to the *N*-oxide group with hydroxyl, while Itai and Kamiya showed that heating of 6-ethoxypyridazine 1-oxide[139] (8-125) or 4-methoxypyridazine 1-oxide[134] with 28% ammonia water or hydrazine hydrate in ethanol solution gave the corresponding amino (8-126) or hydrazino (8-127) derivative, and that heating of 6-azidopyridazine 1-oxide with sodium alkoxide in alcohol gave 6-alkoxypyridazine 1-oxide[139], and the same reaction of 4-azido-3,6-dimethoxypyridazine 1-oxide[166] afforded the corresponding 4-alkoxy compound.

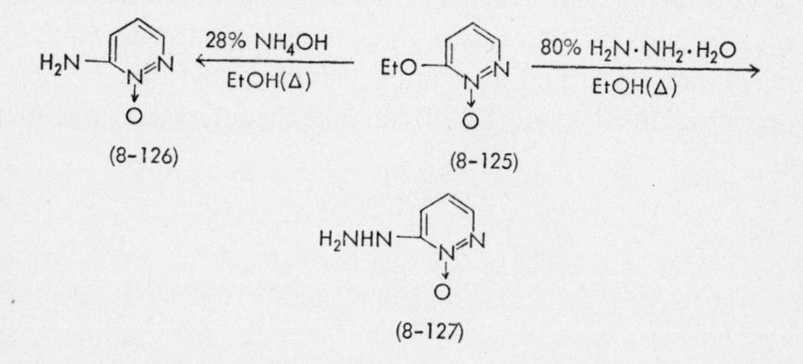

These facts indicate that the alkoxyl and azido groups in the positions *ortho* and *para* to the *N*-oxide group are active towards S_N reaction but, as will be shown in the following section, the same substitution reaction progresses smoothly with an alkoxyl or azido group in the position *meta* to the *N*-oxide group, *i.e.*, *ortho* or *para* to the free tertiary nitrogen. This activity

References p. 418

of the *meta* position has also been recognized with the halogen atom in pyridazine series *N*-oxides (*cf. Section 8.4*, p. 391).

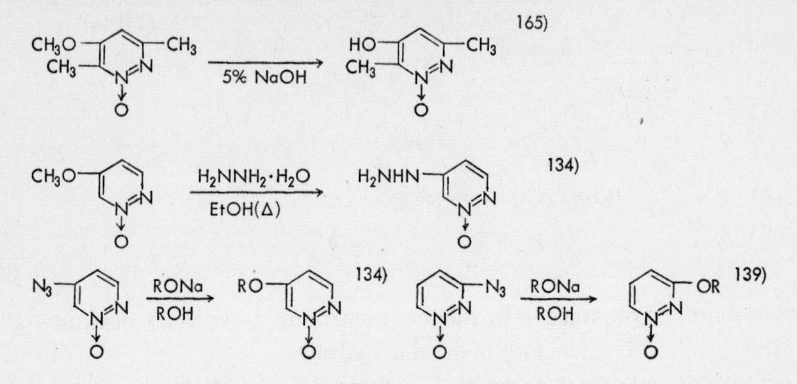

Example. 6-Hydrazinopyridazine 1-Oxide[139]

A mixture of 4.0 g of 6-ethoxypyridazine 1-oxide, 20 ml of 80% hydrazine hydrate, and 20 ml of ethanol was refluxed for 2 h on a water bath, the reaction mixture was evaporated to dryness under a reduced pressure, and a few drops of ethanol were added to the residue. Yellow prisms that separated were collected and washed with ethanol. Recrystallization from ethanol gave 2.0 g (56%) of 6-hydrazinopyridazine 1-oxide, m.p. 160° (decomp.).

8.6.2 *Tautomerism of Hydroxyl Derivatives*

The hydroxyl group in the 2- and 4-positions of pyridine series *N*-oxides forms a prototropic system with the *N*-oxide group and there is a possibility that such compounds can take any of three tautomeric forms; two of benzen, oid and one of quinoid form. For example, 4-hydroxypyridine 1-oxide (8-128- and 2-hydroxypyridine 1-oxide (8-129) can take the following three structures) a, b, and c.

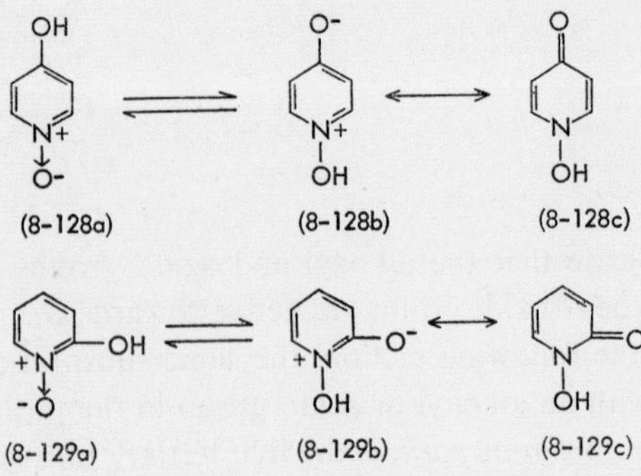

The same may be said of hydroxyquinoline 1-oxide.

Ochiai and Hayashi[167,168] examined the structure of 4-hydroxy derivatives of pyridine and quinoline *N*-oxides and concluded that they are present mainly in the form of (8-128b) from the following facts. Application of diazomethane to 4-hydroxypyridine 1-oxide (8-128) gives two isomeric methyl derivatives. One of them consists of 4-methoxypyridine 1-oxide (8-130), indicating that the hydroxyl group in 4-position had been methylated; the other is an *N*-methoxyl derivative (8-131) because its catalytic reduction over palladium–carbon gives 4-pyridone. This latter derivative is resistant to reaction with Grignard reagents, diene synthesis, and reaction with aniline for azophenine formation, all possible if in quinoid form, and forms a crystalline adduct of 1:1 ratio with phenyl isocyanate. This adduct is easily hydrolyzed at room temperature into the original *N*-methoxyl derivative and triphenyl isocyanurate (8-132), which suggests that the compound is probably the phenylurethan of enol-betain (8-133). The formation of this adduct is possible only with benzenoid compounds.

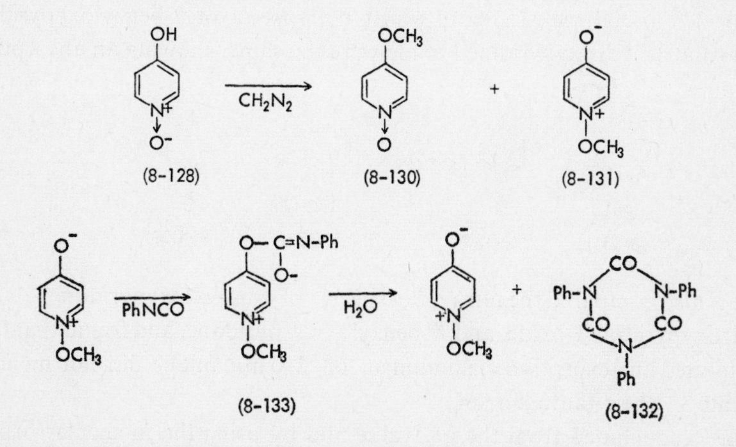

The ultraviolet spectra (in water) of 4-hydroxypyridine 1-oxide and its hydrochloride are similar to that of its sodium salt and the fact indicates that the participation of a quinoid form can be disregarded. It may be assumed from this evidence that of the three tautomeric forms of 4-hydroxypyridine 1-oxide, the compound tends chiefly to take the enol-betaine form (8-128b)[167].

The chemical properties of 4-hydroxyquinoline l-oxide (8-134) are also the same. Application of diazomethane gives 4-methoxyquinoline 1-oxide and

References p. 418

N-methoxy-4-hydroxyquinoline betaine in about 2:9 ratio; the chemical properties of the latter are the same as those of the corresponding pyridine derivative, and it was assumed that the compound mainly takes the enol-betaine form (b) among the three resonance formulae (8-134)[168].

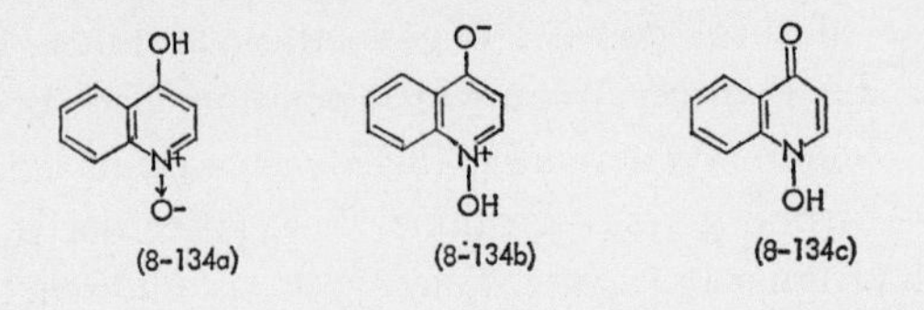

Shaw[169] compared the ultraviolet spectrum of 2-hydroxypyridine 1-oxide (8-135) with those of 2-benzyloxypyridine 1-oxide (8-136), *N*-benzyloxy-2-pyridone (8-137), and 3-hydroxypyridine 1-oxide (8-138), and concluded that 2-hydroxypyridine 1-oxide was present in the form of *N*-hydroxy-2-pyridone since its spectrum [(λ_{max}^{EtOH} mμ (ε): 228 (7,200), 305 (4,600)] was the same as that of *N*-benzyloxy-2-pyridone and similar to that of 2-pyridone [(λ_{max}^{EtOH} mμ (ε): 227 (7,300), 300 (5,000)], whilst the spectra of 2-benzyloxypyridine l-oxide and 3-hydroxypyridine l-oxide were the same, showing an absorption maximum at 263 mμ.

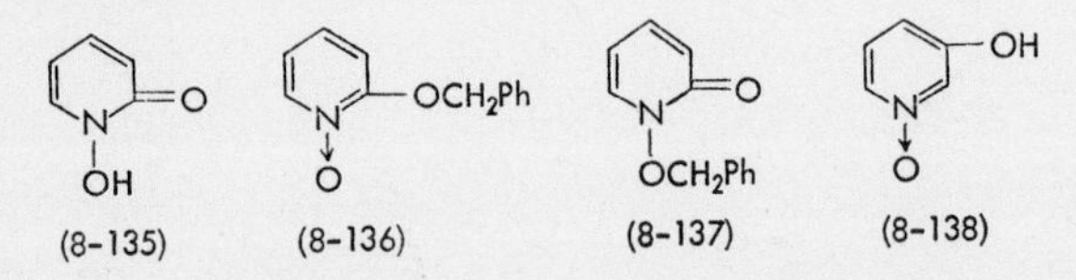

Shaw also compared the ultraviolet spectra of 4-benzyloxypyridine 1-oxide, 4-hydroxypyridine 1-oxide, and *N*-benzyloxy-4-pyridone, and found that they all exhibited an absorption maximum at 268–270 mμ, but he did not mention the state of their tautomerism.

Jaffé[170] concluded from the pK value and by using the molecular orbital method that the *N*-oxide form (8-128a) of 4-hydroxypyridine 1-oxide is more stable than the *N*-hydroxy-4-pyridone form (8-128c). Katritzky[171] concluded from the pK values that 4-hydroxypyridine l-oxide and its alkyl and acyl derivatives are present both in their benzenoid and quinoid forms, and assumed from its ultraviolet and infrared spectra that 2-hydroxypyridine 1-oxide is present in the form of *N*-hydroxy-2-pyridone with a strong hydrogen bonding. Kaneko[172] concluded, from the comparison of infrared spectra of 4- and 2-hydroxyquinoline 1-oxides, their derivatives, and their *N*-acyloxyl derivatives, that while the former took the enol-betaine form, the latter were in the *N*-hydroxycarbostyril form.

In both pyridine and quinoline series *N*-oxides, a 2-hydroxy compound takes the quinoid hydroxamic acid form and the 4-hydroxy compound tends to take the enol-betaine form. This contrasts with the fact that the corresponding amino derivatives, both the 2- and 4-amino compounds, tend to take the benzenoid form (*cf. Section 8.5*, pp. 396, 397).

The same tautomerism is believed to be present in hydroxyl derivatives of other azine and diazine series *N*-oxides, but no detailed examinations have been made.

8.7 Some Rearrangement Reactions

As has already been stated in the section on nucleophilic substitution (*cf. Section 7.4*, p. 259), aromatic *N*-oxide compounds undergo reaction with acid anhydrides and chlorides to give a product which apparently has been formed by rearrangement of the nitrogen-bound oxygen to a position on the aromatic ring or the side chain. Many such examples have been reported and numerous theories have been put forward for its reaction mechanism. The most common theory is that the oxygen atom of the *N*-oxide group is first acylated. In other words, this reaction is triggered by the nucleophilic activity of the oxygen atom and some reactions are known where a group inside the molecule shifts directly to oxygen by this activity.

Shaw[169] found that decomposition of 2-benzyloxypyridine 1-oxide (8-139) by hydrochloric acid gave, besides *N*-hydroxy-2-pyridone (8-135), a small amount of *N*-benzyloxy-2-pyridone (8-140) as a by-product. Tanida[173] found that 4-benzyloxy- or 4-allyloxy-quinoline 1-oxide, when heated in dimethylformamide solution in the presence of boron trifluoride, mainly gave *N*-benzyloxy- or *N*-allyloxy-4-hydroxyquinoline betaine, with 4-hydroxyquinoline 1-oxide as a by-product.

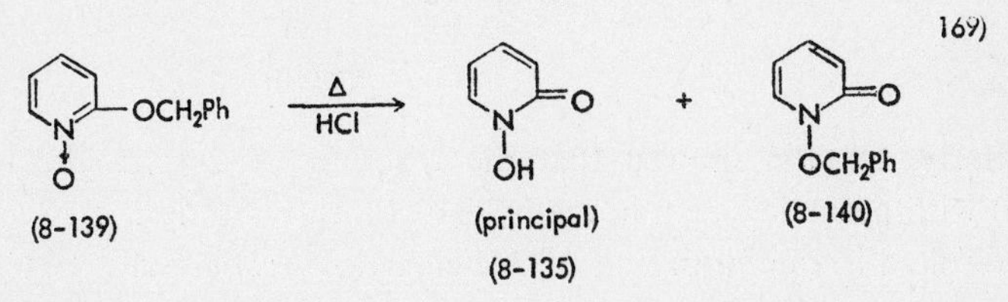

References p. 418

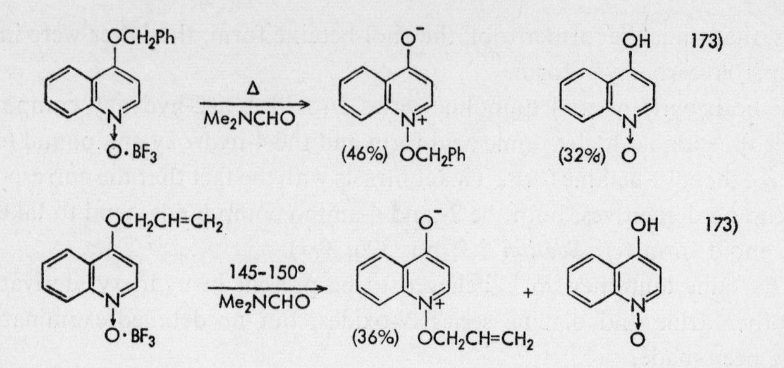

Itai and Kamiya[166] observed that heating of 3,4,6-trimethoxypyridazine 1-oxide (8-141) resulted in the formation of 1,3,4-trimethoxypyridazin-6-one (8-142).

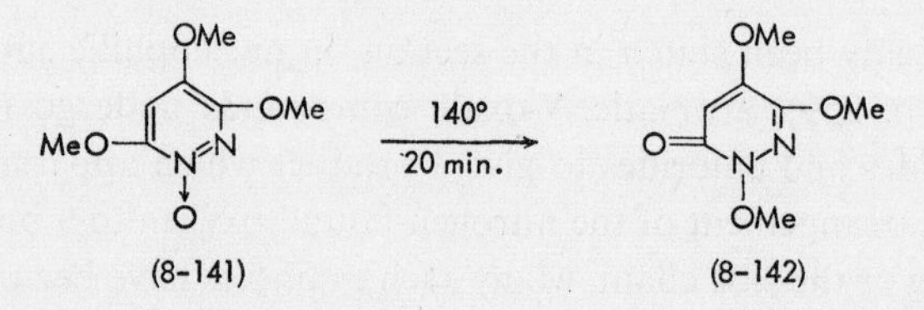

Adams and Reifschneider[8] stated that *N*-oxidation of 2-methoxy-6-methylpyridine (8-143) with peracetic acid exclusively gave the *N*-oxide compound when using fresh peracid, but the reaction was attended with formation of 1-methoxy-6-methyl-2-pyridone (8-144) if peracid older than 1 year since preparation was used.

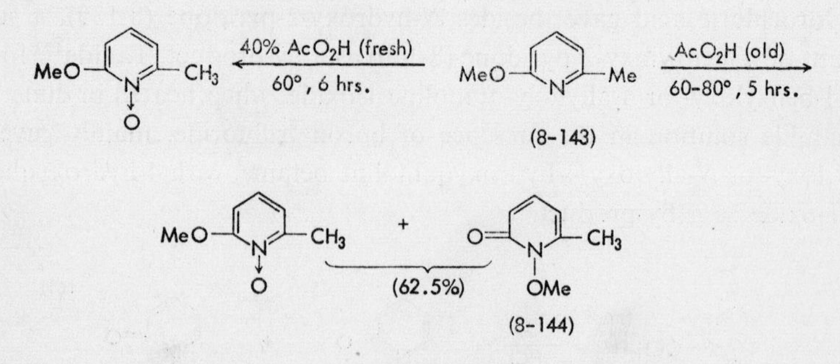

Similarly, Ohta[174] showed that the reaction of 2-benzoyloxyquinoline (8-145) with permaleic acid in methylene chloride solution at room temperature resulted in the formation of *N*-benzoyloxycarbostyril (8-146). He assumed an intermediate formation of 2-benzoyloxyquinoline 1-oxide and proposed the following mechanism for this reaction.

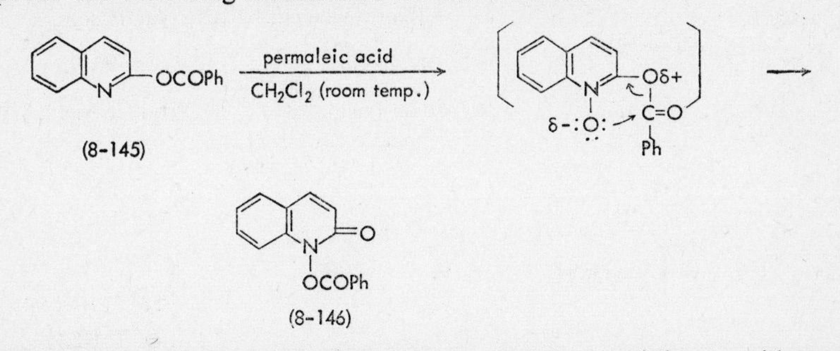

Recently, Dinan and Tieckelmann[175] reported that pyridine 1-oxides substituted with an allyloxy, benzyloxy, or methoxyl group in the 2-position underwent rearrangement to the corresponding *N*-alkoxy-2-pyridone in a good yield at a comparatively low temperature.

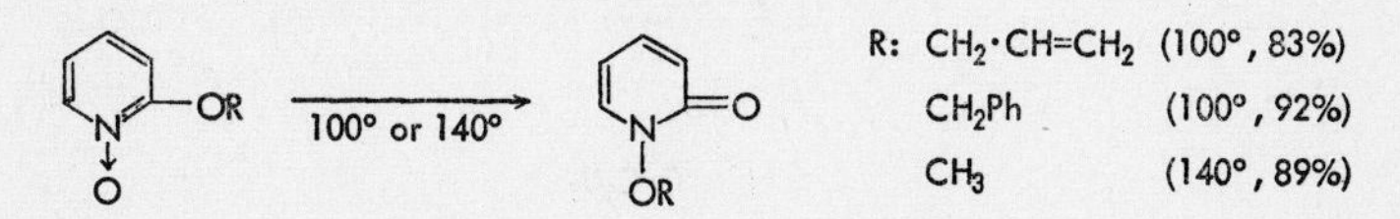

It is known that heating of 2-methoxypyridine to 210° results in its rearrangement to *N*-methyl-2-pyridone and this rearrangement is known to be promoted by the presence of benzoyl peroxide. This evidence suggests that this rearrangement reaction is due to a free radical process[176]. In the corresponding *N*-oxide compounds, however, there has been no evidence to support a radical process for such rearrangement reactions and it has been postulated that such a rearrangement is an ionic process due to the nucleophilic activity of the *N*-oxide group.

A base-catalyzed rearrangement reaction accompanying intermediate nucleophilic addition to the 2- or 4-position of the *N*-oxide group has been discovered. Naito and Dohmori[177a–c] showed that *N*-acetoacetyl, *N*-phenylacetyl, and *N*-cyanoacetyl derivatives of 2- and 4-pyridinesulfonamide 1-oxide underwent a rearrangement similar to the Smiles reaction[179] in the presence of an alkali, as in the case of the corresponding *o*- and *p*-nitrophenylsulfonamide derivatives[178], and resulted in the formation of the compounds listed in Table 8-5, with decomposition of sulfurous acid. A similar rearrangement has been observed with *N*-acetoacetyl-4-quinoline-sulfonamide 1-oxide[177a].

As shown in Table 8-5, the rearrangement of a 2-sulfonamide derivative is accompanied by participation of the oxygen atom in the *N*-oxide group in

References p. 418

TABLE 8-5

REARRANGEMENT OF *N*-ACYL DERIVATIVES OF 2- AND 4-PYRIDINESULFONAMIDE 1-OXIDE

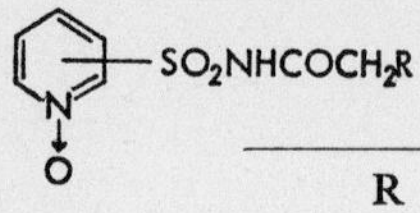 Pyridine 1-oxide–$SO_2NHCOCH_2R$	R	Rearrangement by 10% NaOH soln. at 20–36° Product (yield, %)	Rearrangement by 10% NaOH soln. at 90–95° Product (yield, %)
2-Sulfonamide derivatives	$-COCH_3$	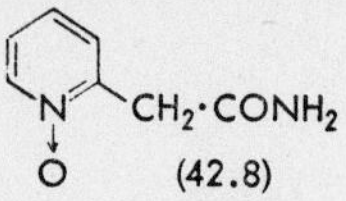 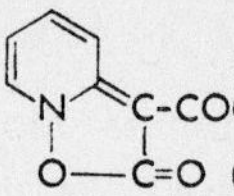 2-$CH_2 \cdot CONH_2$ pyridine 1-oxide (42.8); N—O—C=O ring with C-$COCH_3$ (56.8)	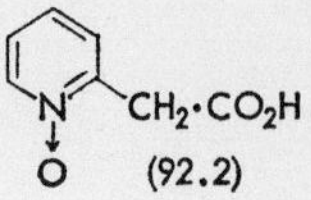 2-$CH_2 \cdot CO_2H$ pyridine 1-oxide (92.2)
	$-C_6H_5$	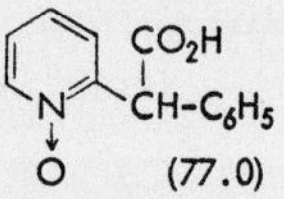 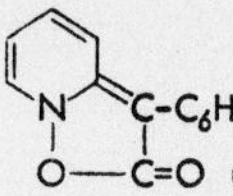 2-$CH(CO_2H)-C_6H_5$ pyridine 1-oxide (77.0); N—O—C=O ring with C-C_6H_5 (6.7)	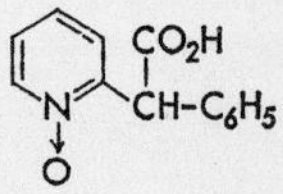 2-$CH(CO_2H)-C_6H_5$ pyridine 1-oxide
	$-CN$	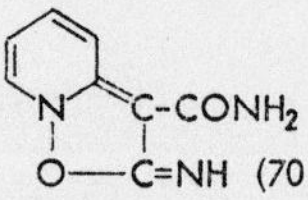 N—O—C=NH ring with C-$CONH_2$ (70.7)	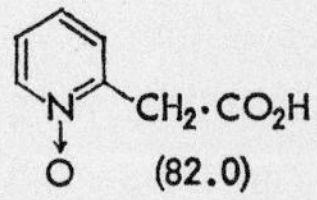 2-$CH_2 \cdot CO_2H$ pyridine 1-oxide (82.0)
4-Sulfonamide derivatives	$-COCH_3$	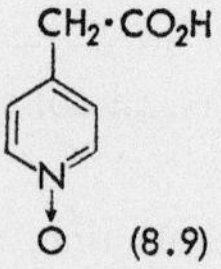 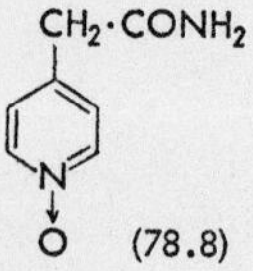 4-$CH_2 \cdot CO_2H$ pyridine 1-oxide (8.9); 4-$CH_2 \cdot CONH_2$ pyridine 1-oxide (78.8)	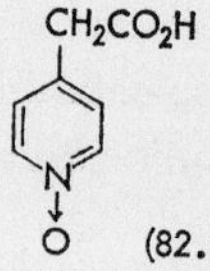 4-CH_2CO_2H pyridine 1-oxide (82.0)
	$-C_6H_5$	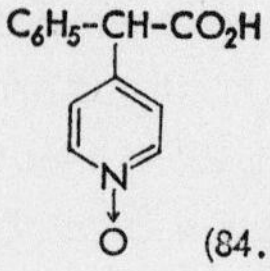 4-($C_6H_5-CH-CO_2H$) pyridine 1-oxide (84.8)	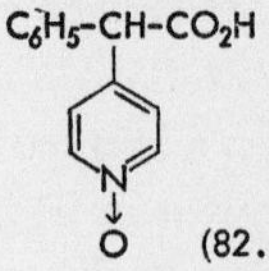 4-($C_6H_5-CH-CO_2H$) pyridine 1-oxide (82.0)

this reaction to form an isoxazolone ring and results in the simultaneous formation of a 3-substituted pyrido[*b*]isoxazol-2-one derivative, the yield being greater, the smaller the volume of the substituent, R[177c]. Dohmori[177d] found that this rearrangement reaction does not take place in the corresponding deoxygenated pyridinesulfonamide derivatives. He also measured the reaction rate of this rearrangement by the determination of sulfurous acid formed by decomposition at the time of rearrangement by converting it to sulfuric acid and showed that this rearrangement is a first-order reaction and that the reaction rate is 10–100 times faster than that of the corresponding *p*-nitrobenzenesulfonamide derivatives. By consideration of these facts, he proposed the following reaction mechanism for this rearrangement[177d].

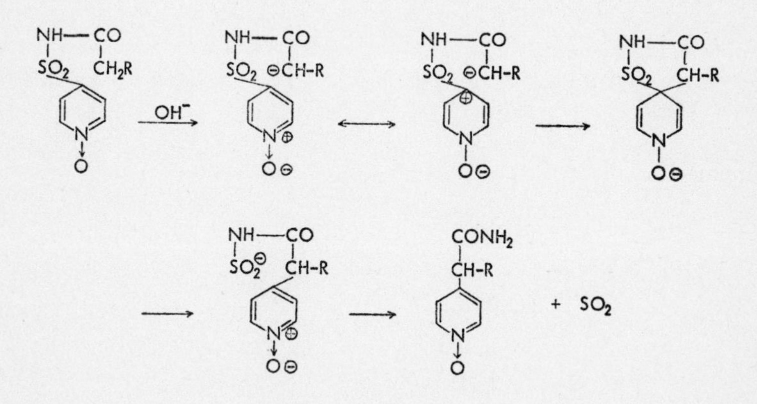

Some acid- or base-catalyzed rearrangements, accompanying a similar intermediate nucleophilic addition to the 2-position of the *N*-oxide group or of a tertiary ring-nitrogen, are known to take place. Clark-Lewis and Katekar[180] showed that the addition of 1-methyl-3-(*N*-methyl-*N*-phenylcarbamoyl)-2-oxo-1,2-dihydroquinoxaline 4-oxide (8-147) to conc. sulfuric acid chilled to 0° results in the formation of 1-methyl-3-(2-methylaminophenyl)-2-oxo-1,2-dihydroquinoxaline (8-148) with evolution of carbon dioxide, and Habib and Rees[181] proposed the following mechanism for this reaction.

References p. 418

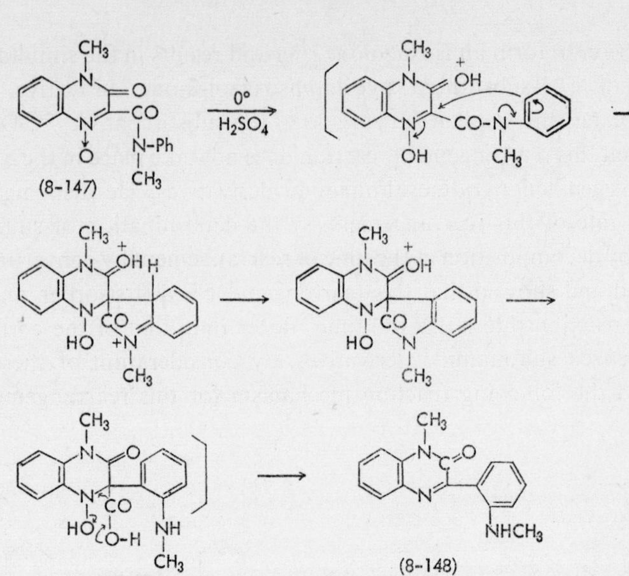

Similarly

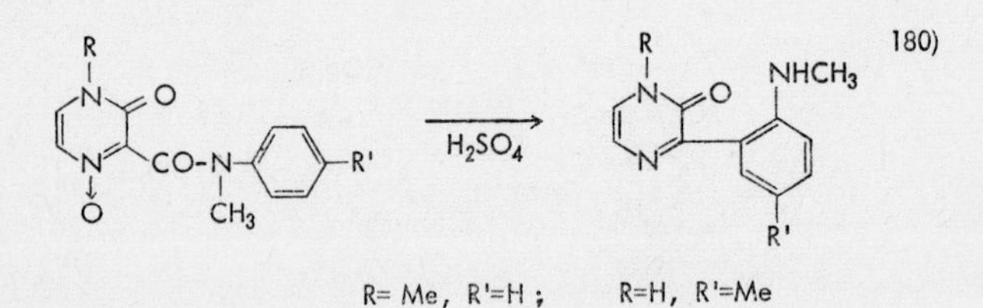

Carbon[182] discovered the rearrangement of benzotriazine *N*-oxides into benzotriazole, as shown below, and assumed cleavage of the triazine ring following intermediate addition of a hydroxyl ion.

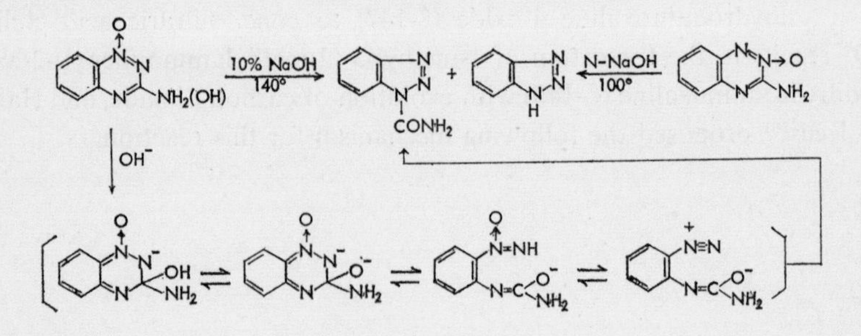

Pinkus and others[183] proved that the structure of 2-phenylisoisatogen, formed by heating 2-phenylisatogen (8-149) with sulfuric acid in methanol at 100°, is not of the oxaziridine type (8-150) as presumed by Ruggli and others[184] but is 3-benzoylanthranil (8-151); they assumed the following reaction mechanism.

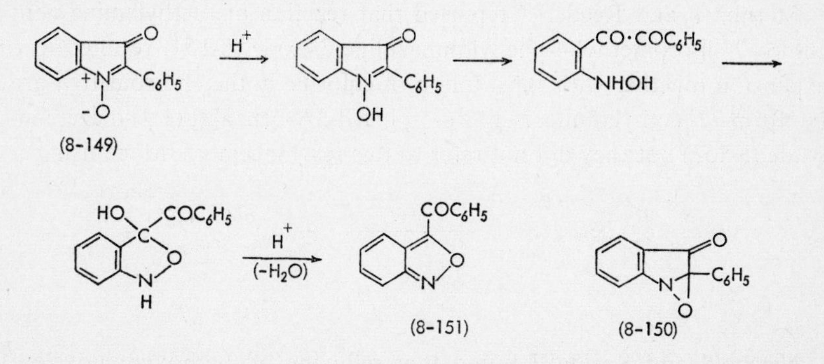

On the other hand, Noland and Jones[185] found, contrary to the foregoing example, that heating of 2-phenylisatogen (8-149) in ethanolic ammonia resulted in intermediate formation of an ammonia adduct and subsequent ring expansion by N-insertion to 4-hydroxy-3-phenylcinnoline l-oxide (8-152). They explained this reaction by the following sequence.

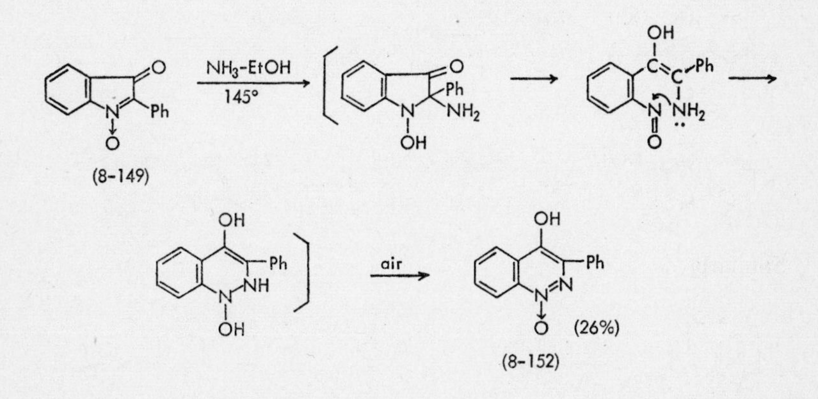

These workers found that heating of 2-phenyl- or 2-methoxycarbonyl-isatogen with tetracyanoethylene in xylene gave the corresponding 2-substituted 3,4-dihydro-4-quinazolone (8-153). It is certain that tetracyanoethylene takes part in this reaction but its mechanism is still unknown.

References p. 418

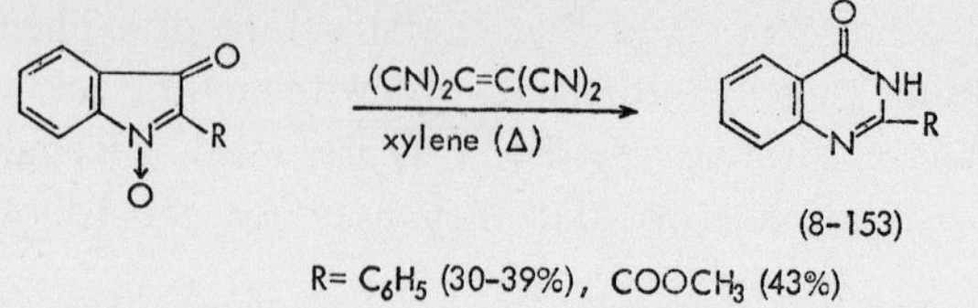

Sternbach and Reeder[186] reported that reaction of methylamine with 6-chloro-2-chloromethyl-4-phenylquinazoline 3-oxide (8-154) resulted in ring expansion together with substitution of chlorine in the chloromethyl group to form 2-methylamino-7-chloro-5-phenyl-3*H*-benzo[*e*]-(1,4)-diazepine 4-oxide (8-155) but they did not refer to this rearrangement mechanism.

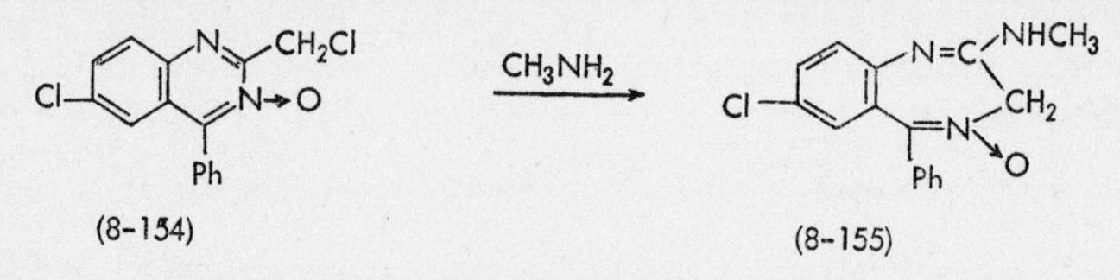

Newbold and Spring[187] found that refluxing of 2-ethoxyquinoxaline 4-oxide in ethanolic hydrochloric acid results in rearrangement of the oxygen of the *N*-oxide group to the 3-position, 2,3-dihydroxyquinoxaline being formed, accompanied by hydrolysis of the ethoxyl group. Cheeseman[188] carried out this reaction with 2-methoxyquinoxaline 4-oxide, obtained the same result, and proposed the following mechanism for this reaction.

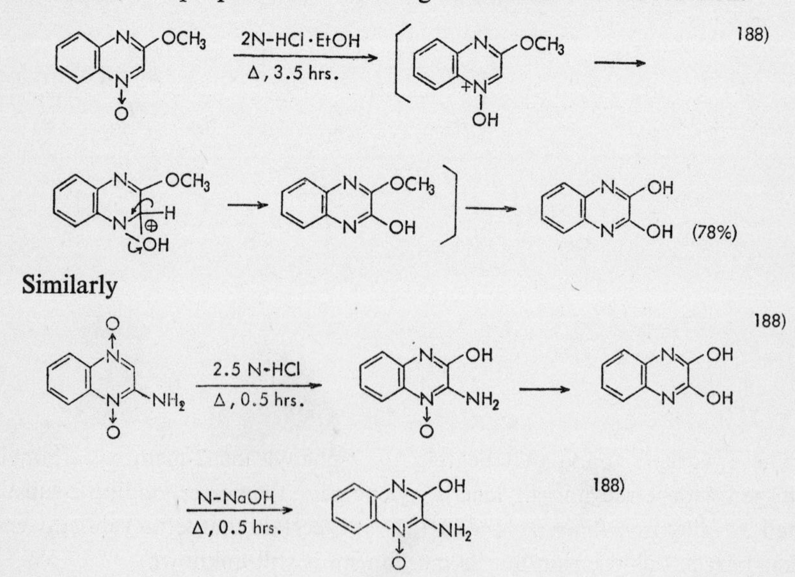

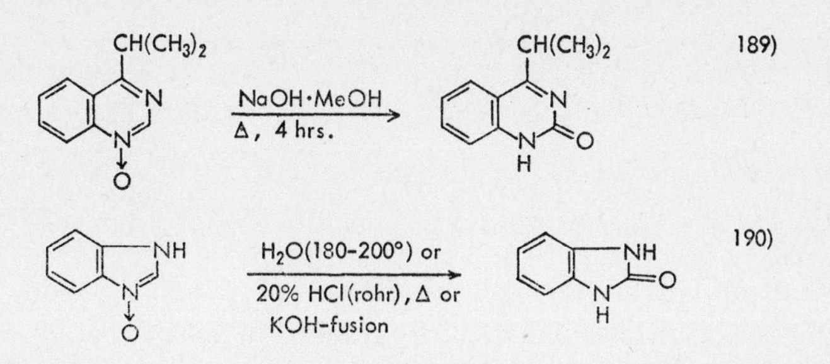

This rearrangement of benzimidazole *N*-oxide proceeds more easily with its 1-methyl derivatives. Takahashi and Kanô[191] found that heating an acetone or chloroform solution of 1-methylbenzimidazole 3-oxide for a few hours or allowing the solution to stand at room temperature for a long period of time resulted in the rearrangement of its oxygen to form 1-methyl-2(3*H*)-benzimidazolone (8-156), and proposed either of the following two mechanisms for this reaction.

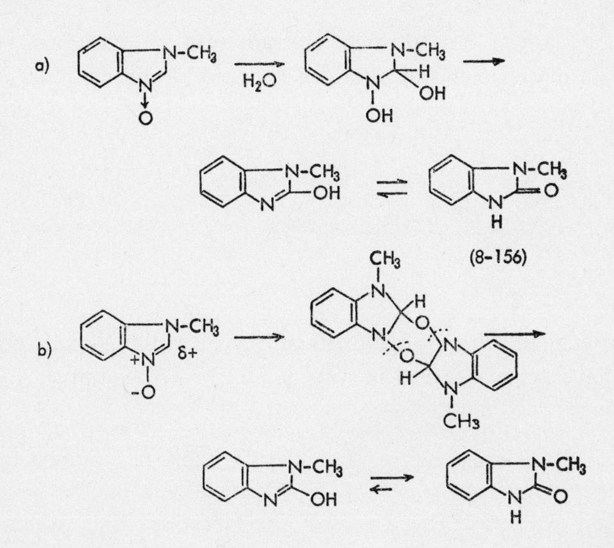

Landquist[192] reported that ultraviolet or sunlight irradiation of the solution of quinoxaline 1,4-dioxide (8-157) in water or dilute sulfuric acid results in the formation of 2-hydroxyquinoxaline 4-oxide (8-158), and that the reaction progresses with quinoxaline 1-oxide, though very slowly, to form 2-hydroxyquinoxaline.

References p. 418

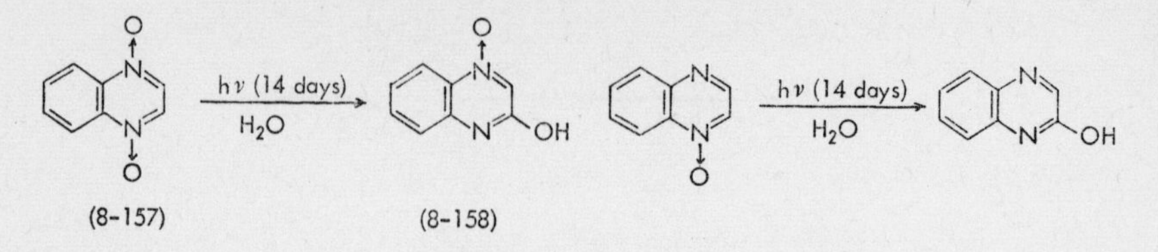

Hata[193] found that ultraviolet irradiation of pyridine 1-oxide and 3-picoline l-oxide in the gas phase chiefly resulted in deoxygenation only, but that irradiation of 2-picoline l-oxide with ultraviolet rays of 3261 Å, corresponding to its n–π^* transition, gave 2-pyridylmethanol.

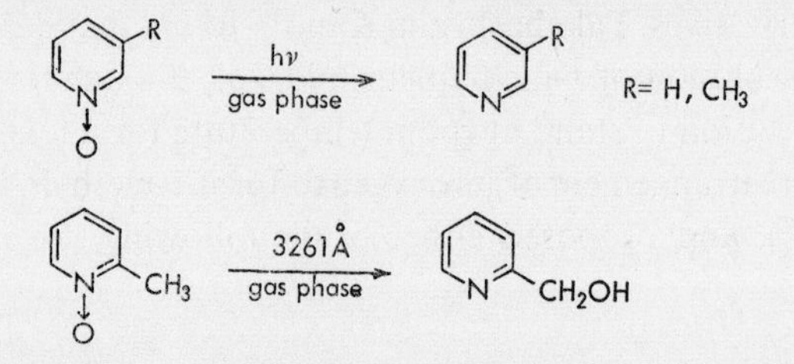

Buchardt[194] carried out ultraviolet irradiation of quinoline 1-oxide in absolute ethanol or water and obtained carbostyril and its dimer in a good yield.

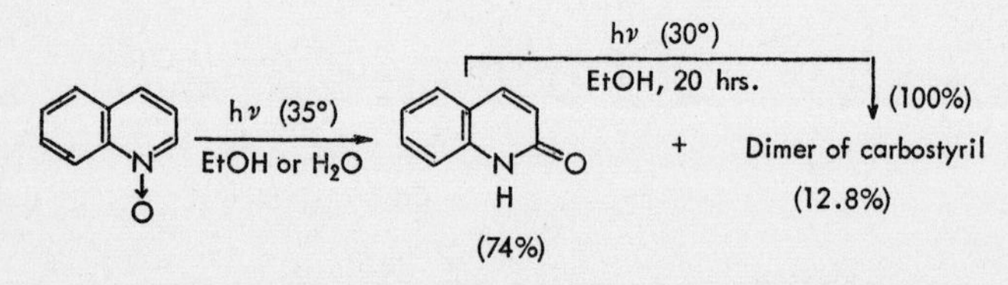

Such photochemical rearrangements of oxygen in the *N*-oxide group to its α-position have been observed in nitrones[195–198]. For example,

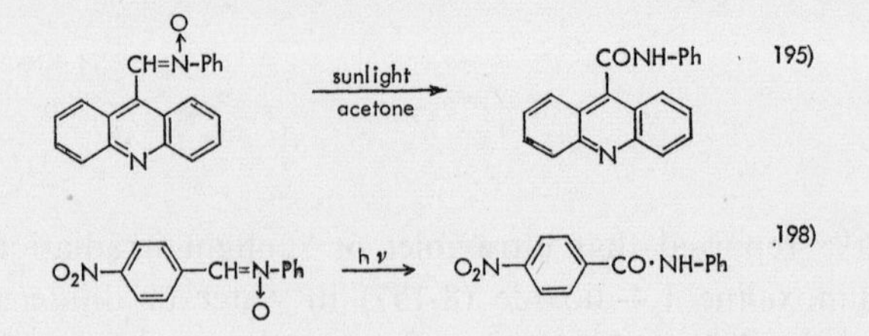

As mechanism for this reaction, it has been proved that the nitrone (8-159) undergoes isomerization to the corresponding oxaziridine (8-160) by light

irradiation[198] and that the latter compound is so unstable that it is decomposed into an acid amide (8-161).

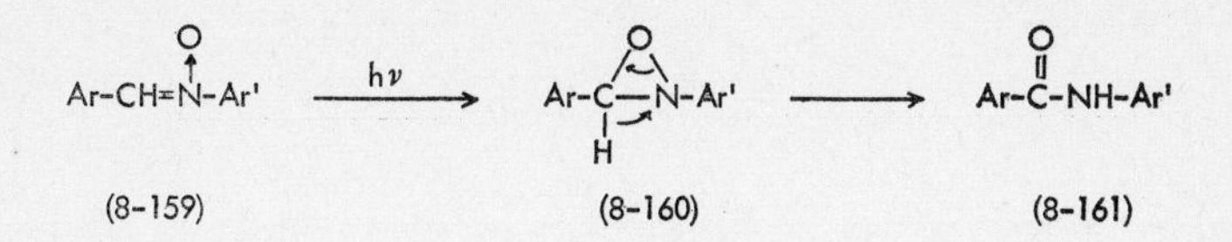

In certain instances, the aryl group is known to undergo rearrangement from the carbon to the nitrogen atom during this reaction.

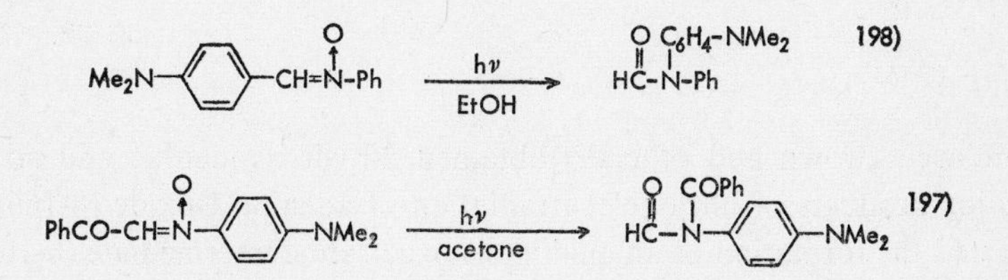

Kröhnke[197] explained these rearrangements as the cleavage of the nitrogen-oxygen bond in the above oxaziridine (8-160) and the passage through the diradical-like state thereby formed.

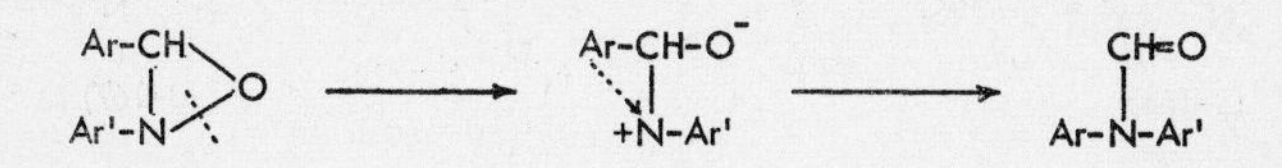

It can be assumed that the rearrangement of the foregoing quinoxaline and quinoline series aromatic *N*-oxides to cyclic acid amides by irradiation also proceeds through similar oxaziridine-type compounds but, in contrast to the nitrones, the intermediate oxaziridines formed have not been isolated as yet. Ishikawa, Kaneko, and Yamada[199] obtained 3-methylcarbostyril (8-163), *N*-methylcarbostyril (8-164), and *N*-acetylindole (8-165) in 22, 10, and 8% yield, respectively, besides a small amount of quinaldine, by ultraviolet irradiation of quinaldine 1-oxide (8-162). The formation of these rearrangement products may be explained as that of an oxaziridine-type compound (8-166) as an intermediate which undergoes rearrangement in three different ways, as shown below.

References p. 418

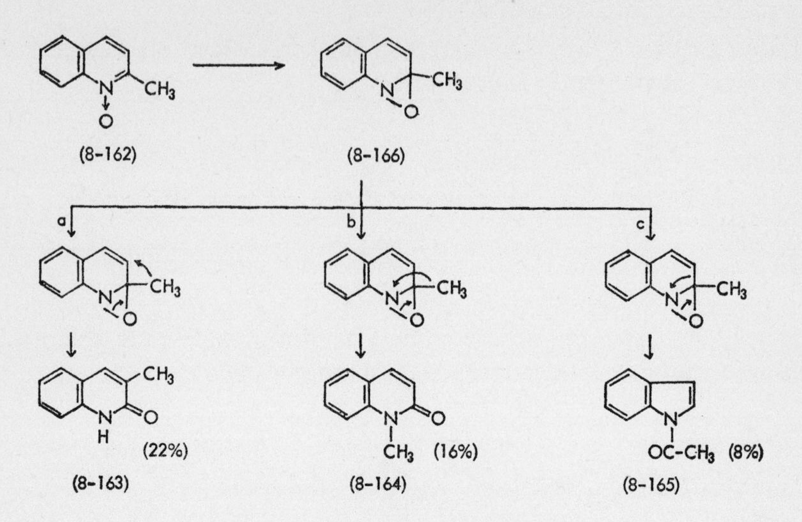

Similarly, Brown and others[200] obtained 2-hydroxyadenine and adenine as major products of ultraviolet irradiation of adenine 1-oxide (8-166) and suggested the formation of an analogous oxaziridine intermediate (8-167).

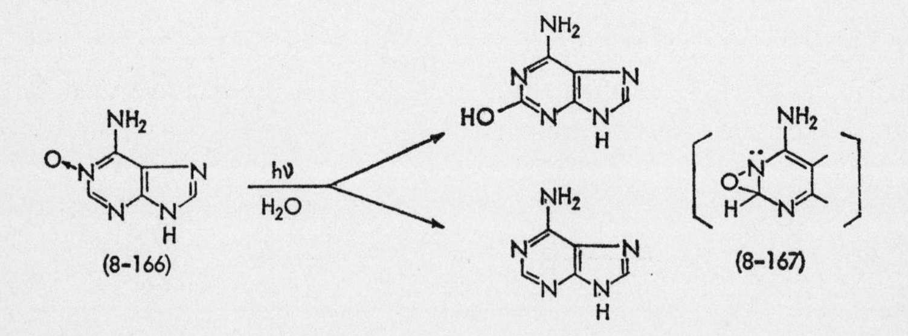

REFERENCES

1 T. Okamoto, H. Hayatsu, and Y. Baba, *Chem. Pharm. Bull. (Tokyo)*, 8 (1960) 892.
2 G. Coppens, F. Declerck, C. Gillet, J. Nasielski, *Bull. Soc. Chim. Belges*, 70 (1961) 480 [*Chem. Abstr.*, 57 (1962) 5880].
3 M. Liveris and J. Miller, *J. Chem. Soc.*, (1963) 3486.
4 S. Sako, *Chem. Pharm. Bull. (Tokyo)*, 11 (1963) 337.
5 Y. Kawazoe and S. Natsume (née Suzuki), *Yakugaku Zasshi*, 83 (1963) 523.
6 R. Adams and S. Miyano, *J. Am. Chem. Soc.*, 76 (1954) 3168.
7 R. L. Bixler and C. Niemann, *J. Org. Chem.*, 23 (1958) 575.
8 R. Adams and W. Reifschneider, *J. Am. Chem. Soc.*, 81 (1959) 2537.
9 V. Boekelheide and R. J. Windgassen, Jr., *J. Am. Chem. Soc.*, 81 (1959) 1456.

10 D. Jerchel and H. F. Heck, *Ann.*, 613 (1958) 171. *cf.* K. Ramaiah and V. R. Srinivasan, *Indian J. Chem.*, 1 (1963) 351 [*Chem. Abstr.*, 60 (1964) 500].
11 E. Ochiai and T. Kato, *Yakugaku Zasshi*, 71 (1951) 156.
12 E. D. Parker and A. Furst, *J. Org. Chem.*, 23 (1958) 201.
13 L. Pentimalli, *Tetrahedron*, 14 (1961) 151.
14 A. R. Katritzky and A. M. Monro, *J. Chem. Soc.*, (1958) 150.
15 W. Herz and D. R. K. Murty, *J. Org. Chem.*, 26 (1961) 418.
16 M. Hamana and H. Noda, *Yakugaku Zasshi*, 83 (1963) 342.
17 L. Pentimalli, *Gazz. Chim. Ital.*, 93 (1963) 1093.
18 M. Colonna, *Gazz. Chim. Ital.*, 90 (1960) 1197.
19 M. Colonna and L. Zamparella, *Gazz. Chim. Ital.*, 72 (1962) 301 [*Chem. Abstr.*, 57 (1962) 9815].
20 C. F. Koelsch and W. H. Gumprecht, *J. Org. Chem.*, 23 (1958) 1603.
21 T. Itai, S. Sako, and G. Okusa, *Chem. Pharm. Bull. (Tokyo)*, 11 (1963) 1146.
22 M. Colonna, *Gazz. Chim. Ital.*, 90 (1960) 1178 [*Chem. Abstr.*, 56 (1962) 5930.]
23 T. M. Mishina and L. S. Efros, *Zh. Obshch. Khim.*, 32 (1962) 2217.
24 G. R. Bedford, A. R. Katritzky, and H. M. West, *J. Chem. Soc.*, (1963) 4600.
25 J. Hamer and A. Macaluso, *Chem. Revs.*, 64 (1964) 480.
26 F. Sachs and R. Kemp, *Ber.*, 35 (1902) 1224.
27 I. Tanacescu and I. Nanu, *Ber.*, 72 (1939) 1083.
28 A. Kaufmann and L. G. Vallette, *Ber.*, 45 (1912) 1736.
29 H. L. de Waal and M. Brink, *Chem. Ber.*, 89 (1956) 636.
30 *(a)* E. C. Taylor, A. J. Crovetti, and N. E. Boyer, *J. Am. Chem. Soc.*, 79 (1957) 3549. *(b)* E. C. Taylor and J. S. Driscoll, *J. Org. Chem.*, 26 (1961) 3796.
31 H. Tanida, *Chem. Pharm. Bull. (Tokyo)*, 7 (1959) 540.
32 M. Hamana, S. Saeki, Y. Hatano, and M. Nagakura, *Yakugaku Zasshi*, 83 (1963) 348.
33 T. Kato, Y. Goto, and M. Kondo, *Yakugaku Zasshi*, 84 (1964) 290.
34 F. E. Cislack (to reilly tar and chemical corp.), *U.S. Pat.* 2,989,534, June 20, 1961.
35 T. Kato and Y. Goto, *Chem. Pharm. Bull. (Tokyo)*, 11 (1963) 461.
36 M. Ogata, *Chem. Pharm. Bull. (Tokyo)*, 11 (1963) 1517.
37 *(a)* M. Hamana, B. Umezawa, Y. Gotoh, and K. Noda, *Chem. Pharm. Bull. (Tokyo)*, 8 (1960) 692. *(b)* M. Hamana, B. Umezawa, and K. Noda, *Chem, Pharm. Bull. (Tokyo)*, 11 (1963) 694.
38 M. Colonna, *Gazz. Chim. Ital.*, 91 (1961) 34.
39 G. Ortoleva, *Gazz. Chim. Ital.*, 29(I) (1899) 503; 30(I) (1900) 509.
40 L. C. King, *J. Am. Chem. Soc.*, 66 (1944) 894.
41 *(a)* F. Kröhnke, *Angew. Chem.*, 75 (1963) 317. *(b)* F. Kröhnke and K. F. Gross, *Chem. Ber.*, 92 (1959) 22.
42 B. Umezawa, *Chem. Pharm. Bull. (Tokyo)*, 8 (1960) 698, 918.
43 M. Hamana, B. Umezawa, and Y. Goto, *Yakugaku Zasshi*, 80 (1960) 1519.
44 D. Jerchel, J. Heider, and H. Wagner, *Ann.*, 613 (1958) 153.
45 C. A. Buehler, L. A. Walker, and P. Garcia, *J. Org. Chem.*, 26 (1961) 1410.
46 M. Katada, *Yakugaku Zasshi*, 68 (1948) 123.
47 E. Ochiai and M. Katada, *Yakugaku Zasshi*, 63 (1943) 186.
48 A. Kirpal and E. Reiter, *Ber.*, 58 (1925) 699; 60 (1927) 664.
49 A. Kirpal and W. Böhm, *Ber.*, 65 (1932) 680.
50 M. G. Bysstitzkaja and A. W. Kirssanow, *Zh. Obshch. Khim.*, 10 (1940) 1107 [*Chem. Zentr.*, II (1940) 4374].
51 E. Ochiai, T. Teshigawara, and T. Naito, *Yakugaku Zasshi*, 65 (1945) 429.
52 H. J. den Hertog and J. Overhoff, *Rec. Trav. Chim.*, 69 (1950) 468.

53 H. J. DEN HERTOG, C. H. HENKENS, AND J. H. VAN ROON, *Rec. Trav. Chim.*, 71 (1952) 1145.
54 E. OCHIAI AND I. SUZUKI, *Yakugaku Zasshi*, 67 (1947) 158.
55 T. KATO AND F. HAMAGUCHI, *Pharm. Bull. (Japan)*, 4 (1956) 174.
56 E. V. BROWN, *J. Am. Chem. Soc.*, 79 (1957) 3565.
57 *(a)* E. OCHIAI AND H. MITARASHI, *Chem. Pharm. Bull. (Tokyo)*, 11 (1963) 1084. *(b)* E. OCHIAI AND H. MITARASHI, *Ann. Rept. Itsuu Lab.*, 13 (1963) 22.
58 F. PARISI, P. BOVINA, AND A. QUILICO, *Gazz. Chim. Ital.*, 90 (1960) 903.
59 E. OCHIAI AND H. MITARASHI, *Ann. Rept. Itsuu Lab.*, 14 (1964) 17.
60 F. PARISI, P. BOVINA, AND A. QUILICO, *Gazz. Chim. Ital.*, 92 (1962) 1138.
61 P. YATES, M. J. JORGENSON, AND S. K. ROY, *Can. J. Chem.*, 40 (1962) 2146.
62 E. OCHIAI AND T. NAITO, *Yakugaku Zasshi*, 64 (1944) 206.
63 E. OCHIAI, A. OHTA, AND H. NOMURA, *Pharm. Bull. (Japan)*, 5 (1957) 310.
64 E. HAYASHI, H. YAMANAKA, AND K. SHIMIZU, *Chem. Pharm. Bull. (Tokyo)*, 7 (1959) 143.
65 T. ITAI AND S. KAMIYA, *Chem. Pharm. Bull. (Tokyo)*, 9 (1961) 87.
66 E. OCHIAI, CH. KANEKO, AND J. INOMATA, *Yakugaku Zasshi*, 78 (1958) 584.
67 E. C. TAYLOR, A. J. CROVETTI, AND N. E. BOYER, *J. Am. Chem. Soc.*, 79 (1957) 3549.
68 E. C. TAYLOR AND W. A. EHRHART, *J. Am. Chem. Soc.*, 82 (1960) 3141.
69 Z. TALIK, *Roczniki Chem.*, 35 (1961) 475 [*Chem. Abstr.*, 57 (1962) 15065].
70 R. F. EVANS AND H. C. BROWN, *J. Org. Chem.*, 27 (1962) 1665.
71 *(a)* T. ITAI AND S. NATSUME (née SUZUKI), *Chem. Pharm. Bull. (Tokyo)*, 11 (1963) 83. *(b)* T. ITAI AND S. NATSUME (née SUZUKI), *Chem. Pharm. Bull. (Tokyo)*, 11 (1963) 342.
72 *(a)* T. ISHII, *Yakugaku Zasshi*, 72 (1952) 1315. *(b)* T. ISHII, *Yakugaku Zasshi*, 72 (1952) 1317.
73 F. ALLISON, J. L. COMTE, AND H. E. FIERZ-DAVID, *Helv. Chim. Acta*, 34 (1951) 818.
74 E. OCHIAI AND M. IKEHARA, *Pharm. Bull. (Japan)*, 2 (1954) 72.
75 E. OCHIAI AND Y. KAWAZOE, *Pharm. Bull. (Japan)*, 5 (1957) 606.
76 E. OCHIAI, *Yakugaku Zasshi*, 61 (1941) 88.
77 *(a)* M. KATADA, *Yakugaku Zasshi*, 67 (1947) 56. *(b)* M. KATADA, *Yakugaku Zasshi*, 67 (1947) 59. *(c)* E. OCHIAI AND M. KATADA, *Yakugaku Zasshi*, 63 (1943) 265.
78 E. OCHIAI, M. O. ISHIKAWA, AND Z. SAI, *Yakugaku Zasshi*, 63 (1943) 280.
79 E. OCHIAI, T. ITAI, AND K. YOSHINO, *Proc. Imp. Acad. (Tokyo)*, 20 (1944) 141.
80 M. O. ISHIKAWA, *Proc. Imp. Acad. (Tokyo)*, 20 (1944) 599.
81 M. O. ISHIKAWA, *Yakugaku Zasshi*, 65B (1945) 105.
82 *(a)* T. ITAI, *Yakugaku Zasshi*, 65B (1945) 4. *(b)* T. ITAI, *Eisei Shikensho Hokoku*, 67 (1950) 138.
83 I. SUZUKI, *Yakugaku Zasshi*, 68 (1948) 126.
84 T. WIELAND AND H. BIENER, *Chem. Ber.*, 96 (1963) 266.
85 M. HAMANA, Y. HOSHIDE, AND K. KANEDA, *Yakugaku Zasshi*, 76 (1956) 1337.
86 R. JÛJO, *Yakugaku Zasshi*, 66B (1946) 49.
87 *(a)* T. ITAI AND H. OGURA, *Yakugaku Zasshi*, 75 (1955) 292. *(b)* A. R. KATRITZKY, *J. Chem. Soc.*, (1956) 2404.
88 E. C. TAYLOR AND A. J. CROVETTI, *J. Am. Chem. Soc.*, 78 (1956) 214.
89 S. YOSHIDA, *Yakugaku Zasshi*, 66 (1946) 158.
90 T. KATO AND H. HAYASHI, *Yakugaku Zasshi*, 83 (1963) 352.
91 I. SUZUKI AND T. NAKAJIMA, *Chem. Pharm. Bull. (Tokyo)*, 12 (1964) 619.
92 T. ITAI AND S. NATSUME (née SUZUKI), *Chem. Pharm. Bull. (Tokyo)*, 12 (1964) 228.
93 T. OKAMOTO, *Yakugaku Zasshi*, 71 (1951) 297.
94 H. J. DEN HERTOG AND W. P. COMBÉ, *Rec. Trav. Chim.*, 70 (1951) 581.
95 E. OCHIAI, S. OKUDA, AND T. ITOH, *Yakugaku Zasshi*, 71 (1951) 591.

96 H. J. DEN HERTOG AND W. P. COMBÉ, *Rec. Trav. Chim.*, 71 (1952) 745.
97 R. M. ACHESON, B. ADCOCK, G. M. GLOVER, AND L. E. SUTTON, *J. Chem., Soc.*, (1960) 3367.
98 H. W. KRAUSE AND W. LANGENBECK, *Chem. Ber.*, 92 (1959) 155.
99 E. PROFFT AND G. SCHULZ, *Arch. Pharm.*, 294 (1961) 292.
100 M. OGATA, H. KANÔ, AND K. TÔRI, *Chem. Pharm. Bull. (Tokyo)*, 11 (1963) 1527.
101 M. KATADA, *Yakugaku Zasshi*, 67 (1947) 61.
102 E. OCHIAI AND M. IKEHARA, *Yakugaku Zasshi*, 68 (1948) 139.
103 M. ENDO AND T. NAKASHIMA, *Yakugaku Zasshi*, 80 (1960) 875.
104 T. WIELAND AND H. BIENER, *Chem. Ber.*, 96 (1963) 266.
105 T. ITAI, *Yakugaku Zasshi*, 66 (1946) 170.
106 *(a)* T. OKABAYASHI, *Yakugaku Zasshi*, 73 (1953) 949. *(b)* T. OKABAYASHI, *Hakko Kôgaku Zasshi*, 35 (1957) 21.
107 T. OKAMOTO AND M. ITOH, *Chem. Pharm. Bull. (Tokyo)*, 11 (1963) 785.
108 H. IGETA, *Chem. Pharm. Bull. (Tokyo)*, 8 (1960) 368.
109 H. IGETA, *Chem. Pharm. Bull. (Tokyo)*, 8 (1960) 550.
110 H. OTOMASU, *Pharm. Bull. (Tokyo)*, 4 (1956) 117.
111 E. OCHIAI AND T. TESHIGAWARA, *Yakugaku Zasshi*, 65 (1945) 435.
112 E. OCHIAI AND T. NAITO, *Yakugaku Zasshi*, 65 (1945) 441.
113 E. HAYASHI, *Yakugaku Zasshi*, 70 (1950) 145.
114 E. OCHIAI AND H. OGURA, *Yakugaku Zasshi*, 72 (1952) 767.
115 E. HAYASHI, private communication.
116 T. ITAI, *Yakugaku Zasshi*, 66B (1946) 170.
117 T. ITAI AND M. SEKIJIMA, *Eisei Shikensho Hokoku*, 74 (1956) 121.
118 A. R. KATRITZKY, *J. Chem. Soc.*, (1957) 191.
119 M. HAMANA AND M. YAMAZAKI, *Yakugaku Zasshi*, 81 (1961) 612.
120 E. C. TAYLOR AND J. S. DRISCOLL, *J. Am. Chem. Soc.*, 82 (1960) 3141.
121 G. BUCHMANN AND B. BRAEUER, *J. prakt. Chem.*, 16 (1962) 225 [*Chem. Abstr.*, 58 (1963) 9022].
122 J. K. LANDQUIST, *J. Chem. Soc.*, (1956) 2550.
123 R. M. CRESSWELL AND G. B. BROWN, *J. Org. Chem.*, 28 (1963) 2560.
124 *(a)* T. ITAI, *Yakugaku Zasshi*, 69 (1949) 542. *(b)* T. ITAI, *Yakugaku Zasshi*, 69 (1949) 545.
125 E. SHAW, J. BERNSTEIN, K. A. LOSEE, AND W. A. LOTT, *J. Am. Chem. Soc.*, 72 (1950) 4362.
126 F. LEONARD AND A. WAJNGURT, *J. Org. Chem.*, 21 (1956) 1077.
127 H. G. MAUTNER, SHIH-HSI CHU, AND C. M. LEE, *J. Org. Chem.*, 27 (1962) 3671.
128 C. W. REES, *J. Chem. Soc.*, (1956) 3684; C. E. MAXWELL AND P. N. GORDON, *U.S. Pat.* 3,056,798 (Cl 260-294.8), Oct. 2, 1962 [*Chem. Abstr.*, 58 (1963) 5654].
129 G. WAGNER, H. PISCHEL, AND R. SCHMIDT, *Z. Chem.*, 2 (1962) 86.
130 G. PALAMIDESSI AND L. BERNARDI, *Gazz. Chim. Ital.*, 93 (1963) 339.
131 M. FUJIMOTO, *Pharm. Bull. (Japan)*, 4 (1956) 1.
132 Y. SUZUKI, *Yakugaku Zasshi*, 81 (1961) 917.
133 E. OCHIAI AND Y. SUZUKI, *Pharm. Bull. (Japan)*, 2 (1954) 247.
134 T. ITAI AND S. KAMIYA, *Chem. Pharm. Bull. (Tokyo)*, 11 (1963) 1059.
135 J. G. MURRAY AND C. R. HAUSER, *J. Org. Chem.*, 19 (1954) 2008.
136 T. KATO, T. NIITSUMA, AND N. KUSAKA, *Yakugaku Zasshi*, 84 (1964) 432.
137 M. J. PIETERSE AND H. J. DEN HERTOG, *Rec. Trav. Chim.*, 80 (1961) 1376.
138 H. IGETA, *Chem. Pharm. Bull. (Tokyo)*, 8 (1960) 559.
139 T. ITAI AND S. KAMIYA, *Chem. Pharm. Bull. (Tokyo)*, 11 (1963) 348.
140 S. SAKO, *Chem. Pharm. Bull. (Tokyo)*, 10 (1962) 956.
141 T. ITAI AND S. SAKO, *Chem. Pharm. Bull. (Tokyo)*, 14 (1966) 269.

142 I. Nakayama, *Yakugaku Zasshi*, 71 (1951) 1391.
143 M. Takahashi, *Ann. Rept. Itsuu Lab.*, 13 (1963) 25.
144 R. Adams and W. Reifschneider, *J. Am. Chem. Soc.*, 79 (1957) 2236.
145 T. Itai, *Eisei Shikensho Hokoku*, 67 (1950) 141.
146 T. Itai, *Eisei Shikensho Hokoku*, 67 (1950) 149.
147 A. Tschitschibabin and M. Rjasanzew, *Zh. Fiz. Khim.*, 47 (1915) 1571 [*Chem. Zentr.*, II (1916) 228].
148 W. Marckwald, *Ber.*, 27 (1894) 1317.
149 T. Naito, *Yakugaku Zasshi*, 65 (1945) 446.
150 H. Tanida, *Yakugaku Zasshi*, 79 (1959) 1063.
151 T. Itai and T. Nakajima, *Chem. Pharm. Bull. (Tokyo)*, 10 (1962) 936.
152 T. Kato, F. Hamaguchi, and T. Oiwa, *Pharm. Bull. (Japan)*, 4 (1956) 178.
153 N. Ikekawa, *Chem. Pharm. Bull. (Tokyo)*, 6 (1958) 266.
154 J. T. Adams, C. K. Bradsher, D. S. Breslow, S. T. Amore, and C. R. Hauser, *J. Am. Chem. Soc.*, 68 (1946) 1317.
155 T. S. Gardner, E. Wenis, and J. Lee, *J. Org. Chem.*, 22 (1957) 984.
156 K. Oda, *Yakugaku Zasshi*, 64B (1944) 76.
157 E. Ochiai, T. Teshigawara, K. Oda, and T. Naito, *Yakugaku Zasshi*, 65 (1945) 431.
158 E. Ochiai, T. Naito, and M. Katada, *Proc. Imp. Acad. Tokyo*, 19 (1943) 574; *Yakugaku Zasshi*, 64 (1944) 210.
159 H. Tanida, *Chem. Pharm. Bull. (Tokyo)*, 7 (1959) 887.
160 A. R. Katritzky, *J. Chem. Soc.*, (1956) 2063.
161 T. Itai, *Eisei Shikensho Hokoku*, 67 (1950) 123.
162 N. M. Voronina, Z. V. Pushkareva, L. B. Radena, and N. V. Bavikova, *Zh. Obshch. Khim.*, 30 (1960) 3476.
163 H. Yamanaka, *Chem. Pharm. Bull. (Tokyo)*, 7 (1959) 505.
164 T. Higashino, *Chem. Pharm. Bull. (Tokyo)*, 9 (1961) 635.
165 S. Sako, *Yakugaku Zasshi*, 82 (1962) 1208.
166 T. Itai and S. Kamiya, *Chem. Pharm. Bull. (Tokyo)*, 11 (1963) 1073.
167 E. Ochiai and E. Hayashi, *Yakugaku Zasshi*, 67 (1947) 151.
168 E. Ochiai and E. Hayashi, *Yakugaku Zasshi*, 67 (1947) 154.
169 E. Shaw, *J. Am. Chem. Soc.*, 71 (1949) 67.
170 H. H. Jaffé, *J. Am. Chem. Soc.*, 77 (1955) 4445, 4448.
171 A. R. Katritzky, *J. Chem. Soc.*, (1957) 4375.
172 Ch. Kaneko, *Yakugaku Zasshi*, 79 (1959) 428.
173 H. Tanida, *Yakugaku Zasshi*, 78 (1958) 613.
174 A. Ohta, *Chem. Pharm. Bull. (Tokyo)*, 11 (1963) 1586.
175 F. J. Dinan and H. Tieckelmann, *J. Org. Chem.*, 29 (1964) 1650.
176 K. B. Wiberg, J. M. Shryne, and R. R. Kintner, *J. Am. Chem. Soc.*, 79 (1957) 3160.
177 *(a)* T. Naito and R. Dohmori, *Chem. Pharm. Bull. (Tokyo)*, 3 (1955) 38. *(b)* T. Naito, R. Dohmori, and T. Kotake, *Chem. Pharm. Bull. (Tokyo)*, 12 (1964) 588. *(c)* R. Dohmori, *Chem. Pharm. Bull. (Tokyo)*, 12 (1964) 595. *(d)* R. Dohmori, *Chem. Pharm. Bull. (Tokyo)*, 12 (1964) 601.
178 T. Naito, R. Dohmori, and O. Nagase, *Yakugaku Zasshi*, 74 (1954) 593; T. Naito, R. Dohmori, and M. Sano, *Yakugaku Zasshi*, 74 (1954) 596; T. Naito, R. Dohmori, and M. Shinoda, *Chem. Pharm. Bull. (Tokyo)*, 3 (1955) 34.
179 S. Smiles and L. A. Warren, *J. Chem. Soc.*, (1930) 956; J. F. Bunnett and R. E. Zahlen, *Chem. Revs.*, 49 (1951) 362.
180 J. W. Clark-Lewis and G. F. Katekar, *J. Chem. Soc.*, (1959) 2825.
181 M. S. Habib and C. W. Rees, *J. Chem. Soc.*, (1960) 3371.
182 J. A. Carbon, *J. Org. Chem.*, 27 (1962) 185.

183 J. L. Pinkus, T. Cohen, M. Sundaralingam, and G. A. Jeffrey, *Proc. Chem. Soc.*, (1960) 70.
184 P. Ruggli, E. Caspar, and B. Hegedüs, *Helv. Chim. Acta*, 22 (1934) 140.
185 W. E. Noland and D. A. Jones, *J. Org. Chem.*, 27 (1962) 341.
186 *(a)* L. H. Sternbach and E. Reeder, *J. Org. Chem.*, 26 (1961) 1111. *(b)* L. H. Sternbach, E. Reeder, and W. Metlesics, *J. Org. Chem.*, 26 (1961) 4488.
187 G. T. Newbold and F. S. Spring, *J. Chem. Soc.*, (1948) 519.
188 G. H. W. Cheeseman, *J. Chem. Soc.*, (1961) 1246.
189 E. Hayashi and T. Higashino, *Chem. Pharm. Bull. (Tokyo)*, 12 (1964) 43.
190 St. von Niementowski, *Ber.*, 43 (1910) 3018.
191 S. Takahashi and H. Kanô, *Chem. Pharm. Bull. (Tokyo)*, 12 (1964) 785.
192 J. K. Landquist, *J. Chem. Soc.*, (1953) 2830.
193 N. Hata, *Bull. Chem. Soc. Japan*, 34 (1961) 1440, 1444; *J. Chem. Phys.*, 36 (1961) 2072.
194 O. Buchardt, *Acta Chem. Scand.*, 17 (1963) 1461.
195 L. Chardonnes and P. Heinrich, *Helv. Chim. Acta*, 32 (1949) 656.
196 M. J. Kalmet and L. A. Kaplan, *J. Org. Chem.*, 22 (1957) 576.
197 F. Kröhnke, *Ann.*, 604 (1957) 203.
198 J. S. Splitter and M. Calvin, *J. Org. Chem.*, 23 (1958) 651.
199 M. Y. Ishikawa, Ch. Kaneko, and S. Yamada, *Chem. Pharm. Bull. (Tokyo)*, 13 (1965) 747.
200 G. B. Brown, G. Levin, and S. Murphy, *Biochemistry*, 3 (1964) 880; G. Levin, R. B. Setlow, and G. B. Brown, *Biochemistry*, 3 (1964) 883.

Chapter 9

Biological Properties of Aromatic Amine Oxides

9.1 General

The first sign of interest in the biological properties of aromatic amine oxides probably came with the isolation, in 1938 from *Chromobacterium iodinum*, of iodinin, a compound having a strong antibacterial activity and the structure of dihydroxyphenazine *N,N*-dioxide[1]. In 1943, an antibiotic with marked bacteriostatic action, aspergillic acid, produced by *Aspergillus flavus*, was found to be a derivative of 2-hydroxypyrazine 1-oxide[2]. In subsequent years, a considerable number of aromatic amine oxides were synthesized with the skeleton of these antibiotics as model, and their action on micro-organisms examined. One outcome of these studies is that a number of patents have been taken out on the use of derivatives of 2-hydroxy- and 2-mercapto-pyridine 1-oxides as agricultural pesticides and fungicides.

In 1952, Arai and Nakayama[3] examined the antibacterial activity of numerous derivatives of 4-substituted pyridine and quinoline 1-oxides and found that 4-nitroquinoline 1-oxide had the most potent activity. This compound was shown by Okabayashi[4] to have a strong antifungal activity.

Okabayashi observed that this compound had an extraordinarily potent mutagenic effect on *Aspergillus niger*[5], while Sakai and Fukuoka[6] found that it had a certain anticancer action. In 1957, Nakahara and his associates[7] found that 4-nitroquinoline 1-oxide possessed a powerful carcinogenicity. Since then, active biochemical studies have been instituted on the carcinogenic action and mutagenic properties of 4-nitroquinoline 1-oxide and its derivatives.

There are but few reports on systematic studies made of other biological properties of aromatic amine oxides, though some patents mention the action of an individual compound or groups of compounds against micro-organisms or describe compounds having some pharmacological effect.

9.2 Antimicrobial and Antiprotozoal Activities

The antibacterial action of iodinin (9-1) is antagonized to some extent by anthraquinone or naphthoquinone derivatives like quinizarin (9-2). The similarity in structural and polar features of (9-1) and (9-2) suggests that there is a competition for receptor sites of the affected organism, analogous to the case of sulfanilamide and *p*-aminobenzoic acid[8].

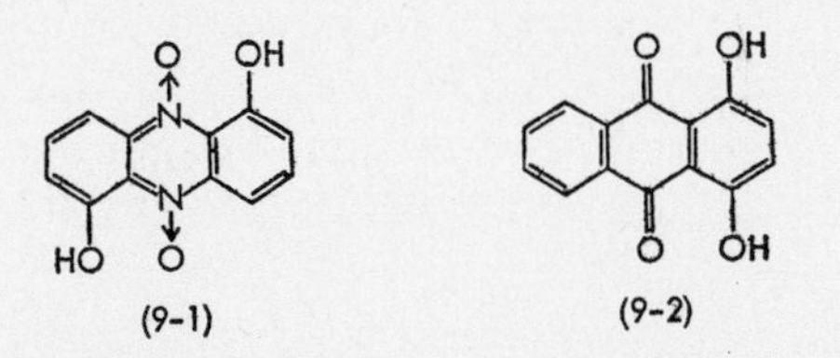

From such an aspect, McIlwain[9] synthesized quinoxaline and its *N,N*-dioxide derivatives, and examined their bacteriostatic activity *in vitro*. The results he obtained, as shown in Table 9-1 in comparison with those of iodinin, indicated that the antibacterial activity of the *N*-oxide compounds was markedly greater than that of the corresponding tertiary bases.

Later, Akabori and Nakamura[10] isolated 1,6-dihydroxyphenazine from the culture broth of *Streptomyces thioluteus* and reported that this compound is active mainly against phytopathogenic fungi and yeast, but did not compare its activity with that of iodinin. Subsequently, King and others[11] carried out the synthesis of hydroxyquinoxaline *N,N*-dioxide derivatives, and Silk[12] that of 2,3-dimethylquinoxaline *N,N*-dioxide derivatives, both for the same purpose, but neither mentioned the antibacterial activity of these compounds.

Landquist and others[13] observed that whereas 2,3-dimethylquinoxaline *N,N*-dioxide (9-3) had a powerful antibacterial activity against gram-negative bacteria, such as *Salmonella dublin* and *Clostridium welchii* of mice, its activity *in vitro* was far weaker and suggested the possibility that the compound might be converted into an active metabolite *in vivo*. They proved that the urine and blood of sheep dosed with (9-3) became activated within an hour of dosing. These workers isolated 2-hydroxymethyl-3-methylquinoxaline 1,4-dioxide (9-4) as the metabolite of (9-3) and confirmed that this compound is the active principle by the *in vitro* experiment indicated in Table 9-2.

In view of these facts, Landquist and Silk[14] synthesized various derivatives of 2-(hydroxymethyl)quinazoline 1,4-dioxides but did not report their antibacterial activity.

References p. 440

TABLE 9-1

ANTIBACTERIAL ACTIVITY OF PHENAZINE AND QUINOXALINE DERIVATIVES[9]

(Tested with *Streptococcus haemolyticus* grown on a peptone medium)

	Concn. (M) of base causing				Concn. (M) of N,N-dioxide causing			
	no inhibition	prevention of visible growth for (days) 1	2–4	5 or more	no inhibition	prevention of visible growth for (days) 1	2–4	5 or more
Iodinin					10^{-7}	5×10^{-7}	10^{-6}	$1.5–2\times10^{-6}$
Phenazine					4×10^{-6}	$2–4\times10^{-5}$	$4–8\times10^{-5}$	$6\times10^{-5}-10^{-4}$
1,2,3,4-Tetrahydrophenazine	10^{-3}				4×10^{-5}	2×10^{-4}	$5\times10^{-4}-10^{-3}$	10^{-3}
2-Methyl-3-amylquinoxaline	5×10^{-4}				2×10^{-4}	5×10^{-4}		
2-Methylquinoxaline	$2\times10^{-4}-2\times10^{-3}$	2×10^{-3}			$2\times10^{-5}-2\times10^{-4}$	$2\times10^{-4}-10^{-3}$		$5\times10^{-4}-10^{-3}$

TABLE 9-2

In vitro ANTIBACTERIAL ACTIVITY OF 2,3-DIMETHYLQUINOXALINE 1,4-DIOXIDE (9-3) AND ITS METABOLITE (9-4)

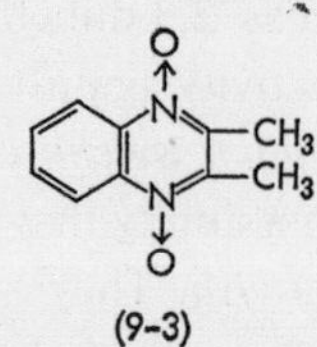

(9-3)

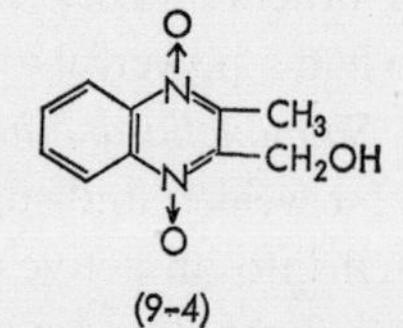

(9-4)

Compound	Max. dilution inhibiting the growth of *Salmonella dublin*	*Clostridium welchii*
9-3	1:1,000	1:100,000
9-4	1:9,000	1:2,700,000

Aspergillic acid isolated from a culture of *Aspergillus flavus* by White and Hill[2] in 1943 is an antibiotic that markedly inhibits the growth of a large number of gram-negative and -positive organisms. Its structure is either 2-hydroxy-3,6-diisobutylpyrazine 1-oxide (9-5) or 1-hydroxy-3,6-diisobutyl-2-pyrazone (9-6)[15], and its deoxygenation results in weaker antibacterial activity[16]. The α-hydroxy *N*-oxide or hydroxamic acid grouping, therefore, seems to be an essential feature of the molecule for biological activity.

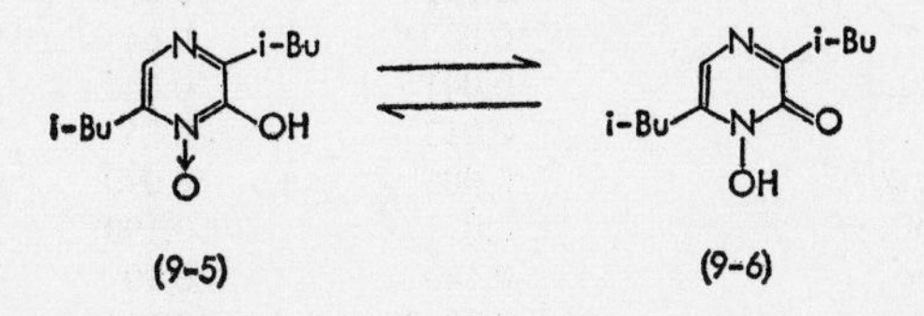

Dutcher[15b] observed that simple hydroxamic acids unrelated to pyrazine also showed some antibacterial activity. Newbold, Spring, and others[16,17], showed that the 2-hydroxy derivatives of pyridine and quinoline 1-oxides, and their methyl derivatives, and 4-hydroxy-2-methylquinazoline 3-oxide possessed a significant antibacterial activity *in vitro*. Such an activity was also found in hydroxamic acids of simple amino acids. At about the same time, Shaw, with the same view in mind, synthesized 2-, 3-, and 4-hydroxypyridine 1-oxides, examined their antibacterial activity, and found that only 2-hydroxypyridine 1-oxide possessed a significant activity[18]. He and his colleagues synthesized this compound, its derivatives, and 1-hydroxy-2-quinolone and 1-hydroxy-4,6-dimethyl-2-pyridone, having the same cyclic hydroxamic acid skeleton, and compared their antibacterial activity with that of aspergillic acid[19]. As shown by the results listed in Table 9-3, some of the synthesized compounds showed antibacterial activity *in vitro* comparable to or better than that of aspergillic acid.

Safir and Williams[20] synthesized five kinds of 2-hydroxypyrazine 1-oxide derivatives (9-7a–e) (Table 9-4) and found (7a) and (7b) to possess antibacterial activity

Shaw and others[21] further synthesized 2-mercaptopyridine 1-oxide (9-8) or *N*-hydroxy-2-pyridinethione (9-9) and its derivatives, and found that these cyclic thiohydroxamic acids had a far more potent antibacterial activity *in vitro* against some organisms than the corresponding *N*-hydroxy-2-pyridone derivatives, as shown in Table 9-5.

2-Mercaptopyridine 1-oxide also shows a strong antifungal activity and numerous patents have been taken out by Olin Mathieson Chemical Corpor-

References p. 440

TABLE 9-3

In vitro ANTIBACTERIAL ACTIVITY OF SOME CYCLIC HYDROXAMIC ACIDS[19]

(Measured by a dilution method)

N-Hydroxy derivatives	Minimum concentration inhibiting growth (mg/ml)		
	Staph. aureus	*Kleb. pneumoniae*	*Myc. smegmatis*
Aspergillic acid	0.021	0.030	0.023
2-Pyridone	0.003	0.043	0.035
4-Methyl-2-pyridone	0.041	0.045	0.041
3-Bromo-2-pyridone	0.013	>0.25	0.050
5-Bromo-2-pyridone	0.005	0.090	0.017
6-Bromo-2-pyridone	0.011	0.190	0.190
5-Nitro-2-pyridone	0.140	0.90	0.450
Carbostyril	0.0012	0.019	0.013
4,6-Dimethyl-2-pyrimidone	0.210	0.95	0.44

TABLE 9-4

DERIVATIVES OF 2-HYDROXYPYRAZINE 1-OXIDE

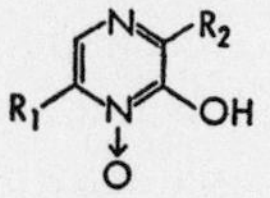

(9-7)

	R_1	R_2		R_1	R_2
a	CH_3	$CH_2CH_2(CH_3)_2$	d	H	CONHOH
b	CH_3	$CHCH_3C_2H_5$	e	C_6H_5	$CH_2CH(CH_3)_2$
c	CH_3	CONHOH			

ation and others on the use of this compound and a variety of its derivatives as pesticides, fungicides, and preservatives. Chief among these derivatives are the heavy metal salts such as manganese, nickel, copper, and zinc salts of the SH group[22], organic salts such as cetyltrimethylammonium and dodecylbenzyldimethylammonium salts[23], disulfides[24], allyl, methallyl, or 2-chlorallyl derivatives[25], alkoxymethyl and alkylthiomethyl derivatives[26], carbamoyl and thiocarbamoyl derivatives[27], the trichloromethyl derivative[28], and the molybdic acid complex[29] or stannous chloride complex[30] of the disulfide. Chas. Pfizer Company has taken out a patent[31] on the use of 2-mercaptopyridine 1-

TABLE 9-5

In vitro ACTIVITY OF SOME 2-MERCAPTOPYRIDINE 1-OXIDES[21]

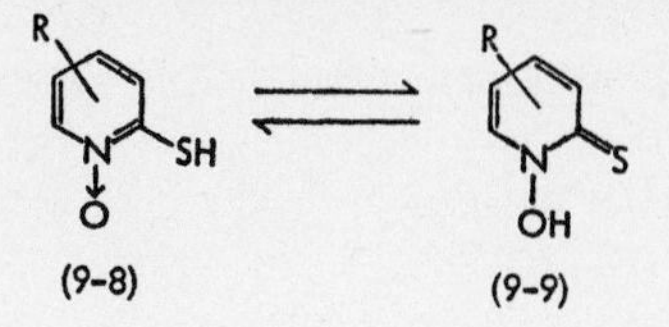

R	Minimum concentration inhibiting growth (γ/ml)		
	Staph. aureus	*Kleb. pneumoniae*	B.C.G.
H	0.06	1.5	0.006
3-CH_3	0.06	0.6	0.004
4-CH_3	0.08	1.5	0.001
5-CH_3	0.07	1.5	0.005
6-CH_3	0.1	3.5	0.003
3-OC_2H_5	0.08	1.5	0.03
5-Br	0.1	2.0	0.008
N-Hydroxy-2-pyridone	3	40	2

oxide, 2-mercaptoquinoline 1-oxide, and their derivatives, substituting the mercapto–hydrogen with an *N*-heterocycle group like imidazolinyl and tetrahydropyrimidinyl groups, for treating superficial mycoses and as active anti-infection agents.

Leonard and others synthesized 2-thiocyanatopyridine 1-oxide and its 4-nitro derivative[32a], and reported that these compounds possessed a broad spectrum against bacteria and fungi in a very low concentration[32b].

In 1954, Lightbown[33] isolated a substance from the culture filtrate of *Pseudomonas pyocyanea* and found the substance to antagonize the growth-inhibitory action of dihydrostreptomycin against *Bacillus subtilis* and *Staphylococcus aureus*. Cornforth and James[34] isolated and purified this substance and showed that it chiefly consisted of an approximately 2:1 mixture of the 2-heptyl and 2-nonyl derivatives of 4-hydroxyquinoline 1-oxide, with a small amount of the 2-undecyl derivative and traces of the 2-octyl derivative and its higher homologs. Lightbown and Jackson[35] found, as shown in Table 9-6, that this antagonistic action of these 2-alkyl-4-hydroxyquinoline 1-oxides was affected by the length of their alkyl chain and became maximum with the nonyl group, and that the *N*-oxide grouping was essential for this activity.

TABLE 9-6

RELATIVE ACTIVITY OF 2-ALKYL-4-HYDROXYQUINOLINE DERIVATIVES IN ANTAGONIZING DIHYDROSTREPTOMYCIN ACTIVITY AGAINST *Bacillus pumilus*[35]

(determined by a modified cup-plate assay)

4-Hydroxyquinoline 1-oxide	Relative activity	4-Hydroxyquinoline	Relative activity
Natural crystalline antagonist	100	2-Heptyl-3-bromo 1-oxide	420
2-Methyl	<0.1	2-Heptyl	0
2-Heptyl	33	2-Nonyl	0
2-Nonyl	206	2-Heptyl-3-carboxy 1-oxide	0
2-Undecyl	66		

In 1952, Arai and Nakayama[3] examined the *in vitro* antibacterial activity of approximately 60 kinds of pyridine and quinoline 1-oxide derivatives substituted with a hydroxyl, alkoxyl, amino, nitro, halogen, mercapto, and arylthio group, using *Salmonella typhosa*, but none of them inhibited the growth of the micro-organism in a concentration below 10^{-3} mole/l, except the derivatives possessing a nitro or azo group in the 4-position. The activity of 4-nitroquinoline 1-oxide was outstanding, as will be seen from Table 9-7.

Okabayashi[4] examined the antifungal activity of pyridine and quinoline 1-oxides substituted in their 4-position and found a very strong activity in 4-nitroquinoline 1-oxide. He further examined its antimicrobial spectrum and obtained the results given in Table 9-8. As will be seen, 4-nitroquinoline 1-oxide inhibits the growth of all the bacteria tested, except *Pseudomonas aeruginosa*, in 400,000 to 1,000,000 dilution, and yeast and fungi in 100,000 to 800,000 dilution. This activity was confirmed by the follow-up test of Leonard and others[32b]. Later workers proved that 6-methyl-4-nitroquinoline 1-oxide[36] and 4-nitro-4′-nitrobenzalquinaldine 1-oxide[37] also possessed fungistatic and bacteriostatic activity.

Hart[38] found that 1,8-naphthyridine and its *N,N*-dioxide possessed a potent antibacterial activity against *Streptococcus haemolyticus*, *Staphylococcus aureus*, and *Escherichia coli*, and Itai and Kamiya[39] reported that 4-azido derivatives of pyridine and quinoline 1-oxides, and 2,2′-dimethyl-4,4′-azopyridine 1,1′-dioxide had a strong bacteriostatic activity against *Staph. aureus*, *Esch. coli*, and *Candida albicans*.

Nishimura, Kanô, and their associates[40] examined the antimicrobial activity of pyridazine 1-oxide derivatives and stated that some of the 3,6-disubsti-

TABLE 9-7

ANTIBACTERIAL EFFECT OF 4-NITROQUINOLINE 1-OXIDE AGAINST VARIOUS STRAINS OF BACTERIA[3]

(Basal medium, bouillon)

Test bacteria	Minimum concentration inhibiting bacterial growth for 6 days (mole $\times 10^{-3}$/l)
Staphylococcus aureus	1/4
Salmonella typhosa	1/8
Salmonella paratyphi A	1/4
Shigella dysenteriae	1/8
Escherichia coli	1/4

TABLE 9-8

ANTIMICROBIAL ACTIVITY OF 4-NITROQUINOLINE 1-OXIDE[4]

(Koji extract-agar, dilution method; incubated at 30° for 14 days)

Test organism	Maximum dilution inhibiting growth ($\times 10^3$)	Test organism	Maximum dilution inhibiting growth ($\times 10^3$)
Esch. coli	1,000	*Torula utilis*	200
Staph. aureus	1,000	*Zygosacch. soja*	400
B. subtilis	4,000	*Asperg. oryzae*	100
B. vulgaris	1,000	*Asperg. niger*	800
Pseud. aeruginosa	4	*Rhod. japonicus*	256
Sacch. sake	400	*Trichophyton sp.*	200

tuted 4-nitro and 4-chloro derivatives had a strong antifungal activity *in vitro*, as shown in Table 9-9. The most noteworthy of these data is the fact that the antitrichophyton activity of 3-halogeno-6-methyl-4-nitropyridazine 1-oxides and 3,4-dichloro-6-(hydroxyiminomethyl)pyridazine 1-oxide is stronger than that of griseofulvin, and the antitrichomonas activity of 3-methoxy-6-methyl-4-nitropyridazine 1-oxide is stronger than that of Trichomycin and Flagyl.

Okabayashi[41] found that the antifungal activity of 4-nitroquinoline 1-oxide is markedly lowered by the addition of SH compounds like cysteine, thioglycolic acid, and glutathione, at a physiological pH. He recognized such

References p. 440

TABLE 9-9

ANTIMICROBIAL ACTIVITY OF SOME 3,6-DISUBSTITUTED PYRIDAZINE 1-OXIDE DERIVATIVES[40a]

	4-Nitro derivatives							4-Chloro deriv.			Control	
Substituent in 3	Cl	I	Br	OCH_3	OCH_3	OCH_3	OCH_3	Cl	Cl	Cl	Griseo-fulvin	Tricho-mycin
Substituent in 6	CH_3	CH_3	CH_3	CH_3	Cl	H	NO_2	CH=N –OH	CN	CH=N –OAc		
Test organism	Minimum concentration inhibiting growth[a] (γ/l.)											
Salmonella typhosa	20	20	20	20	2	5	>50	50	>50	>50		
Escherichia coli	50	20	20	20	10	10	>50	50	>50	>50		
Pseudomonas aeruginosa	>50	>50	>50	>50	>50	>50	>50	>50	>50	>50		
Bacillus subtilis	5	1	5	1	2	1	20	>50	>50	>50		
Staphylococcus aureus	20	5	20	10	10	2	>50	50	>50	>50		
Trichophyton purpureum	< 1.0	< 1.0	2	5	5	5	20				3.1	50
Trichophyton mentagraphytes	0.8	0.8	1.6					2	5	5	6.2	50
Aspergillus niger	50	50	50	50	20	10	>50	>50	20	>50	>100	3.1
Penicillium digitatum	5	2	5	5	5	5	50	1	10	1	>100	0.8
Candida albicans	50	>50	>50	>50	20	>50	>50	>50	50	>50	>100	1.6
Saccharomyces cerevisiae	50	50	20	50	20	>50	>50				>100	0.8
Trichomonas vaginalis	12.5	12.5	12.5	0.8	1.6	1.6	6.3	3.2	6.3	6.3		1.6

[a] *Bacteria*, readings after 20 h at 37°; *Trichophyton*, readings after 10 days at 28°; *Aspergillus* and *Penicillium*, readings after 2 days at 28°; *Trichomonas*, readings after 2 days at 37°.

antagonism to exist between these two kinds of compounds and proved the substitution of the nitro group with SH compound during their reaction under such conditions by the formation of 4-cysteinylquinoline 1-oxide, 4-thioglycolylquinoline 1-oxide, and nitrous acid[42].

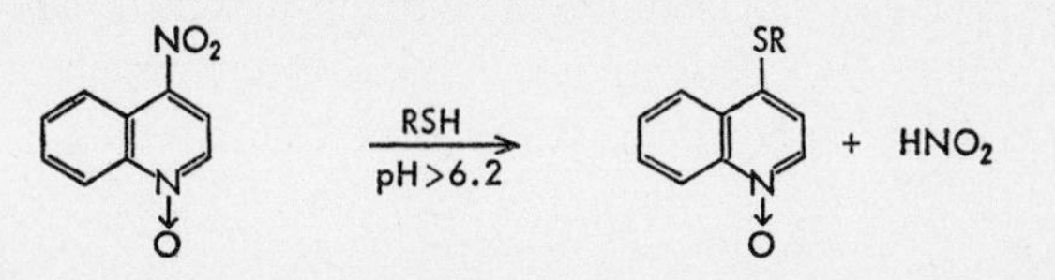

Childress and Scudi[43] synthesized 2-(sulfanilamido)pyridine 1-oxide and its methyl homologs and reported that they all possessed a potent antimicrobial activity and that they did not cause urolith formation in animal experiments. Later, a patent was taken out for the use of 3-(sulfanilamido)pyridazine 1-oxide, and its 6-methyl and 6-methoxy derivatives as antibacterial agents[44]. It has also been reported that some derivatives (9-10) of 4-(arylsulfonyl)pyridine 1-oxide are active against *Escherichia coli* and *Staphylococcus aureus*[45].

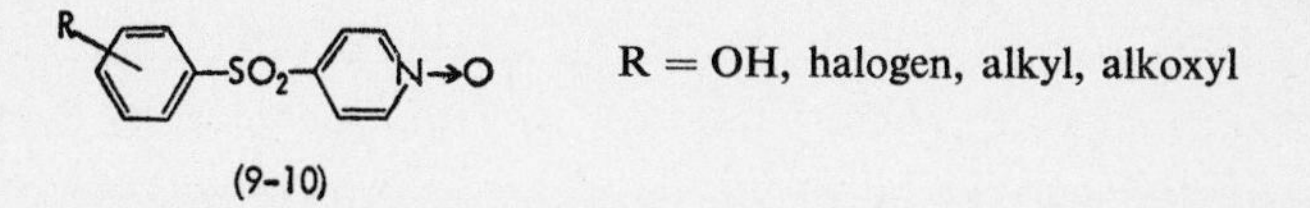

(9-10)

R = OH, halogen, alkyl, alkoxyl

Elslater and Tendick[46] reported that acridine *N*-oxide derivatives of the structural formulae (9-11)[47] and (9-12)[48] had antibacterial property as well as an antiparasitic activity against malaria and entameba, and took out patents for their syntheses.

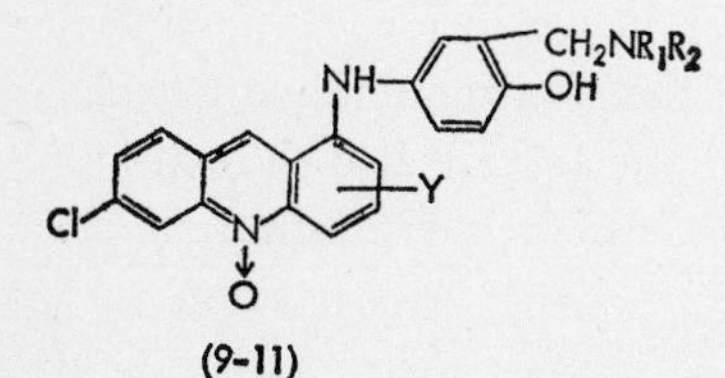

(9-11)

Y = H, Cl. CH_3, NO_2, or Ph in 2- or 3-position
R_1, R_2 = H, alkyl or hydroxyalkyl with C_1 to C_{10}, and N-heterocycle

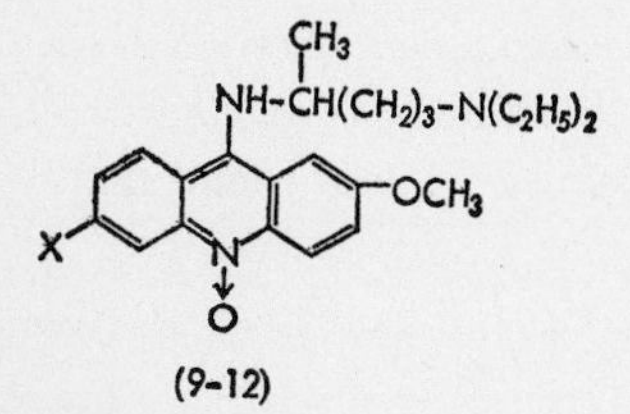

(9-12)

X = Cl, CH_3

References p. 440

Some examples of the action of these *N*-oxide compounds against larger organisms are found in a patent[49] for the use of 2,3-dimethylquinoxaline *N,N*-dioxide and its 6-chloro derivative, and 2,3-bis(hydroxymethyl)quinoxaline *N,N*-dioxide as an anthelmintic, and a patent[50] on the use of pyridine and quinoline 1-oxides, and isoquinoline 2-oxide, and their halogen, alkoxyl, nitro, alkyl, and aralkyl derivatives as attracting materials for birds and rodents, or repellants to be applied to food or surface areas.

9.3 Mutagenic, Carcinogenic, and Carcinostatic Activity of 4-Nitroquinoline 1-Oxide and its Derivatives

9.3.1 Mutagenic Activity

In 1955, Okabayashi[5,51] found that 4-nitroquinoline 1-oxide had an outstanding mutagenic activity against *Aspergillus niger* in a concentration of 7×10^{-5} to 7×10^{-6} mole/l. The mutagenic activity of this compound was later confirmed by Ando and others[52] against *Aspergillus oryzae*, by Mashima and Ikeda[53] against certain biochemical mutants of *Streptomyces griseoflavus*, and by Endo and others[54] against tobacco mosaic virus. Szynalski[55] reported, however, that 4-nitroquinoline 1-oxide had only a weak or doubtful mutagenic activity against *Escherichia coli*, but Okabayashi and others[56] proved by the use of a reverse mutant test that the compound is a potent mutagen for bacteria like *Esch. coli* WN-22 (Pro−) and *Brevibacterium liquefaciens*. Mifuchi[57] was able to induce a respiration-deficient mutant from yeast (*Saccharomyces cerevisiae*) by the action of this 4-nitroquinoline 1-oxide.

Okabayashi[58] observed that 4-nitroquinaldine 1-oxide also showed mutagenic activity and suggested, as its action mechanism, the substitution of the SH group in living materials, with the nitro group in 4-position as has already been indicated (*cf. Section 9.2* p. 433), and concurrent liberation of nitrous acid. He followed the change of 4-nitroquinoline 1-oxide in the culture medium of micro-organisms like *Escherichia coli*, *Brevibacterium liquefaciens*, *Pseudomonas aeruginosa*, *Candida utilis*, and *Aspergillus niger*, by the use of paper chromatography[59] and colorimetry[60], and found that 4-(hydroxyamino)quinoline 1-oxide accumulates first, which is then reduced via 4-aminoquinoline 1-oxide to 4-aminoquinoline, and that these steps of reduction differ according to the kind of micro-organism present.

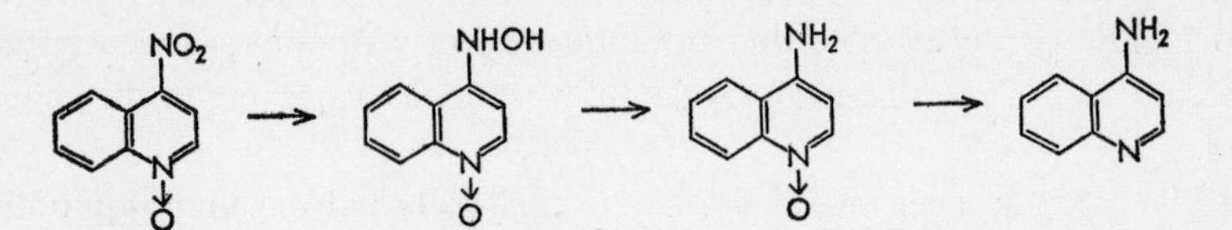

Aspergillus niger, which is the most sensitive to 4-nitroquinoline 1-oxide and easily mutated by treatment with this agent, can reduce it only to 4-hydroxyaminoquinoline 1-oxide, while *Esch. coli*, against which 4-nitroquinoline 1-oxide has only a slight mutagenic affect, caused formation of additional reduction products.

In view of these facts, Okabayashi[61] examined the mutagenic activity of 4-hydroxyaminoquinoline 1-oxide against *Aspergillus niger* and found that its mutagenic activity was of the same potency as that of 4-nitroquinoline 1-oxide. It may be assumed that the hydroxyamino group in 4-(hydroxyamino)-quinoline 1-oxide has a much more weaker activity to nucleophilic reagents than the nitro group in the corresponding 4-nitro derivative, and the foregoing assumption that mutation is caused by the substitution of the nitro group with an SH compound or by production of nitrous acid has thus lost one of its bases unless *in vivo* oxidation of the hydroxyamino group can be proved.

9.3.2 Carcinogenic Activity[62]

In 1957, Nakahara, Fukuoka, and Sugimura[7] found that 4-nitroquinoline 1-oxide has a very powerful carcinogenic activity. They applied this compound as a 0.25% benzene solution to the skin of mice, three times a week, and found that papillomas developed in all the mice in 80–140 days and malignant tumors in 120–200 days. After 150 days, 100% of the mice bore tumors and the tumors were already malignant in 36% of them[62]. Nakahara and Fukuoka[63] found that subcutaneous injection of 0.5–2 mg of 4-nitroquinoline 1-oxide as a propylene glycol solution, with an interval of 7 or 10 days to a total dose of 5–6 mg, produced sarcoma at the site of the injection in 25–66% of the mice. Mori and others[64] used a 100:5 mixture of olive oil and cholesterol in place of propylene glycol as a vehicle for 4-nitroquinoline 1-oxide and succeeded in inducing adenoma and adenocarcinoma in the lungs of mice, besides local sarcoma, by its subcutaneous injection. The same result was obtained with rats[65]. Mori[66] further experimented with the change of this vehicle to a 100:5 mixture of olive oil and lecithin; injection of 4-nitroquinoline 1-oxide in this vehicle into rats was found to induce uterine cancer.

References p. 440

It is highly worthy of note that the induction of a tumor by subcutaneous injection of 4-nitroquinoline 1-oxide differs in locus according to the vehicle used.

As has already been stated (*cf. Section 9.3.1*), 4-(hydroxyamino)quinoline 1-oxide has a mutagenic activity comparable to that of 4-nitroquinoline 1-oxide. Shirasu and Ohta[67] examined the carcinogenic activity of this 4-hydroxyamino derivative in mice and found that the tumor could not be induced by painting its solution on the skin, twice a week for 6 months, but that subcutaneous injection of the compound as a suspension in physiological salt solution, mixed with Tween 80, induced a malignant fibrosarcoma at the site of injection. The same result was obtained with rats[68].

In connection with the mutagenic activity of 4-nitro- and 4-hydroxyamino-quinoline 1-oxides, Endo and his co-workers[69] observed that these compounds caused bacteriophage induction in lysogenic bacteria.

Nakahara and his collaborators[70] examined the relationship between the carcinogenic effect and chemical structure of 4-nitroquinoline 1-oxides and found that the 2-methyl, 2-ethyl, and 6-chloro derivatives also induced papillomas and that most of these progressed to a malignant state, while quinoline 1-oxide, 4-nitroquinoline, and 6-nitroquinoline did not show such an effect. They revealed that the presence of an *N*-oxide group and a nitro group in the 4-position was indispensable for carcinogenic activity of these compounds. This fact seems to suggest, as was assumed by Okabayashi in connection with the mutagenic effect of 4-nitroquinoline 1-oxide, that the reaction between the nitro group and SH compound takes a part in the carcinogenic activity of these compounds. Endo[71] carried out the reaction of the quinoline compounds with cysteine or glutathione at the physiological pH, estimated the amount of nitrous acid formed, and proved that the reactivity of the nitro group decreased in the following order.

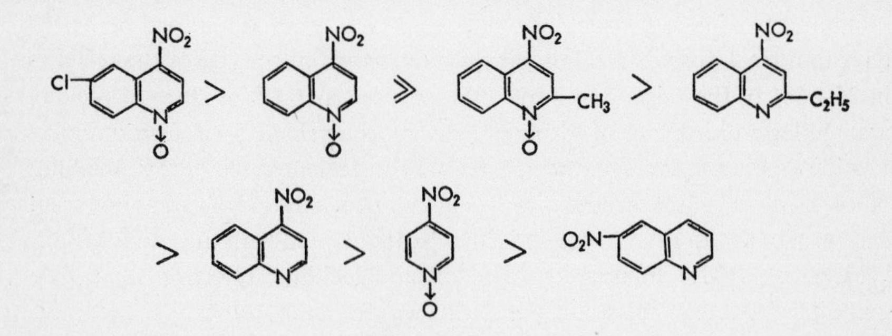

Further, the assumption that 4-nitroquinoline 1-oxide actually reacts with the SH group in the skin *in vivo* was verified by Hayashi[72], who demonstrated both histochemically and by spectrophotometry that there is a definite decrease in intraepithelial content of SH in the skin area within 48 hours after a single application of 4-nitroquinoline 1-oxide. No such decrease occurred after an application of non-carcinogenic quinoline derivatives.

In connection with this problem, Fukuoka and others[73] proved that 4-nitroquinoline 1-oxide markedly inhibited the glycolysis of Ehrlich ascites carcinoma, Yoshida sarcoma, and ascites hepatoma. Endo and others[74] found that application of 4-nitroquinoline 1-oxide in 10^{-5} molar concentration in tissue culture cells produced characteristic intranuclear inclusion bodies surrounded by a clear halo in the resting cells, but that quinoline derivatives not possessing a nitro group in the 4-position or the *N*-oxide group did not behave in this manner.

From these data, Nakahara and Fukuoka[63] assumed that the reaction of the nitro group and SH compounds *in vivo* might be the proximate cause of carcinogenic action of 4-nitroquinoline 1-oxide and suggested that the energy released by this reaction might produce a misconstruction of the genetic determinant of the somatic cell (duplicant) by disturbing its normal replication process. This hypothesis requires re-examination in the light of the fact that 4-(hydroxyamino)quinoline 1-oxide, which should have little reactivity with SH compounds, has been found to have a carcinogenic effect comparable to that of 4-nitroquinoline 1-oxide.

9.3.3 Carcinostatic Activity

In 1955, Sakai and Fukuoka[6] found that mouse sarcoma cells *in vitro* are completely killed by 4-nitroquinoline 1-oxide in a concentration of 0.002% and their viability is greatly reduced at 0.001% concentration. In *in vivo* experiments using Ehrlich ascites carcinoma, they found that the compound prolonged the average survival period in appropriate doses. This observation was confirmed and extended by Sugiura[75], and by Moore and Mannering[76]. Similar antitumor activity has also been found in the 2-alkyl derivatives of 4-nitroquinoline 1-oxide[6]. Taking these facts into consideration, Ochiai and his school[77] synthesized the compounds of formulae (9-13) and (9-14) as the more water-soluble derivatives of 4-nitro quinoline 1-oxide.

References p. 440

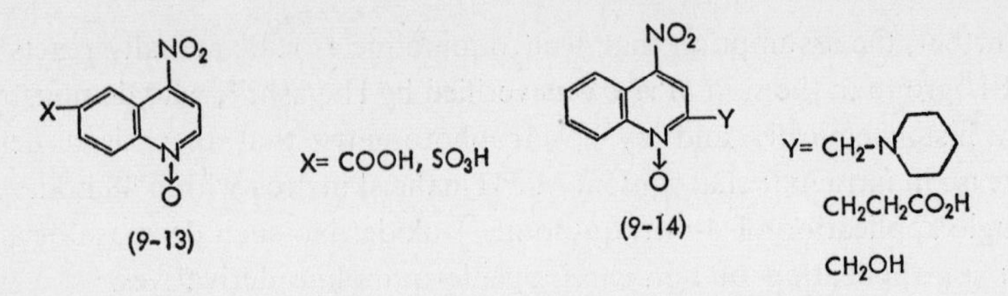

Of these compounds, the anticancer effect of the potassium salt of 6-carboxy-4-nitroquinoline 1-oxide was examined by the same method as above by Fukuoka and others[78]. According to them, the carcinocidal activity of this compound *in vitro* was about one-tenth of that of 4-nitroquinoline 1-oxide, and its toxicity was about one-tenth. The compound completely destroyed Ehrlich ascites carcinoma *in vivo* and its optimum dose was about 20 times that of 4-nitroquinoline 1-oxide. They also found that the compound inhibited glycolytic action of Ehrlich carcinoma cells *in vitro* at about 20 times the concentration necessary for 4-nitroquinoline 1-oxide.

Aromatic amine oxides other than 4-nitroquinoline 1-oxide compounds that have been tested for carcinostatic activity include phenazine *N*,*N*-dioxide, which was found to be effective against Ehrlich ascites carcinoma in mice by intraperitoneal injection, resulting in 90% survival after 30 days, but its 2,3-dimethyl derivatives and phenazine were reported to be ineffective[79]. It has also been reported[80] that 3-methoxy-6-methylpyridazine 1-oxide has antitumor activity both against experimental rodent tumors and in tissue culture.

9.4 Miscellaneous

As has already been stated (*cf. Section 8.7*), Sternbach and others[81] developed a rearrangement reaction for the synthesis of 2-amino-7-chloro-5-phenyl-3*H*-benzo[*e*]-(1,4)-diazepine 4-oxide by the reaction of 6-chloro-2-chloromethyl-4-phenylquinazoline 3-oxide and amines. They found that its 2-methylamino derivative (9-15) and the *N*-acetylated compound of the latter showed muscle relaxing, sedative, and anticonvulsion activities in animals[81d], and patents[82] have been taken out by Hoffmann–La Roche Inc. for such pharmacological activities and for the synthesis of compounds (9-16) possessing psychosedative activity.

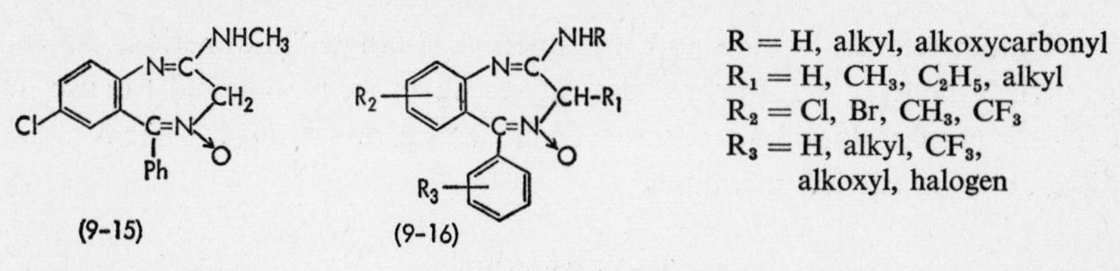

(9-15) (9-16)

Patents of this nature are found in that of Mathes and du Vanzo[83] on the use of *O*-alkyl (C_1 to C_{12}) derivatives of the 3- or 4-aldoxime of pyridine 1-oxide as a tranquilizer, anticonvulsant, muscle relaxant, analgesic, antipyretic, or antiinflamator, and that of Carbon[84] on the use of benzotriazine 1-oxide derivatives (9-17) as a tranquilizer and antibacterial agent.

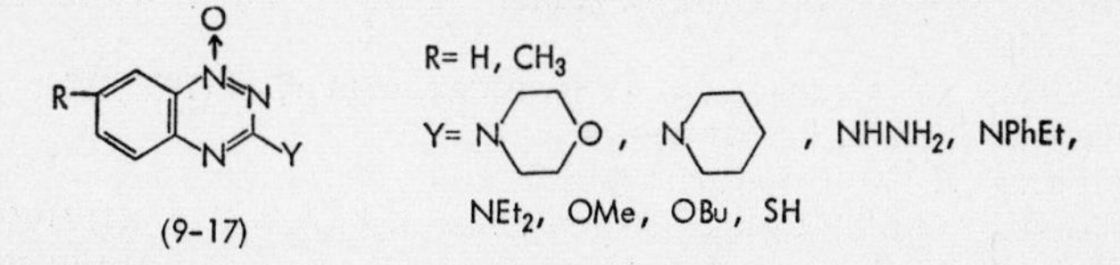

(9-17)

Petering and Van Giessen[85] observed the diuretic action of 8-chloro-1,2,3,4-tetrahydrobenzo[*g*]pteridine-2,4-dione 5,10-dioxide (9-18).

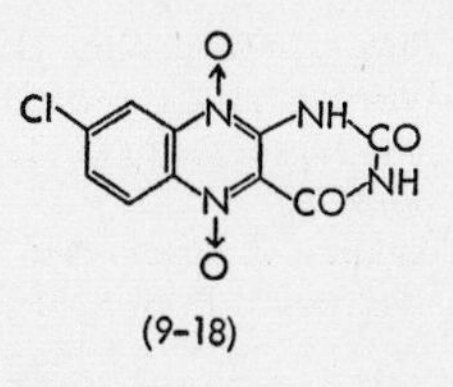

(9-18)

A patent has been granted for the use of nicotinic acid *N*-oxide, either as free acid or as magnesium salt, as a hypocholesteremic agent[86]. For the same purpose, 2-(*m*-methoxyphenyl)-1-(1-oxido-2-pyridyl)-1-phenyl-2-propanol (9-19) has also been mentioned in a patent[87] and this compound is said to have less estrogenic side-effect.

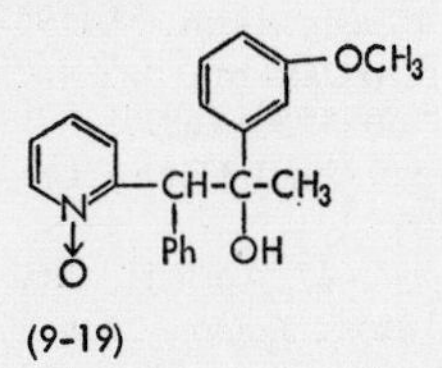

(9-19)

Haley and others[88] reported that quinoxaline *N,N*-dioxide had a protective effect against radiation damage to mice and that 2,3-dimethylquinoxaline *N,N*-dioxide and its 6-chloro derivative also had a similar but weaker effect[89].

References p. 440

4-Cyanopyridine 1-oxide and its 2-methyl, 2-ethyl, 2,6-dimethyl, 2,5-dimethyl, 3-methyl, and 3-propyl derivatives have been suggested for use in cosmetic preparations for preventing sunburn as a ray-absorbing agent in an oil-containing carrier medium[90].

REFERENCES

1 G. R. Clemo and H. McIlwain, *J. Chem. Soc.*, (1938) 479.
2 E. C. White and J. H. Hill, *J. Bacteriol.*, 45 (1943) 433.
3 I. Arai and I. Nakayama, *Yakugaku Zasshi*, 72 (1952) 167.
4 T. Okabayashi, *Hakko Kogaku Zasshi*, 31 (1953) 373.
5 T. Okabayashi, *Hakko Kogaku Zasshi*, 33 (1955) 513.
6 S. Sakai, K. Minoda, G. Saito, S. Akagi, A. Ueno, and F. Fukuoka, *GANN*, 46 (1955) 605.
7 W. Nakahara, F. Fukuoka, and T. Sugimura, *GANN*, 48 (1957) 129.
8 H. McIlwain, *Nature*, 148 (1941) 628.
9 H. McIlwain, *J. Chem. Soc.*, (1943) 322.
10 H. Akabori and M. Nakamura, *J. Antibiotics (Tokyo), Ser. A*, XII (1959) 17.
11 F. E. King, N. G. Clark, and P. M. H. Davis, *J. Chem. Soc.*, (1949) 3012.
12 J. A. Silk, *J. Chem. Soc.*, (1956) 2058.
13 J. Francis, J. K. Landquist, A. A. Levi, J. A. Silk, and J. M. Thorp, *Biochem. J.*, 63 (1956) 455.
14 J. K. Landquist and J. A. Silk, *J. Chem. Soc.*, (1956) 2052.
15 *(a)* J. D. Dutcher and O. Wintersteiner, *J. Biol. Chem.*, 155 (1944) 359. *(b)* J. D. Dutcher, *J. Biol. Chem.*, 171 (1947) 321. *(c)* G. T. Newbold, W. Sharp, and F. S. Spring, *J. Chem. Soc.*, (1951) 2679.
16 G. T. Newbold and F. S. Spring, *J. Chem. Soc.*, (1948) 1684.
17 K. G. Cunningham, G. T. Newbold, F. S. Spring, and J. Stark, *J. Chem. Soc.*, (1949) 2091.
18 E. Shaw, *J. Am. Chem. Soc.*, 71 (1949) 67.
19 W. A. Lott and E. Shaw, *J. Am. Chem. Soc.*, 71 (1949) 70.
20 S. R. Safir and J. H. Williams, *J. Org. Chem.*, 17 (1952) 1298.
21 E. Shaw, J. Bernstein, K. A. Losee, and W. A. Lott, *J. Am. Chem. Soc.*, 72 (1950) 4362.
22 Olin Mathieson Chem. Corp., *Brit. Pat.* 761,171, Nov. 14, 1956 [*Chem. Abstr.*, 51 (1957) 13327].
23 J. Bernstein, W. A. Lott, and K. A. Losee (to Olin Mathieson Chem. Corp.), *U.S. Pat.* 742,393, April 17, 1956 [*Chem. Abstr.*, 51 (1957) 494].
24 J. Bernstein and K. A. Losee (to Olin Mathieson Chem. Corp.), *U.S. Pat.* 2,742,476, April 17, 1956 [*Chem. Abstr.*, 50 (1956) 16877].
25 J. Rockett (to Olin Mathieson Chem. Corp.), *U.S. Pat.* 2,922,793, January 26, 1960 [*Chem. Abstr.*, 54 (1960) 8857].
26 *(a)* J. Rockett, *U.S. Pat.* 2,932,647, April 12, 1960 [*Chem. Abstr.*, 54 (1960) 18557]. *(b)* R. E. McClure and A. Ross, *J. Org. Chem.*, 27 (1962) 304.
27 B. B. Brown (to Olin Mathieson Chem. Corp.), *U.S. Pat.* 2,940,978, June 14, 1960 [*Chem. Abstr.*, 54 (1960) 21140].
28 J. Rockett (to Olin Mathieson Chem. Corp.), *U.S. Pat.* 2,922,790, January 26, 1960 [*Chem. Abstr.*, 54 (1960) 8857].
29 B. A. Starrs (to Olin Mathieson Chem. Corp.), *U.S. Pat.* 3,027,371, March 27, 1962 [*Chem. Abstr.*, 57 (1962) 12442].

30 B. A. Starrs, *U.S. Pat.* 3,027,372, March 27, 1962 [*Chem. Abstr.*, 57 (1962) 12441].
31 L. H. Conover, A. R. English, and C. E. Larrabee (to Chas. Pfizer Co.), *U.S. Pat.* 2,921,073, January 12, 1960 [*Chem. Abstr.*, 54 (1960) 8860].
32 *(a)* F. Leonard and A. Wajngurt, *J. Org. Chem.*, 21 (1956) 1077. *(b)* F. Leonard, F. A. Barkley, E. V. Brown, F. E. Anderson, and D. M. Green, *Antibiot. Chemotherapy*, VI-4 (1956) 261.
33 J. W. Lightbown, *J. Gen. Microbiol.*, 11 (1954) 477.
34 J. W. Cornforth and A. T. James, *Biochem. J.*, 63 (1956) 124.
35 J. W. Lightbown and F. L. Jackson, *Biochem. J.*, 63 (1956) 130.
36 E. Profft, G. Buchmann, and N. Wackrow, *Wiss. Z. Tech. Hochsch. Chem. Leuna-Merseburg*, 2 (1959-1960) 93 [*Chem. Abstr.*, 55 (1961) 1616].
37 G. Buchmann and D. Kirstein, *J. prakt. Chem.*, 18 (1962) 175.
38 E. P. Hart, *J. Chem. Soc.*, (1954) 1879.
39 T. Itai and S. Kamiya, *Chem. Pharm. Bull. (Tokyo)*, 9 (1961) 87.
40 *(a)* H. Nishimura, H. Kanô, K. Tawara, M. Ogata, and Y. Tanaka, *Shionogi Kenkyusho Nempo*, 14 (1964) 86. *(b)* H. Kanô and M. Ogata (to Shionogi & Co., Ltd.) *Jap. Pat.* 22,970/1964, October 15, 1964.
41 T. Okabayashi, *Hakko Kôgaku Zasshi*, 31 (1953) 416.
42 T. Okabayashi, *Yakugaku Zasshi*, 73 (1953) 946.
43 S. J. Childress and J. V. Scudi, *J. Org. Chem.*, 23 (1958) 67.
44 W. E. Taft (to American Cyanamid Co.), *French Pat.* M883, January 19, 1961.
45 S. P. Acharya and K. S. Nargund, *J. Sci. Ind. Res. (India)*, 21B (1962) 451.
46 E. F. Elslater and F. H. Tendick, *J. Med. Pharm. Chem.*, 5 (1962) 1153, 1159.
47 E. F. Elslater and F. H. Tendick (to Parke Davis Co.), *Brit. Pat.* 829,728 (March 9, 1960). [*Chem. Abstr.*, 54 (1960) 18559].
48 E. F. Elslater and F. H. Tendick, (to Parke Davis Co.) *U.S. Pat.* 2,880,210, March 31, 1959 [*Chem. Abstr.*, 53 (1959) 15100].
49 J. J. Urspring (to Chas. Pfizer Co.), *U.S. Pat.* 2,890,981, June 16, 1959 [*Chem. Abstr.*, 53 (1959) 17441].
50 L. D. Goodhue and K. E. Cantrel (to Phillips Petroleum Co.), *U.S. Pat.* 3,044,930, July 7, 1962 [*Chem. Abstr.*, 57 (1962) 15079].
51 T. Okabayashi, *Hakko Kôgaku Zasshi*, 33 (1945) 513.
52 K. Yamagata, M. Oda, and T. Ando, *Hakko Kôgaku Zasshi*, 34 (1956) 378.
53 S. Mashima and Y. Ikeda, *Appl. Microbiol.*, 6 (1958) 45.
54 H. Endo, W. Wada, K. Miura, Z. Hidaka, and C. Hiruki, *Nature*, 190 (1961) 833
55 W. Szybalski, *Ann. N.Y. Acad. Sci.*, 76 (1958) 475.
56 T. Okabayashi, M. Ide, A. Yoshimoto, and M. Otsubo, *Chem. Pharm. Bull. (Tokyo)*, 13 (1965) 610.
57 I. Mifuchi, *GANN*, 54 (1963) 205.
58 T. Okabayashi, *Hakko Kôgaku Zasshi*, 35 (1957) 17.
59 T. Okabayashi, *Chem. Pharm. Bull. (Tokyo)*, 10 (1962) 1221.
60 T. Okabayashi and A. Yoshimoto, *Chem. Pharm. Bull. (Tokyo)*, 12 (1964) 262.
61 T. Okabayashi, *Chem. Pharm. Bull. (Tokyo)*, 10 (1962) 1127; T. Okabayashi, A. Yoshimoto, and M. Ide, *Chem. Pharm. Bull. (Tokyo)*, 12 (1964) 257.
62 *cf.* the review W. Nakahara, *Critique of Carcinogenic Mechanism* in *Progr. Exptl. Tumor Res.*, 2 (1961) 158.
63 W. Nakahara and F. Fukuoka, *GANN*, 50 (1959) 1.
64 K. Mori and A. Yasuno, *GANN*, 52 (1961) 149; K. Mori, *GANN*, 52 (1961) 265; 53 (1962) 303; 55 (1964) 315; K. Mori and I. Hirafuku, *GANN*, 55 (1964) 205.
65 K. Mori, *GANN*, 54 (1963) 415.
66 K. Mori, *GANN*, 55 (1964) 277.

67 Y. Shirasu and A. Ohta, *GANN*, 54 (1963) 221.
68 Y. Shirasu, *GANN*, 54 (1963) 487.
69 H. Endo, M. Ishizawa, and T. Kamiya, *Nature*, 198 (1963) 195.
70 W. Nakahara, F. Fukuoka, and S. Sakai, *GANN*, 49 (1958) 33.
71 H. Endo, *GANN*, 49 (1958) 151.
72 Y. Hayashi, *GANN*, 50 (1959) 219.
73 T. Ono, T. Tomaru, and F. Fukuoka, *GANN*, 50 (1959) 189; F. Fukuoka, T. Ono, M. Ohashi, and S. Nishimura, *GANN*, 50 (1959) 23.
74 H. Endo, M. Aoki, and Y. Aoyama, *GANN*, 50 (1959) 209.
75 K. Sugiura, *Ann. N.Y. Acad. Sci.*, 76 (1958) 575; *Progr. Exptl. Tumor Res.*, 2 (1961) 332.
76 R. E. Moore and G. J. Mannering, *Proc. Am. Ass. Cancer Res.*, 2 (1958) 329.
77 E. Ochiai, S. Suzuki, Y. Utsunomiya, T. Ohmoto, K. Nagatomo, and M. Itoh, *Yakugaku Zasshi*, 80 (1960) 339.
78 F. Fukuoka, T. Sugimura, and S. Suzuki, *Gann*, 48 (1957) 263.
79 A. Furst, C. Klausner, and W. Cutting, *Nature*, 184 (1959) 908. *cf.* K. Hano, H. Iwata, and K. Nakajima, *Chem. Pharm. Bull. (Tokyo)*, 13 (1965) 107.
80 D. L. Aldous and R. N. Castle, *Arzneimittel-Forsch.*, 13 ,1963) 878.
81 *(a)* L. H. Sternbach, S. Kaiser, and E. Reeder, *J. Am. Chem. Soc.*, 82 (1960) 475. *(b)* L. H. Sternbach and E. Reeder, *J. Org. Chem.*, 26 (1961) 1111. *(c)* L. H. Sternbach, E. Reeder, O. Keller, and W. Metlesics, *J. Org. Chem.*, 26 (1961) 4488. *(d)* L. H. Sternbach and E. Reeder, *J. Org. Chem.*, 26 (1961) 4936.
82 E. Reeder and L. H. Sternbach (to Hoffmann–La Roche Inc.), *U.S. Pat.* 858,597, December 10, 1959; L. H. Sternbach and G. Saucy (to Hoffmann–La Roche Inc.), *U.S. Pat.* 2,605, January 15, 1960; E. Reeder and L. H. Sternbach (to Hoffmann–La Roche Inc.), *U.S. Pat.* 3,051,701, August 28, 1962 [*Chem. Abstr.*, 57 (1962) 1664]; L. H. Sternbach (to Hoffmann–La Roche Inc.), *U.S. Pat.* 2,893,992, July 7, 1959 [*Chem. Abstr.*, 54 (1960) 597].
83 W. Mathes and J. P. du Vanzo (to Dr. R. Raschig GmbH), *French Pat.* M. 2092, November 18, 1963.
84 J. A. Carbon (to Abott Laboratories), *U.S. Pat.* 3,137,693, June 16, 1964 [*Chem. Abstr.*, 61 (1964) 5670].
85 H. G. Petering and G. J. van Giessen, *J. Pharm. Sci.*, 52 (1963) 1192.
86 Samip Société Anilloise de Produits Chimique, *French Pat.* 1256, May 11, 1961.
87 J. R. Dice and R. D. Westland (to Parke Davis Co.), *U.S. Pat.* 3,128,281, Apirl 7, 1964 [*Chem. Abstr.*, 60 (1964) 15842].
88 T. J. Haley, A. M. Flesher, R. Veomett, and J. Vincent, *Proc. Soc. Exptl. Biol. Med.*, 96 (1957) 597 [*Chem. Abstr.*, 52 (1958) 5633].
89 T. J. Haley, A. M. Flesher, and N. Komesu, *Nature*, 184 (1959) 198.
90 Rohm and Haas Co., *German Pat.* 1,134,799, March 9, 1961.

Author Index

Subject Index

(Page numbers with E refer to Examples)